Computational Approach for Air Pollution

Akhilesh Kumar Yadav
Editor

Computational Approach for Air Pollution

Springer

Editor
Akhilesh Kumar Yadav
Department of Mining Engineering
Indian Institute of Technology (Banaras Hindu University)
Varanasi, India

ISBN 978-3-032-09214-4 ISBN 978-3-032-09215-1 (eBook)
https://doi.org/10.1007/978-3-032-09215-1

This Springer imprint is published by the registered company Springer Nature Switzerland AG
The registered company address is: Gewerbestrasse 11, 6330 Cham, Switzerland

Dedicated
to
Environmental Engineers and Scientists
around the World

Foreword

Air pollution is a silent global crisis that impacts millions of lives and ecosystems every year. As we grapple with rapid urbanization, industrial growth, and climate change, innovative solutions grounded in computational and data-driven approaches have become indispensable. The *Computational Approach to Air Pollution* emerges as a timely and critical contribution to this discourse, offering a multidisciplinary road map to tackle one of the most persistent challenges.

This book synthesizes cutting-edge research, bridging the gap between theoretical models and real-world applications. From foundational assessments of pollution's health and environmental impacts to advanced AI-driven forecasting and sustainable mitigation strategies, the chapters collectively present a holistic vision for cleaner air. The inclusion of case studies that span regions such as India, Spain, and urban centers around the world adds practical relevance, demonstrating how data and technology can drive actionable policy and innovation. What sets this volume apart is its integration of diverse methodologies, including machine learning, probabilistic modeling, nature-based solutions, and digital tools (e.g., BIM), into a cohesive framework. It not only advances academic understanding but also equips policymakers, researchers, and technologists with evidence-based strategies.

As I stand at the intersection of environmental science and digital transformation, this book serves as both a call to action and a beacon of hope. Its insights underscore the power of collaboration across disciplines to forge a sustainable and breathable future.

Philip Karl Hopke

Bayard D. Clarkson Distinguished
Professor Emeritus
Clarkson University
Potsdam, NY, USA

Department of Public Health Sciences
University of Rochester School of Medicine and Dentistry
Rochester, NY, USA

Preface

Air pollution remains one of the most pressing global challenges of our time, with far-reaching consequences for public health, ecosystems, and climate stability. As urbanization and industrialization accelerate, the need for innovative, data-driven, and sustainable solutions has never been greater. This book, *Computational Approach to Air Pollution*, brings together cutting-edge research, computational methodologies, and real-world case studies to address the multifaceted dimensions of air pollution.

This book is structured into four cohesive parts, each addressing a critical aspect of air pollution research and control: (a) Fundamentals and Assessment of Air Pollution: This section laid the groundwork by introducing key concepts, computational approaches, and comprehensive assessments of health and environmental impacts; (b) Computational and Data-Driven Approaches: This section explores machine learning, AI-based modeling, and predictive analytics as transformative tools for air quality forecasting and emission control; (c) Emerging Strategies and Innovations: This part highlights nature-based solutions, digital technologies (BIM and digital twins), and novel strategies to mitigate air pollution in urban and industrial emissions; and (d) Specialized Applications and Case Studies: The final section presents real-world applications, including COVID-19 emission analyses, fuzzy decision-making techniques, and probabilistic health risk assessments, offering actionable insights for policymakers and researchers.

This compilation bridges the gap between theoretical research and practical implementation, catering to environmental scientists, data analysts, policymakers, and engineers. By integrating computational intelligence with sustainability frameworks, our aim is to contribute to the global discourse on cleaner air and healthier communities.

I extend my gratitude to the contributing authors, reviewers, and institutions whose expertise has shaped this compilation. Of the 49 abstracts submitted, 40 were selected for full chapter contributions and 21 chapters were ultimately included based on the scope and quality of the submissions. Their commitment to advance air quality science underscores the collaborative spirit needed to combat this global crisis. As we navigate the intersection of technology, ethics, and sustainability, this book aims to inspire innovative solutions and foster interdisciplinary dialogue because clean air is not just an environmental goal, but a fundamental human right.

The editor welcomes any feedback, comments, or suggestions to improve future endorsement.

Key features of the book:

- Interdisciplinary Approach: Combines environmental science, AI, and policy studies.
- Real-World Case Studies: Includes applications from India, Spain, and developed nations.
- Cutting-Edge Techniques: Covers machine learning, fuzzy logic, and digital twin technologies.
- Sustainability Focus: Emphasizes nature-based and inclusive evaluation frameworks.

Varanasi, India Akhilesh Kumar Yadav

Contents

About the Editor

Akhilesh Kumar Yadav graduated with a Bachelor of Technology in Electronics and Communication Engineering from the Chaudhary Charan Singh University, Meerut, India. He subsequently obtained a Master of Technology in Environmental Engineering from Madan Mohan Malaviya Engineering College, Gorakhpur, India (affiliated with Dr. A. P. J. Abdul Kalam Technical University, Lucknow). His doctoral degree, with a research focus on air pollution, was awarded by the Indian Institute of Technology (Banaras Hindu University), Varanasi, India. Dr. Yadav has served at various institutions, including Indian Institute of Technology (Banaras Hindu University), Varanasi, India; Indian Institute of Technology Bombay, Mumbai, India; Bhabha Atomic Research Centre, Mumbai, India (as a research collaborator); and Chaoyang University of Technology, Taichung, Taiwan. He has authored/co-authored several research and review articles, book chapters, and edited/authored books, and holds a patent. Dr. Yadav received the Young Engineer Award from the Institution of Engineers (India) and the Young Scientist Award from VDGOOD Technology Factory, Kolkata. His research interests include air, water, and soil pollution; climate change; vulnerability; human health risk assessments; and GIS applications in environmental pollution and management. He is an active member of several reputed international professional bodies, including the IEI (Kolkata), ECI (New Delhi), AEACI (Mumbai), MSI (New Delhi), and IASTA (Mumbai).

Part I
Fundamentals and Assessment of Air Pollution

Chapter 1
Introduction to Computational Approaches in Air Pollution

Umalaxmi Thingujam, Animesh Ghosh Bag, Victor Thingujam, Sourish Pramanik, Ankan Das, Kishore Chandra Swain, Sushil Kumar, and Sandip Debnath

1.1 Introduction

An environmental issue that poses serious risks to ecosystems, human health, and the climate at large is air pollution. All man-made activity, such as industrial pollutants, vehicle exhaust, and agricultural practices, as well as natural events such as

U. Thingujam
Department of Soil Science and Agricultural Chemistry, Institute of Agriculture, Visva-Bharati, Santiniketan, India

A. Ghosh Bag
Department of Agriculture, Swami Vivekananda University, Barrackpore, India

V. Thingujam
ICAR Research Complex for Eastern Region, Farming System Research Centre for Hill and Plateau Region, Plandu, India

S. Pramanik
Department of Plant Pathology, Punjab Agricultural University, Ludhiana, India

Ethophilia Research Foundation, Santiniketan, India

A. Das
Institute of Agriculture, Visva-Bharati University, Sriniketan, India

K. C. Swain
Department of Agricultural Engineering, Institute of Agriculture, Visva-Bharati, Sriniketan, India

S. Kumar
Department of Environmental Studies, Siksha-Bhavana (Institute of Science), Visva-Bharati, Santiniketan, India

S. Debnath (✉)
Department of Genetics and Plant Breeding, Institute of Agriculture, Visva-Bharati, Sriniketan, India
e-mail: sandip.debnath@visva-bharati.ac.in

A. K. Yadav (ed.), *Computational Approach for Air Pollution*, https://doi.org/10.1007/978-3-032-09215-1_1

wildfires and volcanic eruptions, can contribute to it. Particulate matter (PM_{10}), nitrogen oxides (NO_x), sulfur dioxide (SO_2), and volatile organic compounds (VOCs) are among the hazardous gases and particles released by these activities that progressively find their way into the atmosphere (Manisalidis et al., 2020). According to Onivefu and Imarhiagbe (2024), these compounds can harm living things and also react chemically to produce secondary pollutants such as fine particulate matter ($PM_{2.5}$) and ozone (O_3). Computational modeling of air pollution has become an essential tool for assessing the dynamics and formulating mitigation plans to measure these contaminants efficiently. In essence, computational modeling of air pollution involves simulating the physical and chemical processes that control various particle movement mechanisms, such as dispersion, transformation, and deposition of pollutants in the atmosphere, using mathematical and numerical techniques. These models shed light on how pollutants move through the atmosphere and interact with other elements to affect air quality. They are essential for determining the causes of air quality issues, forecasting pollutant amounts under various illnesses, and evaluating the efficacy of regulatory actions (Bai et al., 2018; Yadav et al., 2022).

The first category of computational models that are frequently employed in air quality management is dispersion models, which are able to calculate pollutant concentrations at particular sites taking into account topography, climatic conditions, and emission sources (Snoun et al., 2023; Yadav, 2015). They are especially helpful for local evaluations, including determining how industrial facilities affect the people around them. The second category is photochemical models, which model the production and movement of chemically reactive pollutants on wide geographical scales, such as secondary PM and ground-level ozone (Baker et al., 2016). Finally, receptor models, which examine recorded pollutant concentrations in particular places to trace them back to their sources, are frequently used in policy evaluations for assessing regional or national air quality management programs. For source apportionment investigations, this model is crucial and recent scientific developments have greatly improved the capability and precision of air pollution modeling. To increase forecast accuracy and computing efficiency, machine learning (ML) approaches have also been included into conventional models (Frank et al., 2020; Rastogi et al., 2025). Pollutant concentrations are being predicted using ML methods including support vector machines (SVM), random forests (RF), and artificial neural networks (ANNs) based on real-time input and historical data. These techniques may manage intricate correlations between variables, such as nonlinear relationships among emissions and atmospheric conditions, that are difficult for standard models to capture. The nonlinear reactions of pollutants to changes in emissions were also captured using deep learning techniques. Convolutional neural networks (CNNs), for instance, have been used to improve spatial forecasts by maintaining patterns of pollution dispersion throughout metropolitan regions. These developments are perfect for real-time applications in smart cities because they reduce processing needs while maintaining excellent precision.

To inform policymakers on the possible results of emission reduction methods, computational models are essential. It makes it possible for decision-makers to assess options such as cutting vehicle emissions or switching to greener energy sources. Furthermore, by showing how lowering greenhouse gas emissions can also enhance air quality, these tools emphasize the connection between climate change and air quality. Computational modeling is a crucial method for measuring air quality in recent days. These simulations offer practical suggestions for reducing the detrimental impacts of air pollution on ecosystems, human health, and environmental sustainability by combining conventional methodologies with cutting-edge machine learning approaches.

1.2 Sources of Air Pollution

There are many different types of air pollution, but they can be essentially divided into two categories: anthropogenic (man-made) and natural. Numerous dangerous chemicals are released into the surrounding environment by these sources, affecting the health of people, the environment, and the climate as a whole (Manisalidis et al., 2020; Yadav, 2021; Yadav et al., 2023, 2025). An explanation of these categories based on findings from scientific research is provided below.

1.2.1 Natural Sources of Air Pollution

According to Manisalidis et al. (2020), a "natural source" refers to air polluting agents that do not need human involvement and release pollutants into the atmosphere, including gases, volatile organic compounds (VOCs), and particle matter ($PM_{2.5}$ and PM_{10}). The following is a list of sources of natural pollution.

1. Volcanic eruptions: Eruptions of volcanoes produce a lot of gases, including ash particles, carbon dioxide (CO_2), sulfur dioxide (SO_2), and nitrous oxide (N_2O). According to Hansell and Oppenheimer (2004), these emissions have the potential to cause acid rain and particle matter pollution. By retaining sunlight on the Earth's surface, sulfate aerosols are created when released SO_2 progressively combines with different molecules in the atmosphere.
2. Wildfires: Wildfires are a major source of air pollution, mainly releasing particulate matter ($PM_{2.5}$), carbon dioxide (CO_2), carbon monoxide (CO), and nitrogen oxides (NO_x). These emissions can account for up to 75% of native air pollution during wildfires. Scientists also gathered enough medical data to demonstrate that wildfire smoke exacerbates respiratory conditions and impairs vision in wide regions (Reid et al., 2016).
3. Dust storms: Fine particles are transported over great distances by dust storms that originate in dry regions. These particles have the potential to worsen air quality and aggravate respiratory conditions. For example, Saharan dust storms

are closely associated with worse air quality indices in North American and European nations (Prospero et al., 2014).

4. Biological emissions: The anaerobic breakdown of organic materials in submerged or waterlogged places releases gases, including methane (CH_4), a possible greenhouse gas (Singh et al., 2018). In addition, trees such as poplar and oak emit volatile organic compounds that combine with NO_x to create ground-level ozone that is regarded as a toxic secondary pollutant.

1.2.2 Anthropogenic Sources of Air Pollution

The worldwide air pollution process is being accelerated by air pollutants influenced by anthropogenic or human activity (Fig. 1.1). The main causes of air pollution, which are explained below, are human activities such as industry, traffic, agriculture, and deforestation.

1. Industrial emissions: When fossil fuels were burned and various items were manufactured, industries began to release pollutants, including SO_2, NO_x, and heavy metals throughout the beginning of the industrial revolution. According to a Chinese example study, equipment production accounts for the majority of

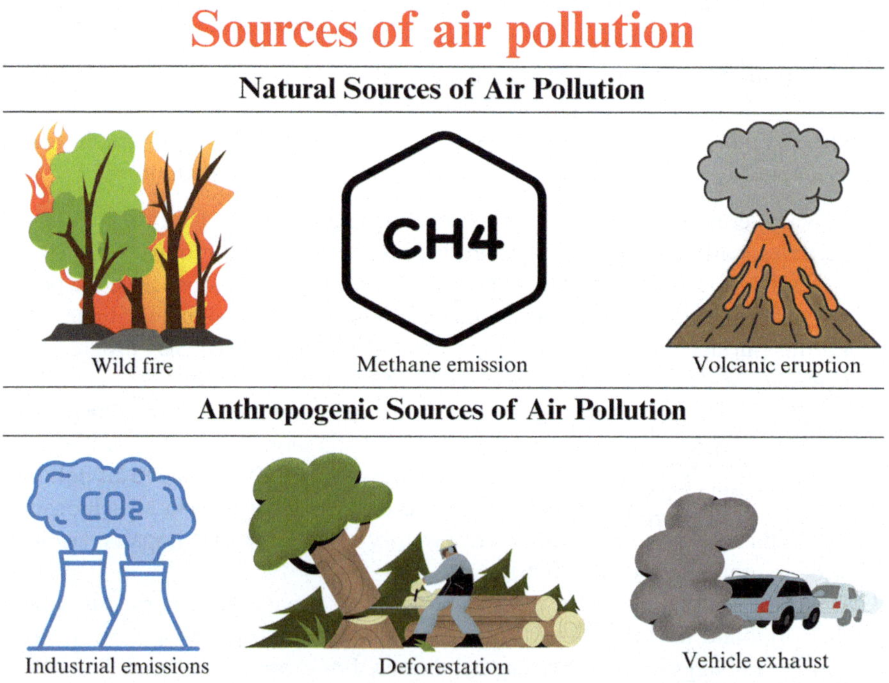

Fig. 1.1 Various sources of air pollution

industrial operations' over 50% contribution to air pollution emissions (Huo et al., 2014).
2. Vehicle exhaust: Transportation-related emissions are an important air quality hazard. Through fuel combustion, cars release harmful gases such as NO_x, CO, and particulate matter ($PM_{2.5}$ and PM_{10}). Road dust and tire non-exhaust pollutants also play a role in this process. The U.S. Environmental Protection Agency (EPA) claims that more than half of all air pollution in America originates from mobile sources (Reitze Jr., 1999).
3. Agricultural activities: Cattle and other livestock undergo enteric fermentation, which produces methane. Through soil microbial activities, fertilized cropland releases nitrogen oxides and ammonia (NH_3). By producing CO_2 and particles during deforestation, slash-and-burn agriculture makes an additional contribution.
4. Deforestation: When forests are cut for urbanization or agriculture, CO_2 emissions rise and the potential to sequester carbon is diminished. Forest fires increase air pollution by releasing significant volumes of greenhouse gases and particles. It is obvious that both natural and man-made factors greatly contribute to air pollution, but due to their growing scope and frequency, human-induced activities predominate. The design of successful strategies that address local pollution issues and global issues such as climate change requires a scientific understanding of these sources.

1.2.3 *Primary and Secondary Air Pollutants*

Depending on where they come from, air pollutants are also often divided into main and secondary pollutants. These classifications help us to understand the origins, characteristics, and effects of pollutants on the environment and human health.

1.2.3.1 The Primary Contaminants

Primary pollutants are released into the atmosphere directly from a variety of sources. These gases are linked to natural or human processes and possess certain chemical characteristics, namely particulate matter (PM), nitrogen oxides (NO_x), sulfur dioxide (SO_2), carbon monoxide (CO), and volatile organic chemicals. These principal pollutants are mostly caused by human activities such as power plants and the burning of fossil fuels. For example, the burning of coal and oil in power plants releases SO_2, while automobile exhaust is the main source of CO emissions. Furthermore, wildfires produce PM and CO, and volcanic eruptions release SO_2 and ash particles, which are powerful sources of main pollutants. These contaminants can have negative health impacts right away. For instance, PM may enter the lungs deeply and cause respiratory and cardiovascular disorders, whereas CO disrupts the bloodstream's ability to carry oxygen.

1.2.3.2 Pollutants that Are Secondary

Secondary pollutants are produced by chemical interactions between main pollutants and other substances, such as sunlight or water vapor, rather than being released directly into the atmosphere. Peroxyacetyl nitrates (PANs), smog, ground-level ozone (O_3), and secondary particle matter are a few examples of important secondary pollutants. These ozone gases, which mostly contribute to photochemical smog, are created when NO_x and VOCs react in the presence of sunlight. PANs are created when NO_x and hydrocarbons react in the presence of ultraviolet (UV) light. Compared to the main pollutants, these contaminants have more widespread impacts. Ozone, for instance, is a far more potent oxidant that may disrupt ecosystems, lower agricultural production, and damage lung tissue. Ozone and particle matter combine to create smog, which impairs vision and makes respiratory conditions worse.

1.2.4 Major Air Pollutants and their Effects

A serious environmental problem caused by the discharge of toxic materials into the atmosphere is air pollution. These contaminants, which come from both man-made and natural sources, have a great effect on ecosystems, infrastructure, and human health. Some significant pollutants are covered in detail here, along with their sources and effects on health.

1. Carbon monoxide (CO): CO is an odorless, colorless gas that is created when fossil fuels are not completely burned. Vehicle emissions, industrial operations, and stoves in homes have been the primary causes of this. The poisonous gas can attach to blood hemoglobin with an affinity 210 times higher than oxygen, which lowers the amount of oxygen that reaches tissues (Pan et al., 2020). Symptoms such as fatigue, light headedness, and cardiovascular problems may result from this. Asphyxia and even death can be caused by elevated levels of exposure. CO contributes to the production of smog, although it has less direct negative effects on ecosystems than other pollutants.
2. Sulfur dioxide (SO_2): Burning coal and oil in power plants is the main source of SO_2 emissions. Additionally, it comes from natural sources, such as volcanic eruptions. According to Khalaf et al. (2024), the pollutant gas irritates the respiratory system, making it harder for people to breathe, especially those who have asthma or other long-term lung diseases. When SO_2 and water vapor combine to create acid rain, soils and water bodies become more acidic, harming ecosystems. Additionally, it causes haze to build, affecting visibility.

3. Nitrogen oxides (NO_x): During high-temperature combustion, NO_x gases, such as nitrous oxide, nitric oxide, and nitrogen dioxide, are released mostly from automobiles, power plants, and industrial settings. According to Manisalidis et al. (2020), these NO_x gases have the potential to promote respiratory tract inflammation, increase the risk of respiratory diseases, including asthma, and, in the event of prolonged exposure, cause lung cancer. These harmful gases have the potential to produce both photochemical smog and ground-level ozone (O_3). In addition, it produces acid rain, which damages aquatic life and flora.
4. PM, or particulate matter: PM is mostly made up of microscopic, suspended particles in the air that come from industrial operations, automobile emissions, building sites, and natural events such as wildfires. By creating secondary pollutants such as ozone, these substances indirectly cause health problems. Long-term exposure to some VOCs can result in neurological damage or cancer. These substances contribute significantly to the creation of smog and have the potential to damage ecosystems by producing reaction products such as ozone.

1.3 Atmospheric Dispersion and Transport of Air Pollutants

The abundance and distribution of pollutants in the natural environment are determined by the movement of air pollutants, which are extremely important processes. These processes, which affect air quality on local, regional, and even global scales, are primarily controlled by meteorological elements such as wind, turbulence, and temperature inversions. Forecasting pollution trends and reducing their negative effects on the environment and human health require understanding of these mechanisms.

1.3.1 Turbulence and Winds

The primary force behind the advection mechanism of pollution transfer is the wind. It has the capacity to transport contaminants across great distances, spanning regional, metropolitan, and occasionally even international borders. For instance, based on wind direction and speed, contaminants from wildfires or industrial emissions can reach tens of thousands of kilometers (Diamond et al., 2015). The atmosphere is extremely turbulent and close to the Earth's surface because of waves of different sizes and timeframes. Turbulence in the process improves the mixing of pollutants with fresh air, lowering their concentration locally while dispersing them over a wider region. Compared to molecular diffusion, this mixing method is far more efficient.

1.3.2 Inversions of Temperature

Whenever a layer of warm air covers colder air close to the ground surface, temperature inversions typically occur, interfering with the vertical mixing process (Pal et al., 2021). The warm air rises as a result of this circumstance, causing the contaminants to spread further. The warm layer, on the other hand, serves as a cover during the inversion phenomena, trapping pollutants close to the ground. Inversions are frequent in steep metropolitan areas during beautiful weather, which causes a serious pollution problem. The best illustration of temperature inversion is when large amounts of nitrogen oxides (NO_x) or particulate matter (PM) accumulate in cities and reduce visibility due to smog production.

1.4 Health and Environmental Impacts of Air Pollution

The ecology and human health are severely harmed by air pollution, a major worldwide problem (Manisalidis et al., 2020). According to scientific data, it may play a significant role in chronic diseases, human mortality, and the quick deterioration of the environment. The effects of air pollution on the ecosystem and human health are explained in more detail in the following.

1.4.1 Health Effects of Air Pollution

1. Immediate health effects: According to Izah et al. (2024), brief contact with air pollutants such as nitrogen dioxide (NO_2), ozone (O_3), and particulate matter (PM_{10} and $PM_{2.5}$) might result in a number of respiratory illnesses, eye irritation, coughing, and headaches. In metropolitan areas, the impacts of these pollutants are especially noticeable in winter. Children under 5 years of age are far more susceptible to respiratory diseases due to household air pollution, which is produced by incomplete burning of fuels such as wood, coal, or kerosene. In this age range, it accounts for 44% of pneumonia-related deaths.
2. Long-term health effects: Long-term or chronic exposure to pollutants such as $PM_{2.5}$ increases the risk of lung cancer, asthma, chronic obstructive pulmonary disease (COPD), and cardiovascular disease (Fig. 1.2). Systemic inflammation is caused by particulate matter particles that penetrate the bloodstream and lungs deeply. Due to oxidative stress caused by pollutants, neurological conditions including Alzheimer's disease, dementia, and cognitive decline can also be brought on by air pollution. Children and the elderly are among the highest exposed to these contaminants, and those with preexisting medical issues are disproportionately impacted. According to medical research, children who are

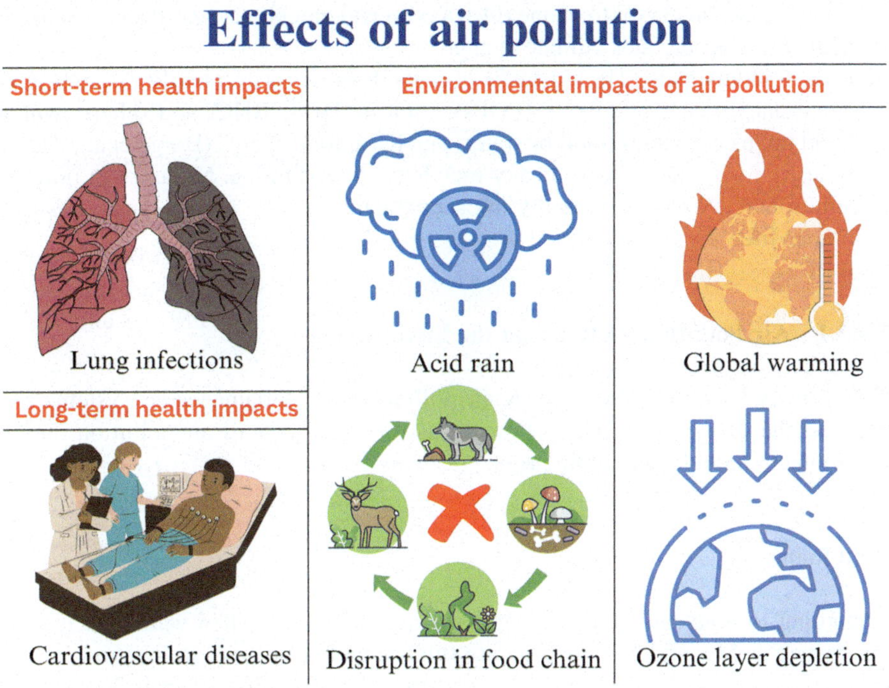

Fig. 1.2 Various effects of air pollution on human health and the environment

exposed to high levels of pollution are more likely to acquire asthma and have compromised lung development compromised.

3. Evidence from epidemiology: According to research by the European Environment Agency (EEA), $PM_{2.5}$ or fine particulate matter caused more than 238,000 premature deaths in the EU in 2020 (Chevet et al., 2024). A total of 3.2 million premature deaths are caused by air pollution each year, with a large proportion of women living in lower- or middle-income nations.

1.4.2 Environmental Impacts of Air Pollution

1. Degradation of ecosystems: Acid rain, which damages soils and water bodies and endangers aquatic life or plants, is the result of pollutants such as sulfur dioxide (SO_2) and nitrogen oxides (NO_x) (Izah et al., 2024). Food security in nations such as India that rely largely on agriculture may be threatened by ground-level ozone interference with photosynthesis, which destroys crops and lowers agricultural production.
2. Connections to climate change: By absorbing sunlight, air pollutants such as black carbon, a significant component of PM, contribute to global warming.

Methane is a precursor to ozone (O_3), a powerful greenhouse gas that worsens air quality and accelerates climate change.
3. Loss of biodiversity: The existence of several species is eventually threatened by persistent organic pollutants (POPs), such as DDT, BHC, and others, which build up in ecosystems (soil or water) and harm food chains (Borgå et al., 2022). Industrial activities release heavy metals (such as Pb, Hg, Ni, and Cr) into the atmosphere, which eventually bioaccumulate in fish bodies and endanger human life.

1.4.2.1 Air Pollution's Effects on the Environment

The ecology is severely affected by air pollution, which damages ecosystems and destroys the ozone layer. The following list of the effects of air pollution on the environment includes scientific proof.

Acid Rain

According to Sedyaaw et al. (2024), acid rain is created when water reacts with gases such as SO_2 and NO_x to generate sulfuric and nitric acid. Rain, snow, or fog can carry these acids to the ground, affecting the ecology. Acid rain has several environmental impacts that are listed as follows:

1. Aquatic ecosystems: Acid rain affects aquatic life by lowering the pH of lakes and rivers. As a result, aluminum from soils may leak into water bodies, which is harmful to aquatic life, such as fish and algae.
2. Soil fertility: By removing vital components such as calcium, magnesium, phosphorus, and nitrogen, acid rains lower soil fertility and have a negative impact on plant development.
3. Damage to infrastructure: It can erode materials used in infrastructure, such as marble, concrete, and steel, raising building maintenance expenses.

Changes in Climate

Three main greenhouse gases that trap heat in the atmosphere and cause global warming are carbon dioxide (CO_2), methane (CH_4), and nitrous oxide (N_2O). Numerous human endeavors, such as the burning of fossil fuels, agriculture, and industrial operations, release these gases into the atmosphere. Here is a summary of the main ways air pollution contributes to climate change.

1. Temperature increase: As the world's temperature rises, extreme weather phenomena such as heat waves, cyclones, and droughts become more frequent.
2. Increase in sea level: Coastal ecosystems are at risk due to the increase brought on by melting glaciers and ice sheets.

3. Ecosystem disruption: As a result of habitat changes caused by climate change, species distribution is changing, and extinction risk is rising, which lowers biodiversity.

Depletion of Ozone

In general, solvents, aerosol sprays, and refrigeration all employ chlorofluorocarbons (CFCs) and other halogenated gases. These chemicals have the ability to enter the stratosphere, especially the ozone layer, where they can destroy ozone (O_3) molecules by releasing chlorine and bromine atoms (Molina, 2020).

1. Increased UV radiation: By absorbing damaging ultraviolet (UV) radiation, the ozone layer protects all life on Earth. Skin cancer rates increase as a result of increased light to UV exposure to the human body caused by depletion of the ozone layer.
2. Impacts of the ecosystems: In a similar vein, increased UV radiation can harm marine ecosystems and plants, which might have an impact on the ecosystem's food chains and biodiversity.

1.5 Air Quality Standards and Regulations

To reduce the negative consequences of air pollution on the environment and human health, air quality standards and regulations are essential. Numerous national and international organizations have established these guidelines to direct the formulation and implementation of policies. An outline of the major groups and agreements that helped create these standards may be seen below.

1.5.1 Global Air Quality Guidelines (AQGs) Issued by the World Health Organization (WHO)

To protect public health worldwide, the WHO offers evidence-based recommendations for air pollutant concentrations. The 2021 update contains recommendations for carbon monoxide (CO), nitrogen dioxide (NO_2), sulfur dioxide (SO_2), ozone (O_3), and particle pollution ($PM_{2.5}$ and PM_{10}). These recommendations work as a worldwide standard for nations looking to improve air quality and lessen its negative effects on human health. Although not legally binding, they have an impact on national norms and policies and this is represented in Table 1.1.

Table 1.1 Global air quality guideline by WHO (Krzyzanowski & Cohen, 2008)

Pollutant	Averaging period	AQG (μg/m^3)
$PM_{2.5}$	Annual mean	5
	24 h	15
NO_2	Annual mean	10
	24 h	25
O_3	Peak season, 8-h mean	60

1.5.2 The Environmental Protection Agency

Particulate matter (PM), ozone (O_3), nitrogen dioxide (NO_2), sulfur dioxide (SO_2), carbon monoxide (CO), and lead (Pb) are the six common air pollutants for which the EPA in the United States establishes legally binding regulations. To ensure that these standards safeguard public health and welfare, the EPA periodically analyzes and revises them in light of the most recent scientific findings. State implementation plans, which include methods to meet and upholding these criteria, are used to enforce adherence to NAAQS.

1.5.3 International Treaties

The Kyoto Protocol is a global agreement to cut greenhouse gas emissions. It set legally enforceable goals for wealthy nations to cut emissions between 2008 and 2012 by an aggregate of 5% below 1990 levels. A worldwide framework was established by the 2015 Paris Agreement to keep global warming well below 2 °C and strive to keep it to 1.5 °C beyond preindustrial levels was established by the 2015 Paris Agreement. It pushes nations to submit Nationally Determined Contributions (NDCs) detailing their goals for reducing greenhouse gas emissions. These treaties focus on preventing climate change by lowering carbon dioxide (CO_2), methane (CH_4), and other greenhouse gases emissions. To reduce the effects of air pollution, air quality standards and regulations are essential. While national organizations such as the EPA create legally binding rules, the WHO offers worldwide guidance. Global environmental issues such as climate change are addressed by international agreements such as the Paris Agreement. Public participation, technology innovation, and policy enforcement must all be combined for these requirements to be implemented effectively.

1.6 Mitigation and Control Strategies for Air Pollution

An integrated strategy that incorporates public participation, regulatory regulations, and technology improvements is necessary for air pollution mitigation. The following is a discussion of air pollution reduction strategies.

1.6.1 Technologies for Emission Control

In industrial settings, technologies including fabric filters, scrubbers, and electrostatic precipitators are frequently used to lower particle emissions. While vehicles are equipped with catalytic converters, which transform toxic gases such as carbon monoxide (CO) and nitrogen oxides (NO_x) into less toxic substances such as carbon dioxide (CO_2) and nitrogen, scrubbers can eliminate contaminants such as sulfur dioxide (SO_2) from flue gases using a liquid spray (Wilailak et al., 2021). The tiny particulate matter (PM) from industrial pollutants can be captured using baghouses, which are fabric filters.

1.6.2 Adoption of Renewable Energy

Renewable energy sources, such as solar, wind, and hydro power, can replace fossil fuels, lowering greenhouse gas emissions and atmospheric air pollutants. To cut emissions, developed nations such as Canada are providing incentives for sustainable energy projects in isolated and rural areas (Agu et al., 2023). The Paris Agreement (2015) global climate targets are also in line with the use of renewable energy. Infrastructure modifications and a large investment are needed to make the switch to renewable energy. However, international initiatives to decarbonize energy systems are gathering steam and several countries are putting into place that will phase down coal-fired power plants by the 2030s.

1.6.3 Planning for the City

Public transit is being promoted by developed nations to reduce automobile emissions, which are responsible for up to 70% of urban air pollution. By serving as organic carbon sinks and mitigating the impacts of heat islands, green space expansion improves air quality. Nitrogen dioxide (NO_2) levels in urban areas have already decreased by 94% as a result of London's implementation of Ultra Low Emission Zones (ULEZ), which impose fees on cars that do not meet emission requirements (Ma et al., 2021). Walking and bicycling are encouraged by investments in bicycle infrastructure and pedestrian-friendly urban planning, which further reduces dependency on vehicular transportation.

1.6.4 Policies and Public Awareness

Behavior change can be achieved by teaching communities' new ways to combat pollution and how to minimize it. Campaigns to promote energy-efficient cooking stoves, for instance, have decreased exposure to air pollution in households in low-income areas.

1.7 Computational Modeling of Air Pollution

In order to comprehend and forecast the behavior of various contaminants in the atmosphere, computational modeling is crucial. These models replicate the dispersion of pollutants, chemical reactions, and transport processes in a range of environmental and climatic scenarios. Chemical transport models (CTM) and atmospheric dispersion models (Fig. 1.3) are the two main types of models that are frequently used to evaluate air quality and direct mitigation and regulatory actions.

1.7.1 Models of Atmospheric Dispersion

The primary goal of these models is to replicate the way pollutants propagate through the atmosphere following their release from a source. To forecast pollutant concentrations at certain sites, they take into account emission characteristics, topographical factors, and meteorological data. Typical kinds of model for atmospheric dispersion include the following.

1. Gaussian plume model: Among the first and most popular models is the Gaussian plume model, which postulates the dispersion of pollutants from the source and follows a normal or Gaussian distribution in both horizontal and vertical directions (Abdel-Rahman, 2008). It is mainly used for the dispersion of near-field pollutants, such as those from limited sources or industrial stacks. Due to its oversimplified assumptions, it is not very successful for long-distance transportation or complicated terrain.
2. Lagrangian particle models: These models use random turbulence effects to monitor individual pollutants as they travel through the atmosphere. It offers thorough depictions of the dispersion of pollutants, which makes it appropriate

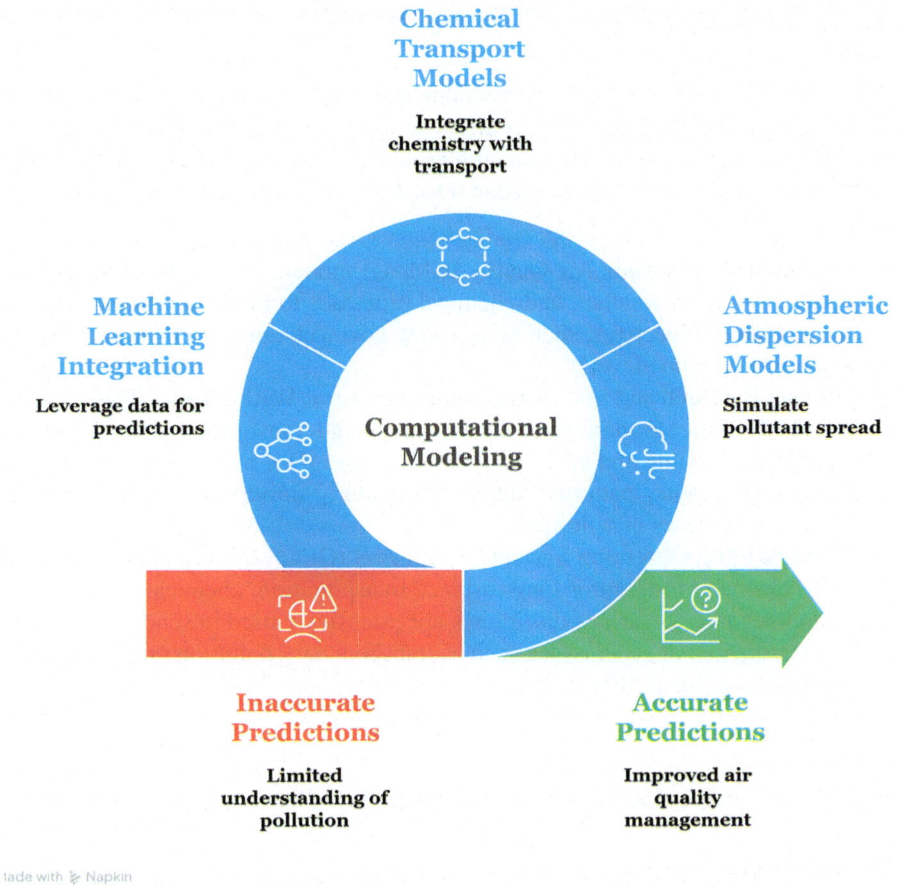

Fig. 1.3 Types of computational modeling to predict the behavior of different air pollutants in the atmosphere to improve air quality management

for intricate situations such as metropolitan settings or fluctuating weather patterns (Yang et al., 2021). A popular Lagrangian puff model that takes into account long-distance transit and weather into consideration is the CALPUFF model.

3. Models of Eulerian grids: In order to simulate the transport, dispersion, and transformation of pollutants, grid models typically include dividing the atmosphere into grid structures and solving mass balance equations for each grid (Wania & Mackay, 1999). These models work well in analyzing regional air quality when there are several sources and chemical processes at play. The adherence of both urban and rural regions to air quality norms is frequently assessed using models such as AERMOD.

1.7.2 Models of Chemical Transport (CTM)

These models replicate chemical reactions and long-term movement of contaminants through the atmosphere by combining atmospheric chemistry with transport mechanisms. These models are crucial for understanding secondary pollutants such as particulate matter ($PM_{2.5}$) and ozone (O_3). A discussion of some significant CTM is given as follows.

1. Community Multiscale Air Quality (CMAQ) model: This regional CTM integrates emissions, weather, and chemical processes to model air quality utilized for regulatory objectives, such as assessing how well emission control measures affect PM or ozone levels.
2. Weather Research and Forecasting with Chemistry (WRF-Chem): This program creates dynamic simulations of the movement and transformation by combining weather forecasting and air quality modeling useful for researching how air pollution and weather patterns interact, particularly during extreme occurrences such as heat waves or wildfires.
3. Goddard Earth Observing System—Chemistry (GEOS-Chem): An international CTM designed to monitor long-distance transportation, chemical changes, and pollution sources. Various disciplines are extensively used in scholarly research to examine global-scale issues such as transboundary pollution or how air quality is affected by climate change.

1.7.3 Machine Learning in Air Pollution Prediction

Machine learning (ML) has become a powerful instrument for forecasting air pollution levels, locating pollution sources, and understanding intricate connections between weather patterns and pollutant concentrations. ML models use large datasets to produce precise and timely predictions that support policymaking and air quality management. Here are a few machine learning techniques that are frequently applied to forecasting of air pollution.

1.7.3.1 Supervised Learning Approaches

Labeled data are used in supervised learning to train models that predict air quality indices (AQI) or pollutant concentrations. Common methods include the following.

1. Regression models: The concentrations of pollutants are predicted using both linear and nonlinear regression techniques, which are based on historical data and meteorological factors. For instance, $PM_{2.5}$ and PM_{10} levels may be well predicted using regression-based models.

2. Artificial neural networks (ANNs): ANNs are able to capture nonlinear correlations between temperature, humidity, and pollution levels and meteorological data. When estimating the amounts of particular pollutants and the AQI in metropolitan areas, these models have proven to be quite accurate.
3. Random forests and gradient boosting: These ensemble learning techniques combine several decision trees to achieve great accuracy. To assist determine the most important aspects that influence air quality, they also provide feature significance analysis. The effect of weather factors, including wind velocity and atmospheric pressure, on the amount of pollution has been evaluated using random forests.

1.7.3.2 Unsupervised Learning for Air Pollution Clustering

Patterns in unlabeled data are analyzed using unsupervised learning techniques, frequently to find pollution hotspots or simplify datasets.

1. K-means clustering: This technique identifies hotspots where mitigation efforts should be concentrated by grouping sites with comparable pollutant characteristics. For example, clustering of K-means has been used to categorize urban regions according to changes in PM 2.5 concentrations.
2. Principal component analysis (PCA): PCA improves visualization and comprehension of the main elements that cause pollution by reducing the dimensionality of large datasets of air pollution. Seasonal changes in pollutant concentrations and their relationship to meteorological factors have been examined using it.

1.7.3.3 Deep Learning for Air Quality Forecasting

Deep learning methods are perfect for forecasting air quality because they are excellent at handling complicated datasets with temporal or geographic correlations.

1. Long short-term memory (LSTM) and recurrent neural networks (RNNs): Because they are made for time series data, LSTMs and RNNs are useful for predicting future pollution levels based on past patterns. According to studies, LSTMs have a goodness-of-fit value of over 90% and can accurately predict AQI, as well as particular pollutants such as NO_2 and O_3.
2. Convolutional neural networks (CNNs): CNNs are used to identify regional pollution trends by identifying spatial patterns in sensor data or satellite images. $PM_{2.5}$ concentrations have been predicted using hybrid models that integrate spatial and temporal variables by combining CNN and LSTMs.

1.7.4 Big Data Analytics for Air Pollution Management

In recent times, with the advent of big data tools along with IoT (Internet of Things) and remote sensing, the management of air pollution got a new direction. These tools and techniques in an integrated manner analyze the massive amount of big data that enables monitoring, prediction, and management in real-time.

1.7.4.1 Sources of Air Pollution Big Data

Satellite Imagery

Satellites such as MODIS (Moderate Resolution Imaging Spectroradiometer), Sentinel-5P, and TROPOMI monitor pollutants such as NO_2, SO_2, and particulate matter ($PM_{2.5}$) and provide global datasets on air quality. These datasets are prerequisite for tracking long-term trends and identifying transboundary pollution.

Ground-Based Sensors

At local scale, AQI stations and IoT-enabled sensors collect real-time data on pollutants. This data collection can be seen in cities such as Delhi, where sensor networks are used to monitor $PM_{2.5}$ levels and inform policy interventions.

Databases of Traffic and Industrial Emissions

Industrial and vehicular emissions data often help to identify primary pollution sources. These databases are used to design targeted mitigation strategies, such as emission caps or traffic restrictions.

1.7.4.2 Techniques in Big Data Analytics

When working with large-scale air pollution data, several computational frameworks play a key role in processing and interpreting the information effectively. Distributed systems such as Hadoop and Spark are particularly useful for handling massive datasets. Hadoop's MapReduce, for instance, has been applied to environmental monitoring data to support predictive modeling, while Spark's real-time processing is great for instant updates, such as calculating the Air Quality Index (AQI) on the fly. Beyond raw data processing, geospatial analysis tools such as Geographic Information Systems (GIS) and remote sensing help track pollution patterns across different regions. These methods have proven to be valuable in mapping pollution hotspots and understanding how urban expansion affects air quality over time.

Another important approach is data fusion, which brings together multiple sources, such as satellite images, ground-level sensor readings, and weather data, to create a more complete picture of air pollution. For example, merging IoT sensor data with satellite observations can significantly improve the precision of pollution assessments. By combining these techniques, researchers and policymakers can make better informed decisions about air quality management.

1.7.4.3 Application of Big Data Analysis in Air Pollution Management

Big data is changing the game when it comes to managing air pollution, giving us smarter ways to track, predict, and even prevent dangerous air quality levels, e.g., real-time monitoring. IoT sensors now provide constant updates on pollution, helping authorities alert the public faster when smog or other hazards spike. But it does not stop there. Thanks to machine learning, we can now predict air quality with impressive accuracy. Hybrid models such as ARIMA-NN (which blends traditional forecasting with neural networks) and deep learning tools such as RNN (which are great at spotting patterns in historical data) are making these forecasts sharper than ever.

The real power comes from combining different data sources, such as satellite imagery, ground sensors, industrial emissions reports, and crunching them with tools such as Hadoop, geospatial mapping, and data fusion. This helps cities monitor pollution in real time, make better predictions, and shape policies that work. Of course, there are still hurdles. Data is not always perfect, and not every city has the expertise to use these tools effectively. But as technology continues to improve, big data will be the key to achieving global clean air goals, such as the UN Sustainable Development Goals (SDGs). The bottom line? Smarter data means cleaner air, and that is something everyone can breathe easier about.

1.7.4.4 Geographic Information Systems (GIS) in Air Pollution Analysis

GIS has become an important component in the analysis and management of air pollution. Integrating spatial variability data with advanced analytical techniques, GIS provides proper insights into pollution patterns, sources, and their effect on human population and environment. The application of GIS techniques to detect air pollution is discussed below.

1.7.4.5 Spatial Mapping of Pollution Sources

GIS enables the visualization of pollution sources such as traffic emissions, industrial hotspots, and power plants on geospatial maps of specific locality. This spatial representation helps identify areas with high pollutant concentrations and prioritize mitigation efforts.

Industrial Hotspots

Air pollution can be prepared in industrial regions such as Gujarat and Maharashtra in India with the help of GIS where areas identified with cumulative pollutants such as SO_2, NO_2, and $PM_{2.5}$ (Guttikunda & Jawahar, 2012).

Traffic Emissions

Urban areas often use GIS to monitor vehicular emissions. For example, GIS integrates traffic databases with emission inventories to locate high-emission zones and assess their impact on urban air quality.

1.7.4.6 Interpolation Techniques for Pollution Estimation

GIS uses interpolation methods to estimate pollutant concentrations in unsampled locations, providing a continuous spatial representation of air quality.

Inverse Distance Weighting (IDW)

IDW estimates pollution levels by assigning higher weights to nearby data points. This method has been used in Delhi to spatially predict $PM_{2.5}$ concentrations and their association with health outcomes such as emergency room visits for respiratory symptoms.

Kriging Methods

Kriging is a geostatistical technique that provides more accurate spatial predictions by considering both the distance and the spatial autocorrelation of data points. Studies have used kriging to model the spatial variability of pollutants such as $PM_{2.5}$ and NO_2 in industrial cities, enabling better understanding of their distribution patterns.

1.7.5 GIS-Based Risk Assessment and Mitigation Planning

Geographic Information Systems (GIS) have become an indispensable tool for understanding how air pollution impacts communities and developing targeted solutions. By combining real-time pollution measurements with population data, health records, and urban infrastructure details, GIS helps paint a clear picture of who is most at risk and where interventions are needed most urgently. We have seen powerful examples of this in cities such as New Delhi, where researchers mapped particulate matter levels against emergency room visits for respiratory issues, revealing disturbing patterns—especially during winter months when pollution peaks. The same approach has proven valuable in rural areas near industrial zones, where GIS analysis helps identify vulnerable populations that are exposed to dangerous levels of pollution without even realizing it.

What makes GIS truly valuable is how it turns complex data into actionable plans for cleaner air. Urban planners now routinely use these systems to determine optimal locations for green spaces that can filter polluted air, identify neighborhoods that would benefit most from traffic reduction measures, and even model how proposed infrastructure changes might affect local air quality before shovels hit the ground. Technology has gone beyond simple mapping to become a dynamic planning tool that considers everything from weather patterns to commuting routes. For instance, by layering school locations with real-time pollution data, some cities have implemented alternate walking routes during high-smog days to protect children's health. This practical application of spatial analysis represents a major leap forward in how we approach environmental health challenges, transforming raw data into life-saving policies and community protections. The true power lies in GIS's ability to make invisible threats visible and provide concrete ways to address them.

1.7.6 Computational Approaches in Air Pollution Control Strategies

The field of air pollution control is undergoing a radical transformation as computational methods bring unprecedented precision to emission reduction strategies. Traffic management systems now rely heavily on sophisticated modeling techniques that analyze real-time vehicle movement patterns through IoT sensor networks, allowing cities to dynamically adjust traffic light sequences and significantly cut down on congestion-related emissions. Predictive capabilities have grown remarkably sophisticated—where traditional fluid dynamics models once struggled with processing delays, machine learning approaches such as artificial neural networks now deliver faster, more accurate forecasts of particulate matter concentrations along busy roadways. The latest hybrid systems combine these neural networks with time series analysis methods such as LSTM, creating emission prediction tools that are both computationally efficient and reliable enough for urban planning applications (Singh & Suthar, 2025).

The industrial sectors are seeing similar advancements through computational optimization. Manufacturing facilities employ genetic algorithms to simulate countless production scenarios, identifying configurations that maintain output while reducing energy use and associated emissions. The same plants use computational fluid dynamics simulations to model how pollutants disperse from smokestacks, allowing engineers to strategically position scrubbers and filtration systems where they will have maximum impact. Even more impressive are emerging deep learning applications such as the Deep-learning-based Response Surface Model, which merges convolutional neural networks with chemical transport simulations. These systems can predict with astonishing accuracy how complex atmospheric pollutants such as $PM_{2.5}$ and ozone will respond to specific emission control measures, giving regulators powerful tools for policy testing before implementation.

Smart city infrastructures are amplifying these benefits through widespread integration of the Internet of Things. Dense networks of air quality sensors across urban centers continuously feed data on nitrogen dioxide, sulfur dioxide, and fine particulate matter to cloud-based analysis platforms. Singapore's approach demonstrates this capability in action: its systems detect emerging pollution hotspots in real time, triggering adaptive responses such as redirecting traffic flows or temporarily scaling back industrial operations in affected zones. The true power emerges when these ground-level sensor readings combine with satellite observations and weather data through advanced fusion techniques, creating comprehensive pollution maps that inform dynamic policy adjustments. This integrated approach enables measures such as restricting high-emission vehicles during critical periods, with enforcement precisely targeted to where and when it will make the most difference.

1.7.6.1 Challenges in Computational Air Pollution Analysis

Data Quality and Availability

For efficient working of the models, two factors are critical: (1) completeness and (2) consistency of datasets obtained from sensors and emission inventories. Data processing techniques such as imputation or outlier detection are essential but resource-intensive.

Computational Complexity

High-resolution models, such as chemical transport models (CTMs), require substantial computational power and time. Simplified models, such as ANN-based predictors, offer faster alternatives, but not much reliable.

Model Uncertainty

Variability in input parameters such as meteorological conditions introduces uncertainties in predictions. Techniques such as ensemble modeling can mitigate this issue but add complexity to model development.

1.7.7 Remote Sensing and GIS in Air Pollution Mitigation

Traditionally, ground-based instrumentation has been used to monitor air pollution and its quality, which has largely limited the geographical area that can be monitored. Many of the ground-based equipment cannot accurately describe the overall concentration of all pollutants at a city-wide level because they are designed to measure individual contaminants. In this regard, it is crucial to employ remote sensing satellites to support the air quality monitoring. The process of collecting data about an object without coming into direct contact with it is known as remote sensing. To gather data to be integrated into a Geographic Information System (GIS), it is a valuable tool. Geographic Information Systems are crucial instruments for combining and evaluating spatial data from many fields and sources. All types of spatially linked data can be captured, managed, analyzed, and displayed using this integrated system, which consists of hardware, software, and data. This computer-based application is used to map and analyze objects and events that occur on Earth. Maps, tables, satellite remote sensing, and the Global Positioning Satellite (GPS) system are some of the data sources used in GIS. GIS combines the special visualizations and advantages of spatial analysis provided by maps with standard database functions such as query and statistical analysis. A GIS is a tool that helps decision-makers solve problems by offering several approaches to issues pertaining to people, the environment, planning, and the planet as a whole. Numerous atmospheric studies have been conducted using on-site measurements and a variety of mathematical models, including particulate matter and gaseous pollutant concentrations, measurements, and evaluations of air quality and dispersion. Remote sensing is used in places where fundamental infrastructure is far away and cannot be found. Remote sensing is the ability to sense the Earth's surface from space by utilizing the characteristics of electromagnetic waves emitted, reflected, or diffracted by sensed including models of air quality in remote sensing: The first step involves using a standalone software program to simulate air quality models, complete with input and outputs. Other software systems such as GIS, RDBMS, and so on can use all of the data on their own. To assess air quality of the air in cities, four main satellite remote sensing methods are employed. The first and most well-known approach is measuring the thickness of aerosols in the atmosphere; the second is detecting black particles; the third is visually inspecting satellite imagery; and the fourth is analyzing changes in land-use and land-cover. Choosing the right satellites for pollution monitoring requires careful consideration of each mission's specific capabilities. To track black carbon particles within atmospheric aerosols, satellites equipped with high-resolution sensors prove particularly valuable, as they can detect these fine particulates with greater precision. The advent of satellite remote sensing has fundamentally changed environmental monitoring, providing an unprecedented ability to observe pollution patterns across vast regions from orbit. This space-based perspective enables continuous tracking of key air pollutants such as carbon dioxide, sulfur dioxide, nitrogen dioxide, and ground-level ozone, while also monitoring changes in aerosol concentrations over time. The technology's ability to deliver near-real-time data allows for rapid identification of pollution hotspots, supporting more responsive environmental management decisions. Beyond atmospheric monitoring, these systems also track water and land pollution through various spectral bands—infrared sensors detect heat signatures from industrial emissions, while visible spectrum instruments identify oil slicks on water surfaces. This multispectral capability provides comprehensive environmental intelligence that helps bridge the gap between scientific research and practical policy implementation.

Although satellite systems offer significant advantages over ground-based monitoring in terms of coverage and cost-effectiveness, they do face certain operational limitations. Spatial resolution constraints sometimes make it difficult to detect hyperlocal pollution variations, particularly in urban areas with complex microclimates. The substantial expenses involved in satellite development, launch, and ongoing operation can limit access for some nations and organizations. Data interpretation requires specialized expertise that may not be readily available in all regions, and atmospheric conditions, such as persistent cloud cover, can temporarily obstruct measurements in certain locations. Despite these challenges, the unique benefits of global-scale monitoring, combined with the improvement of sensor technologies and data processing capabilities, continue to make satellite remote sensing an indispensable tool for environmental protection. The ability to provide both policymakers and citizens with accessible pollution data through web platforms and mobile applications represents a significant advancement in environmental transparency and public engagement. As sensor technologies advance and data processing techniques improve, these orbital observation systems are becoming increasingly sophisticated in distinguishing between different pollution and sources, though they still work best when combined with targeted ground measurements for calibration and verification purposes. Integration of satellite data with other monitoring systems creates a more complete picture of environmental quality that supports everything from international climate agreements to local air quality warnings (Fig. 1.4).

1.7.8 Management of air Pollution through Spatial Data

The process of managing air quality involves choosing key locations and using ground-based instruments to measure the concentration of pollutants there. These instruments are expensive and lack spatial continuity and, because of the limited number of air pollutant locations in each area, they are unable to provide a good spatial distribution of air pollutant readings over a city or area. Furthermore, measurement stations might not be easily accessible in developing countries or isolated places. Satellite photos of the Earth's surface in the solar spectrum are impacted by the atmosphere. Therefore, the signal that the satellite sensor saw was the total of the atmospheric and ground impacts. Because atmospheric aerosol distributions vary widely in terms of type, size, space, and time, it is challenging to quantify the change in radiation caused by tropospheric aerosols (Mishra et al. 2020). Remote sensing capabilities are being researched in the area of controlling and tracking air

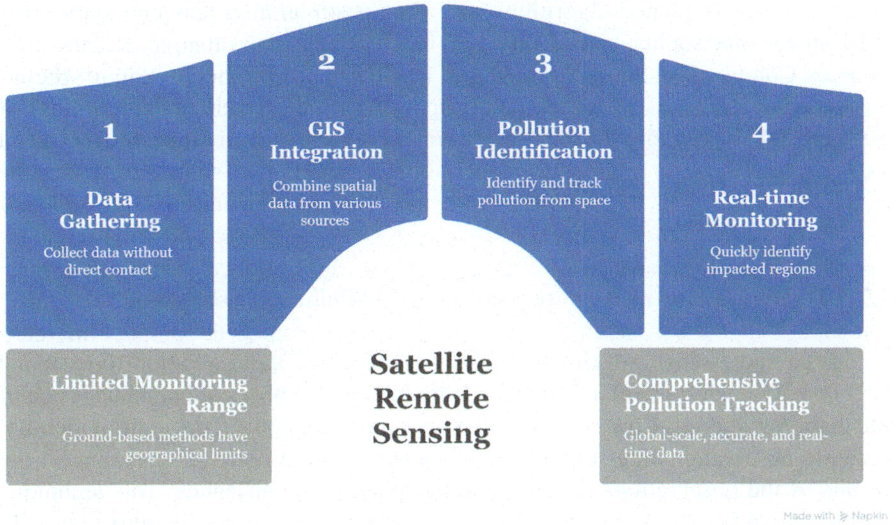

Fig. 1.4 Satellite remote sensing for air-quality intelligence—modern Earth-observing satellites complement limited ground stations by capturing multispectral data on aerosols, trace gases (NO_2, SO_2, O_3), black-carbon hotspots, and land-use change. Integrated with GIS, these spaceborne measurements generate city to global scale, enable real-time alerts, guide policy, and enable citizens to check local air quality from any smartphone—transforming how we detect, track, and combat environmental pollution despite challenges of resolution, cost, and cloud cover

pollution. It offers a structure for creating an inventory database and a themed map for air pollution studies.

1.7.9 Use of Dispersion Modeling in Air Pollution Management

Using specific emission sources, meteorological scenarios, land use, terrain, meteorological data, and a measure of the total atmospheric concentration of the metrics in the model domain, air quality dispersion modeling is a computer simulation that provides a solution by estimating the impact of point, line, volume, and area sources on surface air quality in the atmosphere. To determine concentrations, it uses meteorological information including temperature, mixing height, wind direction, and wind speed. This is applied to a variety of airborne gaseous and particle contaminants. It helps to classify an area as very, fairly, or moderately polluted based on the temporal and spatial variance of PM. A number of models are in use, including the American Meteorological Society and the U.S. Environmental Protection Agency Regulatory Model (AERMOD), the Industrial Source Complex Short Term Version 3 (ISCST3), the Fugitive Dust Model, and the Point, Area, and Lines Sources Model (PAL2). The most popular model among all of them is AERMOD. It was created using more sophisticated algorithms and ideas, such as planetary boundary layer (PBL) theory and sophisticated techniques for complicated terrains, by the Industrial Sources Complicated Short-Term Model (ISCST3). To provide a state-of-the-art dispersion model for everyday regulatory applications, the American Meteorological Society/EPA Regulatory Model Improvement Committee created AERMOD, a steady-state Gaussian plume dispersion model targeted at short range (Landis et al., 2021).

1.7.9.1 Limitations of Remote Sensing in Air Pollution Estimation

Satellite remote sensing technology has a few drawbacks in addition to its many benefits, of which spectral interference from other atmospheric inhabitants is one of the main restrictions. Low concentrations of pollutants make it difficult to identify. Satellite observations are limited to atmospheric window wavelength ranges; the results of the observations depend on atmospheric circumstances. This technique will require a highly skilled workforce and the procedure is exceedingly costly. In order to monitor environmental conditions and protect human health, air quality measurement is essential. This initiative uses a variety of tools that follow global guidelines established by agencies such as the Environmental Protection Agency (EPA) and the World Health Organization (WHO). These devices, intended for research and regulatory compliance, include portable and fixed air quality monitors that detect PM ($PM_{2.5}$, PM_{10}), gases (O_3, NO_2, SO_2, and CO), and volatile organic compounds (VOCs), all at the same time (Higgins et al., 2024). To facilitate real-time monitoring and analysis, a lot of equipment is connected to data loggers. Although gas analyzers measure atmospheric gases including CO_2, SO_2, NO_2, and O_3, specific instruments such as PM monitors, gravimetric (weight-based), or optical measure small particles ($PM_{1.0}$, $PM_{2.5}$, PM_{10}) (Moazami et al., 2023). Like regular weather stations, meteorological stations help predict the dispersion of pollutants by providing information on temperature, humidity, wind direction, and speed.

Passive sampling devices offer an affordable solution for long-term pollution monitoring, although they sacrifice real-time data capabilities. For broader spatial coverage, remote sensing technologies such as satellite systems and LiDAR (light detection and ranging) have become indispensable tools, despite their substantial upfront costs (Liang et al., 2023). Instruments such as the Moderate Resolution Imaging Spectroradiometer (MODIS) demonstrate this trade-off—while providing valuable large-scale data, they require careful selection based on regional regulations, target pollutants, and specific monitoring needs. In the realm of gas detection, electrochemical sensors such as Alphasense NO2-B4F3 for nitrogen dioxide and SPEC Sensors 110–601 for sulfur dioxide deliver impressive sensitivity, but their performance can be compromised by cross-reactivity with other gases and environmental factors such as temperature fluctuations (Thingujam et al. 2025). These sensors also face durability challenges, which often require frequent recalibration, especially in variable outdoor conditions. Recent technological advances have introduced more sophisticated options such as photo-ionization detectors (PIDs), which excel at detecting volatile organic compounds at minute concentrations, although their higher costs and maintenance requirements limit widespread adoption (Maeso-García et al., 2021). The field has seen significant progress with Micro-Electro-Mechanical Systems (MEMS) technology, where compact devices such as Sensirion's SGP30 sensor enable portable, multigas detection ideal for indoor air quality monitoring. However, these microsensors still struggle to maintain consistent performance in outdoor environments exposed to changing weather patterns (Algamili et al., 2021). Researchers are increasingly turning to machine learning algorithms to address these limitations, using advanced data processing techniques to compensate for sensor drift and environmental interference. This approach shows particular promise for enhancing the reliability of lower-cost sensor networks, potentially bridging the gap between affordable monitoring solutions and traditional high-end instrumentation. Effective air quality measurement continues to evolve through these technological innovations, but requires ongoing refinement to address the complex challenges of pollution source identification and environmental variability in different monitoring scenarios.

1.7.9.2 Understanding Air Pollution in Agriculture

Monitoring air pollution in agricultural settings presents unique challenges that require specialized sensor technologies to track emissions from livestock operations and chemical applications. Livestock facilities release levels of ammonia (NH_3), methane (CH_4), volatile organic compounds (VOCs), and particulate matter (PM), while pesticide and herbicide applications contribute additional airborne contaminants that impact both environmental and human health (Traiwatcharanon et al., 2024). Modern sensor networks help researchers pinpoint pollution sources by correlating emission spikes with specific activities such as feeding schedules or waste management practices at animal facilities. These monitoring systems provide real-time data that, when combined with agricultural productivity metrics and health statistics, reveal the complex relationships between farming practices and their environmental consequences. Recent advances in sensor technology have introduced innovative nanomaterials—including conducting polymers, metal oxide semiconductors, and carbon-based structures—that significantly improve detection capabilities for key agricultural pollutants such as ammonia and nitrogen oxides (Choudhari & Jagtap, 2023). These next-generation sensors benefit from precise engineering controls on factors such as doping concentrations and annealing temperatures, enabling more flexible and intelligent monitoring solutions. Field studies demonstrate the practical value of these technologies, such as research documenting seasonal variations in cattle farm emissions and their connection to microclimate conditions (Popa et al., 2022). Compared to established modeling approaches such as the EMEP methodology, sensor-derived data provide farm managers with actionable insights to reduce environmental impacts. Comparative evaluations of sensor performance reveal important nuances: while CO and CO_2 detectors demonstrate reliable accuracy, NO_2 sensors tend to show weaker correlation with reference instruments (Whitehill et al., 2022). Spatial analysis of monitoring data consistently shows predictable pollution patterns, with elevated NO_2 levels near roadways and higher ammonia concentrations surrounding livestock operations, though interestingly, these measurable impacts do not always correlate directly with community odor complaints. The variation in stakeholder concerns and preferred mitigation approaches highlights the importance of combining technical monitoring data with inclusive community engagement processes to develop effective solutions.

1.7.10 Air Microbiome and Metagenomic Analysis in Air Pollution Studies

Metagenomics has revolutionized the way we study airborne microbes by allowing direct DNA analysis of air samples without traditional culture methods. Key techniques such as 16S rRNA amplicon sequencing and ITS regions reveal the composition of bacterial and fungal communities floating in our air. More comprehensive approaches include shotgun metagenomics, which examines all genetic material in an air sample, and meta-transcriptomics, which identifies which microbial genes are actively being expressed. These powerful tools help researchers track microbial diversity across different environments—from city streets to industrial sites to natural areas—while also showing how pollution alters airborne microbial communities. The health implications are significant since we are constantly breathing in bioaerosols containing everything from allergens to dangerous pathogens. Metagenomic analysis proves particularly valuable for several critical applications: detecting how air pollution changes the microbiome, identifying airborne disease-causing organisms and antibiotic resistance genes, and uncovering links between air quality and respiratory illnesses such as asthma and COPD. By providing this molecular-level view of what is in the air we breathe, these techniques offer new insights into environmental health risks that were impossible to see with older methods.

1.7.11 Microbial Interactions with Air Pollutants

The relationship between airborne microbes and common pollutants reveals a complex biological–chemical interaction in our atmosphere. Certain bacteria and fungi actively engaged with pollutants such as nitrogen oxides (NO_x), particulate matter (PM), and volatile organic compounds (VOCs), sometimes transforming these substances in surprising ways. Some microbial species, including various strains of *Pseudomonas*, demonstrate the ability to break down hazardous VOCs through natural metabolic processes. Meanwhile, diverse microbial communities frequently attach themselves to airborne particulate matter, effectively creating living coatings that alter the chemical properties. These biological modifications can significantly influence the way that particulate matter interacts with human respiratory systems, potentially changing the health impacts when inhaled. The microorganisms do not just passively coexist with pollution—they actively participate in atmospheric chemistry through these interactions, creating a dynamic system where biological and chemical processes continuously influence each other. This microbial activity adds another layer of complexity to understanding air pollution's effects, as the same chemical pollutant might behave differently depending on which microbes it encounters during its atmospheric journey. Antibiotic-resistant microbial strains can be found in highly polluted air environments, which can have an impact on public health and interfere with regular social interactions. Another promising strategy to reduce air pollution is microbial bioremediation. Examples include the use of genetically modified microbes with improved pollutant degradation capabilities, biofilters that use microbial consortiums to break down industrial emissions, and phytoremediation, which uses improved microbe interactions in which plant-associated microbes help purify the air.

Notwithstanding notable progress, several issues remain, such as the need to standardize DNA extraction and air sampling techniques, interpret metagenomic data in intricate air matrices, and address ethical concerns with the use of genetically modified bacteria for air purification. Future studies are required to investigate bioengineering for pollution mitigation, create microbial-based air quality monitoring instruments, and integrate multigenomic techniques.

Numerous genes are as follows: (1) Glutathione-S-transferases (GSTs): Variants in GSTM1 and GSTT1 impact antioxidant defense mechanisms, influencing susceptibility to damage induced by oxidative stress. These genes have been found to contribute to differential responses to air pollution and play a role in the regulation, inflammation, and detoxification of oxidative stress. (2) Cytochrome P450 (CYP) enzymes: Variants in CYP1A1 and CYP2E1 alter how harmful air pollutants are metabolized, adjusting personal risk levels. (3) Mutations in Nrf2 Pathway Genes (NFE2L2) affect cellular defense systems against oxidative damage caused by contaminants. (4) Inflammatory response genes: Variations in the levels of inflammatory reactions to pollutants are caused by genetic variations in TNF-α, IL-6, and IL-1β.

Exposure to pollutants and epigenetic changes without changing the DNA sequence, air contaminants can cause epigenetic modifications that change how genes are expressed. Among these changes are:

1. *NA methylation*: Changes in the methylation patterns of genes relevant to inflammation and immunological response have been linked to exposure to $PM_{2.5}$ and NO_x.
2. *Histone modifications*: Variations in histone acetylation and methylation can affect the expression of genes linked to lung function and oxidative stress.
3. *MicroRNAs (miRNAs)*: People exposed to high levels of pollution have variable expression of several miRNAs, which affect pathways linked to lung and cardiovascular disorders.

1.7.12 Gene–Environment Interactions in Pollution-Related Diseases

An important part of disease susceptibility is the interaction of environmental exposure and genetic factors, for instance COPD with asthma, i.e., when exposed to pollutants; people with particular GSTM1 or IL-4 gene variations are more prone to develop asthma or COPD. Cardiovascular disease as variants in the NOS_3 and APOE genes alter how exposure to air pollution affects the heart. In neurodegenerative disorders, $APOE_4$ carriers' genetic predisposition has been connected to a higher risk of neuroinflammatory and pollution-induced cognitive decline. Targeted interventions and policies, such as Risk Assessment Models, which integrate genetic data into exposure risk assessments to identify vulnerable populations, and Precision Medicine Approaches, which develop tailored interventions for people with high genetic susceptibility to air pollution, including policy recommendations for enforcing stricter air quality regulations for populations at greater genetic risk, can be informed by knowledge of genetic susceptibility to air pollutants. It is necessary to properly comprehend the complex connections between contaminants and genes. The generalizability of the results is restricted by variations in population genetics and research methodology. Concerns about discrimination and privacy are raised by ethical considerations when using genetic data for public health initiatives. To lessen the negative health effects of air pollution, researchers should concentrate on extensive genetic analyses, multi-omics techniques, and customized approaches.

1.7.13 Metagenomic Approaches for Studying Polluted Environments

In the microbial communities without the need for culture, metagenomics offers a deeper understanding of their makeup and functional roles. Important techniques consist of the following:

1. *Shotgun metagenomics*: Offers a thorough examination of microbial genomes, making it possible to identify the genes responsible for the breakdown of pollutants.
2. *Amplicon sequencing (16S rRNA, ITS)*: Evaluates the taxonomic makeup and diversity of microorganisms in polluted areas. The study of gene expression in microbial communities, or meta-transcriptomics, reveals metabolic pathways that are active in response to pollution.
3. *Metaproteomics and metabolomics*: Examine proteins and metabolites to understand the roles that microorganisms play in contaminated settings.

1.7.13.1 Microbial Interactions with Pollutants

In contaminated environments, microorganisms react to pollutants in a variety of ways such as adaptation, resistance, and deterioration. These interactions include heavy metal resistance and biotransformation by microbes such as *Pseudomonas* and *Bacillus* possess genes for metal resistance, bioaccumulation, and detoxification; biodegradation of organic pollutants, wherein specific bacterial and fungal species can break down hydrocarbons, pesticides, and plastics; and antibiotic resistance in polluted environments, where industrial and agricultural waste contributes to the spread of antibiotic resistance genes among microbial populations. In a variety of polluted environments, metagenomic studies have been used such as oil spill sites to identify hydrocarbon-degrading bacteria and their metabolic pathways, heavy metal-contaminated soils to characterize microbial communities involved in metal resistance and detoxification, industrial wastewater to analyze microbial consortia that contribute to the breakdown of pollutants and the cycling of nutrients, and airborne pollution to evaluate airborne microbial communities impacted by airborne emissions from urban and industrial sources.

Using the knowledge and findings of metagenomic analyses, microbial-based solutions for pollution mitigation can be developed. These include genetic engineering for biodegradation using synthetic biology to enhance microbial degradation capabilities, bioaugmentation for introducing pollutant-degrading microbes to contaminated sites, and bio-stimulation for increasing native microbial activity through nutrient supplementation. Microbial monitoring holds promise for developing biomarkers that assess environmental health and pollution impacts, but significant challenges remain. Understanding microbial interactions is particularly complex, as communities exhibit antagonistic and cooperative relationships that influence their response to pollutants. Standardizing metagenomic workflows presents another hurdle, requiring consistent protocols for sampling, DNA sequencing, and data analysis to ensure reliable comparisons across studies. Ethical and ecological concerns also arise, especially when considering the intentional release of engineered microorganisms for bioremediation: Potential risks must be carefully evaluated before large-scale applications. Future research should prioritize integrating multi-omics approaches (combining metagenomics, meta-transcriptomics, and metabolomics) to gain a more comprehensive view of microbial responses to pollution. Improved computational tools are needed to handle the large datasets generated by these techniques, and here machine learning and AI could prove invaluable. These technologies may help identify key genes involved in pollution adaptation and predict how contaminants affect biological systems, bridging the gap between microbial data and actionable environmental insights. Sustainable bioremediation strategies should also be explored, balancing effectiveness with ecological safety to avoid unintended consequences.

1.7.14 Computational Genomics in Microbial Pollution Studies

Computational genomics is revolutionizing our understanding of how microbes interact with air pollution by revealing their survival strategies at the genetic level. By analyzing microbial communities in urban, industrial, and natural environments, researchers can track how different levels of pollution shape these invisible ecosystems. The technology goes beyond simple identification—it pinpoints specific genes such as those coding for monooxygenases and dioxygenases that help microbes break down pollutants. One particularly fascinating discovery involves horizontal gene transfer, where microbes share resistance genes such as trading cards, rapidly spreading the tolerance to pollution through entire communities. On the human side, these genomic tools are equally valuable, helping identify why some people's genetics make them more vulnerable to pollution-related illnesses than others. The same sequencing methods that show us how bacteria adapt to dirty air can also reveal genetic variants that increase the risks of asthma or other conditions in polluted areas. This dual perspective—understanding both microbial adaptations and human susceptibility—gives us a much clearer picture of pollution's full biological impact.

1.7.15 Applications of Bioinformatics in Air Pollution Monitoring and Mitigation

Bioinformatics is transforming air pollution research through several innovative approaches that bridge environmental science and human health. AI-driven predictive models now analyze genetic datasets to predict how different pollution levels might impact population health, helping identify at-risk communities before problems emerge. On the detection front, scientists are engineering microorganisms to function as living biosensors—these biologically tuned detectors can pinpoint specific pollutants in real time, offering a dynamic alternative to traditional monitoring equipment. Perhaps most impactful is how genetic research informs personalized health strategies. By screening individuals for genetic markers of susceptibility to pollution, clinicians can develop targeted prevention plans for those most vulnerable to asthma, cardiovascular disease, or other pollution-related conditions. Together, these bioinformatics applications create a more nuanced understanding of air pollution's effects, from the molecular level up to population-wide health outcomes.

1.8 Conclusions

As air pollution continues to challenge communities worldwide, the fusion of computational methods and environmental science offers powerful solutions. Advanced machine learning algorithms and predictive models now enable more accurate pollution tracking and forecasting, transforming raw data into actionable insights for policymakers. The marriage of remote sensing with GIS technology provides an invaluable big-picture view, mapping pollution sources and movement patterns while identifying high-risk areas vulnerable to exposure. Meanwhile, breakthroughs in metagenomics reveal how microbial communities interact with pollutants, uncovering natural bioremediation potential through genetic adaptations. However, these promising approaches face practical hurdles—from the need for standardized genomic databases to the challenges of processing massive multi-omics datasets while maintaining data privacy. Looking ahead, the development of integrated AI platforms could bridge these gaps, combining predictive modeling with multi-omics analysis to create comprehensive pollution management systems. These technological advances, when paired with traditional environmental research, are reshaping how we monitor air quality, assess health risks, and develop targeted interventions, moving us toward smarter pollution control strategies that protect both ecosystems and public health. The path forward lies in strengthening these computational tools while ensuring that they remain accessible and actionable for communities that need them most.

Acknowledgments The authors express their sincere appreciation to all individuals whose direct or indirect contributions facilitated the successful completion of this work.

References

Abdel-Rahman, A. A. (2008). On atmospheric dispersion and the Gaussian plume model. In *proceedings of the 2nd international conference on waste management, water pollution, air pollution, indoor climate* (Vol. 26, pp. 1-10). Corfu, Greece.

Agu, O. S., Tabil, L. G., & Mupondwa, E. (2023). Actualization and adoption of renewable energy usage in remote communities in Canada by 2050: A review. *Energies, 16*(8), 3601. https://doi.org/10.3390/en16083601

Algamili, A. S., Khir, M. H. M., Dennis, J. O., Ahmed, A. Y., Alabsi, S. S., Ba Hashwan, S. S., & Junaid, M. M. (2021). A review of actuation and sensing mechanisms in MEMS-based sensor devices. *Nanoscale Research Letters, 16*(1), 1–21. https://doi.org/10.1186/s11671-021-03481-7

Bai, L., Wang, J., Ma, X., & Lu, H. (2018). Air pollution forecasts: An overview. *International Journal of Environmental Research and Public Health, 15*(4), 780. https://doi.org/10.3390/ijerph15040780

Baker, K. R., Woody, M. C., Tonnesen, G. S., Hutzell, W., Pye, H. O. T., Beaver, M. R., & Pierce, T. (2016). Contribution of regional-scale fire events to ozone and PM2.5 air quality estimated by photochemical modeling approaches. *Atmospheric Environment, 140*, 539–554. https://doi.org/10.1016/j.atmosenv.2016.06.032

Borgå, K., McKinney, M. A., Routti, H., Fernie, K. J., Giebichenstein, J., Hallanger, I., & Muir, D. C. (2022). The influence of global climate change on accumulation and toxicity of persistent organic pollutants and chemicals of emerging concern in Arctic food webs. *Environmental Science: Processes & Impacts, 24*(10), 1544–1576. https://doi.org/10.1039/D1EM00469G

Chevet, E., Boiron, O., & Anselmet, F. (2024). Modeling of air pollution due to marine traffic in Marseille. *Atmospheric Environment, 329*, 120542. https://doi.org/10.1016/j.atmosenv.2024.120542

Choudhari, U., & Jagtap, S. (2023). A panoramic view of NO_x and NH_3 gas sensors. *Nano-Structures & Nano-Objects, 35*, 100995. https://doi.org/10.1016/j.nanoso.2023.100995

Diamond, M. L., de Wit, C. A., Molander, S., Scheringer, M., Backhaus, T., Lohmann, R., & Arvidsson, R. (2015). Exploring the planetary boundary for chemical pollution. *Environment International, 78*, 8–15. https://doi.org/10.1016/j.envint.2015.02.001

Frank, M., Drikakis, D., & Charissis, V. (2020). Machine-learning methods for computational science and engineering. *Computation, 8*(1), 15. https://doi.org/10.3390/computation8010015

Guttikunda, S. K., & Jawahar, P. (2012). Application of SIM-air modeling tools to assess air quality in Indian cities. *Atmospheric Environment, 62*, 551–561. https://doi.org/10.1016/j.atmosenv.2012.08.074

Hansell, A., & Oppenheimer, C. (2004). Health hazards from volcanic gases: A systematic literature review. *Archives of Environmental Health: An International Journal, 59*(12), 628–639. https://doi.org/10.1080/00039890409602947

Higgins, C., Kumar, P., & Morawska, L. (2024). Indoor air quality monitoring and source apportionment using low-cost sensors. *Environmental Research Communications, 6*(1), 012001. https://doi.org/10.1088/2515-7620/ad1cad

Huo, H., Zhang, Q., Guan, D., Su, X., Zhao, H., & He, K. (2014). Examining air pollution in China using production- and consumption-based emissions accounting approaches. *Environmental Science & Technology, 48*(24), 14139–14147. https://doi.org/10.1021/es503959t

Izah, S. C., Ogwu, M. C., Etim, N. G., Shahsavani, A., & Namvar, Z. (2024). Short-term health effects of air pollution. In *Air pollutants in the context of one health: Fundamentals, sources, and impacts* (pp. 249–278). Springer. https://doi.org/10.1007/698_2024_1132

Khalaf, E. M., Mohammadi, M. J., Sulistiyani, S., Ramírez-Coronel, A. A., Kiani, F., Jalil, A. T., & Derikondi, M. (2024). Effects of sulfur dioxide inhalation on human health: A review. *Reviews on Environmental Health, 39*(2), 331–337. https://doi.org/10.1515/reveh-2022-0237

Krzyzanowski, M., & Cohen, A. (2008). Update of WHO air quality guidelines. *Air Quality, Atmosphere & Health, 1*(1), 7–13. https://doi.org/10.1007/s11869-008-0008-9

Landis, M. S., Long, R. W., Krug, J., Colón, M., Vanderpool, R., Habel, A., & Urbanski, S. P. (2021). The US EPA wildland fire sensor challenge: Performance and evaluation of solver submitted multi-pollutant sensor systems. *Atmospheric Environment, 247*, 118165. https://doi.org/10.1016/j.atmosenv.2020.118165

Liang, M., Chao, Y., Tu, Y., & Xu, T. (2023). Vehicle pollutant dispersion in the urban atmospheric environment: A review of mechanism, modeling, and application. *Atmosphere, 14*(2), 279. https://doi.org/10.3390/atmos14020279

Ma, L., Graham, D. J., & Stettler, M. E. (2021). Has the ultra low emission zone in London improved air quality? *Environmental Research Letters, 16*(12), 124001. https://doi.org/10.1088/1748-9326/ac30c1

Maeso-García, M. D., Esteve-Turrillas, F. A., & Verdú-Andrés, J. (2021). Applications of the photoionization detector (PID) in occupational hygiene: Estimation of air changes per hour in premises with natural ventilation. *Chemosensors, 9*(12), 331. https://doi.org/10.3390/chemosensors9120331

Manisalidis, I., Stavropoulou, E., Stavropoulos, A., & Bezirtzoglou, E. (2020). Environmental and health impacts of air pollution: A review. *Frontiers in Public Health, 8*, 14. https://doi.org/10.3389/fpubh.2020.00014

Mishra, M. K., Rathore, P. S.,Misra, A., & Kumar, R. (2020). Atmospheric correction of multispectral VNIR remote sensing data: Algorithm and inter-sensor Thara Seesaard, Kamonrat Kamjornkittikoon, Chatchawal Wongchoosuk

Moazami, T. N., Svendsen, K. V. H., Buhagen, M., & Jorgensen, R. B. (2023). Comparing PM2.5, respirable dust, and total dust fractions using real-time and gravimetric samples in an exposure chamber study. *Heliyon, 9*(6), e16127. https://doi.org/10.1016/j.heliyon.2023.e16127

Molina, L. T. (2020). Ozone layer. In *Managing air quality and energy systems* (pp. 235–260). CRC Press.

Onivefu, A. P., & Imarhiagbe, O. (2024). Types of air pollutants. In *Air pollutants in the context of one health: Fundamentals, sources, and impacts* (pp. 123–160). Springer. https://doi.org/10.1007/698_2024_1128

Pal, S., Clark, N. E., Lee, T. R., Conder, M., & Buban, M. (2021). When and where horizontal advection is critical to alter atmospheric boundary layer dynamics over land: The need for a conceptual framework. *Atmospheric Research, 264*, 105825. https://doi.org/10.1016/j.atmosres.2021.105825

Pan, K. T., Leonardi, G. S., & Croxford, B. (2020). Factors contributing to CO uptake and elimination in the body: A critical review. *International Journal of Environmental Research and Public Health, 17*(2), 528. https://doi.org/10.3390/ijerph17020528

Popa, R. A., Popa, D. C., Pogurschi, E. N., Vidu, L., Marin, M. P., Tudorache, M., & Vulpe, A. (2022). Comparative evaluation of the dynamics of animal husbandry air pollutant emissions using an IoT platform for farms. *Agriculture, 13*(1), 25. https://doi.org/10.3390/agriculture13010025

Prospero, J. M., Collard, F. X., Molinié, J., & Jeannot, A. (2014). Characterizing the annual cycle of African dust transport to the Caribbean Basin and South America and its impact on the environment and air quality. *Global Biogeochemical Cycles, 28*(7), 757–773. https://doi.org/10.1002/2013GB004802

Rastogi, U., Kumar, S., Maurya, S. P., & Yadav, A. K. (2025). Air pollution forecasting using genetic algorithm and LSTM deep learning technique. *Aerosol Science and Engineering*. https://doi.org/10.1007/s41810-025-00343-y

Reid, C. E., Brauer, M., Johnston, F. H., Jerrett, M., Balmes, J. R., & Elliott, C. T. (2016). Critical review of health impacts of wildfire smoke exposure. *Environmental Health Perspectives, 124*(9), 1334–1343. https://doi.org/10.1289/ehp.1409277

Reitze, A. W., Jr. (1999). Mobile source air pollution control. *Environmental Law, 6*, 309–354.

Sedyaaw, P., Bhaladhare, D., Kawade, S., & Pandey, A. (2024). A review on acid rain its causes, effects and management measures. Bhartiya Krishi Anusandhan Patrika. 12. 2320–2882. 10.18805/BKAP732

Singh, N. K., Patel, D. B., & Khalekar, G. D. (2018). Methanogenesis and methane emission in rice/paddy fields. In *Sustainable agriculture reviews 33: Climate impact on agriculture* (pp. 135–170). Springer. https://doi.org/10.1007/978-3-319-99076-7_5

Singh, S., & Suthar, G. (2025). Machine learning and deep learning approaches for $PM_{2.5}$ prediction: A study on urban air quality in Jaipur, India. *Earth Science Informatics, 18*(1), 97. https://doi.org/10.1007/s12145-024-01648-1

Snoun, H., Krichen, M., & Chérif, H. (2023). A comprehensive review of Gaussian atmospheric dispersion models: Current usage and future perspectives. *Euro-Mediterranean Journal for Environmental Integration, 8*(1), 219–242. https://doi.org/10.1007/s41207-023-00354-6

Thara, S., Kamonrat, K. & Chatchawal W. (2024). A comprehensive review on advancements in sensors for air pollution applications, Science of The Total Environment. 951:175696. https://doi.org/10.1016/j.scitotenv.2024.175696

Thingujam, U., Prabha, D., Ghosh Bag, A. , Thingujam, V., Darshan, N.P, Dutta,S. & Gorain, S. (2025). From point sensing to intelligent systems: a comprehensive review on advanced sensor technologies for soil health monitoring. *Discov. Sens. 1*, 27 (2025). https://doi.org/10.1007/s44397-025-00028-8

Traiwatcharanon, P., Pon-On, W., Zacharias, M., & Wongchoosuk, C. (2024). Electrochemical copper oxide nanoparticles-based sensor for butachlor plus propanil herbicide detection. *Journal of Materials Science: Materials in Electronics, 35*(9), 1–13. https://doi.org/10.1007/s10854-024-12419-5

Wania, F., & Mackay, D. (1999). The evolution of mass balance models of persistent organic pollutant fate in the environment. *Environmental Pollution, 100*(1–3), 223–240. https://doi.org/10.1016/S0269-7491(99)00093-7

Whitehill, A. R., Long, R. W., Urbanski, S. P., Colón, M., Habel, A., & Landis, M. S. (2022). Evaluation of Cairpol and Aeroqual air sensors in biomass burning plumes. *Atmosphere, 13*(6), 877. https://doi.org/10.3390/atmos13060877

Wilailak, S., Yoo, B. H., Kim, Y., & Lee, C. J. (2021). Parametric analysis and design optimization of wet SO_x scrubber system in marine industry. *Fuel, 304*, 121369. https://doi.org/10.1016/j.fuel.2021.121369

Yadav, A. K. (2015). Elemental composition and source apportionment of suspended particulate matters and health risk assessment in mining and nonmining areas of Odisha, India. *Journal of Hazardous, Toxic, and Radioactive Waste, 19*(3), 04014037.

Yadav, A. K. (2021). Human health risk assessment in opencast coal mines and coal-fired thermal power plants surrounding area due to inhalation. *Environmental Challenges, 3*, 100074.

Yadav, A. K., Gupta, K., Bodhale, R., & Phuleria, H. C. (2025). Metals and trace element associated health risk in size-segregated household dust of an industrial town in India. *Air Quality, Atmosphere & Health*. https://doi.org/10.1007/s11869-025-01794-5

Yadav, A. K., Sahoo, S. K., Kumar, A. V., Shirin, S., Jamal, A., Patra, A. C., Dubey, J. S., Thakur, V. K., Lenka, P., Singh, S., Jha, V., & Tripathi, R. M. (2023). Identifying size-segregated particulate matter ($PM_{2.5}$, PM_{10} and SPM) sources in an industrial town of India. *Aerosol Science and Engineering, 7*(4), 455–473.

Yadav, A. K., Shirin, S., Emmanouil, C., & Jamal, A. (2022). Effect of seasonal and meteorological variability of air pollution in Singrauli coalfield. *Aerosol Science and Engineering, 6*, 61–70.

Yang, T., Yin, Y., Zhou, H., & Ren, Z. (2021). Review of Lagrangian stochastic models for turbulent combustion. *Acta Mechanica Sinica, 37*(10), 1–22. https://doi.org/10.1007/s10409-021-01142-7

Chapter 2
Understanding the Health and Environmental Impacts of Air Pollutants

Salil Bhargava, Shailesh Agrawal, Kashish Grover, and Anay Kodape

Abbreviations

8-OHdG	8-hydroxydeoxyguanosine
AERMOD	Atmospheric dispersion modeling system
AHA	American Heart Association
AI	Artificial intelligence
AP-1	Activator protein-1
AQI	Air quality index
ASD	Autism spectrum disorder
BC	Black carbon
BMI	Body mass index
CIMT	Carotid intima-media thickness
COPD	Chronic obstructive pulmonary disease
CRP	C-reactive protein
DTI	Diffusion tensor imaging
ECG	Electrocardiography
EDGAR	Emissions Database for Global Atmospheric Research
ESC	European Society of Cardiology
FeNO	Fractional exhaled nitric oxide
FEV_1	Forced expiratory volume in 1 s
FVC	Forced vital capacity
GAD-7	Generalized anxiety disorder-7 scale
GHQ	General Health Questionnaire
GM-CSF	Granulocyte-macrophage colony-stimulating factor
HADS	Hospital Anxiety and Depression Scale
HDRS	Hamilton Depression Rating Scale
IARC	International Agency for Research on Cancer
IL-6	Interleukin-6

S. Bhargava (✉) · S. Agrawal · K. Grover · A. Kodape
Department of Respiratory Medicine, Mahatma Gandhi Memorial Medical College, Indore, India

A. K. Yadav (ed.), *Computational Approach for Air Pollution*,
https://doi.org/10.1007/978-3-032-09215-1_2

iNOS	Inducible nitric oxide synthase
K6	Kessler 6 non-specific psychological distress scale
LDH	Lactate dehydrogenase
MMSE	Mini-mental state examination
MoCA	Montreal cognitive assessment
MODIS	Moderate resolution imaging spectroradiometer
NAMP	National Air Quality Monitoring Program
NO_x	Nitrogen oxides
OR	Odds ratio
PAHs	Polycyclic aromatic hydrocarbons
PD-CFRS	Parkinson's Disease Cognitive Functional Rating Scale
PEFR	Peak expiratory flow rate
PET	Positron emission tomography
PHQ-9	Patient health questionnaire-9
$PM_{2.5}$	Particulate matter ($\leq$2.5 μm)
qPCR	Quantitative polymerase chain reaction
SO_2	Sulfur dioxide
STAT-1	Signal transducer and activator of transcription-1
TC	Total cholesterol
TG	Triglycerides
TNF-α	tumor necrosis factor alpha
UPP-age-R	Urinary peptidome aging ratio
VOCs	Volatile organic compounds
WRF-Chem	Weather research and forecasting model with chemistry

2.1 Introduction

Air pollution over the years has emerged as a dual threat, endangering both human health and the environment that sustains us. Pollution, both indoors and outdoors, has detrimental effects on health. The joint impact of outdoor air pollution and household origin contributes to approximately 6.7 million early deaths each year (Cohen et al., 2017). India and China have the highest mortality rates related to air pollution. In 2021, air pollution claimed 2.1 million lives in India and 2.4 million lives in China (Burnett et al., 2023).

Air pollution has been a major global concern, affecting both human health and the environment. To fully understand its effects, it is essential to assess the harm caused by pollutants such as particulate matter (PM), carbon dioxide (CO_2) and monoxide (CO), nitrogen oxides (NO_x), and sulfur dioxides (SO_2). Assessing the impacts of air pollution on health and the environment requires a multidisciplinary approach. This article focuses on exploring various methods and tools used to evaluate health risks, including respiratory, cardiovascular, and neurological problems

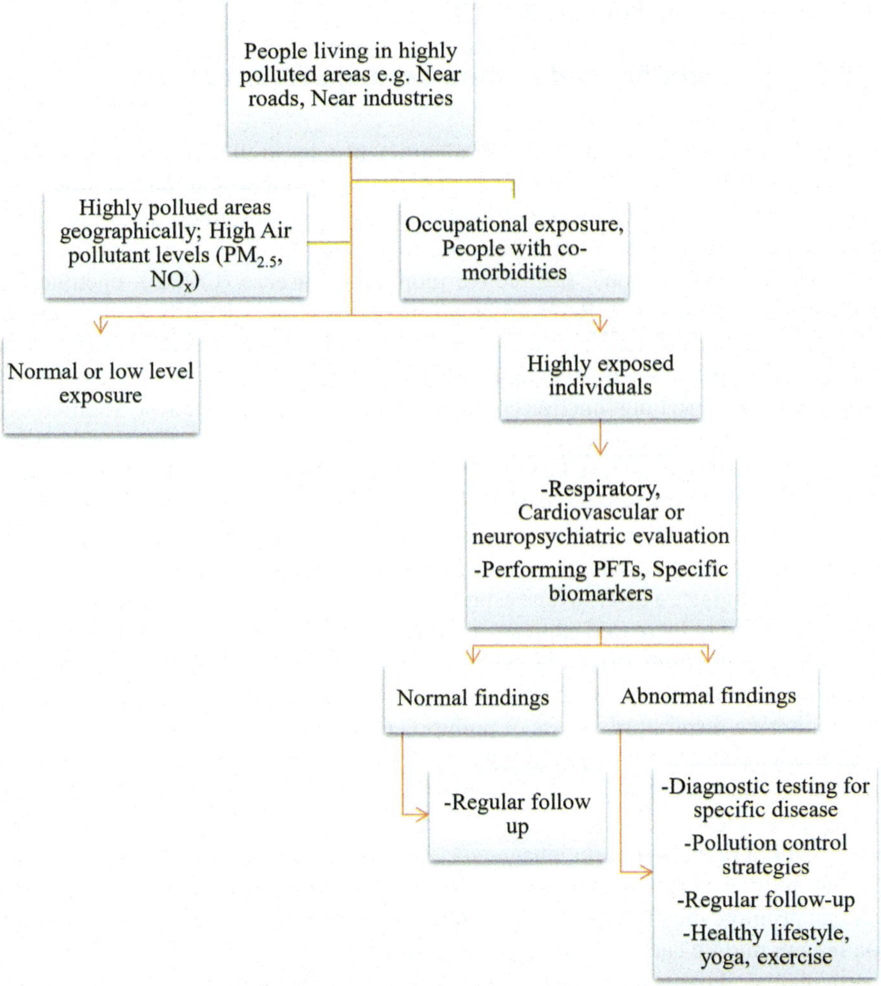

Fig. 2.1 An approach to assessment of air pollution impact

(Fig. 2.1), as well as environmental damage such as ecosystem decline and climate change. This will help not only reduce the disease burden due to air pollution but also identify serious diseases sooner and thereby reduce diseases and deaths due to air pollution. By assessing environmental impacts, strategies and programs can be planned to mitigate pollution and its impact. By focusing on assessment techniques, this aims to shed light on the magnitude of the problem and support the development of effective solutions.

2.2 Health Impact Assessment

2.2.1 Air Pollution and Respiratory Impact Assessment

Air pollutants have been found to significantly harm the respiratory system. Prolonged exposure to air pollutants can cause inflammation, impaired lung function, and an increased incidence of lung diseases. In recent times, air pollution has been associated with decreased lung function, allergic rhinitis, asthma incidence and exacerbations, chronic obstructive pulmonary disease (COPD), pneumonias, and even carcinoma lung. Reduced lung function is seen especially in children and elderly in highly polluted areas; worsened for traffic police exposed to $PM_{2.5}$ and black carbon (BC); nonsmokers more affected; women living within 100 m of major roads show reduced lung function compared to those living farther away (Schikowski et al., 2005; Shakya et al., 2016).

A study highlighted that FeNO (fractional exhaled nitric oxide) measurement, alongside symptom assessment and lung function testing, can effectively identify asthma and other inflammatory conditions. Factors such as allergen exposure, air pollution, and smoking can influence FeNO levels. Elevated ozone levels in air pollution seem to raise FeNO levels, especially in individuals with asthma. This effect may be related to increased expression of inducible nitric oxide synthase (iNOS) in the airway epithelium, driven by AP-1 and STAT-1-dependent pathways (Flamant-Hulin et al., 2009). FeNO testing is noninvasive, easy to perform, and can be useful in monitoring respiratory effects of pollution exposure.

Particles settle in the respiratory pathway and trigger inflammation, cause oxidative stress, and disrupt the body's natural defense mechanisms, such as impaired mucociliary clearance and compromised epithelial tight junctions. These problems, along with a reduction in the phagocytic activity of macrophages, can increase the vulnerability to respiratory infections. Research has suggested that particulate matter can promote the growth and colonization of *Streptococcus pneumoniae* biofilms on human middle ear epithelial cells. Additionally, animal models have shown that particulate pollutants may increase pneumococcal colonization in the nasopharynx. These findings indicate that air pollutants could enhance the pathogenicity of infectious agents in our respiratory tract, in part by weakening immune responses (Kirwa et al., 2021). Oxidative stress and persistent state of inflammation also result in rise of systemic biomarkers such as IL-6, CRP, and TNF-α over prolonged exposure to air pollutants. These biomarkers can be elevated in blood and used to assess the impact of air pollution (Suhaimi & Jalaludin, 2015; Yadav, 2021). Table 2.1 shows an overview of respiratory diseases impacted by air pollution and their possible assessment options to consider when combating the problem.

Table 2.1 Respiratory effects of air pollution and its assessment

Disease/condition	Magnitude of problem	Assessment tools
Chronic obstructive pulmonary disease (COPD)	PM_{10} had a significant effect: a 7 μg/m^3 increase in 5-year average PM_{10} linked to 5.1% FEV_1 reduction, 3.7% FVC reduction [OR:1.33 (95% CI: 1.03–1.72)] (Santos et al., 2016)	• Symptom assessment • Spirometry (FEV_1, FVC, FEV1/FVC ratio) • Chest X-ray • Inflammatory biomarkers (IL-6, CRP)
Asthma	Increased prevalence due to long-term exposure to pollutants; more frequent and severe asthma attacks, higher hospitalizations, and reduced quality of life (Guarnieri & Balmes, 2014)	• FeNO testing (to detect inflammation) • PEFR • Spirometry (FEV_1, FVC) • Symptom assessment
Respiratory tract infections	Greater risk in heavily polluted areas; higher prevalence in children and elderly populations (Kirwa et al., 2021)	• Symptom assessment • Chest X-rays • Sputum cultures

2.2.2 *Air Pollution and Cardiovascular Health Assessment*

Air pollution poses a relatively lower risk of cardiovascular diseases compared to traditional risk factors such as smoking or high cholesterol, but its omnipresence and involuntary impact on billions of people make it a critical global health issue. Consequently, the American Heart Association (AHA) and the European Society of Cardiology (ESC) recognize $PM_{2.5}$ as a significant and causal factor in cardiovascular disease and associated mortality (Chaudhary et al., 2024). Research shows the utility of conventional biochemical and novel biomarkers as indicators of inflammation and oxidative stress, as well as DNA damage in the body due to exposure to pollutants. Exposure to $PM_{2.5}$ has been found to lower antioxidant defenses in mice and impair cardiomyocyte contractility. Furthermore, oxidative stress and inflammation are intricately linked, triggering molecular cascades that further influence vascular function and blood clotting (de Bont et al., 2022; Rajagopalan et al., 2018). High levels of C-reactive protein (CRP) have been linked to a greater risk of developing cardiovascular disease due to pollution-induced oxidative stress and inflammatory responses. Similarly, interleukin-6 (IL-6), tumor necrosis factor-alpha (TNF-α), and 8-hydroxydeoxyguanosine (8-OHdG) have also been identified as indicators of systemic inflammation due to exposure to air pollution (Kelly & Fussell, 2017; Suhaimi & Jalaludin, 2015). Studies have revealed that fine PM_{10} and nitrogen dioxide (NO_2) is associated with elevated triglyceride (TG) levels. Specifically, for every 10 μg/m^3 increase in exposure, TG levels increased by 3.14% for PM_{10} and 4.24% for NO_2. These findings emphasize the potential role of air pollution in the impact of cardiovascular health through lipid metabolism (Table 2.2).

Exposure to air pollution particles has also been associated with a decrease in red blood cell levels and fluctuations in white blood cell levels, reflecting inflammatory responses. Exposure to PM can stimulate the release of GM-CSF and IL-6,

Table 2.2 CVDs associated with air pollution and their assessment

Health impact	Magnitude/findings	Assessment tools/indicators
Cardiovascular diseases (CVDs)		
– Hypertension and stroke	Short-term exposure to PM and NO_x increases risks of hypertension, stroke, and heart attacks (Rajagopalan et al., 2018)	• Blood pressure monitoring • ECG (electrocardiography) • Biomarkers (e.g., CRP, IL-6)
– Atherosclerosis and CAD	• Long-term $PM_{2.5}$ exposure contributes to atherosclerosis and myocardial infarction • Every 10 μg/m^3 increase in exposure, TG levels rose by 3.14% for PM_{10} and 4.24% for NO_2 (Zhang et al., 2023)	• TG levels, TC levels • CRP levels • Carotid intima-media thickness (CIMT) using ultrasound
– Heart failure and arrhythmias	Chronic exposure leads to heart failure and arrhythmias, with higher impacts in low- and middle-income countries (LMICs) (de Bont et al., 2022)	• ECG • Biomarkers: CRP, IL-6, TNF-α (SpringerLink, n.d.)

promoting leukocyte maturation and mobilization. A decrease in white blood cells after exposure to PM may result from neutrophil migration into the lungs, aligning with the observed inflammatory responses. Furthermore, lactate dehydrogenase (LDH) levels decreased post-exposure, possibly due to interactions with airborne metals. Elevated concentrations of PM and metals could contribute to this effect, indicating a potential systemic impact of pollution on blood biomarkers (Ghio et al., 2003). Evaluating red blood cell and neutrophil counts, along with GM-CSF and IL-6 levels in the blood, can serve as an effective method for investigating the hematological effects of air pollution.

2.2.3 Air Pollution and Neuropsychiatric Disorders

Pollution could serve as a persistent carrier, introducing harmful substances into the brain over a lifetime. Growing research highlights neuroinflammation and cerebral oxidative stress as critical factors linking air pollution to neurological and psychological disorders. Ultrafine particles (UFPs) in air can penetrate the alveolar epithelial barriers in the lungs, enter the bloodstream, and circulate to different organs, including the brain. Additionally, these particles, laden with toxic substances, can enter the body through nasal inhalation and travel directly to the brain along the olfactory nerves, effectively bypassing the blood–brain barrier. Once in the brain, contaminants, such as metals, can accumulate with limited mechanisms for removal. This suggests that air pollution could serve as a persistent carrier, introducing harmful substances into the brain over a lifetime (Lin et al., 2021). Heavy metals in the environment, e.g., lead (vehicle exhaust, industrial waste) and mercury (coal plants) have shown neurotoxic side effects in the long run (Yadav, 2015; Yadav et al., 2019,

2023). These mechanisms can contribute to the development of cerebrovascular and neurological disorders, such as stroke, dementia, Alzheimer's disease, and Parkinson's disease.

The greener environment has shown a positive impact on an individual's overall well-being and mental health. Research on the psychological effects remains limited but indicates potential links to depression, anxiety, suicide risk, and emergency visits. However, the findings are mixed due to small sample sizes, inconsistent methods, and restricted scopes. While a US study did not find any link between air pollution and depression among older adults, others reported connections between pollution and anxiety or depression symptoms. A recent study explored psychological distress, a broad mental health indicator that includes depression, anxiety, and mood disorders, using more than a decade of national data combined with $PM_{2.5}$ exposure measurements. Psychological distress, linked to chronic diseases and mortality, worsened with higher concentrations of $PM_{2.5}$ (95% CI: 0.35–0.56). Furthermore, higher levels of PM_{10}, NO_2, and O_3 were significantly associated with increased depressive symptoms among the elderly. Higher concentrations of sulfur dioxide above the median, as well as acute episodes of air pollution, were linked to a rise in emergency visits for depression among women, with more pronounced effects observed in older age groups (Szyszkowicz, 2011). It is important to assess not only the cardiorespiratory effects but also the neuropsychiatric effects of air pollution (Table 2.3).

2.2.4 Air Pollution and Cancer

The polluted atmosphere has emerged as a significant environmental factor in cancer, with increasing evidence linking it to the development and progression of various cancers.

Exposure to prolonged air pollution has been strongly linked to lung cancer, even in people who have never smoked, underscoring its widespread health impact (Fig. 2.2). The International Agency for Research on Cancer (IARC) has classified both outdoor air pollution and the particulate matter it contains as Group 1 human carcinogens for lung cancer. Additionally, the IARC classified household coal burning as a carcinogen of Group 1 and biomass fuel combustion as Group 2A, which means that it is likely carcinogenic for lung cancer (Diver et al., 2024).

A study of workers in different environments examined exposure to genotoxic air pollutants using specific biomarkers. Bus drivers operating in the central city exhibited increased levels of bulky carcinogen–DNA adducts and increased concentrations of 2-amino-adipic semialdehyde (AAS) in plasma proteins, while suburban drivers demonstrated elevated levels of AAS in hemoglobin, malondialdehyde in plasma and PAH–albumin adducts. The findings suggest that selected biomarkers effectively differentiate between high and low environmental exposure (Autrup et al., 1999). The magnitude of cancer risk with air pollution and its assessment is described in Table 2.4.

Table 2.3 Neuropsychiatric impacts of air pollution and their assessment

Disease/ condition	Key findings	Assessment tools
Autism spectrum disorder (ASD)	$PM_{2.5}$ levels correlated with ASD risk during prenatal development and early childhood (Lin et al., 2021)	• Developmental history, behavioral observations • Autism diagnostic observation schedule (ADOS), M-CHAT (modified checklist for autism in toddlers)
Dementia and Alzheimer's disease	Persistent pollutant accumulation, neuroinflammation, and oxidative stress pathways	• Cognitive assessments using mini-mental state examination (MMSE); Montreal cognitive assessment (MoCA) • Neuroimaging for atrophy or damage • Biomarker analysis of inflammatory pathways • Lead levels (blood) • Mercury levels (blood/urine)
Parkinson's disease	Chronic exposure to air pollutants and neurotoxic substances is linked to an increased risk of Parkinson's disease due to oxidative stress and neural damage	• MMSE • Parkinson's disease cognitive functional rating scale (PD-CFRS) • Brain imaging (e.g., PET scans, DTI)
Psychological impact		
Psychological distress	Statistically significant association with $PM_{2.5}$ exposure; K6 scores worsen with higher $PM_{2.5}$ levels (95% CI: 0.35–0.56) (Sass et al., 2017)	• Kessler 6 (K6) non-specific psychological distress scale
Depressive symptoms	• Increased concentrations of PM_{10}, NO_2, and O_3 significantly associated with depressive symptoms • Stronger effects with age [OR = 1.12 (95% CI: 0.99–1.25)] for all ages; [OR = 1.27 (95% CI: 1.10–1.47)] for females 35+ with rise in SO_2 (Szyszkowicz, 2011)	• Hamilton depression rating scale (HDRS) • Patient health questionnaire (PHQ-9) • General health questionnaire (GHQ)
Anxiety	Air pollution (including $PM_{2.5}$ and SO_2) linked to higher anxiety symptoms among women and elderly populations (Szyszkowicz, 2011)	• Generalized anxiety Disorder-7 (GAD-7) scale • Hospital anxiety and depression scale (HADS)

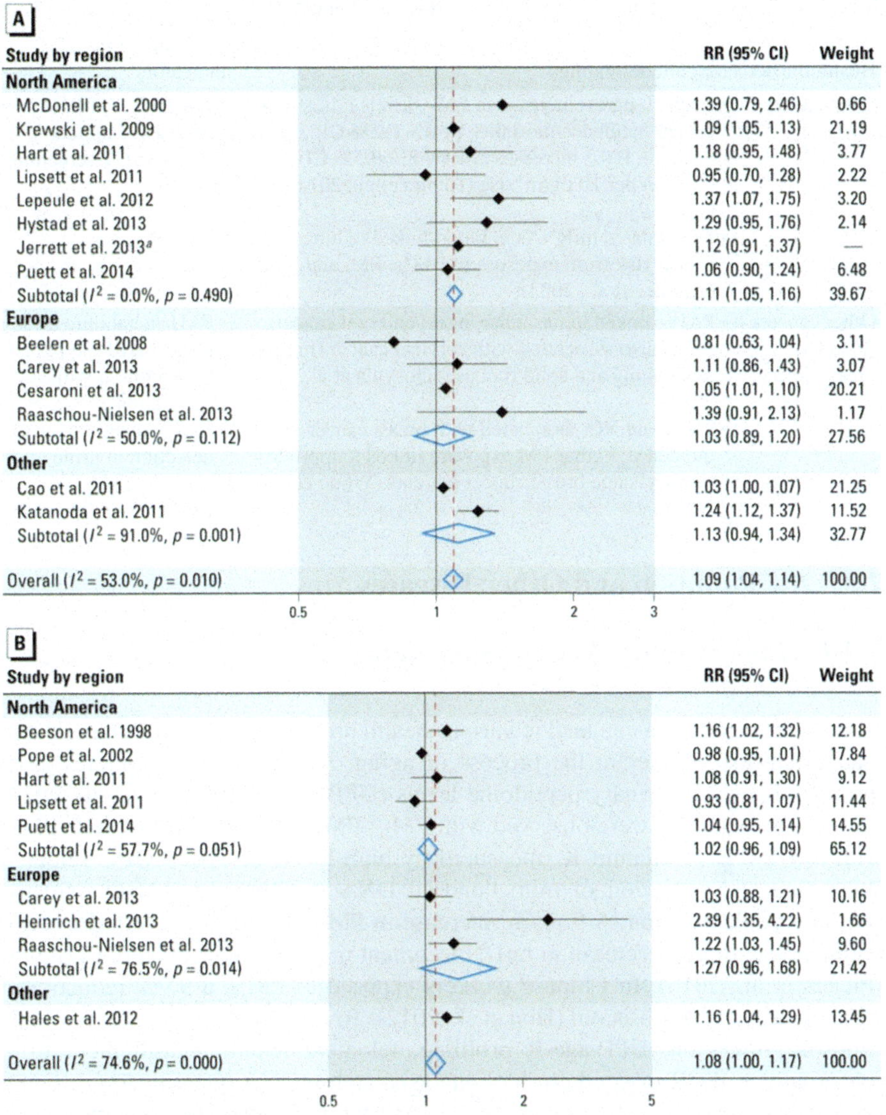

Fig. 2.2 Estimates of lung cancer risk linked to a 10-μg/m³ variation in exposure to $PM_{2.5}$ (**a**) and PM_{10} (**b**), both in general and in geographic study regions (Hamra et al., 2014)

Table 2.4 Risk of cancer due to air pollution and its assessment

Health impact	Magnitude/findings	Assessment tools/ indicators
Lung cancer risk	• Each 10 μg/m³ increase in $PM_{2.5}$ levels raises lung cancer incidence/mortality by 9% (95% CI: 4–14%). PM_{10} increases risk by 8% (95% CI: 0–17%) per 10 μg/m³ rise (Hamra et al., 2014) (Fig. 2.2) • Deregulated miRNAs in blood linked to lung cancer risk from exposure to PAHs, PM, and VOCs (Letelier et al., 2023)	Biomarkers: miRNAs as noninvasive indicators Low-dose CT screening (LDCT)
Other cancers (bladder, breast, liver)	• PAHs, linked to nasopharyngeal and oral cancers. PAHs also associated with cervical cancer (linked to smoking and solid fuel use) (Josyula et al., 2015) • $PM_{2.5}$ and NO_2 associated with breast cancer incidence; higher PM exposure linked to mortality in early-stage breast cancer patients (White et al., 2019)	• PAH-metabolite urine tests: 1-hydroxypyrene • Screening in vulnerable population • Heavy metal (arsenic) detection in urine

2.3 Air Pollution and Other Diseases

2.3.1 Impact of Air Pollution on Aging

Atmospheric pollution can lead to various health problems, and studies have shown that it can also accelerate the process of aging. Air pollution accelerates aging through changes in urinary peptidome levels (UPP-age-R) (Martens et al., 2023). Stronger associations were observed with PM_{10}, $PM_{2.5}$, BC, and NO_2 in individuals with low levels of vitamin K (measured by high levels of dpucMGP). Telomere shortening and reduction of mitochondrial DNA (mtDNA) related to oxidative stress and inflammation. A 5 μg/m³ increase in $PM_{2.5}$ was linked to a reduction in telomere length and decrease in mtDNA content in elderly nonsmokers (Belgium) (Pieters et al., 2016). In Chinese workers exposed to PM_{10}, a 9.9% reduction in telomere length was observed (Hou et al., 2012). To assess the effects of aging in the exposed population, UPP-age-R profiling, telomere length measurement using (quantitative) qPCR or southern blot analysis, and mtDNA quantification through sequencing or biochemical assays can be considered. Further studies with specific pollutant levels ($PM_{2.5}$, PM_{10}) that examine DNA damage and biological aging indicators are needed.

2.3.2 Air Pollution and Obesity: Does there Exist a Correlation?

Long-term exposure to PM_1, $PM_{2.5}$, PM_{10}, and NO_2 associated with obesity, gestational weight gain, and cardiometabolic risks, especially with a family history of heart disease. Childhood obesity was linked to NO_2 in the Netherlands, but not $PM_{2.5}$ or PM_{10} (Bloemsma et al., 2019; Furlong & Klimentidis, 2020). Population studies correlating pollution levels (PM_{10}, NO_2) with BMI, obesity, and cardiometabolic risks are needed. Body mass index (BMI) calculations, exposure mapping with air quality monitors, and longitudinal health surveys can be done to study the ramifications of polluted air on health.

2.4 AI as a Tool to Understand Air Pollution

AI and computer-aided tools predict risks using networks of cause and effect. Systems such as health apps and image interpretation identify at-risk populations and environmental triggers for diseases such as lung disease. AI enhances early intervention by detecting risks in defined timeframes. Applications of AI include disease prediction, environmental trigger analysis, and health monitoring systems. Using AI-driven imaging analysis, predictive modeling, and wearable health apps for real-time exposure tracking will help not only assess the impact but also plan pollution control measures.

2.5 Environmental Impact of Airborne Pollution and Its Assessment

Air pollution has significant effects on the environment. For instance, pollutants such as SO_2 and NO_x can lead to acid rain, which harms plants and disrupts the natural balance of ecosystems. Greenhouse gases, such as CO_2 and methane, contribute to global warming by trapping heat in the atmosphere, leading to changes in climate patterns. Furthermore, air pollution reduces visibility and hinders photosynthesis due to the occurrence of fine particles in the atmosphere. It also impacts biodiversity by introducing toxins into ecosystems, lowering soil quality, and damaging habitats for animals and plants. Additionally, chemicals such as CFCs can weaken the ozone layer, exposing the Earth to higher levels of harmful ultraviolet rays (Table 2.5).

Table 2.5 Summary of tools and methods to study the environmental impact of airborne pollution

Method/tool	Purpose	Probable use
Air quality monitoring – AQI (air quality index) – National air Quality Monitoring Program (NAMP) in India	Measures pollutant levels (e.g., $PM_{2.5}$, NO_x) in the air	Identifying pollution hotspots in a city
Dispersion models – AERMOD	Simulate how pollutants spread based on emission sources and weather patterns	Forecasting air quality in urban areas
Emission inventories – EDGAR	Track pollution sources and quantify their emissions	Assessing the contribution of industries to pollution
Remote sensing – MODIS, sentinel	Use satellite imagery to monitor pollutants on a large scale	Mapping ozone or CO_2 levels globally
Environmental impact assessment (EIA)	Evaluate potential environmental effects of proposed projects, including air quality impact	Approving or modifying development projects
WHO tools and guidelines	Provide frameworks for assessing air quality and its health consequences	Planning clean air initiatives globally
Simulation models – WRF-Chem	Use historical and real-time data to model air pollution impacts	Predicting long-term effects of industrial emissions
Portable air quality monitors	Measure personal exposure to air pollutants in real-time	Devices such as Airveda and Prana air used in India

2.6 Conclusion

Advancements in technology, including AI-driven analysis, biomarker assessments, and satellite-based monitoring, have significantly enhanced our ability to assess air pollution and its effects. A comprehensive and multidisciplinary approach, which integrates air quality assessments, epidemiological research, and policy interventions, is important to formulate effective control measures. Addressing this challenge requires coordinated global efforts, stringent regulatory measures, and the promotion of sustainable practices. The commitment to proactive intervention and scientific innovation will be crucial to addressing the lasting consequences of air pollution and ensuring a healthier future for generations to come.

Acknowledgments The authors sincerely thank our prestigious institute and other people who are involved in this research.

References

Autrup, H., Daneshvar, B., Dragsted, L. O., Gamborg, M., Hansen, M., Loft, S., et al. (1999). Biomarkers for exposure to ambient air pollution—comparison of carcinogen-DNA adduct levels with other exposure markers and markers for oxidative stress. *Environmental Health Perspectives, 107*(3), 233–238. https://doi.org/10.1289/ehp.99107233

Bloemsma, L. D., Wijga, A. H., Klompmaker, J. O., Janssen, N. A. H., Smit, H. A., Koppelman, G. H., et al. (2019). The associations of air pollution, traffic noise and green space with overweight throughout childhood: The PIAMA birth cohort study. *Environmental Research, 169*, 348–356. https://doi.org/10.1016/j.envres.2018.11.015

Burnett, R. T., Pope, C. A., Ezzati, M., et al. (2023). Air pollution deaths attributable to fossil fuels. *BMJ, 383*, 077784. https://doi.org/10.1136/bmj-2023-077784

Chaudhary, F., Ahmed, Z., Fraix, M. P., & Agrawal, D. K. (2024). Air pollution and the pathogenesis of cardiovascular disease. In D. M. Djuric & D. K. Agrawal (Eds.), *Environmental factors in the pathogenesis of cardiovascular diseases* (pp. 133–166). Springer Nature. https://doi.org/10.1007/978-3-031-62806-1_4

Cohen, A. J., Brauer, M., Burnett, R., Anderson, H. R., Frostad, J., Estep, K., et al. (2017). The global burden of disease due to outdoor air pollution. *Environmental Science & Technology, 51*(12), 6457–6466. https://doi.org/10.1021/acs.est.6b03736

de Bont, J., Jaganathan, S., Dahlquist, M., Persson, Å., Stafoggia, M., & Ljungman, P. (2022). Ambient air pollution and cardiovascular diseases: An umbrella review of systematic reviews and meta-analyses. *Journal of Internal Medicine, 291*(6), 779–800. https://doi.org/10.1111/joim.13492

Diver, W. R., Teras, L. R., Deubler, E. L., et al. (2024). Outdoor air pollution and risk of incident adult haematologic cancer subtypes in a large US prospective cohort. *British Journal of Cancer*. https://doi.org/10.1038/s41416-024-02718-3

Flamant-Hulin, M., Caillaud, D., Sacco, P., Pénard-Morand, C., & Annesi-Maesano, I. (2009). Air pollution and increased levels of fractional exhaled nitric oxide in children with no history of airway damage. *Journal of Toxicology and Environmental Health. Part A, 73*(4), 272–283. https://doi.org/10.1080/15287390903388948

Furlong, M. A., & Klimentidis, Y. C. (2020). Associations of air pollution with obesity and body fat percentage, and modification by polygenic risk score for BMI in the UK biobank. *Environmental Research, 185*, 109364. https://doi.org/10.1016/j.envres.2020.109364

Ghio, A. J., Hall, A., Bassett, M. A., Cascio, W. E., & Devlin, R. B. (2003). Exposure to concentrated ambient air particles alters hematologic indices in humans. *Inhalation Toxicology, 15*(14), 1465–1478. https://doi.org/10.1080/08958370390242041

Guarnieri, M., & Balmes, J. R. (2014). Outdoor air pollution and asthma. *Lancet, 383*(9928), 1581–1592. https://doi.org/10.1016/S0140-6736(14)60617-6

Hamra, G. B., Guha, N., Cohen, A., Laden, F., Raaschou-Nielsen, O., Samet, J. M., et al. (2014). Outdoor particulate matter exposure and lung cancer: a systematic review and meta-analysis. *Environmental Health Perspectives, 122*(9), 906–911. https://doi.org/10.1289/ehp.1408092

Hou, L., Wang, S., Dou, C., Zhang, X., Yu, Y., Zheng, Y., et al. (2012). Air pollution exposure and telomere length in highly exposed subjects in Beijing, China: A repeated-measure study. *Environment International, 48*, 71–77. https://doi.org/10.1016/j.envint.2012.06.020

Josyula, S., Lin, J., Xue, X., Rothman, N., Lan, Q., Rohan, T. E., et al. (2015). Household air pollution and cancers other than lung: A meta-analysis. *Environmental Health, 14*, 24. https://doi.org/10.1186/s12940-015-0005-8

Kelly, M. S., & Fussell, J. A. (2017). Biomarkers of the health outcomes associated with ambient air pollution. *Science of the Total Environment, 579*, 1–10. https://doi.org/10.1016/j.scitotenv.2016.10.074

Kirwa, K., Eckert, C. M., Vedal, S., Hajat, A., & Kaufman, J. D. (2021). Ambient air pollution and risk of respiratory infection among adults: evidence from the multiethnic study of atherosclerosis (MESA). *BMJ Open Respiratory Research, 8*(1), e000866. https://doi.org/10.1136/bmjresp-2020-000866

Letelier, P., Saldías, R., Loren, P., Riquelme, I., & Guzmán, N. (2023). MicroRNAs as potential biomarkers of environmental exposure to polycyclic aromatic hydrocarbons and their link with inflammation and lung cancer. *International Journal of Molecular Sciences, 24*(23), 16984. https://doi.org/10.3390/ijms242316984

Lin, C. K., Chang, Y. T., Lee, F. S., Chen, S. T., & Christiani, D. (2021). Association between exposure to ambient particulate matters and risks of autism spectrum disorder in children: A systematic review and exposure-response meta-analysis. *Environmental Research Letters, 16*(6), 063003. https://doi.org/10.1088/1748-9326/abfd08

Martens, D. S., An, D. W., Yu, Y. L., Chori, B. S., Wang, C., Silva, A. I., et al. (2023). Association of air pollution with a urinary biomarker of biological aging and effect modification by vitamin K in the FLEMENGHO prospective population study. *Environmental Health Perspectives, 131*(12), 127011. https://doi.org/10.1289/EHP12722

Pieters, N., Janssen, B. G., Dewitte, H., Cox, B., Cuypers, A., Lefebvre, W., et al. (2016). Biomolecular markers within the core axis of aging and particulate air pollution exposure in the elderly: A cross-sectional study. *Environmental Health Perspectives, 124*(7), 943–950. https://doi.org/10.1289/ehp.1509728

Rajagopalan, S., Al-Kindi, S. G., & Brook, R. D. (2018). Air pollution and cardiovascular disease: JACC state-of-the-art review. *Journal of the American College of Cardiology, 72*(17), 2054–2070. https://doi.org/10.1016/j.jacc.2018.07.099

Santos, U. P., Garcia, M. L. S. B., Braga, A. L. F., Pereira, L. A. A., Lin, C. A., André, P. A., et al. (2016). Association between traffic air pollution and reduced forced vital capacity: A study using personal monitors for outdoor workers. *PLoS One, 11*(10), e0163225. https://doi.org/10.1371/journal.pone.0163225

Sass, V., Kravitz-Wirtz, N., Karceski, S., Hajat, A., Crowder, K., & Takeuchi, D. (2017). The effects of air pollution on individual psychological distress. *Health & Place, 48*, 72–79. https://doi.org/10.1016/j.healthplace.2017.09.002

Schikowski, T., Sugiri, D., Ranft, U., Gehring, U., Heinrich, J., Wichmann, H. E., et al. (2005). Long-term air pollution exposure and living close to busy roads are associated with COPD in women. *Respiratory Research, 6*, 152. https://doi.org/10.1186/1465-9921-6-152

Shakya, K. M., Rupakheti, M., Aryal, K., & Peltier, R. E. (2016). Respiratory effects of high levels of particulate exposure in a cohort of traffic police in Kathmandu, Nepal. *Journal of Occupational and Environmental Medicine, 58*(6), e218–e225. https://doi.org/10.1097/JOM.0000000000000748

Suhaimi, N. F., & Jalaludin, J. (2015). Biomarker as a research tool in linking exposure to air particles and respiratory health. *BioMed Research International, 2015*, 962853. https://doi.org/10.1155/2015/962853

Szyszkowicz, M. (2011). Ambient sulfur dioxide and female ED visits for depression. *Air Quality, Atmosphere and Health, 4*(3–4), 259–262. https://doi.org/10.1007/s11869-010-0100-4

White, A. J., Keller, J. P., Zhao, S., Carroll, R., Kaufman, J. D., & Sandler, D. P. (2019). Air pollution, clustering of particulate matter components, and breast cancer in the sister study: A U.S.-wide cohort. *Environmental Health Perspectives, 127*(10), 107002. https://doi.org/10.1289/EHP5138

Yadav, A. K. (2015). Elemental composition and source apportionment of suspended particulate matters and health risk assessment in mining and nonmining areas of Odisha, India. *Journal of Hazardous, Toxic, and Radioactive Waste, 19*(3), 04014037.

Yadav, A. K. (2021). Human health risk assessment in opencast coal mines and coal-fired thermal power plants surrounding area due to inhalation. *Environmental Challenges, 3*, 100074.

Yadav, A. K., Sahoo, S. K., Dubey, J. S., Kumar, A. V., Pandey, G., & Tripathi, R. M. (2019). Assessment of particulate matter, metals of toxicological concentration, and health risk around a mining area, Odisha, India. *Air Quality, Atmosphere & Health, 12*(7), 775–783.

Yadav, A. K., Sahoo, S. K., Kumar, A. V., Shirin, S., Jamal, A., Patra, A. C., Dubey, J. S., Thakur, V. K., Lenka, P., Singh, S., Jha, V., & Tripathi, R. M. (2023). Identifying size-segregated particulate matter ($PM_{2.5}$, PM_{10} and SPM) sources in an industrial town of India. *Aerosol Science and Engineering, 7*(4), 455–473.

Zhang, Y., Shi, J., Ma, M., Yu, N., Zheng, P., Chen, Z., et al. (2023). Association between air pollution and lipid profiles. *Toxics, 11*(11), 894. https://doi.org/10.3390/toxics11110894

Chapter 3
Assessing the Health and Environmental Consequences of Air Pollution

Pankaj Kumar and Parimal Kumar Khan

3.1 Introduction

Air pollution can be characterized as environmental contamination caused by an abiotic and biotic factors that have the potential to affect the inherent characteristics of the atmosphere (whether indoor or outdoor), leading to unpleasant odors, decreased visibility, and several health hazards. The main constituents of air pollution include particulate matter (PM), ground-level ozone (O_3), carbon monoxide (CO), carbon dioxide (CO_2), sulfur oxides (SO_x) and nitrogen oxides (NO_x) that arise from both natural and human activities. Common sources of pollution include emissions from vehicles, industries, brake and tire wear, oil refineries, biomass combustion, thermal power stations, metal and coal extraction, waste disposal sites, as well as from construction and demolition activities. Population explosion coupled with rapid industrialization, expansion of urbanization, and increased number of vehicles have led to an increase in several problems that particularly include air pollution. As a result, cities turn into frequently high-risk zones, and the people living there become susceptible to negative health effects such as chronic obstructive lung disease (COPD), cardiovascular mortality (CM), and respiratory mortality (RM), and certainly it is going to be an emergency for both human and environmental health. Air pollution poses the greatest environmental threat to human health, leading to approximately eight million premature deaths each year globally, which is still on the rise (State of Global Air, 2024). Nearly the entire global population (99%) breathes air that exceeds the air quality standards set by the WHO (2022). In countries with low and middle income, outdoor air pollution accounts for nine out of ten fatalities. According to the Air Quality Life Index (AQLI) report, the presence of PM exceeding the WHO threshold of 5 μg/m^3 in the atmosphere decreases the

P. Kumar (✉) · P. K. Khan
Department of Zoology, Patna University, Patna, India

A. K. Yadav (ed.), *Computational Approach for Air Pollution*,
https://doi.org/10.1007/978-3-032-09215-1_3

global average lifespan by 2.2 years, contributing to a total loss of 17 billion life-years worldwide (Lee & Greenstone, 2021). The impact of this pollution on lifespan is three times greater than the effects of alcohol consumption and unsafe drinking water, six times greater than the risk of HIV/AIDS, and 89 times greater than that of conflict and terrorism (Lee & Greenstone, 2021). Air pollution poses a serious threat to various forms of life, the environment, sites of cultural heritage, and regional climate patterns. Presently, India ranks eighth globally in terms of $PM_{2.5}$ pollution, with an average concentration of 53.3 $\mu g/m^3$, significantly exceeding the WHO guideline. In particular, 12 of the 15 most polluted cities in Central and South Asia are located within India, indicating severe urban air quality problems. Approximately 60% of Indian cities recorded levels of $PM_{2.5}$ that exceeded WHO recommendations by a factor of 7. In 2021, India's $PM_{2.5}$ level was 58.1 $\mu g/m^3$, 4.8 $\mu g/m^3$ higher than the current measurement. In India, approximately 25–30% of $PM_{2.5}$ emissions come from transportation and crop burning, which significantly contributes to higher levels of $PM_{2.5}$, particularly in northern India.

Impact assessment is a systematic approach that examines the possible effects, both beneficial and adverse, of a suggested action or policy on individuals and the environment. It plays a significant role in conveying scientific knowledge to policymakers, enabling them to implement measures to minimize the impact of toxic substances and assess the health advantages. The health impact assessment focuses on assessing the effect of a substance/factor on health. Similarly, environmental impact assessment evaluates the alterations caused by a substance/factor on the environment, such as acid rain, haze, ozone depletion, soil erosion, water contamination, wildlife depletion, and ultimately climate change, which aims to reduce harmful effects and strengthen the beneficial effects by a systematic approach.

Air pollution has numerous harmful impacts on land and aquatic ecosystems, reducing biodiversity and affecting growth rates, yields, and ecosystem services. Nitrogen oxides (NO_x) and ammonia (NH_3) settled on Earth's surface, causing eutrophication and affecting ecosystem structure. In sentient terrestrial ecosystems such as grasslands, surpassing critical nitrogen deposition limits can result in the decline of sensitive species, but can promote the growth of species that thrive in high nitrogen conditions and can alter ecosystem structure and function. The deposition of sulfur dioxide (SO_2), nitrogen oxides (NO_x), and ammonia (NH_3) contributes to acidification, disturbs ecosystems, and leads to a loss of biodiversity (Fig. 3.1). Heavy metals and toxic pollutants accumulate in soils and bioaccumulate through the food chain.

In Europe, Air Quality Limit Values (AQLVs) are established to initiate environmental planning (Kuklinska et al., 2015). These values pinpoint areas at risk by designing monitoring and assessment zones to track emission sources and levels of air pollution. In the United States, the national air quality threshold is given by the National Ambient Air Quality Standards (NAAQS). In India, this task is carried out by the Central Pollution Control Board (CPCB) and the Bureau of Indian Standards (BIS). Although their foundations are based on different frameworks, significant advances have been achieved in reducing total emissions along with their associated health and environmental effects. It also presents worldwide geographic air quality

Fig. 3.1 Wildfire causing massive air pollution in California, the United States (Creator: August Frank Credit: AP, Copyright: ©Lewiston Tribune)

standards according to the severity of air quality problems and lists all sources of pollutants and their precursors (Kuklinska et al., 2015; Yadav, 2021).

This chapter aims to explore the evaluation of health and environmental impacts of air pollution. It discussed the key pollutants, their sources, exposure pathways, health impact (short- and long-term consequences along with mutagenic and carcinogenic effects), environmental impacts, case studies, and finally, impact assessment methodologies.

3.2 Key Pollutants and Their Sources

Air pollution can have a profoundly negative impact on all aspects of the environment and human well-being. It stems from a complex interplay of anthropogenic and natural sources with significant implications for global health and environmental stability. A significant contributor is the combustion of fossil fuels used in transportation, energy generation, and industrial activities, which releases pollutants such as particulate matter, nitrogen, and sulfur oxides. Industrial operations, including manufacturing and construction, discharge a variety of harmful substances, while agricultural activities (livestock rearing and fertilizer use) lead to ammonia and methane. Furthermore, household sources such as the use of wood and coal for heating and cooking significantly contribute to the release of harmful gases,

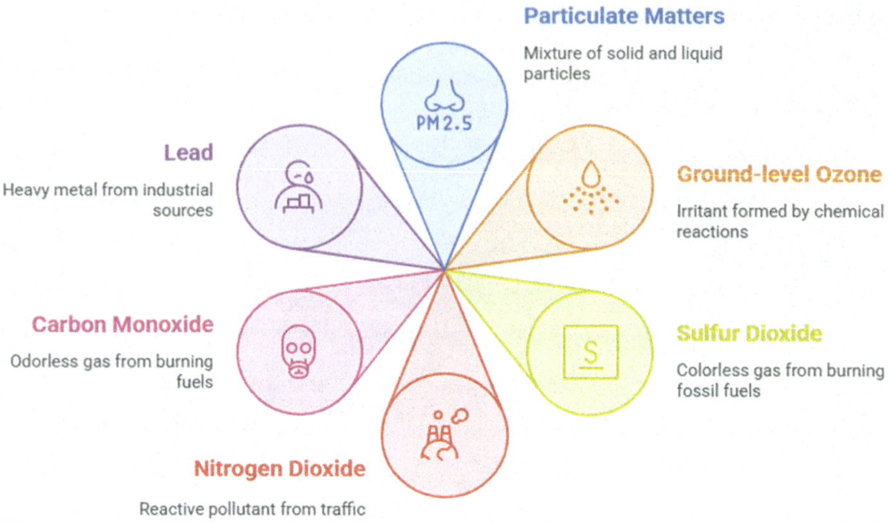

Fig. 3.2 Key air pollutants and their primary sources

especially in developing countries. It is also essential to note that natural events, such as volcanic eruptions and dust storms, contribute to the overall levels of air pollution.

The World Health Organization (WHO) has reported six key air pollutants, namely particulate matter (PM), ground-level ozone (O_3), sulfur dioxide (SO_2), nitrogen dioxide (NO_2), carbon monoxide (CO), and lead (Pb) (Fig. 3.2).

Particulate matter (PM) represents a diverse combination of tiny solid and liquid particles of different sizes suspended in the air, including dust, smoke, soot, and liquid droplets. These particles are hazardous and can penetrate deeply into the bodies of living organisms. Particulate matter consists of particles that are 10 micrometers (μm) in diameter or smaller, referred to as PM_{10}, and very fine particles that are 2.5 micrometers (μm) in diameter or less, called $PM_{2.5}$ (Manisalidis et al., 2020).

Ground-level ozone (O_3) forms through a chemical reaction involving nitrogen oxides and volatile organic compounds (VOCs) emitted from natural and human-made sources situated up to 2 miles above the earth's surface. It acts as an irritant that can adversely affect human health as well as the health of other living organisms. Ground-level ozone is commonly known as "bad ozone" (Yildizhan et al., 2024).

Sulfur dioxide (SO_2) is a colorless, chemically reactive, and highly odorous air pollutant that is hazardous to the environment and human health. It originates from the burning of fossil fuels and is released from vehicles, power plants, and other industrial processes. SO_2 affects the vital function of the lungs, causes eye irritation, coughing, throat irritation, and even chest tightness.

Nitrogen dioxide (NO_2) is highly reactive and a traffic-related pollutant that, upon reaction with volatile organic compounds, forms ground-level ozone (bad ozone). It acts as a respiratory irritant that can reach deep into the lungs, leading to respiratory illnesses, coughing, wheezing, shortness of breath, bronchospasm, and potentially pulmonary edema when inhaled in large amounts.

Carbon monoxide (CO) is a colorless and odorless gas formed primarily from the burning of fossil fuels and is released from vehicular emissions, power stations, as well as from spacecraft and cigarette smoking. Inhaling CO causes poisoning that results in headache, dizziness, fatigue, nausea, vomiting, and sometimes unconsciousness. CO binds more readily to hemoglobin than oxygen does. Consequently, people who are subjected to high levels of carbon monoxide for a prolonged period can suffer from serious poisoning. Nevertheless, both laboratory and field studies have shown that it is beneficial to plants and can enhance plant growth (Soon et al., 1999).

Lead (Pb) is a heavy metal utilized in various industries and released by certain petrol engine motors, batteries, paints, plumbing materials, radiators, waste incineration processes, and also cosmetics. Lead poses significant health risks, especially for children and pregnant women. Children exposed to lead can experience issues with behavior and learning, reduced IQ, increased hyperactivity, stunted growth, hearing difficulties, and anemia (Brabhukumr et al., 2020; Farhat et al., 2013; Shao et al., 2012). In extreme cases, lead ingestion can cause seizures, unconsciousness, and even death. High concentrations of Pb in the ecosystem negatively affect plants and agriculture as well. Neurological effects are seen in diverse living beings exposed to elevated levels of lead (Assi et al., 2016).

3.3 Impacts of Air Pollution on Health

The adverse effect of air pollution on human health occurs through a complex interplay of biological mechanisms. The respiratory system acts as the primary entry point for most pollutants into the body, leading to the onset and progression of various complications. Air pollutants penetrate deep into lung tissue and tend to accumulate in various organs through the circulatory system (Fig. 3.4). The accumulation of pollutants in the human body will ultimately cause various health hazards. The level of harm inflicted by the particulate and gaseous pollutants inhaled, on the respiratory system, also affects the protective mechanisms. The potential mechanisms through which these pollutants affect the cardiovascular and respiratory systems begin with inflammation and oxidative stress mediated by reactive species of oxygen and nitrogen that are generated by inhaled pollutants (Al-Kindi et al., 2020).

Most pollutants are primary contributors to diseases in people. Specifically, small particulate matter such as PM_{10} and $PM_{2.5}$ can infiltrate the lungs and cardiovascular organs, causing respiratory distress, chronic obstructive pulmonary disease (COPD), asthma, lung cancer, cardiovascular diseases, and even strokes (Fig. 3.3) (Thangavel et al., 2022). Outdoor air pollution and urban development are linked to a rise in the occurrence as well as the severity of asthma. A study conducted across ten European cities found that the cases of incident asthma among children and all

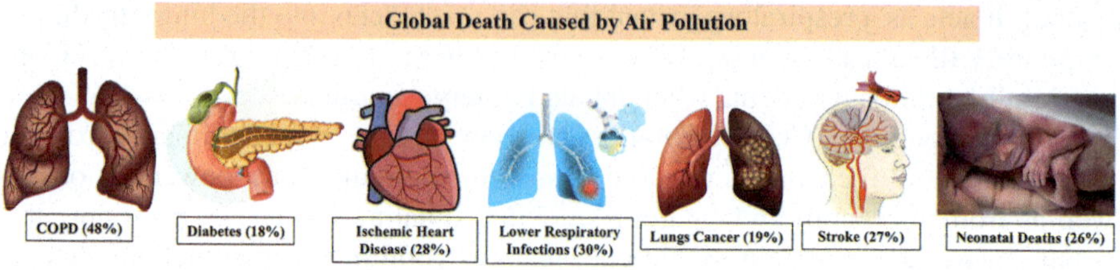

Fig. 3.3 Percentage of global deaths due to specific diseases by total air pollution (State of Global Air, 2024)

exacerbations of childhood asthma following exposure to pollutants from vehicle emissions were 14% and 15%, respectively. Children with asthma exhibit marked deterioration in their health conditions. Furthermore, research has indicated a close relationship between $PM_{2.5}$ and PM_{10} in the atmosphere and asthma symptoms, particularly in atopic children (Altman et al., 2023).

The key factors that affect the interaction between air pollutants and the organ system of human body are outlined as follows:

1. The physicochemical nature of pollutants and their amount.
2. The age, sex, genetic profile, lifestyle, daily exposure, and medical history of individuals as well as their socioeconomic circumstances.
3. Daily dose, route of exposure, intensity, and duration of exposure.
4. Rate of uptake by cells or tissues, physiological status, and rate of clearance from the body (Fig. 3.4).

It is important to mention that quick exposure to very less amount of pollutant does not have any significant impact on human health due to body's own defense mechanism. However, long-term exposure with large amount of pollutants causes chronic health complications if the accumulation is fast coupled with slow clearance or recovery is slow. In contrast, when the substance is rapidly eliminated from the body, its effects are immediately reversed. In general, the type of pollutants, their quantity, route of exposure, retention, and clearance play significant roles in how individuals and pollutants interact with each other. However, the significance of these factors may vary greatly (Lippmann, 2013).

3.3.1 Short-Term Exposure

With short-term exposure, particulate pollutants and ozone are harmful to human health at each stage of life, increasing the risk of premature delivery in females, developing or exacerbating heart and lung diseases, and even reducing life expectancy. However, due to increased exposure and/or higher sensitivity, certain groups of people are more prone to diseases than others. Moreover, children, the elderly, and individuals with preexisting disease conditions are disproportionately affected.

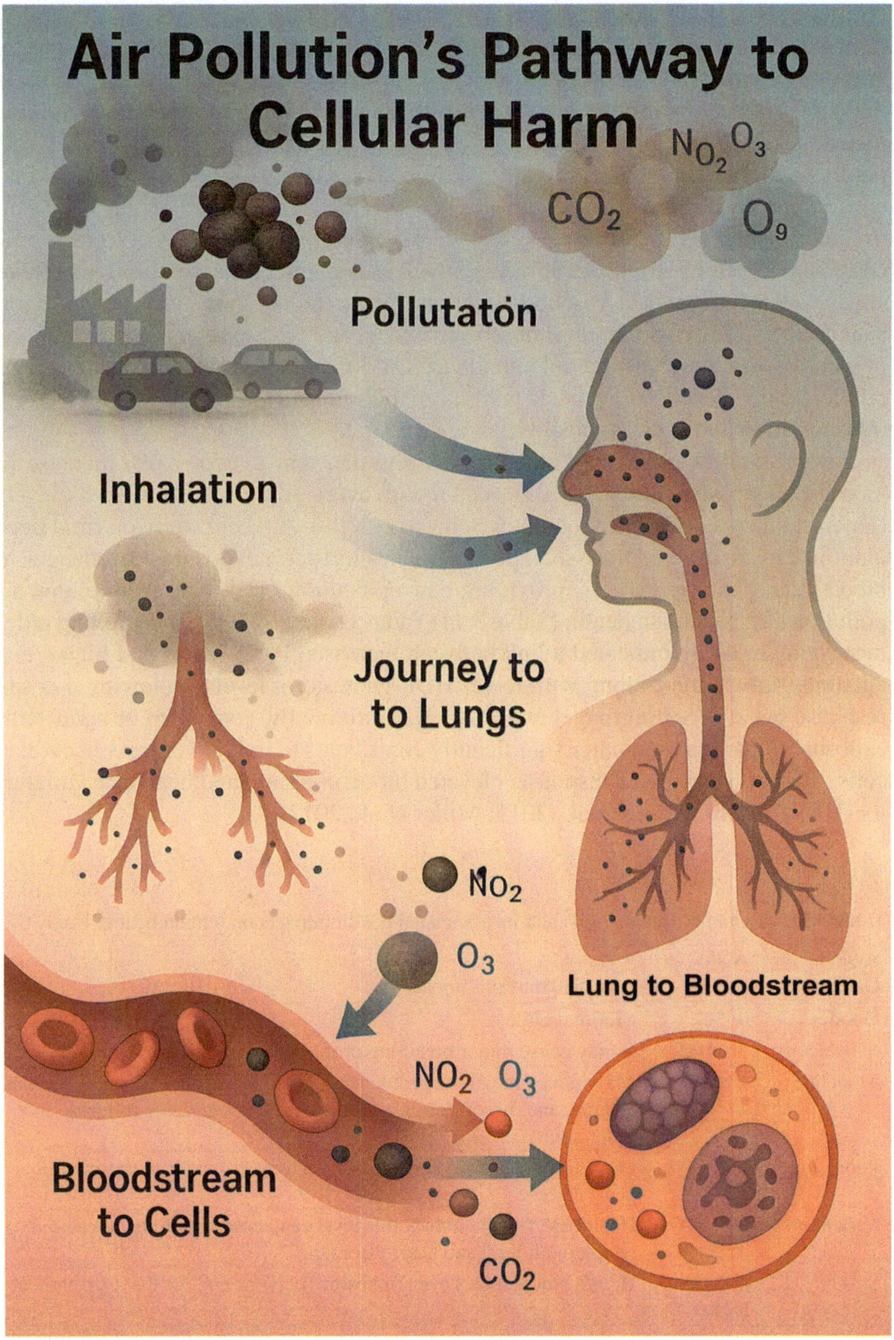

Fig. 3.4 Possible pathways of exposure of air pollution in humans

Air pollution causes a variety of short-term effects on human health and the environment. A brief exposure to high levels of particulate matter ($PM_{2.5}$ and PM_{10}), nitrogen dioxide (NO_2), sulfur dioxide (SO_2), ozone (O_3), and carbon monoxide (CO) may result in instant negative health effects. Common clinical symptoms include respiratory problems such as coughing, irritation of the throat and eyes, difficulty breathing, and worsening asthma. In addition to breathing difficulties, short-term exposure to air pollution has also been linked to heart ailments, including irregular heartbeats, heart attacks, and strokes. The underlying physiological processes contributing to these impacts involve systemic inflammation, oxidative stress, and compromised vascular function. Furthermore, short-term air pollution can affect visibility and hinder outdoor activities, and lead to temporary reductions in work productivity and school attendance. Additionally, air quality warnings and public health alerts during peak periods of pollution can adversely affect urban transportation and economic activities.

Several studies conducted in US cities revealed that a 1.51–2.0% increase in mortality from respiratory diseases occurs with every 10 μg/m^3 rise in $PM_{2.5}$ levels (Fajersztajn et al., 2017; Zanobetti & Schwartz, 2009). A recent study on rural populations further showed an association between short-term exposure and changes in blood inflammatory markers, providing new epidemiological insights into how air pollution affects human health (Table 3.1) (Yadav et al., 2014). A study among older men living in a city indicated a link between increased $PM_{2.5}$ levels and higher circulating triglycerides, along with lower HDL cholesterol levels, following a dose–response pattern (Schraufnagel et al., 2019). Additionally, controlled or short-term exposure to particulate matter significantly contributes to oxidative stress in cardiac cells, reduced vasodilator response, elevated blood pressure, and heart rate variability (HRV) (Hemmingsen et al., 2015; Miller et al., 2012).

Table 3.1 Air Quality Index (AQI) and its associated health impacts on human health

AQI Category	AQI (range)	Associated health impacts
Good	0–50	Minimal effect
Satisfactory	51–100	It may cause minor breathing problems in sensitive people
Moderate	101–200	May induce breathing problems in persons with lung disease, such as asthma, and discomfort in persons with heart disease, children, and older adults
Poor	201–300	Long-term exposure may cause breathing difficulty in those with heart disease
Vary poor	301–400	Health alert: pollutants at this level can cause breathing sickness in people with lung and heart disorders
Severe	401 or higher	Health emergencies even in healthy persons and for the environment

3.3.2 Long-Term Exposure

Long-term exposure to air pollution is strongly associated with an increased incidence of cardiovascular diseases, chronic respiratory diseases, COPD, lung cancer, and adverse neurodevelopmental as well as neurodegenerative outcomes. Inhalation of particulate matter (PM), ozone (O_3), and environmentally persistent free radicals (EPFR), whether chronic or acute, produces reactive species of oxygen and nitrogen (RONS) that may initiate an inflammatory response intensified by the further endogenous generation of RONS. When RONS production surpasses the level of antioxidant defenses, it activates the mitogen-activated protein kinase (MAPK) complex, which plays a role in triggering nuclear transcription factors such as NF-κB and AP-1, leading to increased RNA synthesis and the release of pro-inflammatory cytokines IL-8 and TNF-α, potentially resulting in the creation of DNA adducts (Zhao et al., 2019).

Air pollution has also been found to diminish the functionality of T lymphocytes, elevated levels of IgE, and heighten the release of CD4+ and CD8+ T lymphocytes, as well as an enhanced Th2 response to antigens in a polluted environment, conditions related to rhinitis and asthma (Schraufnagel et al., 2019). Pregnant women and newborns, as the most susceptible groups, are likely to face greater risks from the negative impact induced by air pollution. Various epidemiological studies revealed that exposure to air pollution can negatively influence infant health, resulting in poor birth outcomes, as well as respiratory and neurodevelopmental problems (Shang et al., 2019). Among these, birth weight emerges as a critical indicator of health risks in newborns, and it also plays a significant role in morbidity during childhood and adulthood. Both long and high birth weight is typically associated with the development of numerous chronic diseases in later life, such as cardiovascular disease, type 2 diabetes, and certain types of cancers (Moraitis et al., 2014).

3.3.3 Mutagenic and Carcinogenic Impacts

Mutagenicity is defined as the capacity of a chemical substance to modify genetic material (DNA), which may lead to inheritable mutations. Pollutants can create reactive intermediates that may result in the formation of DNA adducts. If such adducts are not properly repaired/removed, they can cause point mutations, insertions, or deletions during DNA replication. Both in vitro and in vivo tests, such as the Ames test and the micronucleus assay, have consistently shown the mutagenic properties of airborne particulates, particularly $PM_{2.5}$ and PM_{10} (Lepers et al., 2014). Air samples collected from urban and industrial areas often exhibit greater mutagenicity, which aligns with emissions from traffic, industry, and biomass combustion. There is sufficient evidence of lung and urinary bladder cancers caused by outdoor air pollution. The PM present in outdoor air can persist in the respiratory tract as a result of its poor water solubility, leading to inflammation and induction of

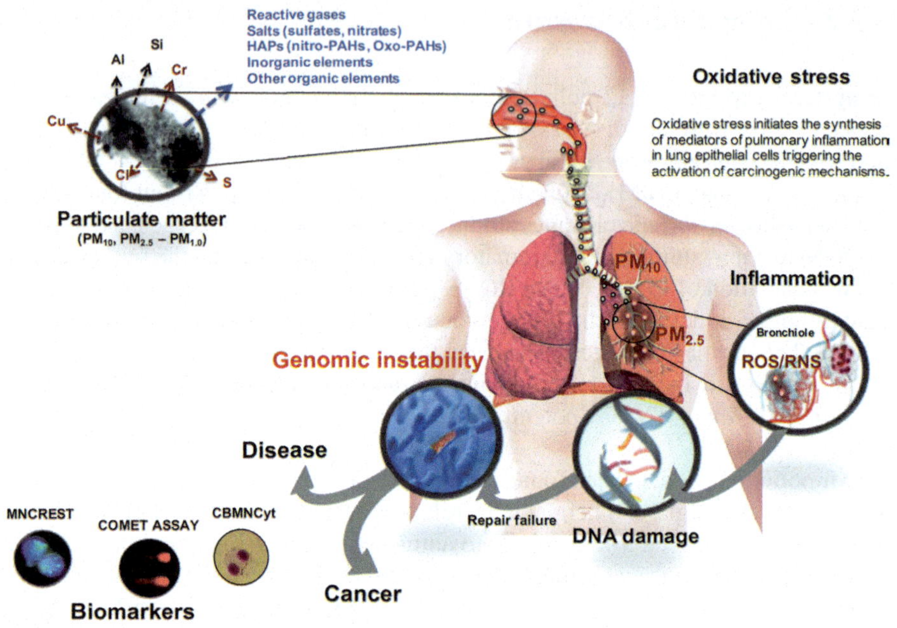

Fig. 3.5 Particulate matter induces oxidative stress, inflammation, and risk of cancer (Espitia-Pérez et al., 2020)

oxidative stress. Variability in particle deposition and clearance occurs among individuals based on factors such as age, sex, lifestyle, and preexisting respiratory diseases. Research involving human subjects has used urinary concentrations of hydroxylated polycyclic aromatic hydrocarbons (PAHs) to assess exposure and metabolism of organic substances adsorbed to PM, revealing that individuals can be exposed to carcinogenic pollutants through the air and excrete their metabolites. Outdoor air pollution, including particulate matter, is reported to be carcinogenic (Group 1) in humans (Forbes & Garland, 2016; Yadav et al., 2019).

However, the studies conducted to assess the mutagenicity and carcinogenicity of heavy metals (such as Pb and Cd) in particulate matter exhibit their lower capacity (Fig. 3.5). Organic compounds such as polycyclic aromatic hydrocarbons (PAHs), nitroarenes, and aromatic amines have greater capacity in this regard (DeMarini & Linak, 2022). Both cancer and noncancer risk due to metals present in the air can be calculated by the final report exposure factors handbook 2011, given by the USEPA.

3.4 Environmental Impacts

The impacts of air pollution are multifaceted on the environment. Air pollutants cause acid rain that modifies soil pH, drains vital nutrients, and harms plant life. Ground-level ozone negatively affects plant functions, reducing agricultural and forest productivity. Long-lasting air pollutants also accumulate in water ecosystems, polluting aquatic environments and endangering species diversity. Furthermore, numerous air pollutants are closely associated with climate change, functioning either as greenhouse gases or as climate influences that affect radiative forcing and atmospheric patterns indirectly.

3.4.1 Acid Rain

Acid rain refers to both wet (such as rain, fog, or snow) and dry (particulates and gases) precipitation, which has harmful contents of nitric and sulfuric acids released from vehicles and industrial emissions, as well as from the burning/exhaust of fossil fuels. In nature, acid rain harms tree and leads to acidification of water bodies and soils, making them unsuitable for aquatic life and other forms of wildlife. Additionally, it accelerates the deterioration of not only buildings but also statues and sculptures that are part of our cultural heritage.

Acid rain impacts ecological systems in aquatic and terrestrial habitats, especially in streams, lakes, and wetlands, resulting in harm to wildlife and forests. It extracts aluminum from clay particles, which adversely affects fish and wildlife. Plants and animals that are sensitive to acid can withstand moderate levels of aluminum, while others may decline as pH levels drop. Young organisms are particularly vulnerable to the pH level of their surroundings, and certain species may not survive in acidic waters. At low pH, the majority of eggs of fishes fail to hatch, and even lead to the death of adult fish.

3.4.2 Impacts on Vegetation

Pollutants such as sulfur and nitrogen penetrate plant leaves through the stomata, disrupting the diffusion of CO_2. NO_x is absorbed by cells, resulting in the formation of toxic nitrite (NO_2^-) and nitrate (NO_3^-) ions, which interfere with nitrogen metabolism as they are absorbed by the roots. In some cases, exposure to toxic gases, especially SO_2, causes stomata to close as a defense against further pollutant entry, but this also decreases the rate of photosynthesis. Inside the cells, SO_2 dissolves to form bisulfite and sulfite ions. While sulfite is toxic, it can be metabolized by chloroplasts into nontoxic sulfate at lower concentrations. Currently, ozone is considered the most harmful phytotoxic air pollutant, as it adheres to the plasma membrane and disrupts metabolic processes. This impairs the regulation of stomatal openings, damages chloroplast thylakoid membranes, and leads to the degradation of Rubisco, ultimately hindering photosynthesis. Interaction with O_2 produces reactive oxygen species such as hydrogen peroxide (H_2O_2), superoxide (O_2^-), singlet oxygen ($1O_2\bullet$), and hydroxyl radicals (OH^-). These molecules cause protein denaturation, damage to nucleic acids (mutations), and lipid peroxidation, which deteriorates the lipoprotein organization of cellular membranes. Air pollution is a significant factor contributing to forest decline in heavily polluted regions of Europe and North America (Florentina & Io, 2011).

3.4.3 Impacts on Soil

The potential impact of acid precipitation leads to the acidification of soils, an increase in loss of plant nutrients, increased weathering of mineral nutrients, and a slowdown of decomposition of organic matter, with a decrease in the pollution of soil organism populations, a decrease in cation exchange capacity, and affects plant development and water quality in many ways. As soils become more acidic, the capacity for critical nutrients and minerals such as calcium (Ca), magnesium (Mg), and potassium (K) decreases. As a result, these nutrients and minerals are leached away from the soil, rendering them less available to land creatures. Similarly, an increase in soil acidity can promote the elimination of heavy metals such as aluminum (Al). Thus, metals are more prone to enter surrounding lakes, rivers, and streams.

3.4.4 Impact on Water Bodies

Air pollutants impact water bodies significantly but indirectly by adding wet and dry depositions of nitrates, sulfates, and the toxic heavy metals such as Pb and Cd. These pollutants affect water bodies in several ways such as water quality degradation, ecological damage, and health risk to aquatic animals. Water bodies can undergo sudden and intense acidification in the short term, particularly when acid precipitation falls as snow. This phenomenon, known as "acid shock," can be fatal to numerous aquatic species, including fish. In some cases, eutrophication can occur when increased nutrient levels, rich in nitrogen, promote algae growth in water, leading to an imbalance in fish diversity and resulting in their mortality.

Heavy metals such as cadmium, lead, mercury, and arsenic infiltrate aquatic ecosystems primarily through agricultural runoff and acid rain, which present a significant danger to aquatic organisms. These metals are persistent, nondegradable, and are prone to bioaccumulate within living organisms, resulting in biomagnification throughout the food web. Even in minimal amounts, they can interfere with essential physiological processes, hinder reproduction, stunt growth, and lead to developmental abnormalities in fish and invertebrates.

3.4.5 *Impacts on Climate*

When long-term alterations occur in average weather conditions, such as temperature and rainfall, in a specific area, they refer to it as climate change. Climate change and air pollution are interconnected issues. Contaminants such as methane, black carbon, tropospheric ozone layers, and aerosols influence the quantum of sunlight that reaches the Earth's atmosphere. As a result, an increase in the temperature of Earth occurs, causing the melting of icebergs and glaciers.

3.5 Assessment Methodologies

Air pollution has emerged as one of the most alarming environmental concerns around the world, resulting in millions of premature deaths each year and inflicting significant ecological damage. A comprehensive and systematic approach is essential to understand, quantify, and manage air pollution, supported by various assessment methods. These methods are vital for measuring pollutant levels, identifying pollution sources, modeling pollutant dispersion, and evaluating human exposure along with the associated health risks. Evaluation of air pollution is a multidisciplinary endeavor, employing tools and techniques from atmospheric sciences, environmental engineering, public health, and data analytics. Conventional approaches, such as ground-level monitoring, provide precise and real-time information on pollutant concentrations, while sophisticated modeling methods aid in forecasting pollution distribution and simulating future scenarios. Satellite remote sensing and affordable sensor technologies have broadened the geographical and temporal scope of air quality assessment, particularly in resource-limited areas (Fig. 3.6).

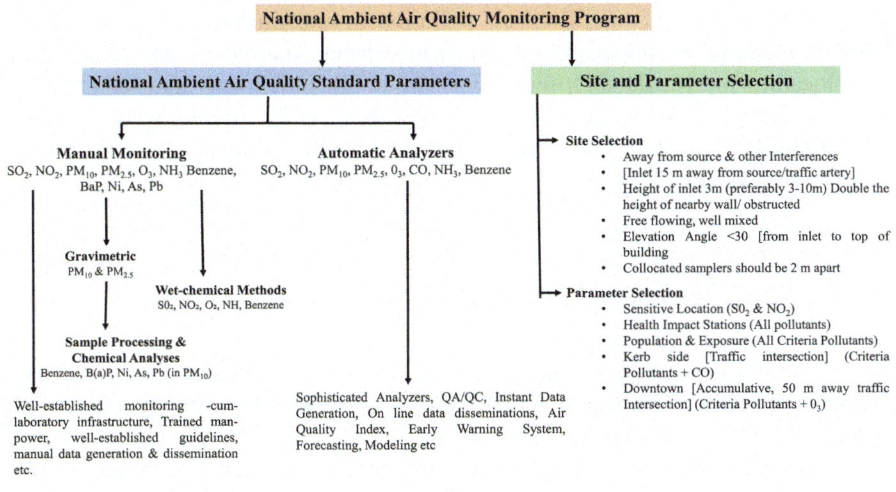

Fig. 3.6 Parameters and site selection in National Ambient Air Quality Monitoring Program

3.5.1 Ground-Level Monitoring of Air Pollution

Ground-level monitoring involves direct assessment of air pollutant levels at or near the Earth's surface using specialized instruments placed in stationary or mobile locations. This method appears to be the most accurate and reliable way to evaluate real-time ambient air quality, since it provides detailed information on pollutant concentrations that directly affect human health and the environment.

3.5.1.1 Stationary Monitoring Stations

Stationary air pollution monitoring systems offer essential, high-quality temporal information that serves as the foundation for air quality management and research, although unmatched in data accuracy, precision, and long-term reliability for designated locations. Typically, monitoring of emissions at stationary sources consists of four components: (1) indicator performance, (2) measurement methods, (3) frequency of monitoring, and (4) average time. Examples of measurement methods include the following four systems:

(a) Continuous Emission Monitoring Systems (CEMS),
(b) Continuous Opacity Monitoring Systems (COMS),
(c) Continuous Parametric Monitoring Systems (CPMS), and.
(d) Manual checks involving documentation of process conditions or work practices.

3.5.1.2 Mobile Monitoring Units

They are dedicated vehicles or portable devices that contain air quality sensors designed to monitor ground-level air pollution in real time as they move. These units are essential in supporting stationary monitoring stations by providing adaptable, location-specific data. Different types of mobile monitoring units can be seen, such as Google's air quality street view cars, CPCB mobile vans in India, ENVIRONBUS in Germany, and EPA's VAMMS (Vehicle Add-on Mobile Monitoring System).

3.5.2 Satellite-Based Monitoring of Air Pollution

Advancements in monitoring technology, remote sensing, and modeling can deliver practical air pollution information more quickly and cost-effectively than traditional approaches. For more than 20 years, satellites have been collecting global air quality information, addressing data deficiencies in various areas. For countries with limited monitoring capabilities, particularly low- and middle-income countries, there is no single method of managing air quality. The use of satellite-derived estimates can help fill data gaps in areas lacking ground monitoring. Satellites acts as platforms for various instruments or sensors. A single satellite can carry multiple sensors, and a specific sensor can be present on several satellites. Each sensor can gather data on different parameters. For example, the MODIS3 sensor, which is found on both the Aqua and Terra satellites of NASA, serves as a vital source for global $PM_{2.5}$ estimates. The Tropospheric Monitoring Instrument (TROPOMI) is mounted on the Sentinel 5 precursor satellite and measures gases such as ozone, NO_2, SO_2, and CO, as well as aerosols including dust, smoke, and black carbon (Holloway et al., 2021).

The Indian Space Research Organization (ISRO) has developed and validated an aerosol optical depth (AOD) product from its Ocean Color Monitor (OCM-3) sensor. This data product allows for understanding the complexities of changes in air quality across the Indian subcontinent. It was developed with the help of the cutting-edge SAC AErosol Retrieval (SAER) algorithm, which uses data from the EOS-6 satellite.

Acknowledgments We express our sincere gratitude to all people who provided guidance during the preparation of this book chapter. Special thanks to Dr. Akhilesh Kumar Yadav, Assistant Professor, Manipal University Jaipur, Jaipur, India, for giving me this opportunity.

References

Al-Kindi, S. G., Brook, R. D., Biswal, S., & Rajagopalan, S. (2020). Environmental determinants of cardiovascular disease: lessons learned from air pollution. *Nature Reviews. Cardiology, 17*(10). https://doi.org/10.1038/s41569-020-0371-2

Altman, M. C., Kattan, M., O'Connor, G. T., Murphy, R. C., Whalen, E., LeBeau, P., Calatroni, A., Gill, M. A., Gruchalla, R. S., Liu, A. H., Lovinsky-Desir, S., Pongracic, J. A., Kercsmar, C. M., Khurana Hershey, G. K., Zoratti, E. M., Teach, S. J., Bacharier, L. B., Wheatley, L. M., Sigelman, S. M., & Jackson, D. J. (2023). Associations between outdoor air pollutants and non-viral asthma exacerbations and airway inflammatory responses in children and adolescents living in urban areas in the USA: a retrospective secondary analysis. *Lancet Planetary Health, 7*(1). https://doi.org/10.1016/S2542-5196(22)00302-3

Assi, M. A., Hezmee, M. N. M., Haron, A. W., Sabri, M. Y. M., & Rajion, M. A. (2016). The detrimental effects of lead on human and animal health. *Veterinary World, 9*(6). https://doi.org/10.14202/vetworld.2016.660-671

Brabhukumr, A., Malhi, P., Ravindra, K., & Lakshmi, P. V. M. (2020). Exposure to household air pollution during first 3 years of life and IQ level among 6–8-year-old children in India—A cross-sectional study. *Science of the Total Environment, 709*, 135110. https://doi.org/10.1016/J.SCITOTENV.2019.135110

DeMarini, D. M., & Linak, W. P. (2022). Mutagenicity and carcinogenicity of combustion emissions are impacted more by combustor technology than by fuel composition: A brief review. *Environmental and Molecular Mutagenesis., 63*(3). https://doi.org/10.1002/em.22475

Espitia-Pérez, L., Jiménez-Vidal, L., & Espitia-Pérez, P. (2020). Particulate matter exposure: Genomic instability, disease, and cancer risk. *Environmental Health - Management and Prevention Practices*. https://doi.org/10.5772/intechopen.86536

Fajersztajn, L., Saldiva, P., Pereira, L. A. A., Leite, V. F., & Buehler, A. M. (2017). Short-term effects of fine particulate matter pollution on daily health events in Latin America: A systematic review and meta-analysis. *International Journal of Public Health., 62*(7). https://doi.org/10.1007/s00038-017-0960-y

Farhat, A., Mohammadzadeh, A., Balali-Mood, M., Aghajanpoor-Pasha, M., & Ravanshad, Y. (2013). Correlation of blood Lead level in mothers and exclusively breastfed infants: A study on infants aged less than six months. *Paediatria Croatica, 2*(4), 150–152.

Florentina, I., & Io, B. (2011). The effects of air pollutants on vegetation and the role of vegetation in reducing atmospheric pollution. In *The impact of air pollution on health, economy, environment and agricultural sources*. IntechOpen. https://doi.org/10.5772/17660

Forbes, P. B. C., & Garland, R. M. (2016). Outdoor air pollution. In *Comprehensive analytical chemistry* (Vol. 73, pp. 73–96). https://doi.org/10.1016/bs.coac.2016.02.004

Hemmingsen, J. G., Rissler, J., Lykkesfeldt, J., Sallsten, G., Kristiansen, J., & Loft, S. (2015). Controlled exposure to particulate matter from urban street air is associated with decreased vasodilation and heart rate variability in overweight and older adults. *Particle and Fibre Toxicology, 12*(1). https://doi.org/10.1186/s12989-015-0081-9

Holloway, T., Miller, D., Anenberg, S., Diao, M., Duncan, B., Fiore, A. M., Henze, D. K., Hess, J., Kinney, P. L., Liu, Y., Neu, J. L., O'Neill, S. M., Odman, M. T., Pierce, R. B., Russell, A. G., Tong, D., West, J. J., & Zondlo, M. A. (2021). Satellite monitoring for air quality and health. *Annual Review of Biomedical data Science, 4*. https://doi.org/10.1146/annurev-biodatasci-110920-093120

Kuklinska, K., Wolska, L., & Namiesnik, J. (2015). Air quality policy in the U.S. and the EU - A review. *Atmospheric Pollution Research, 6*(1). https://doi.org/10.5094/APR.2015.015

Lee, K., & Greenstone, M. (2021). Air quality life index. Annual Update. In *Annual update* (Issue September).

Lepers, C., Dergham, M., Armand, L., Billet, S., Verdin, A., Andre, V., Pottier, D., Courcot, D., Shirali, P., & Sichel, F. (2014). Mutagenicity and clastogenicity of native airborne particulate matter samples collected under industrial, urban or rural influence. *Toxicology In Vitro, 28*(5), 866–874. https://doi.org/10.1016/J.TIV.2014.03.011

Lippmann, M. (2013). National Particle Component Toxicity (NPACT) initiative: integrated epidemiologic and toxicologic studies of the health effects of particulate matter components. *Research Report (Health Effects Institute), 177*, 5–13.

Manisalidis, I., Stavropoulou, E., Stavropoulos, A., & Bezirtzoglou, E. (2020). Environmental and health impacts of air pollution: A review. *Frontiers in Public Health, 8*. https://doi.org/10.3389/fpubh.2020.00014

Miller, M. R., Shaw, C. A., & Langrish, J. P. (2012). From particles to patients: Oxidative stress and the cardiovascular effects of air pollution. *Future Cardiology., 8*(4). https://doi.org/10.2217/fca.12.43

Moraitis, A. A., Wood, A. M., Fleming, M., & Smith, G. C. S. (2014). Birth weight percentile and the risk of term perinatal death. *Obstetrics and Gynecology, 124*(2 PART1). https://doi.org/10.1097/AOG.0000000000000388

Schraufnagel, D. E., Balmes, J. R., Cowl, C. T., De Matteis, S., Jung, S. H., Mortimer, K., Perez-Padilla, R., Rice, M. B., Riojas-Rodriguez, H., Sood, A., Thurston, G. D., To, T., Vanker, A., & Wuebbles, D. J. (2019). Air pollution and noncommunicable diseases: A review by the forum of international respiratory societies' environmental committee, part 1: The damaging effects of air pollution. *Chest, 155*(2). https://doi.org/10.1016/j.chest.2018.10.042

Shang, L., Huang, L., Yang, W., Qi, C., Yang, L., Xin, J., Wang, S., Li, D., Wang, B., Zeng, L., & Chung, M. C. (2019). Maternal exposure to PM2.5 may increase the risk of congenital hypothyroidism in the offspring: A national database based study in China. *BMC Public Health, 19*(1). https://doi.org/10.1186/s12889-019-7790-1

Shao, B., Zhu, L., Dong, M., Wang, J., Wang, J., Xie, H., Zhang, Q., Du, Z., & Zhu, S. (2012). DNA damage and oxidative stress induced by endosulfan exposure in zebrafish (Danio rerio). *Ecotoxicology, 21*(5). https://doi.org/10.1007/s10646-012-0907-2

Soon, W., Baliunas, S. L., Robinson, A. B., & Robinson, Z. W. (1999). Environmental effects of increased atmospheric carbon dioxide. *Climate Research, 13*(2). https://doi.org/10.3354/cr013149

State of Global Air. (2024). *Soga-2024-Report_0.Pdf* (pp. 1–35). https://www.stateofglobalair.org/sites/default/files/documents/2024-06/soga-2024-report_0.pdf

Thangavel, P., Park, D., & Lee, Y. C. (2022). Recent insights into particulate matter (PM2.5)-mediated toxicity in humans: An overview. *International Journal of Environmental Research and Public Health, 19*(12). https://doi.org/10.3390/ijerph19127511

WHO. (2022). *Billions of people still breathe unhealthy air: New WHO data*. World Health Organization.

Yildizhan, H., Udriştioiu, M. T., Pekdogan, T., & Ameen, A. (2024). Observational study of ground-level ozone and climatic factors in Craiova, Romania, based on one-year high-resolution data. *Scientific Reports, 14*(1), 26733. https://doi.org/10.1038/s41598-024-77989-0

Yadav, A. K. (2021). Human health risk assessment in opencast coal mines and coal-fired thermal power plants surrounding area due to inhalation. *Environmental Challenges, 3*, 100074.

Yadav, A. K., Sahoo, S. K., Dubey, J. S., Kumar, A. V., Pandey, G., & Tripathi, R. M. (2019). Assessment of particulate matter, metals of toxicological concentration, and health risk around a mining area, Odisha, India. *Air Quality, Atmosphere & Health, 12*(7), 775–783.

Yadav, A. K., Sahoo, S. K., Patra, A. C., Dubey, J. S., Lenka, P., Sagar, D. V., Kumar, A. V., & Tripathi, R. M. (2014). Source identification of particulate matter and associated intake of elements through inhalation in an industrial area of Odisha, India. *Toxicological & Environmental Chemistry, 96*(3), 410–425.

Zanobetti, A., & Schwartz, J. (2009). The effect of fine and coarse particulate air pollution on mortality: A national analysis. *Environmental Health Perspectives, 117*(6). https://doi.org/10.1289/ehp.0800108

Zhao, J., Li, M., Wang, Z., Chen, J., Zhao, J., Xu, Y., Wei, X., Wang, J., & Xie, J. (2019). Role of $PM_{2.5}$ in the development and progression of COPD and its mechanisms. *Respiratory Research, 20*(1). https://doi.org/10.1186/s12931-019-1081-3

Chapter 4
Assessment on Health and Environmental Impacts Due to Air Pollution

Sourish Pramanik, Subhrajyoti Bhattacharjee, Dibyendu Seth, Ankan Das, Narender Pal, Chayan Munshi, Umalaxmi Thingujam, Kishore Chandra Swain, Sushil Kumar, and Sandip Debnath

S. Pramanik (✉)
Department of Plant Pathology, Punjab Agricultural University, Ludhiana, India

Ethophilia Research Foundation, Santiniketan, India

S. Bhattacharjee
Department of Soil and Water Conservation Engineering, Punjab Agricultural University, Ludhiana, India

D. Seth
Ethophilia Research Foundation, Santiniketan, India

Department of Genetics and Plant Breeding, Chaudhary Charan Singh Haryana Agriculture University, Hisar, India

A. Das
Division of Crop Protection, ICAR-Indian Institute of Horticultural Research, Bengaluru, India

N. Pal
Department of Environmental Engineering, Punjab Engineering College (Deemed to be University), Chandigarh, India

C. Munshi
Ethophilia Research Foundation, Santiniketan, India

U. Thingujam
Department of Soil Science and Agricultural Chemistry, Palli Siksha Bhavana, Visva-Bharati, Sriniketan, India

K. C. Swain
Department of Agricultural Engineering, Institute of Agriculture, Visva-Bharati, Sriniketan, India

S. Kumar
Department of Environmental Studies, Siksha-Bhavana (Institute of Science), Visva-Bharati, Santiniketan, India

S. Debnath
Department of Genetics and Plant Breeding, Institute of Agriculture, Visva-Bharati, Sriniketan, India

A. K. Yadav (ed.), *Computational Approach for Air Pollution*,
https://doi.org/10.1007/978-3-032-09215-1_4

4.1 Introduction

The survival and growth of every living thing on the planet depend on air. It is a core environmental element that maintains the life cycle trend on the earth surface from the drawn of the time. The dependency trend sometimes gets different due to natural phenomena and human activity that make the environment more vulnerable. Air pollution is one of the major disturbances in the core environmental framework. It is currently a global issue due to the rapid development of modern production and transportation, as well as a sharp increase in growing populations and urbanization (Zhou et al., 2019). In addition to affecting the lives of individuals and their careers, poor air quality also hinders economic growth and contributes to environmental degradation (Kampa & Castanas, 2008). According to research by the World Health Organization (WHO), outdoor air pollution, including small particles, is more harmful than previously thought (Aayush et al., 2020). The WHO estimates that air pollution kills almost seven million people annually (Krzyzanowski & Cohen, 2008). In both urban and rural areas, outdoor air pollution, primarily particulate matter, resulted in 5.2 million premature deaths worldwide in 2016 (World Health Organization, 2021). Numerous contaminants are significant contributors to human illness. These include particulate matter (PM), which are particles with varying but extremely small diameters that enter the respiratory system by inhalation and can lead to cancer, reproductive, and cardiovascular disorders, and malfunctions of the central nervous system and reproductive system. Although ozone in the upper atmosphere protects against UV radiation, it is dangerous on the surface when present in large concentrations, as it may also have an adverse effect on the cardiovascular and pulmonary systems. In addition, air pollutants that are harmful to people include nitrogen oxide, dioxins, sulfur dioxide, and polycyclic aromatic hydrocarbons (PAHs). High concentrations of carbon monoxide can potentially cause immediate poisoning (Manisalidis et al., 2020). Although a variety of pollutants resulting from human activity influence air pollution, industrial processes and automotive engines account for the majority of the problem. Significant air pollutants such as ozone (O_3), particulate matter (PM), nitrogen dioxide (NO_2), and sulfur dioxide (SO_2) have been shown to have extremely detrimental impacts (Saini et al., 2019). Other than the primary contaminants, such as carbon monoxide (CO), soot particles, nitrogen dioxide (NO_2), and sulfur oxides (SO_x); emissions into the atmosphere additionally contain photochemical secondary pollutants including ozone (O_3), as well as trace amounts of organic compounds, heavy metals, radioactive isotopes, and toxic chemical gases (Cong et al., 2022; Han et al., 2022; Mahmood et al., 2020; Sharma et al., 2022; Somani et al., 2020; Yadav, 2022; Yadav & Hopke, 2020). Immoderate vegetation damage and excessive crop output declines are attributed to air pollution (Joshi & Swami, 2007). Additionally, O_3 and PM with aerodynamic diameter <2.5 μm ($PM_{2.5}$) are also found to be those with the greatest economic importance for the elimination of air pollutants related to the health of plants and trees (Nowak et al., 2023). Furthermore, air pollution has an impact on the health of soil and water reserves through wet and dry deposition of contaminants; therefore, its effects go beyond the atmosphere (Saxena, 2025). For instance, acid rain from sulfur and nitrogen oxides or heavy metals from chemical manufacturing may pollute surface water, reduce soil fertility, and even result in ultimate soil erosion and land degradation (Manisalidis et al., 2020).

Artificial intelligence is a computer-simulated human intellect that has problem-solving and decision-making ability. Assessment and forecasting of pollution is the major priority in dealing with the increasing complexity of the environment. However, to fulfil the criteria, AI is employed as a key tool for pollutants predicting and helping communities to create more effective regulations to reduce the dangers of pollution-related public exposure in environmental (Cetin et al., 2019; Rastogi et al., 2025). This immersive technology is a superior technique for air quality assessment and forecasts since it can handle complex and nonlinear relationships among spatial–temporal factors (Fernández & Vico, 2013; Yadav et al., 2022). It also provides an exceptional ability to track the present pollutant scenario and illustrates a rapid and precise approach to recognize contamination-prone regions. Additionally, over time, various scientists proposed several models to make this work more effectively and efficiently. The observed data, presumably through satellite or ground-based, is the foundation for developing air quality models, issuing risk alerts, and formulating policies. A variety of methods have been employed for air quality (AQ) tracking, including remote sensing by satellite, wireless sensor network (WSN) systems, low-cost sensors, and ground-based combined systems for surveillance (Bahadur et al., 2023). Large-scale data has been made possible by remote sensing (RS) and geographic information systems (GIS), which regularly observes the atmosphere and surface of the earth (Holloway et al., 2021). The data have far more records and sequence changes on the basis of space–time scenario (Zhou et al., 2021). The use of satellite data, although formerly perceived as a useful complement to ground information, has risen due to improvements in AQ models that utilize remote data, the downsizing of satellite sensors, and increases in the spatiotemporal and spectral accuracy of satellite sensors. Because standard computer systems cannot handle the large amounts of information generated by satellites, it is currently difficult to maintain and analyze the information. ML and AI have made improved comprehension of both satellite and ground-based data.

4.2 Sources and Types of Air Pollutants

4.2.1 Emerging Anthropogenic Sources

Modernization and mechanization are most important for the growth of any community, but they create new and invisible forms of air pollution that pose serious risks to health and ecosystems. Causing pollution is a very dynamic event that becomes more complicate when it is a one-dimensional consideration. Microplastics are tiny plastic particles released from everyday items that play a vital role in causing a non-pointel pollution. Synthetic fabrics such as polyester shed up to 1900 microscopic fibers during a single laundry cycle, which escape into the air through drying or wind, can be considered as an example. Vehicles add to this problem because tire wear produces rubber particles (containing chemicals such as styrene) that make up 1–10% of fine dust ($PM_{2.5}$ and PM_{10}) in cities (Johannessen et al., 2022). Industrial areas alone release more than 90,000 tons of these particles yearly. Even 3D printers, often seen as harmless, contribute by melting plastics such as ABS and PLA, spewing out ultrafine particles (as small as a virus) at levels exceeding 106 particles per cubic centimeter. Indoors, where people spend 90% of their time, microplastics hide in dust and ventilation systems, with everyday actions such as vacuuming stirring 13,000–68,000 particles per person annually (Amato-Lourenço et al., 2020). These plastics are now found all over the world, including in remote areas, but their long-term effects on the lungs, climate, and food chains are uncertain.

3D printing releases volatile organic compounds (VOC)—airborne chemicals such as ethyl acetate, methyl isobutyl ketone (MIBK), and benzene—that degrade indoor air quality (Min et al., 2021). Ethyl acetate, detected at 640.73 $\mu g/m^3$ in flex resin, surpasses safe indoor thresholds, while MIBK in Water Elego resin (462.66 $\mu g/m^3$) and ethylbenzene in flex resin (332.73 $\mu g/m^3$) (Garcia-Gonzalez et al., 2024) pose risks ranging from respiratory irritation to kidney damage in animal studies. Xylene isomers (*m*- and *o*-xylene) and styrene, found in resins and filaments, contribute to headaches and potential cancer risks, with benzene, a confirmed carcinogen—appearing in Water Elego resin (19.99 $\mu g/m^3$) and traces in ABS/PLA filaments (Garcia-Gonzalez et al., 2024). Emissions vary widely: PLA filaments release aromatic hydrocarbons (toluene, *n*-hexadecane) (Garca & Pola, 2022), while PETG emits toluene and *n*-hexadecane, deviating from assumptions about benzene dominance (Bernatikova et al., 2021). Discrepancies arise from the enclosures trapping pollutants, additives of materials, and temperature fluctuations. Cumulative exposure to even low-level VOCs, such as toluene in PLA (9.45 $\mu g/m^3$), can trigger immediate symptoms (dizziness, throat irritation) or long-term damage (organ damage, cancers), highlighting the urgent need to recognize 3D printing as an emerging airborne hazard that requires rigorous air quality scrutiny (Khan et al., 2018).

Volatile organic compounds (VOCs), toxic gases released by everyday products and industrial processes, pose invisible risks as air pollutants, particularly from emerging technologies and modern lifestyles. Extremely fast-evaporating VOCs (VVOCs), such as propane and butane in aerosols or refrigerants, escape quickly into the air, harming health even at low concentrations. Common VOCs, such as formaldehyde in building materials (plywood, synthetic fabrics) and toluene in paints or adhesives, remain indoors, linked to cancer and nerve damage (Woolley, 2024). Less volatile but equally dangerous semi-VOCs (SVOCs), such as pesticides (chlordane) and plasticizers (phthalates) in furniture or electronics, seep slowly from plastics and fire retardants, disrupting hormones and causing asthma (Landeg-Cox et al., 2025). Key sources include industrial solvents in manufacturing, compressed aerosols, and biomass burning, which leaks toxins during incomplete combustion (Yao et al., 2023). Even "green" solutions, such as biofuels or biodegradable products, release VOCs during production or use (Jang et al., 2022). Modern reliance on plastics, synthetic materials, and energy-efficient building designs traps these chemicals indoors, where they accumulate in dust or air; phthalates in cosmetics, flame retardants in electronics, and hexanal in food flavorings silently expose people to chronic health risks. From 3D printed plastics to metalworking fluids, emerging technologies increase these emissions, turning homes and workplaces into reservoirs of airborne toxins. Addressing this growing threat requires rethinking material design and industrial practices to curb VOC release in an increasingly synthetic world.

4.2.2 Satellite-Borne Air Pollutants

Satellite launches contribute significantly to atmospheric pollution, with 103 documented launches in 2019 distributed primarily across China (34), Russia/Kazakhstan (22), and the United States (27, including New Zealand operations) (Ryan et al., 2022). In 2019, the global use of rocket propellants totaled 32 Gg, with kerosene leading at 45%, hypergolic fuels comprising 32%, solid propellants making up 14%, and liquid hydrogen contributing 8%. Different regions showed distinct preferences: Japan, India, and French Guiana relied on solid fuels; China, Kazakhstan, and Iran preferred hypergolic types; whereas New Zealand, Russia, and the United States mainly used kerosene-based propellants (Ryan et al., 2022). These propellants release various pollutants, including nitrogen oxides (NO_x), water vapor (H_2O), black carbon (BC), aluminum oxide (Al_2O_3), and chlorine compounds (Adeniyi et al., 2022). Most emissions occur within the lower atmosphere, with 78–79% of H_2O and BC, and 68–69% of chlorine compounds and Al_2O_3 released above 15 km altitude (Ryan et al., 2022). NO_x emissions are primarily from reentry burn, with 95% emitted above 60 km. The growing space tourism industry poses additional environmental concerns, with projected annual emissions from 782 launches that could reach 30.4 Gg of H_2O (triple the level of 2019), 2.1 Gg of NO_x, and 1.0 Gg of BC (double the amount of 2019). When combined with existing launches, total annual reentry NO_x emissions could reach 4.2 Gg, comparable to the lower range of NO_x emissions from meteorite burn-up, estimated at 2–40 Gg annually (Ryan et al., 2022).

4.2.3 Air Pollution from Agricultural Activities

Intensive animal farming, a growing contributor to air pollution, emits ammonia (NH_3, linked to smog and lung irritation), methane (CH_4, a potent greenhouse gas), fine particulate matter (PM, dust-like particles), and odors from manure storage, housing, and feedlots (Conti, 2022). Emissions vary by season, animal type, and farm practices: for example, peak NH_3 levels of NH_3 are observed in summer (Feng et al., 2022), while poultry operations release stronger odors than dairy barns (Konkol et al., 2022). Ventilation systems, weather, and manure management further influence the spread of the pollutant, with PM concentrations spiking during the day due to higher temperatures and activity. Biomass burning, such as crop residue fires in regions like China's Beijing–Tianjin–Hebei area, adds carbon monoxide (CO), nitrogen oxides (NO_x), and black carbon (soot) to the air, with fire peaks in March, June–July, and October worsening regional air quality (Chen et al., 2017). These emissions travel for long distances and reach densely populated areas. Bioaerosols pose health risks, especially in winter and at night, where the AQI (Air Quality Index) exceeds 100. These include airborne bacteria, fungi, and antibiotic-resistant pathogens such as *Bacillus* and *Staphylococcus* (Gao et al., 2019). Inhalation of these pathogens often leads to several health hazards linked to the respiratory and cardiovascular system (Saber et al., 2014). Village-level operations such as crop processing and improper waste disposal amplify these particles, which often move to the cities and accumulates there.

Despite being banned long ago, highly toxic pesticides such as DDT and HCB persist in the air, especially over regions with intensive farming such as Asia and Europe (Tian et al., 2021). Pollution levels vary dramatically around the world: countries such as Moldova, France, and Armenia suffer from severe contamination, while places such as New Zealand and Finland experience minimal exposure (Huang et al., 2024). Interestingly, although Asia and the Americas currently use about 80% of the world's pesticides, the health risks from these specific old, banned chemicals are not necessarily highest where current pesticide use is greatest, instead, they depend heavily on how much of this lingering legacy pollution exists locally. Asian nations like China, the major importers of hazardous pesticides, struggle with improper disposal: discarded pesticide bottles leak toxins into the air. Meanwhile, the Americas export high-risk pesticides, exemplified by Brazil's increase in airborne toxins after lax regulations in 2016 (Huang et al., 2024). In Europe, patchy enforcement of pesticide laws explains why eastern countries (e.g., Ukraine) suffer worse air quality than central Europe. Although the United States curbs risks through robust monitoring—such as California's pesticide air tracking—outdated data in Russia and Australia masks true levels of pollution, worsened by wind-spreading pesticides across borders. Mixing multiple pesticides in regions such as Turkey creates unpredictable hazards, even with fewer chemicals in use (Huang et al., 2024). Modern farming and global trade silently fuel this crisis, as legacy chemicals such as DDT linger in the atmosphere. Addressing this invisible threat requires unified global policies to track emissions, regulate trade, and phase out high-risk pesticides. Figure 4.1 summarizes the different sources of air pollution and how human innovation and traditional practices collectively impact global air quality.

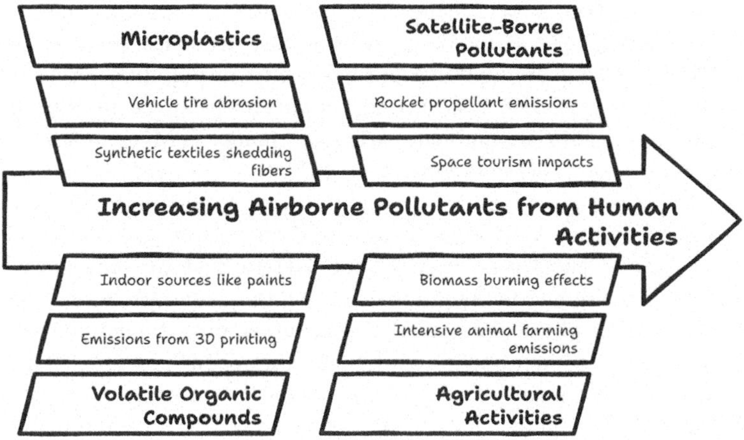

Fig. 4.1 Diverse sources contaminating the atmosphere, from emerging technologies (3D printing, microplastics) and emissions from the space industry to agricultural activities (animal farming, crop burning) and pesticide distribution patterns

4.3 Health Impacts of Air Pollution

There is major health concerns related to the effects of air pollution and climate change on respiratory diseases. The purpose of this chapter is to examine how these problems interact with a focus on respiratory disorders. Extreme weather events are expected to occur more frequently and with greater severity due to climate change, increasing air pollution levels and aggravate respiratory conditions. Lowering greenhouse gas emissions helps mitigate the intensity of extreme weather events and slow down climate change. Reducing the morbidity of respiratory disorders is another benefit of improving air quality through technical and regulatory advances. Furthermore, policies and initiatives that improve social support and healthcare access can help reduce population susceptibility to the harmful health consequences of climate change and air pollution (Tran et al., 2023). Due to climate change, heat extremes and particle air pollution are increasing and pose a health risk to people (Rahman et al., 2022). In vivo studies showing significant exposure-induced changes in the diversity and relative abundance of gut bacterial taxa have recently drawn attention to the effect of air pollution on the gut microbiota as a potential mechanism by which exposure to air pollutants increased the risk of chronic diseases. The state-of-the-art results of research on how air pollution affects the human gut microbiota. In general, there is evidence that both short-term and long-term exposure to air pollutants can change the diversity and composition of the gut microbiota (Filardo et al., 2022). Among all risk factors for human health, air pollution ranks fourth. Information collected from the Global Burden of Disease Study 2019 on overall adjusted life years for ischemic heart disease (IHD), stroke, and disability-adjusted life years (DALYs) for lung, tracheal, and bronchus cancers, as well as years of lost life (YLL) and mortality attributable to air pollution for 43 European countries from 1990 to 2019. The State of Global Air 2020 was used to determine the concentrations of ozone, ambient particulate matter ($PM_{2.5}$), and domestic air pollution from solid fuels (Juginovic et al., 2021).

Due to inadequate planning, design, and traffic in rapidly growing metropolitan areas, many cities around the world suffer from air pollution (Yadav et al., 2014, 2019, 2023, 2025). The Urban Heat Island (UHI) effect worsens these circumstances. Human health and the built environment are related because of air quality and opportunities for physical exercise. To encourage physical activity and reduce air pollution caused by automobile emissions, the recent planning literature has focused on the development of compact, walkable urban zones with green infrastructure scattered throughout (Piracha & Chaudhary, 2022). In addition to reducing occupational health issues such as silicosis and coal workers' pneumoconiosis (black lung disease), this substantial decrease in coal use and the corresponding closure of coal mines and coal-burning power plants will lead to significantly cleaner air, fewer heart attacks, hospitalizations, and other health benefits. However, other forecasts suggest that coal production and consumption would rise in China, India, Russia, and several other Asian nations. Coal is occasionally burnt in outdated and extremely polluting power plants. Coal consumption may have worsening health effects in certain areas. Additionally, millions of people in these areas continue to burn coal in their homes, increasing their exposure to toxins including arsenic, selenium, fluorine, and mercury (Finkelman et al., 2021). Vehicle traffic, coal-fired power plants, hydraulic fracturing, agricultural activity, and forest fires are some of the sources of ambient air pollution. It is made up of secondary pollutants that are created in the atmosphere from precursor gases and primary pollutants that are produced during combustion. Climate change is exacerbated by air pollution, whereas the health impacts of air pollution are exacerbated by climate change. Due to their growing organs and increased air intake per body weight, infants and children are particularly vulnerable to air pollution. In addition to worsening respiratory conditions, air pollution has been associated with decreased development of lung function and an increased prevalence of asthma (Brumberg et al., 2021).

There is substantial evidence that exposure to air pollution has a negative impact on health outcomes, which are often quantified in terms of hospitalizations and deaths. Subclinical routes that still affect behavior, performance, and abilities have received less attention in this research. Economic studies examine causal relationships between pollution and non-health outcomes, such as worker productivity, cognitive function, and other decision-making processes. The subclinical impacts can be harder to notice than those of official medical visits, but they could be more widespread if they affect otherwise healthy individuals (Aguilar-Gomez et al., 2022).

Physical inactivity and air pollution exposure are important risk factors for noncommunicable diseases (NCDs). Empirical data supports the following claims: (1) exposure to air pollution affects behavior related to physical activity; (2) exposure to air pollution occurs during physical activity; and (3) exposure to air pollution has immediate and long-term health effects on individuals who engage in physical activity. Public health modeling studies have measured the combined impact of physical activity and air pollution. Expert conversations, including two public sessions held at health-related conferences, helped identify these main study themes. According to a recent research, air pollution can make people less active during periods of severe air pollution or may even make it impossible for them to exercise in areas with high pollution levels. According to public health modeling studies, the advantages of physical exercise often exceed the hazards of air pollution, at least when it comes to active transportation (Tainio et al., 2021). Environmental epidemiology, which is usually a quantitative area, may benefit from the data produced by qualitative research. A multidisciplinary study using mixed methods is probably going to offer a more comprehensive explanation of the environmental health trends seen in quantitative studies. To effectively manage environmental health and create preventive, mitigation, and communication strategies, these explanations are essential. Environmental epidemiology benefits from the use of qualitative research methods because they: (1) enable triangulation of findings; (2) produce nuanced findings and new research questions; (3) prompt in-depth comprehension of quantitatively identified patterns; (4) produce additional unexpected and/or complex responses; (5) strengthen researcher–respondent relationships; (6) raise awareness of significant context-dependent dynamics or interactions that can produce biases; and (7) understand how perceptions of health risk are influenced by local, contextual, situational, and cultural factors (Noel et al., 2021).

The largest environmental risk factor for premature mortality worldwide is airborne fine particulate matter ($PM_{2.5}$), which is also likely to be the cause of several hundred thousand premature deaths. Here, it measures the mortality burden experienced by urban and rural families by income decile, as well as the consumption-based $PM_{2.5}$ contributions from these households (Rao et al., 2021). The WHO Global Health Observatory provided data on ambient and household air pollution, while the 2019 Global Burden of Disease Study provided statistics on morbidity and death. Noncommunicable diseases account for most of the fatalities brought on by ambient air pollution (Fisher et al., 2021). The identification method takes advantage of the way exogenous wind directions cause air contaminants to disperse spatially. The findings show that while $PM_{2.5}$ emissions from downwind fires do not affect the cognitive abilities of respondents 55 years of age or older, they do from upwind straw burning. The effect is temporary and is caused by concurrent $PM_{2.5}$ emissions on the day of the cognitive test (Lai et al., 2022). Local communities will be affected differently by the health effects of mitigating global climate change. However, only a small number of studies have produced results at the local level; most co-benefits assessments have been conducted at the global level. The importance of both socioeconomic development and climate policy in mitigating the health burden of air pollution is shown by the examination of local $PM_{2.5}$-attributable premature mortality under scenarios in three local populations (Ingole et al., 2022). The connection between poverty and exposure to ambient air pollution in 211 nations and territories. 7.3 billion people worldwide, 80% of whom reside in low- and middle-income nations, are directly exposed to dangerous average annual $PM_{2.5}$

concentrations using the World Health Organization's (WHO) 2021 revised fine particulate matter ($PM_{2.5}$) criteria. Because their economies tend to rely more heavily on polluting businesses and technology, lower-middle-income nations have especially high levels of air pollution. Subnational poverty estimates derived from harmonized household surveys and high-resolution global population and air pollution maps serve as the foundation for these conclusions (Rentschler & Leonova, 2023).

More than 60% of the Air Quality Index (AQI) readings remained at an unsafe level, and both events continued with little variation throughout the day. Source and synoptic meteorological conditions supporting both transboundary pollution episodes were identified using CALIPSO pictures, MERRA-2 AOD data, and backward trajectories (Nguyen et al., 2022). Although the number of environmental studies on ports and shipping has increased dramatically in recent decades, little is known about how these factors affect local health. The many forms of traffic, including cargo, people, and containers, are examined together with information on the environment, air pollution, socioeconomic indicators, and public health (Ducruet et al., 2024). The economy, trade, transportation, health care, social services, and other sectors were all impacted by the rapid spread of the COVID-19 pandemic. Most nations used social distancing measures and nationwide lockdowns to slow the virus's fast spread. This resulted in less commercial, industrial, and human activity, which in turn led to fewer emissions of air pollutants and an increase in air quality. The impact of lockdown policies on changes in air quality before and during COVID-19 lockdown was examined using air pollution monitoring data from the European Environment Agency (EEA) datasets (Skirienė & Stasiškienė, 2021). Numerous epidemiological studies have linked air quality to a wide range of detrimental health consequences, highlighting the large contribution of air pollution to the overall burden of diseases in the population, which can range from subclinical effects to early mortality. At the level of prevention of illness and health promotion on a personal and global scale, the assessment of health risk of air quality can be extremely important. An important instrument for directing public policy choices is the Air Pollution Health Risk Assessment (AP-HRA), which predicts the anticipated health effects of measures affecting air quality in diverse regulatory, environmental, and socioeconomic contexts. The idea of AP-HRA is introduced long with a guide to conduct AP-HRA correctly in various scenarios. It also provides a general explanation of how air pollution-related consequences are assessed and the health risks of air pollutants and their sources are measured.

Despite certain advances, there are still some issues with academic studies on the connection between air pollution and health expenses, such as the failure to adequately account for individual avoidance behavior and rural–urban and international movement (Xu et al., 2022). Recent developments include the creation of community-based air quality monitoring networks using inexpensive gas and particulate sensors, the development of forecasting techniques for ground-level concentrations and related public advisories, the development of risk and impact assessment techniques to investigate the potential health effects of future eruptions, the development of evidence-based, locally specific health protection measures, and standardized ash assessment methods to characterize. However, the fact that the health impacts of several massive and occasionally prolonged eruptions close to dense populations have not been thoroughly observed is still worrisome. Compared to explosive eruptions, the impacts of extended degassing on exposed populations have received very little consideration (Stewart et al., 2022).

4.4 Environmental Impacts of Air Pollution

4.4.1 Impact on Plants

The most vulnerable plant element is the leaf, as they are the primary recipients of air pollution. Major changes in leaf anatomy, decreased root development, and altered transport patterns are all triggered by air pollution. The current O_3 content in the air has a significant detrimental effect on agricultural yield and reduces the photosynthetic ability of plants. O_3 actively contributes to the creation of species of reactive oxygen species (ROS) due to its extremely reactive character. Oxidative stress caused by ROS seriously damages proteins, membrane lipids, and DNA (Saxena, 2025). Dust on the road damages the surface of leaves. Dust deposition also clogs the stomata, which alters gas exchange, lowers carbon uptake, and stunts plant development.

PM contaminants raise the temperature of the leaves and decrease photosynthetically active radiation (PAR). Dust and SO_2 pollution cause a reduction in the area of leaf blades. According to Barthlott et al. (1998), the epicuticular wax serves as a barrier among the plant and road dust and is an indicator of plant health. Road dust deposition breaks down epicuticular wax and alters the wetness of leaves. Necrosis and brown and yellow spots on foliar surfaces are caused by PM from stone dust. Additionally, SO_2, nitrogen oxides, and O_3 gas *pollutants* cause plants to create oxygen radicals. Chlorophyll pigment levels are decreased, and the membranes of the cell organelles are damaged by these oxyradicals. Podda et al. (2019) found that O_3-induced oxidative damage resulting from superoxide anion radical generation modifies the membrane structure of poplar clones exposed to O_3. Similarly, high concentrations of lead, copper, and cadmium prevent the absorption of CO_2 by the Calvin Benson cycle in cucumber leaves (*Cucumis sativus*) and interfere with the flow of electrons in light reactions. The main cause of toxicity is the production of reactive oxygen species (ROS), leading to DNA damage and oxidative stress. In order to counteract the harmful reactive species (RS), plants possess a sophisticated and multilayered network of antioxidative systems (AOS), wherein reactive oxygen species are crucial. In plants, cellular equilibrium is maintained by AOS's superiority over ROS. Plants at antioxidant levels lower their "ROS-neutralizing capacity," oxidize biomolecules, degrade proteins, trigger apoptosis, peroxide lipids, restrict enzyme release, and oxidize nucleotides. In contrast to other pollutants, ozone has a very distinct effect on plants, since it is an oxidative gas. Three steps form the process by which O_3 interacts with plants. The interaction between O_3 and plants occurs in three phases. In the first stage, oxygen enters the plant through the leaf border layer; in the second stage, it enters the plant tissue through the stomata; and in the third stage, it interacts chemically with the plant tissue, creating changes before helping the plant detoxify and repair. O_3 damages organic compounds and causes related disruption by reacting directly with organic double bonds such as an electrophile or a 1,3-dipole. Injury indicators such as white patches that develop into brown necrotic spots are typically caused by ozone exposure. Furthermore poplar-infecting larvae *Chrysomela populi* preferred to eat fresh leaves of the Eridano clone that received exposure to O_3, indicate a sensitivity to O_3 treatment that is individual to their genotype, indicating additional damage. The early maturity of leaves can also be induced by ozone (Mahmood et al., 2020).

4.4.2 Impact on Soil Quality

Acidification: When precipitation exceeds evapotranspiration, a significant paedogenetic process, known as soil acidification, takes place. Atmospheric emissions of acidifying substances such as sulfur dioxide (SO_2) and nitrogen oxides (NO_x), mainly from burning fossil fuels, can persist in the air for up to some days and thus can be transported over thousands of kilometers when they undergo chemical conversion into acids (sulfuric and nitric). Understanding current soil processes in relation to acidification requires a comprehensive evaluation of the interplay between natural processes and human effect. Soil acidification can be caused by several sources of proton, including acidic deposition, dissociation of carbonic and organic acids, nitrification, and excess cation uptake by plants (Fig. 4.2) (Ulrich, 1989).

The input and source of atmospheric deposition are represented by black arrows. Blue and red arrows represent hydrogen consumption and production, respectively. The orange arrow represents the process of consumption without hydrogen generation. Thinner lines represent natural acidification. The buffer system of the soil at various pH levels is displayed by the dotted box (Zhang et al., 2021).

Heavy metal contamination: Metals or metalloids, which are elements with both metal and nonmetal properties, are considered heavy metals. The elements that are most concerning are those found in all areas of the environment and include arsenic, cobalt, cadmium, chromium, copper, manganese, mercury, nickel, lead, thallium, and tin. Most typically, they have densities greater than 5 g/cm^3, making them impervious to degradation or destruction (Garbarino et al., 1995). It has long been known that *air pollution* is a deadly type of *pollution*. Human-induced mobilization of heavy metals in the biosphere has grown to constitute a significant step in the geochemical cycling (Li et al., 2004). Large amounts of heavy metals are released into the environment, soil, and plants by stationary and mobile sources of industrial areas, surpassing the levels of natural emissions (Bilos et al., 2001).

Loss of fertility: By releasing volatile molecules into the atmosphere, the soil also naturally contributes to *air pollution*. Volatilization of ammonia and denitrification

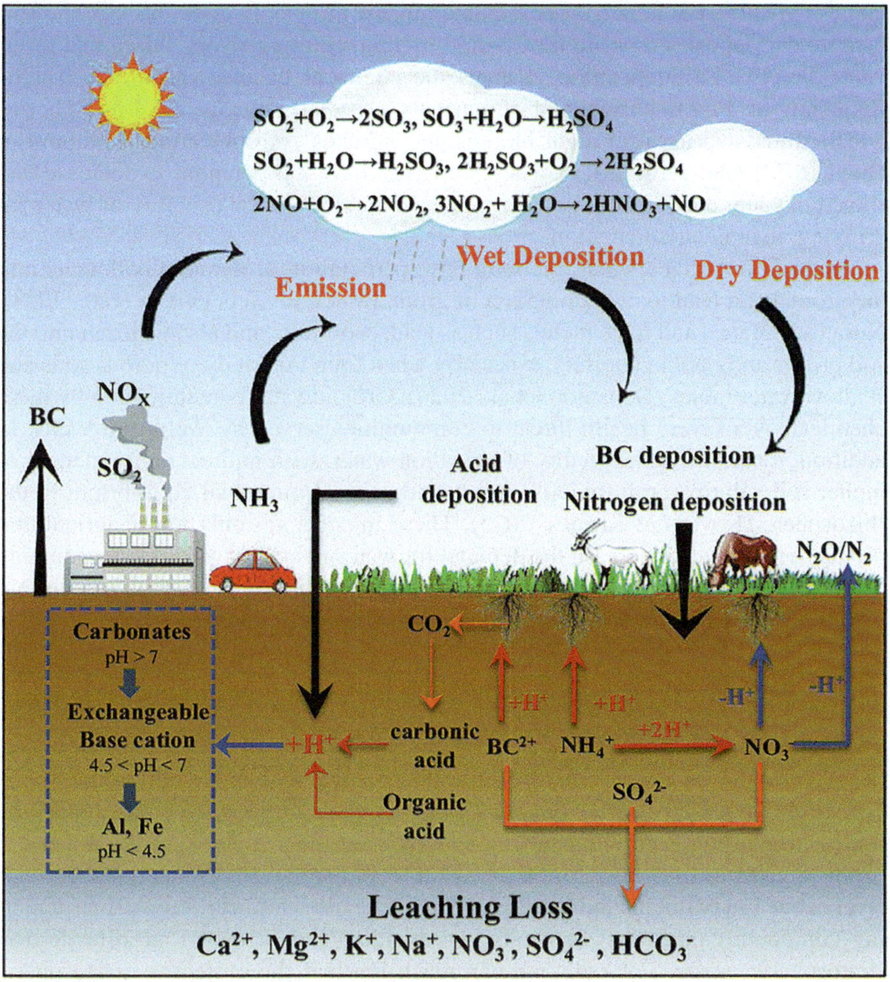

Fig. 4.2 Fundamental mechanisms of soil acidification

allow nitrogen to escape. Due to *pollution*, the soil contains hazardous compounds that can reduce soil fertility and, consequently, soil yield (Mishra et al., 2016).

4.4.3 Effect on Water Resources

Airborne contaminants, when they fall on land or water surfaces, can significantly influence the viability and quality of water resources (Asiminicesei et al., 2024). Industrial process emissions, vehicular exhaust, and agricultural emissions contribute to the atmospheric load of nitrogenous compounds, sulfur gases, and heavy metals (Mishra et al., 2016). These contaminants ultimately come back to the surface by dry deposition or are transported by rain, entering rivers, lakes, and reservoirs directly. The deposition changes the chemical balance of surface waters, frequently leading to enrichment of nutrients, especially nitrogen and phosphorus, which stimulates harmful algal blooms and initiates eutrophication (Ebrahimi & Ojani, 2024; Gao et al., 2019). These activities have the potential to decrease dissolved oxygen, destroy aquatic environments, and impair the usability of water for drinking, fishing, and irrigation.

Aside from surface water, the long-term percolation of dumped pollutants into the ground can lead to contamination of groundwater aquifers (Vivan et al., 2023). Nitrates, sulfates, and trace metals such as lead, cadmium, and arsenic leach into the soil profile and pool in aquifers, especially when found in sandy or porous soils and shallow water tables (Densmore et al., 2024). Groundwater contamination by these chemicals is a severe health threat to communities served by Wells and Wells. In addition, it can lower the quality of irrigation water, with indirect consequences of higher soil salinity, crop toxicity, and disturbance of microbial equilibrium in the rhizosphere (Kovacs & Kovacs, 2023). These impacts not only lower agricultural productivity but also increase the demand for water treatment and soil remediation. Mitigation of the cascading impacts of *air pollution* on water systems is therefore critical to integrated environmental management and sustainable agriculture.

4.4.4 Influence on Soil Erosion and Land Degradation

Vegetation loss due to ozone and acid rain: Ozone is found in the stratosphere, the upper layer of the Earth atmosphere, as well as at the ground level. Stratospheric ozone protects us from the harmful ultraviolet (UV) rays. On the other hand, ground-level ozone is a pollutant and harmful to human health. Unfortunately, ozone depleting compounds (such as chemicals, insecticides, and aerosols) steadily destroy stratospheric ozone. UV radiation can reach Earth if the protective stratospheric ozone layer is compromised, which could have detrimental consequences on human health (skin cancer) and crops (Madronich & de Gruijl, 1993; Teramura, 2006). Ozone enters the plants in the stomata and causes them to close, preventing CO_2 transport and lowering photosynthesis (Singh et al., 2009), and acid rain is any type of precipitation that has an acidic composition. A mixture of deposited material from the atmosphere that contains higher than usual levels of sulfuric and nitric acids is referred to as acid rain. Transportation, industry, and various power plants are the main providers of human resources (Gheorghe & Ion, 2011). Acid rain is defined as dry (particulates and gas) or wet (rain, fog, and snow) precipitation that contains hazardous levels of sulfuric and nitric acids. They can harm trees and plantations, acidify the water and soil, and even destroy buildings and outdoor monuments, sculptures, and constructions (Manisalidis et al., 2020).

Increased soil erosion that affects agricultural productivity: Air pollution indirectly causes soil erosion as a result of a chain of environmental disturbances. Pollutants such as sulfur dioxide (SO_2) and nitrogen oxides (NO_x) once emitted into the atmosphere react with water vapor to create acid rain. Acid precipitation not only deteriorates the structure of the soil by leaching vital nutrients such as calcium and magnesium but also reduces the strength of soil aggregates, thus causing the top soil vulnerable to wind and water erosion (Fu et al., 2017; Nawaz et al., 2020). Moreover, vegetation cover can also be harmed by air pollution, especially leafy parts of plants, due to particulate matter and hazardous gas deposition, lowering plant biomass and root tensile strength, which otherwise contribute to soil binding. The removal of vegetation leaves exposed soil bare, thus hastening the erosion (Wittwer & Castilla, 1995). This deterioration in soil health and greater topsoil loss have the potential to substantially lower agricultural yields by decreasing soil ability to hold water and nutrients, eventually undermining food security and sustainable agriculture.

4.4.5 Impact on Biodiversity and Ecosystems

Decline in microbial activity affecting soil health: Biodiversity may be negatively affected by air pollution at the local, regional, and global levels. Reduced vegetation cover, biodiversity, and ecosystem integrity can occur downstream of main sources due to local plume impacts caused by emissions of sulfur dioxide, hydrogen, fluoride, trace metals, and other toxics from pulp and paper industries, ore smelters, chemical manufacturers, power stations, and so on (Gordon & Gorham, 1963). Air pollution may impact biodiversity if it: (1) changes genetic variety within populations; (2) decreases biota's capacity for reproduction; (3) lowers agricultural or natural vegetation production; or (4) disrupts ecosystem structure and function, and it may have an effect on biodiversity (Barker & Tingey, 2012).

Loss of aquatic life due to water pollution: Eutrophication occurs when high concentrations of nutrients, especially nitrogen, promote the blooming of aquatic algae. This can lead to a disruption in fish variety and even fish mortality. The ability of an ecosystem to neutralize acid nature is unquestionably linked to the critical concentration of pollution that it can withstand without being destroyed (Manisalidis et al., 2020).

4.5 Computational Approaches in Impact Assessment

Air pollution and the problem associated with it are as vulnerable for health as the carcinogenic agent. That is why the calculation method for assessment of the magnitude of it is very important. The impacts of air pollution on human health and the environment are complex and multifaceted, thus we are required to advance the computation and modeling of data-driven assessment methods (Merz et al., 2020; Zhuoxian et al., 2025). Traditionally, assessment methods that were likely to be current foundational do not have the dynamic analytical capability to assess big multidimensional dimensions of the datasets that environmental monitoring systems provide (Dwivedi et al., 2019). The aggregation of computational methods, specifically, machine learning (ML), artificial intelligence (AI), and remote sensing-geographic information systems (RS-GIS), has emerged to address this gap in impact assessment frameworks (Ayanwale et al., 2024; (Kornaros, 2022).

This section explores the evolving role of these computational approaches in quantifying, predicting, and visualizing the consequences of air pollution (Dwivedi et al., 2019). It highlights the use of regression algorithms (e.g., Random Forest, XGBoost, Artificial Neural Networks) and deep learning models (CNNs, LSTMs) in predicting health outcomes such as respiratory illnesses, disease outbreaks, and hospital admission trends under the ML and AI-based assessment category (Çınar et al., 2020; Firouzi et al., 2021). The inclusion of hybrid models that integrate air quality, meteorological, and epidemiological data offers a more holistic risk evaluation platform.

In reviewing computational modeling, they have increasingly prominent role in quantifying, forecasting, and visualizing the impacts of air pollution (Dwivedi et al., 2019). The regression algorithms (e.g., Random Forest, XGBoost, and Artificial Neural Networks) and deep learning models (e.g., CNNs, LSTMs) in the ML and AI-based health assessment category successfully applied to health-related outcomes and disease prediction such as respiratory diseases, disease outbreak predictions, and hospital admission patterns (Çınar et al., 2020; Firouzi et al., 2021). The use of hybrid models that combine air quality, meteorological, and epidemiological data forms an even more comprehensive risk assessment framework.

RS-GIS-based assessment methods show that satellite monitoring and GIS modeling approaches facilitate high-resolution mapping processes for the dispersion of pollutants (Torgersen et al., 2021). Mapping pollution impacts and identification of urban and rural hotspots are second-order impacts on assessment systems. These spatial predictions can be further enhanced through GIS-based geospatial assessment and tools such as kriging and inverse distance weighting (IDW) (Chen et al., 2020). The methods also emphasize the alignment of the ML models with RS-GIS, showing hybrid systems that can provide real-time assessment monitoring support public health planning and decision-making, and provide policy considerations for environmental studies (Kousis & Tjortjis, 2023). Figure 4.3 shows an overview of the models used in the assessment and prediction.

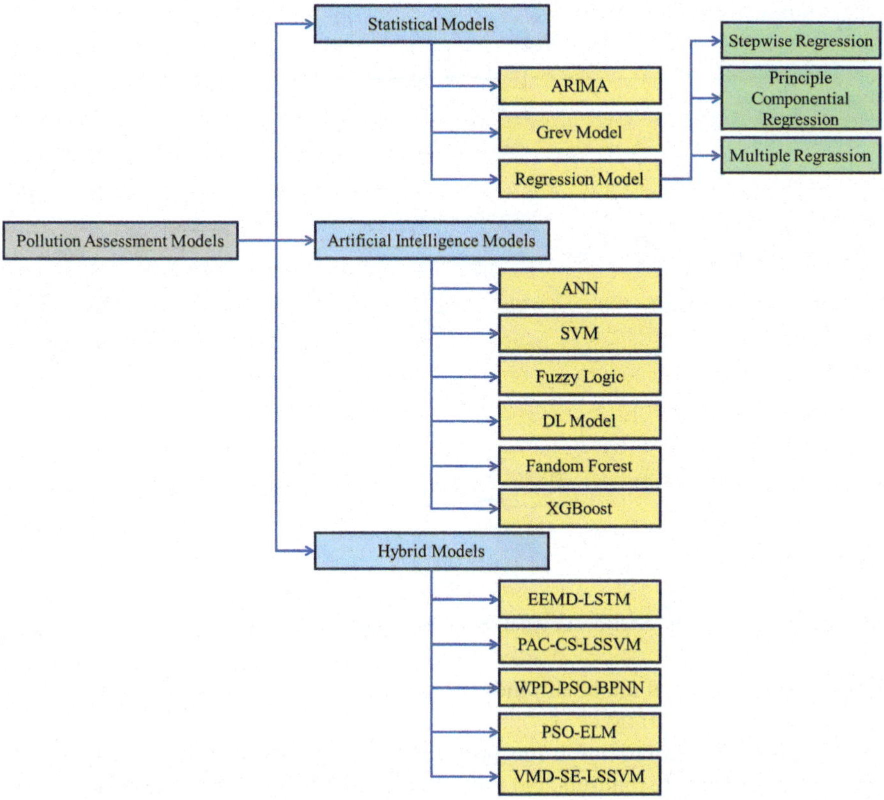

Fig. 4.3 The flow diagram of the models associated with pollution assessment and prediction

4.5.1 ML- and AI-Based Assessment Methods

4.5.1.1 Predicting Air Pollution Impact on Human Health

Air pollution is related to numerous health problems, including asthma, bronchitis, chronic obstructive pulmonary disease (COPD), cardiovascular problems, and premature deaths (Abogunrin-Olafisoye & Adeyi, 2025; Toptaş, 2024). Predictive models such as ML (machine learning) and AI (artificial intelligence) can be used to predict the health impacts of air pollution based on the concentration of pollutants and the exposure of the population.

AI predictive models include Random Forest (RF), XGBoost, and Artificial Neural Networks (ANNs), which are commonly used to predict surface health outcomes of air quality.

Random Forest is an ensemble learning method that works well for large complex datasets. It can detect nonlinear interactions between exposure to pollutants

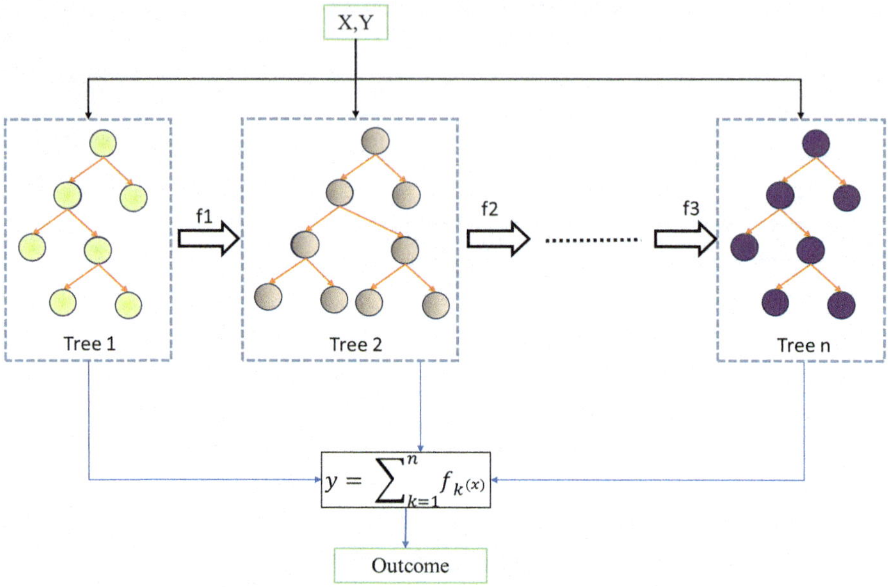

Fig. 4.4 Diagram of the XGBoost model (Source: Author)

($PM_{2.5}$, NO_2, SO_2) and the health indicators studied (e.g., hospital visits, morbidity) (Shayanmehr et al., 2025).

XGBoost has regularization mechanisms to prevent overfitting, so it is suitable for large public health datasets, but it has higher accuracy for prediction and is computationally superior (Fig. 4.4) (Khurshid et al., 2025).

Artificial Neural Networks (ANNs) are proficient in analyzing complex relationships with pollutant concentrations occurring in time and space. Additionally, this model can accommodate high-dimensional input variables and provide accurate output predictions with respect to respiratory illness, cardiovascular disease, and general population health risk (Fig. 4.5) (Nystrøm et al., 2024).

Deep learning approaches such as Convolutional Neural Networks (CNNs) and Long Short-Term Memory Networks (LSTMs) are the long-term prediction models (Park & Kim, 2022).

CNNs are used for the interpretation of remote sensing images to detect pollution-induced changes in the land surface and its features (Jahanbakht et al., 2021). Some examples of these applications include urban air quality mapping and vegetation health monitoring.

LSTMs are a type of RNN suited to time-series analysis and therefore is excellent for predicting future time series of pollutants or for estimating some future health effects from pollutants based on historical input (Zhang et al., 2019).

In addition to pollutant-specific impact assessments, AI models are increasingly being utilized to forecast disease outbreaks associated with air pollution (Popescu et al., 2024). For example, predictive analytics can correlate pollution trends to

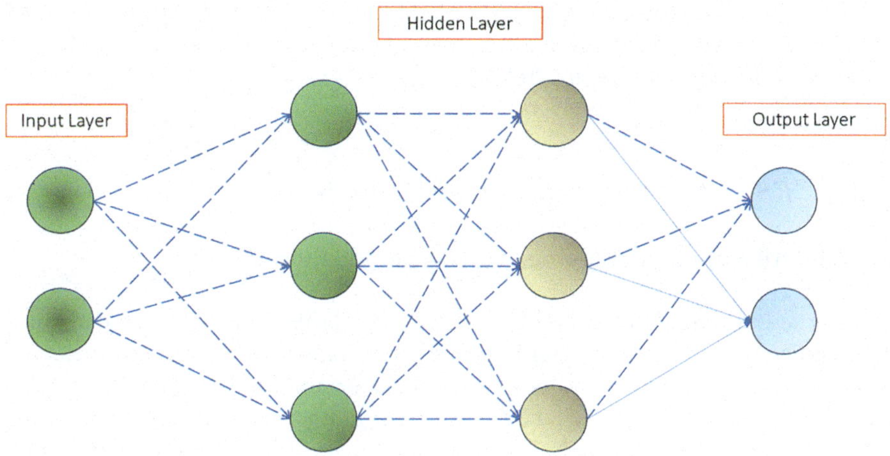

Fig. 4.5 Diagram of the ANN model (Source: Author)

incidence rates of respiratory infections and cardiovascular diseases, even as far-reaching as neurological impairments.

Machine learning classification models (e.g., support vector machines, decision trees) are capable of identifying patterns of disease incidence that correlate with air quality metrics (Filho et al., 2022), ultimately freeing up human resources from the burden of translating data into actionable policy.

The hospitals surrounding metropolitan centers can forecast patient volume during high pollution days, helping health systems mobilize and allocate accountability to protect vulnerable patients by providing timely and relevant support.

These tools not only serve to create early warning systems but also provide an evidence base for making public health policy decisions, emergency planning, and resource allocation for those at-risk populations.

4.5.1.2 Hybrid Models for Health Risk Assessment

Hybrid models work with diverse datasets for pollution concentrations, weather variables, demographic factors, and epidemiological statistics to provide a comprehensive framework for health risk assessment (Sahani et al., 2019).

These models are particularly useful in the following:

- Combining satellite-derived air pollution data with ground-based health statistics to spatially and temporally map disease susceptibility.
- Quantifying cumulative risk by incorporating meteorological influences (e.g., wind speed, humidity, temperature) that affect pollutant dispersion and inhalation risk.

- Leveraging AI techniques alongside established tools such as AirQ+, BenMAP, and GAINS to produce robust estimations of mortality and morbidity attributable to air pollution (Sokhi et al., 2022).

4.5.2 RS-GIS-Based Assessment Methods

4.5.2.1 Air Pollution Mapping and Hotspot Identification

Remote sensing and GIS technologies play a central role in environmental monitoring, particularly in detecting and mapping air pollution patterns across wide geographical areas. Satellite sensors such as those on the *MODIS*, *Sentinel-5P*, and *Landsat* satellites provide critical data for tracking atmospheric pollutants including $PM_{2.5}$, NO_2, and SO_2 (Boy et al., 2019; Edwards et al., 2022). Satellite-derived indices such as *aerosol optical depth* (AOD) can able to identify the pollution levels through inferred even in areas lacking ground-based monitoring stations (Benavente, 2022). Once these values are obtained, GIS-based spatial analysis is used to visualize pollutant distribution. Spatial interpolation methods include: (1) kriging, which takes spatial autocorrelation into account, and (2) inverse distance weighting (IDW), which relies on proximity for estimation (Poterek et al., 2025). These methods are widely applied to create high-resolution maps. These maps are essential for the identification of hotspots, allowing policymakers and environmental managers to prioritize interventions in the most affected regions (Ma et al., 2024).

4.5.2.2 Impact on Soil and Water Systems

The influence of air pollution extends beyond the atmosphere, affecting soil quality and water resources through dry and wet deposition of pollutants (Saxena, 2025). For example, heavy metals from industrial emissions or acid rain resulting from sulfur and nitrogen oxides can degrade soil fertility and contaminate surface water.

Remote sensing tools detect such changes using the following:

(a) Vegetation and water indices (e.g., NDVI, NDWI) to assess plant stress caused by soil contamination.
(b) Spectral reflectance patterns to identify changes in water turbidity or chemical composition.

Geostatistical techniques such as kriging and IDW are also used to estimate the spatial spread of contaminants, mainly in agricultural or peri-urban areas (Alam et al., 2024). These types of models are crucial for understanding long-term environmental degradation caused by sustained air pollution.

4.5.2.3 Integration of ML and RS-GIS

The incorporation of ML with the RS-GIS technique creates powerful hybrid systems that are capable of near real-time monitoring, forecasting, and policy support. For instance:

- Satellite images processed through CNNs can be linked with pollutant data through regression or classification models (Zou et al., 2024).
- Real-time alerts and health risk maps can be generated through web-based GIS platforms augmented by AI algorithms, aiding in rapid response and mitigation (Ali et al., 2025).

These hybrid approaches flexible its limit for a user to analyze spatial–temporal trends in pollution, assess cumulative impacts, and simulate future scenarios under various circumstances. A practical example of air quality monitoring systems deployed in Indian megacities such as Delhi. Sentinel-5P satellite data is integrated with AI-based forecasting models and GIS dashboards for public disciplinary work development and government action (Kalluri et al., 2023).

4.6 Case Studies and Real-World Applications

4.6.1 Health Impact Assessments

Air pollution is responsible for killing millions of people around the world every year. The health effects of air pollution are diverse, complex and disproportionately affect susceptible groups, hence damage multiple organ systems. Studies illuminates the unequal load carried out by particular communities and underscores the primary health impacts of air pollution on the respiratory, cardiovascular, nervous system, and developmental functioning. Especially for respiratory diseases, artificial intelligence (AI) and machine learning (ML) have become powerful tools in recent years to improve our understanding, prediction, and mitigation of such adverse health outcomes.

Respiratory disease is one of the most well-established consequences of exposure to air pollution. The increase in NO_2, IDO2, and $PM_{2.5}$ can be destructive to lung health, while O_3 and $PM_{2.5}$ are worsening. According to Pope et al. (2002), prolonged exposure to $PM_{2.5}$ has been associated with increased risk of lung cancer, asthma attacks, and chronic obstructive pulmonary disease. According to Gauderman et al. (2015), children's lung function development is significantly limited as a result of high TRAP exposure. World Health Organization (2021) has estimated that air pollution is attributable to more than 43% of deaths from lung cancer and COPD worldwide, underscoring the prevalence of pollution and the burden of respiratory diseases.

In this context, artificial intelligence is required to improve the precision and resolution of pollution exposure assessments. Di et al. (2019) established deep learning model for daily $PM_{2.5}$ levels in the United States, which is a great progress in the estimation of exposure in epidemiological studies. Similarly, Zhan et al. (2021) combined satellite with the ground data. They predicted NO_2 exposure in urban areas using convolutional neural networks (CNN). Requia et al. (2020) demonstrated that maps of environmental health risks can be generated at finer levels using artificial intelligence (AI) such as with random forest models to assess the effect of long-term exposure to ozone on pediatric asthma.

AI has also shown potential in the diagnosis and prognosis of respiratory diseases. Machine learning (ML) models can predict exacerbations with a high degree of accuracy when trained on environmental data and electronic health records (EHR). With the XGBoost model, Xia et al. (2022) achieved 89% accuracy in the prediction of asthma exacerbations attributed to PM_2. Bédard et al. (2021) associated exposure to pollution with symptoms of COPD by extracting them from clinical data using natural language processing (NLP). In order to identify early-stage cancer in patients with heavy exposure to $PM_{2.5}$, Liu et al. (2020) employed deep learning on CT images in the context of lung cancer. By modeling the intricate relationships between pollutants, genetics, and lung cancer risk through a graph neural network (GNN), Chen et al. (2023) took it a step further and opened doors to personalized medicine.

In addition, artificial intelligence provides vital assistance to public health initiatives. Using real-time air quality data, Kumar et al. (2021) developed an AI alert system in real time that predicts hospitalizations related to asthma. Healthcare providers can minimize patient risk and prepare ahead of time with the use of such technologies. Meanwhile, Sicard et al. (2022) demonstrated how reinforcement learning can be used to strategically place urban green areas, predicting a 15% reduction in respiratory mortality attributed to PM_2. These applications spotlight AI's possibilities for generating healthier environments and shaping legislative decisions along with its application in diagnostics.

Although it is promising, integrating AI into environmental health research is not without its difficulties. Low-income or rural communities, where exposure risk may be important, are often underrepresented in most models due to lack or biased statistics. Furthermore, there are also problems with the interpretability of AI models in healthcare applications, especially those involving deep learning. Explainable AI (XAI) methodologies are necessary to establish transparency and trust (Loyola-González et al., 2023). Another frontier, as per Yang, Jiang, et al. (2023), is the incorporation of wearable sensors and artificial intelligence, allowing for personalized real-time exposure monitoring. The vast majority of evidence shows that air pollution is an avoidable contributor to the global high burden of disease. Because AI and ML can handle large-scale, better reflect environmental and health data than conventional statistical methods, they dramatically enhance our knowledge of the interconnections between disease and pollution. Early warnings and intelligent urban planning are two other ways that real-time AI systems might significantly reduce risk.

Future studies should focus on longitudinal studies that establish causal correlations, explore federated learning methods to ensure data privacy in health research, and facilitate the integration of AI-driven pollution warnings into public health systems to further this topic. In addition to being an environmental necessity, combating air pollution and its health impacts through technological innovation also offers a significant public health opportunity.

4.6.2 Soil and Water Degradation Studies

Air pollution is a major cause of land and water degradation ecosystems in addition to having an impact on the atmosphere and human respiratory health. Heavy metals, sulfur (S), nitrogen (N) compounds, polycyclic aromatic hydrocarbons (PAHs), and particulate matter (PM) are some of the pollutants that are carried from the air to the soil and aquatic systems by atmospheric deposition. These pollutants pose serious threats to agriculture, biodiversity, and human health, as well as changing soil chemistry and worsening water quality. Creating comprehensive mitigation and policy control measures requires a knowledge of the mechanisms through which air pollutants contaminate soil and water.

Acid deposition, which consists mainly of sulfur dioxide (SO_2) and nitrogen oxides (NO_x), is an important mechanism by which air pollution degrades soil and water. These substances contribute to acid rain by combining with atmospheric moisture to generate nitric and sulfuric acids. Acid deposition, as stated by Likens and Bormann (1995), seriously depletes essential soil nutrients such as calcium (Ca^{2+}) and magnesium (Mg^{2+}), which leads to decreased fertility and compromised plant health. Furthermore, Driscoll et al. (2001) noted that acid rain leads to the release of toxic aluminum ions (Al^{3+}) into the soil, which then seep into freshwater systems, harming aquatic life and disrupting the balance of the environment.

Geographic information systems (GIS) and remote sensing (RS) are becoming fundamental tools for investigating the spatial impacts of acid rains on soils. They make it possible to map patterns of soil acidification on a large scale, pinpoint ecosystems that are at risk, and monitor long-term trends in deposition. Liang et al. (2020), for instance, utilized machine learning approaches to predict soil pH throughout China through the combination of Landsat-8 and Sentinel-2 photos. They discovered that the sulfur deposition resulted in 24% of cultivatable land having pH levels below 5.5. Similarly, Zhou et al. (2021) detected aluminum mobilization in acid soils using hyperspectral imaging and linked these changes to a decrease in NDVI in forested areas.

The sulfur and nitrogen deposition can also be efficiently monitored using RS-GIS methods. Wang et al. (2019) used MODTRAN atmospheric correction models to map SO_2 and NO_x plumes, with 30–50% higher deposition within a 10 km radius downwind of coal plants. EMEP GIS databases have also shown large soil pH gradients over Europe, where Scandinavian podzols have lost up to 40% of base cations since 1980. Vegetation stress, as a surrogate measure of acid rain effects, has also been tracked using RS. In a different study, Nowak et al. (2023) used LiDAR to illustrate that canopies in acidified Adirondack watersheds were 15–30% thinner than unaffected areas.

Heavy metals such as lead (Pb), cadmium (Cd), mercury (Hg), and arsenic (As) are also introduced into the environment through atmospheric deposition from industrial emissions, traffic, and coal combustion. Alloway (2013) reported that airborne particles from industrial and vehicular sources deposit these metals on agricultural soils, where they accumulate and enter the food chain through crops. Similarly, Wuana and Okieimen (2011) found that mercury emissions from fossil fuel combustion contribute to widespread water contamination, with mercury entering aquatic ecosystems and bioaccumulating in fish and other organisms. Persistent organic pollutants such as PAHs and dioxins further compound the problem. According to Wilcke (2007), PAHs from incomplete combustion bind strongly to soil particles and persist for decades, while Guzzella et al. (2016) linked dioxin fallout to freshwater contamination, where it bioaccumulates in fish and poses risks to human consumers.

These contaminants have significant effects on agriculture and the environment. The accumulation of acidic substances and heavy metals in soil reduces the microbial diversity, which is essential for the cycling of nutrients. According to Liu and Diamond (2019), PM-bound metals alter the microbiota in the soil, which has an impact on crop productivity and soil health. According to Khan et al. (2018), ozone (O_3), another airborne contaminant, damages plant tissues and reduces agricultural yields. Nitrogen deposition from the atmosphere causes nutrient overload in aquatic systems, resulting in eutrophication and toxic algae blooms. The usability and freshwater biodiversity are significantly affected by this process, as demonstrated by Camargo and Alonso (2006). Furthermore, Gall and Thompson (2015) highlighted the extensive ecological effects by connecting runoff that carries PAHs from air sources to the destruction of coral reefs.

Case studies also demonstrate the usefulness of RS-GIS tools. China's acid rain control program, for instance, employed HJ-1A/B satellite observation to record a 12% increase in soil pH in Guangdong province after the implementation of SO_2 emission limits in 2015 (Liu et al., 2022). Within the European context, GIS-based overlays of soil types, land use, and deposition data found that 22% of EU soils were above critical acid loads (Schöpp et al., 2021). However, many RS-GIS applications also have restrictions such as cloud cover preventing optical data acquisition in the tropics and the coarse resolution of certain platforms such as MODIS (250–500 m), which can neglect local heterogeneity. New innovations are helping to address these challenges. Sentinel-5P's TROPOMI instrument delivers hourly SO_2/NO_2 measurements with 7 km resolution (Veefkind et al., 2021), whereas AI-based models, such as CNNs that integrate soil spectral data and digital elevation models, improve pH predictions (Zhang, Liu, et al., 2023).

Human health is seriously in jeopardy when food and water sources are contaminated. Zheng et al. (2020) established a link between kidney disease and crops grown in soils contaminated with cadmium, such as rice near smelters. Similarly, Smith et al. (2019) discovered a link between drinking water contaminated with arsenic and a higher chance of bladder and lung cancer. The bioaccumulation of toxins in the human body can also arise due to direct contact with infected soil and water. According to Tóth et al. (2016), kids who had been exposed to Pb-contaminated soils, which are often the outcome of leaded gasoline, exhibited cognitive impairments. Consumption of fish in mercury-contaminated waters is associated with neurological disorders, particularly in susceptible populations (Ali et al., 2022).

To mitigate these effects, several mitigation measures have been suggested and practiced. Lime, for example, is a very successful soil amendment practice. Goulding (2016) showed that the addition of lime to acid soils can restore the pH balance and lower the bioavailability of toxic metals. Precision agriculture is also becoming popular, with multispectral UAV-based imaging to direct lime application on the field scale, as tested in Germany (Bönecke et al., 2020). Phytoremediation, the use of plants such as sunflowers and willows to absorb heavy metals, has shown promise in rehabilitating contaminated soils (Ali et al., 2013). At a larger level, policy initiatives such as emission control legislation are essential. United Nations Environment Programme (2021) emphasized the efficiency of policies such as Euro 6 automotive emission standards in minimizing atmospheric deposition. Furthermore, global conventions such as the UNECE CLRTAP (Convention on Long-Range) increasingly depend on RS-GIS data for the establishment of emission ceilings, as observed in the Göteborg Protocol.

There remain extensive research gaps despite progress. Long-term monitoring initiatives are required to trace the connections among air, water, and soil pollutants throughout ecosystems and geographical areas. Cutting-edge tools such as RS-GIS fusion techniques and AI-based modeling approaches can revolutionize our ability to predict deposition patterns and assess environmental concerns. Furthermore, since pollutants emitted in one country can impact ecosystems and communities in another, international cooperation is essential to address transboundary pollution.

4.6.2.1 Predictive Modeling for Risk Assessment

Machine learning (ML) has been used more and more in urban air pollution forecasting due to its capability of dealing with huge and varied datasets, managing complex, nonlinear relationships, and offer real-time prediction capability. In dynamic urban environments, these methods are particularly valuable for the prediction of concentrations of significant pollutants such as PM_2, PM_{10}, NO_2, O_3, and SO_2. Neural networks, ensemble methods, and novel transformer-based models are some of the ML architectures discussed within this chapter, in addition to the variety of input data that these systems need (such as satellite observations, IoT sensor feeds, traffic concentration, and meteorological conditions). It also analyzes the operational challenges of scalability, interpretability of the model, and data availability in assessing the efficiency of various strategies on standard assessment metrics such as RMSE, MAE, and R^2.

In urban air pollution simulation, neural networks—especially deep learning models—have been at the forefront. Long short-term memory (LSTM) networks, for example, have exhibited a good level of prediction power for temporal changes in pollutants. With the help of traffic and weather data, Li et al. (2020) applied LSTM in Beijing and achieved a mean absolute error (MAE) of 4.7 μg/m^3 for $PM_{2.5}$ predictions. Similarly, Bai et al. (2021) lowered the RMSE by 18% while combining spatial and temporal data with convolutional LSTM (ConvLSTM) compared to traditional ARIMA models. Meanwhile, graph neural networks (GNNs) are gaining popularity because of their ability to model complex metropolitan systems. Based on traffic and road network data, Wang et al. (2022) employed GNNs to model the propagation of pollutants in the city, achieving a high R^2 of 0.89.

Moreover, ensemble models such as Random Forest (RF) and XGBoost have been to be both very precise and robust. When Sahu et al. (2021) merged satellite-derived aerosol optical depth (AOD) and land use data, they found that RF significantly outperformed linear regression (R^2 = 0.82 vs. 0.65) in the prediction of NO_2 in Delhi. A promising new direction is the application of hybrid models, which combine machine learning and physical simulations. Zhang, Zhang, et al. (2023) enhanced ozone predictions by 27% through the integration of neural networks with WRF-Chem simulations. Moreover, transformer-based models such as AirFormer (Qi et al., 2023) provide scalability and contextual knowledge over urban systems utilizing attention mechanisms for multicity $PM_{2.5}$ prediction and reporting an MA of 5.2 μg/m^3.

Urban pollution prediction requires a variety of data sources. Real-time pollutant concentrations are delivered through ground monitoring stations, e.g., those run by the US EPA AirNow program and local IoT-based air quality networks (Liu et al., 2022). With tools such as MODIS offering AOD data and Sentinel-5P TROPOMI offering high-resolution NO_2 measurements, satellite remote sensing greatly increases spatial coverage (Veefkind et al., 2021). Chen et al. (2022) demonstrated that road density metrics and GPS trajectory records are two instances of traffic and urban mobility datasets that are quite helpful for prediction of NO_2.

To improve the accuracy and transparency of the model, feature selection and interpretation are still crucial. A popular XAI method called SHAP (SHapley Additive Explanations) has succeeded in highlighting the influence of a number of climatic conditions. For example, Lundberg et al. (2020) identified wind speed and relative humidity as the main predictors of PM_2 concentrations in the urban environment, offering interpretable knowledge regarding model choices.

The behavior of the various models over contaminants and data types has been benchmarked by comparative studies. For $PM_{2.5}$ forecasting with LSTM using traffic and meteorological inputs, Li et al. (2020) achieved an RMSE of 8.2 μg/m^3. Sahu et al. (2021) employed XGBoost trained on land use and satellite data to forecast NO_2 with an RMSE of 6.5 ppb. While Qi et al. (2023) employed transformer models to predict PM_{10}, with an RMSE of 12.4 μg/m^3 based on historical data, Wang et al. (2022) demonstrated the efficacy of GNNs for O_3 modeling with an RMSE of 9.1 ppb. These results indicate how ML models can be adapted to different contaminants and urban configurations (Table 4.1).

Table 4.1 Performance and comparative studies

Model	Pollutant	Best RMSE	Data used	Study
LSTM	$PM_{2.5}$	8.2 μg/m^3	Meteorological + traffic	Li et al. (2020)
XGBoost	NO_2	6.5 ppb	Satellite + land-use	Sahu et al. (2021)
GNN	O_3	9.1 ppb	Road networks + weather	Wang et al. (2022)
Transformer	PM_{10}	12.4 μg/m^3	Multicity historical data	Qi et al. (2023)

Despite progress, hurdles remain in the road to deploying ML-based air pollution forecasting systems in a fair and scalable way. In low-income and rapidly urbanizing places, where ground monitoring networks are often not available, lack of data is still a major problem (Requia et al., 2020). Similarly, Rudin (2019), who advocates for more transparency in machine learning models, notes the inscrutability of complicated models as an obstruction of regulatory acceptance and policymaking.

These issues are being addressed in a variety of innovative ways. For example, federated learning models provide a model for scalable multiregion forecasting by allowing decentralized training across cities while not sacrificing data privacy (Yang, Liu, et al., 2023). Meanwhile, the concept of "digital twins," virtual constructs of real cities updated in real time to make decisions that simulate their hygienic conditions, is gaining more relevance and offers new opportunities to manage pollution in urban areas (World Health Organization, 2023). These methods close the gaps between scientific modeling and real-world governance in air quality control, in addition to improving technological performance.

4.7 Conclusion and Future Perspectives

Air pollution remains a complicated and multifaceted issue worldwide, impacting public health, environmental sustainability, and socioeconomic growth. This chapter has reviewed the complex nature of air pollution, identifying both traditional, well-known sources (e.g., fossil fuel combustion, industrial contributions) and emerging sources of air pollution (e.g., microplastics, 3D printing, and space-based pollutants). Human health effects range from acute respiratory and cardiovascular diseases to chronic developmental, neurological, and gastrointestinal diseases that disproportionately impact vulnerable communities, including children, older adults, and low-income communities. Additionally, environmental impacts include implications for soil quality, water quality, biodiversity loss, and contributions to climate phenomena—e.g., acid rain and eutrophication.

The sophistication of computational technologies including machine learning (ML), artificial intelligence (AI), and remote sensing-geographic information systems (RS-GIS) allow for better monitoring, assessment, and prediction of air quality impacts. Integrating environmental, meteorological, and health datasets improves the ability to provide a more comprehensive perspective on the dynamics of pollution and their cascading effects. Thus, computational technologies make it possible to map data at higher resolutions, track temporal–spatial trends in pollution, conduct real-time forecasts of air pollution, assess health risks, and support evidence-based policymaking and targeted interventions.

Although steps have been taken to improve, there are still significant gaps in the availability of data, public health preparedness, and the implementation of global policy. We still face inequalities in the availability of monitoring infrastructure to allow for timely and fair responses and actions for mitigation, and many national strategies are still lacking considerations of socioeconomic impact from pollution-related morbidity and mortality.

Looking ahead, future research and policy should focus on the following key areas:

1. *Development of integrated monitoring networks*: Combining satellite-based and ground-level data with citizen science and low-cost sensor technologies to build inclusive, scalable, and real-time air quality surveillance systems.
2. *Personalized exposure assessment*: Advancing wearable and mobile-based pollution exposure trackers that incorporate individual health data for dynamic risk prediction and personalized healthcare response.
3. *Climate–pollution nexus*: Deepening research on how climate change and air pollution interact, especially in terms of extreme weather events, wildfire emissions, and transboundary pollutant transport, which could compound global health and environmental crises.
4. *Policy–technology convergence*: Encouraging the codesign of pollution mitigation policies with AI-powered decision support systems to ensure transparency, adaptability, and scalability across different socioeconomic contexts.
5. *Capacity building and education*: Strengthening interdisciplinary education, community engagement, and workforce training to promote environmental literacy and empower local stakeholders in pollution management.

In summary, effectively addressing air pollution requires a multidisciplinary approach that integrates scientific developments with inclusive governance and engaged public participation. By taking advantage of new technologies and fostering collaboration and research across sectors, the global community can work toward sustainable solutions for air quality and the betterment of planetary and human health into the future.

Acknowledgments The authors sincerely thank our prestigious institute and other people who are involved in this work.

References

Aayush, K., Vishal, D., Hammad, N., & Manu, K. S. (2020). Application of artificial intelligence in curbing air pollution: The case of India. *Asian Journal of Management, 11*(3), 285–290.

Abogunrin-Olafisoye, O. B., & Adeyi, O. (2025). Environmental and health impacts of unsustainable waste electrical and electronic equipment recycling practices in Nigeria's informal sector. *Discover Chemistry, 2*(1), 4.

Adeniyi, G. O., Fakinle, B. S., Sholiyi, O. S., Sonibare, J. A., & Akeredolu, F. A. (2022). Gaseous emissions characterization from potassium nitrate-based rocket composite propellant combustion. *International Journal of Energy for a Clean Environment, 23*(5).

Aguilar-Gomez, S., Dwyer, H., Graff Zivin, J., & Neidell, M. (2022). This is air: The "nonhealth" effects of air pollution. *Annual Review of Resource Economics, 14*(1), 403–425.

Alam, P., Khan, A. H., Islam, R., Sabi, E., Khan, N. A., & Zargar, T. I. (2024). Identification of prevalent leachate percolation of municipal solid waste landfill: A case study in India. *Scientific Reports, 14*(1), 8910.

Ali, G., Mijwil, M. M., Adamopoulos, I., & Ayad, J. (2025). Leveraging the internet of things, remote sensing, and artificial intelligence for sustainable forest management. *Babylonian Journal of Internet of Things, 2025*, 1–65.

Ali, H., Khan, E., & Ilahi, I. (2022). Environmental chemistry and ecotoxicology of hazardous heavy metals: Environmental persistence, toxicity, and bioaccumulation. *Journal of Chemistry, 2019*, 1–14.

Ali, H., Khan, E., & Sajad, M. A. (2013). Phytoremediation of heavy metals-concepts and applications. *Chemosphere, 91*(7), 869–881.

Alloway, B. J. (2013). *Heavy metals in soils: Trace metals and metalloids in soils and their bioavailability* (3rd ed.). Springer.

Amato-Lourenço, L. F., dos Santos Galvão, L., de Weger, L. A., Hiemstra, P. S., Vijver, M. G., & Mauad, T. (2020). An emerging class of air pollutants: Potential effects of microplastics to respiratory human health? *Science of the Total Environment, 749*, 141676.

Asiminicesei, D., Fertu, D. I., & Gavrilescu, M. (2024). Impact of heavy metal pollution in the environment on the metabolic profile of medicinal plants and their therapeutic potential. *Plants, 13*(6), 913. https://doi.org/10.3390/plants13060913

Ayanwale, M. A., Adelana, O. P., Molefi, R. R., Adeeko, O., & Ishola, A. M. (2024). Examining artificial intelligence literacy among pre-service teachers for future classrooms. *Computers and Education Open, 6*, 100179.

Bahadur, F. T., Shah, S. R., & Nidamanuri, R. R. (2023). Applications of remote sensing Vis-à-Vis machine learning in air quality monitoring and modelling: A review. *Environmental Monitoring and Assessment, 195*(12), 1502.

Bai, L., Wang, J., Ma, X., & Lu, H. (2021). Air pollution forecasting using ConvLSTM with spatiotemporal correlations. *Environmental Pollution, 284*, 117115.

Barker, J. R., & Tingey, D. T. (Eds.). (2012). *Air pollution effects on biodiversity*. Springer Science & Business Media.

Barthlott, W., Neinhuis, C., Cutler, D., Ditsch, F., Meusel, I., Theisen, I., & Wilhelmi, H. (1998). Classification and terminology of plant epicuticular waxes. *Botanical Journal of the Linnean Society, 126*(3), 237–260. https://doi.org/10.1111/j.1095-8339.1998.tb02529.x

Bédard, A., Basille, D., Pourcher, V., & Laviolette, M. (2021). Natural language processing for the identification of COPD in electronic health records. *Journal of Biomedical Informatics, 118*, 103792.

Benavente, N. R. (2022). Analysis of the spatial and temporal variability of gases and particulate material concentrations in the São Paulo metropolitan region and its transport to other regions using satellite images and numerical modeling.

Bernatikova, S., Dudacek, A., Prichystalova, R., Klecka, V., & Kocurkova, L. (2021). Characterization of ultrafine particles and VOCs emitted from a 3D printer. *International Journal of Environmental Research and Public Health, 18*(3), 929.

Bilos, C., Colombo, J. C., Skorupka, C. N., & Presa, M. J. P. (2001). Source, distribution and variability of airborne trace metals in La Plate City area, Argentina. *Environmental Pollution, 111*, 149–159.

Bönecke, E., Franko, U., Gebbers, R., Ruehlmann, J., & Lück, K. (2020). Delineating soil acidity management zones using UAV-based data and geostatistical techniques. *Precision Agriculture, 21*(5), 1029–1049.

Boy, M., Thomson, E. S., Navarro, J. A., Arnalds, O., Batchvarova, E., Bäck, J., Berninger, F., Bilde, M., Brasseur, Z., Dagsson-Waldhauserova, P., Castarède, D., Dalirian, M., De Leeuw, G., Dragosics, M., Duplissy, E., Duplissy, J., Ekman, A. M. L., Fang, K., Gallet, J., & Kulmala, M. (2019). Interactions between the atmosphere, cryosphere, and ecosystems at northern high latitudes. *Atmospheric Chemistry and Physics, 19*(3), 2015–2061.

Brumberg, H. L., Karr, C. J., Bole, A., Ahdoot, S., Balk, S. J., Bernstein, A. S., Byron, L. G., Landrigan, P. J., Marcus, S. M., Nerlinger, A. L., & Pacheco, S. E. (2021). Ambient air pollution: Health hazards to children. *Pediatrics, 147*(6), e2021051484.

Camargo, J. A., & Alonso, Á. (2006). Ecological and toxicological effects of inorganic nitrogen pollution in aquatic ecosystems: A global assessment. *Environment International, 32*(6), 831–849.

Cetin, M., Onac, A. K., Sevik, H., & Sen, B. (2019). Temporal and regional change of some air pollution parameters in Bursa. *Air Quality, Atmosphere & Health, 12*, 311–316.

Chen, J., Li, C., Ristovski, Z., Milic, A., Gu, Y., Islam, M. S., et al. (2017). A review of biomass burning: Emissions and impacts on air quality, health and climate in China. *Science of the Total Environment, 579*, 1000–1034.

Chen, X., Lee, R. M., Dwivedi, D., Son, K., Fang, Y., Zhang, X., Graham, E., Stegen, J., Fisher, J. B., Moulton, D., & Scheibe, T. D. (2020). Integrating field observations and process-based modeling to predict watershed water quality under environmental perturbations. *Journal of Hydrology, 602*, 125762. https://doi.org/10.1016/j.jhydrol.2020.125762

Chen, X., Liu, R., Liu, X., Peng, Y., Zhang, L., & Li, J. (2023). Graph neural networks for modeling the joint effects of air pollution and genetic factors on lung cancer risk. *Scientific Reports, 13*(1), 5678.

Chen, X., Liu, X., Wu, J., & Zhang, Y. (2022). Urban traffic and air pollution: High-resolution modeling for city planning. *Sustainable Cities and Society, 76*, 103438.

Çınar, Z. M., Nuhu, A. A., Zeeshan, Q., Korhan, O., Asmael, M., & Safaei, B. (2020). Machine learning in predictive maintenance towards sustainable smart manufacturing in industry 4.0. *Sustainability, 12*(19), 8211.

Cong, L., Zhou, S., Niyogi, D., Wu, Y., Yan, G., Dai, L., & Hu, Y. (2022). Concentrations and isotopic analysis for the sources and transfer of lead in an urban atmosphere-plant-soil system. *Journal of Environmental Management, 311*, 114771.

Conti, C. (2022). *Evaluation of different abatement systems for emissions of ammonia, PM and odour in intensive livestock farms*.

Densmore, J. N., Thayer, D. C., Dick, M. C., Swarzenski, P. W., Ball, L. B., Rosecrans, C. Z., & Johnson, C. (2024). *Evaluation of the characteristics, discharge, and water quality of selected springs at Fort Irwin National Training Center, San Bernardino County, California*. Scientific Investigations. https://doi.org/10.3133/sir20235142

Di, Q., Kloog, I., Koutrakis, P., Lyapustin, A., Wang, Y., & Schwartz, J. (2019). Assessing PM2.5 exposures with high spatiotemporal resolution across the continental United States. *Environmental Science & Technology, 53*(18), 10592–10601.

Driscoll, C. T., Lawrence, G. B., Bulger, A. J., Butler, T. J., Cronan, C. S., Eagar, C., & Weathers, K. C. (2001). Acid rain revisited: Advances in scientific understanding since the passage of the 1970 clean air act. *Environmental Science & Policy, 4*(1), 1–14.

Ducruet, C., Martin, B. P., Sene, M. A., Prete, M. L., Sun, L., Itoh, H., & Pigné, Y. (2024). Ports and their influence on local air pollution and public health: A global analysis. *Science of the Total Environment, 915*, 170099.

Dwivedi, Y. K., Hughes, L., Ismagilova, E., Aarts, G., Coombs, C., Crick, T., Duan, Y., Dwivedi, R., Edwards, J., Eirug, A., Galanos, V., Ilavarasan, P. V., Janssen, M., Jones, P., Kar, A. K., Kizgin, H., Kronemann, B., Lal, B., Lucini, B., & Williams, M. D. (2019). Artificial intelligence (AI): Multidisciplinary perspectives on emerging challenges, opportunities, and agenda for research, practice and policy. *International Journal of Information Management, 57*, 101994.

Ebrahimi, E., & Ojani, M. R. (2024). *Phosphorus dynamics in soil-water-sediment environment.* IntechOpen. https://doi.org/10.5772/intechopen.113225

Edwards, M. R., Holloway, T., Pierce, R. B., Blank, L., Broddle, M., Choi, E., Duncan, B. N., Esparza, Á., Falchetta, G., Fritz, M., Gibbs, H. K., Hundt, H., Lark, T., Leibrand, A., Liu, F., Madsen, B., Maslak, T., Pandey, B., Seto, K. C., & Stackhouse, P. W. (2022). Satellite data applications for sustainable energy transitions. *Frontiers in Sustainability, 3*, 910924.

Feng, K., Wang, Y., Hu, R., & Xiang, R. (2022). Continuous measurement of ammonia at an intensive pig farm in Wuhan, China. *Atmosphere, 13*(3), 442.

Fernández, J. D., & Vico, F. (2013). AI methods in algorithmic composition: A comprehensive survey. *Journal of Artificial Intelligence Research, 48*, 513–582.

Filardo, S., Di Pietro, M., Protano, C., Antonucci, A., Vitali, M., & Sessa, R. (2022). Impact of air pollution on the composition and diversity of human gut microbiota in general and vulnerable populations: a systematic review. *Toxics, 10*(10), 579.

Filho, W. L., Wall, T., Mucova, S. A. R., Nagy, G. J., Balogun, A., Luetz, J. M., Ng, A. W., Kovaleva, M., Azam, F. M. S., Alves, F., Guevara, Z., Matandirotya, N. R., Skouloudis, A., Tzachor, A., Malakar, K., & Gandhi, O. (2022). Deploying artificial intelligence for climate change adaptation. *Technological Forecasting and Social Change, 180*, 121662.

Finkelman, R. B., Wolfe, A., & Hendryx, M. S. (2021). The future environmental and health impacts of coal. *Energy Geoscience, 2*(2), 99–112.

Firouzi, F., Farahani, B., Daneshmand, M., Grise, K., Song, J., Saracco, R., Wang, L. L., Lo, K., Angelov, P., Soares, E., Loh, P., Talebpour, Z., Moradi, R., Goodarzi, M., Ashraf, H., Talebpour, M., Talebpour, A., Romeo, L., Das, R., & Luo, A. (2021). Harnessing the power of smart and connected health to tackle COVID-19: IoT, AI, robotics, and Blockchain for a better world. *IEEE Internet of Things Journal, 8*(16), 12826–12846.

Fisher, S., Bellinger, D. C., Cropper, M. L., Kumar, P., Binagwaho, A., Koudenoukpo, J. B., Park, Y., Taghian, G., & Landrigan, P. J. (2021). Air pollution and development in Africa: impacts on health, the economy, and human capital. *The Lancet Planetary Health, 5*(10), e681–e688.

Fu, B., Wang, S., Liu, Y., Liu, J., Liang, W., & Miao, C. (2017). Hydrogeomorphic ecosystem responses to natural and anthropogenic changes in the loess plateau of China. *Annual Review of Earth and Planetary Sciences, 45*(1), 223–243. https://doi.org/10.1146/annurev-earth-063016-020552

Gall, S. C., & Thompson, R. C. (2015). The impact of debris on marine life. *Marine Pollution Bulletin, 92*(1-2), 170–179.

Gao, K., Beardall, J., Häder, D., Hall-Spencer, J. M., Gao, G., & Hutchins, D. A. (2019). Effects of ocean acidification on marine photosynthetic organisms under the concurrent influences of warming, UV radiation, and deoxygenation. *Frontiers in Marine Science, 6*. https://doi.org/10.3389/fmars.2019.00322

Garbarino, J. R., Hayes, H., Roth, D., Antweider, R., Brinton, T. I., & Taylor, H. (1995). Contaminants in the Mississippi River, U.S Geological survey circular 1133, Virginia, U. S. A. http://pubs.usgs.gov/circ/circ1133

Garca, H., & Pola, T. L. (2022). Health and safety in 3D printing. *International Journal of Occupational and Environmental Safety, 6*(1), 14–25.

Garcia-Gonzalez, H., Lopez-Pola, T., Fernandez-Rubio, P., & Fernandez-Rodriguez, P. (2024). Analysis of volatile organic compound emissions in 3D printing: Implications for indoor air quality. *Buildings, 14*(11), 3343.

Gauderman, W. J., Urman, R., Avol, E., Berhane, K., McConnell, R., Rappaport, E., & Gilliland, F. (2015). Association of improved air quality with lung development in children. *New England Journal of Medicine, 372*(10), 905–913.

Gheorghe, I. F., & Ion, B. (2011). The effects of air pollutants on vegetation and the role of vegetation in reducing atmospheric pollution. In *The impact of air pollution on health, economy, environment and agricultural sources* (Vol. 29, pp. 241–280). IntechOpen.

Gordon, A. G., & Gorham, E. (1963). Ecological aspects of air pollution from an iron-sintering plant at Wawa, Ontario. *Canadian Journal of Botany, 41*(7), 1063–1078.

Goulding, K. W. T. (2016). Soil acidification and the importance of liming agricultural soils with particular reference to the United Kingdom. *Soil Use and Management, 32*(3), 390–399.

Guzzella, L., Poma, G., De Paolis, A., & Roscioli, C. (2016). Air-water exchange of anthropogenic and natural organochlorine compounds in the Venice lagoon. *Chemosphere, 144*, 201–208.

Han, Y., Lee, J., Haiping, G., Kim, K. H., Wanxi, P., Bhardwaj, N., & Brown, R. J. (2022). Plant-based remediation of air pollution: a review. *Journal of Environmental Management, 301*, 113860.

Holloway, T., Miller, D., Anenberg, S., Diao, M., Duncan, B., Fiore, A. M., & Zondlo, M. A. (2021). Satellite monitoring for air quality and health. *Annual Review of Biomedical Data Science, 4*(1), 417–447.

Huang, Y., & Li, Z. (2024). Assessing pesticides in the atmosphere: A global study on pollution, human health effects, monitoring network and regulatory performance. *Environment International, 187*, 108653.

Ingole, V., Dimitrova, A., Sampedro, J., Sacoor, C., Acacio, S., Juvekar, S., Roy, S., Moraga, P., Basagaña, X., Ballester, J., & Antó, J. M. (2022). Local mortality impacts due to future air pollution under climate change scenarios. *Science of the Total Environment, 823*, 153832.

Jahanbakht, M., Xiang, W., Hanzo, L., & Azghadi, M. R. (2021). Internet of underwater things and big marine data analytics—A comprehensive survey. *IEEE Communications Surveys & Tutorials, 23*(2), 904–956.

Jang, M., Yang, H., Park, S. A., Sung, H. K., Koo, J. M., Hwang, S. Y., et al. (2022). Analysis of volatile organic compounds produced during incineration of non-degradable and biodegradable plastics. *Chemosphere, 303*, 134946.

Johannessen, C., Liggio, J., Zhang, X., Saini, A., & Harner, T. (2022). Composition and transformation chemistry of tire-wear derived organic chemicals and implications for air pollution. *Atmospheric Pollution Research, 13*(9), 101533.

Joshi, P. C., & Swami, A. (2007). Physiological responses of some tree species under roadside automobile pollution stress around city of Haridwar, India. *The Environmentalist, 27*, 365–374.

Juginovic, A., Vukovic, M., Aranza, I., & Bilos, V. (2021). Health impacts of air pollution exposure from 1990 to 2019 in 43 European countries. *Scientific Reports, 11*, 22516.

Kalluri, B., Vishnupriya, V., Arjunan, P., & Dhariwal, J. (2023). Net-zero energy campuses in India: Blending education and governance for sustainable and just transition. *Sustainability, 16*(1), 87.

Kampa, M., & Castanas, E. (2008). Human health effects of air pollution. *Environmental Pollution, 151*(2), 362–367.

Khan, M. I. R., Fatma, M., Per, T. S., Anjum, N. A., & Khan, N. A. (2018). Salicylic acid-induced abiotic stress tolerance and underlying mechanisms in plants. *Frontiers in Plant Science, 6*, 462.

Khurshid, M. R., Manzoor, S., Sadiq, T., Hussain, L., Khan, M. S., & Dutta, A. K. (2025). Unveiling diabetes onset: Optimized XGBoost with Bayesian optimization for enhanced prediction. *PLoS One, 20*(1), e0310218.

Konkol, D., Popiela, E., Skrzypczak, D., Izydorczyk, G., Mikula, K., Moustakas, K., et al. (2022). Recent innovations in various methods of harmful gases conversion and its mechanism in poultry farms. *Environmental Research, 214*, 113825.

Kornaros, G. (2022). Hardware-assisted machine learning in resource-constrained IoT environments for security: Review and future prospective. *IEEE Access, 10*, 58603–58622.

Kousis, A., & Tjortjis, C. (2023). Investigating the key aspects of a Smart City through topic Modeling and thematic analysis. *Future Internet, 16*(1), 3.

Kovacs, E. D., & Kovacs, M. H. (2023). *Global change drivers impact on soil microbiota: Challenges for maintaining soil ecosystem services*. IntechOpen. https://doi.org/10.5772/intechopen.111585

Krzyzanowski, M., & Cohen, A. (2008). Update of WHO air quality guidelines. *Air Quality, Atmosphere & Health, 1*, 7–13.

Kumar, P., Morawska, L., Birmili, W., Paasonen, P., Hu, M., Kulmala, M., & Britter, R. (2021). Ultrafine particles in cities. *Environmental Health Perspectives, 129*(4), 047003.

Lai, W., Li, S., Li, Y., & Tian, X. (2022). Air pollution and cognitive functions: Evidence from straw burning in China. *American Journal of Agricultural Economics, 104*(1), 190–208.

Landeg-Cox, C., Middleton, A., Halios, C. H., Marczylo, T., & Dimitroulopoulou, S. (2025). Chemicals in European residences—Part II: A review of emissions, concentrations, and health effects of semi-volatile organic compounds (SVOCs). *Environments, 12*(2).

Li, X., Peng, L., Yao, X., Cui, S., Hu, Y., You, C., & Chi, T. (2020). Long short-term memory neural network for air pollutant concentration predictions: Method development and evaluation. *Environmental Pollution, 231*, 997–1004.

Li, X. D., Lee, S. L., Wong, S. C., Shi, W., & Thornton, L. (2004). The study of metal contamination in urban soils of Hong Kong using a GIS-based approach. *Environmental Pollution, 129*, 113–124.

Liang, Z., Chen, S., Yang, Y., Zhou, Y., & Shi, Z. (2020). Prediction of soil pH using spatially distributed soil spectra and machine learning in a subtropical hilly watershed of China. *Remote Sensing of Environment, 237*, 111540.

Likens, G. E., & Bormann, F. H. (1995). *Biogeochemistry of a forested ecosystem* (2nd ed.). Springer-Verlag.

Liu, J., & Diamond, J. (2019). China's environment in a globalizing world. *Nature, 435*(7046), 1179–1186.

Liu, Y., Gadepalli, K., Norouzi, M., Dahl, G. E., Kohlberger, T., Boyko, A., & Stumpe, M. C. (2020). Detecting cancer metastases on gigapixel pathology images. *IEEE Journal of Biomedical and Health Informatics, 24*(5), 1507–1515.

Liu, Y., Zhou, Y., & Lu, J. (2022). Exploring the relationship between air pollution and meteorological variables in Beijing during 2014-2017. *Environmental Research, 204*(Pt A), 111887.

Loyola-González, O., Martínez-Trinidad, J. F., Carrasco-Ochoa, J. A., & García-Borroto, M. (2023). Explainable artificial intelligence (XAI): Concepts, taxonomies, opportunities and challenges toward responsible AI. *Journal of Artificial Intelligence in Medicine, 136*, 102485.

Lundberg, S. M., Erion, G., Chen, H., DeGrave, A., Prutkin, J. M., Nair, B., & Lee, S.-I. (2020). From local explanations to global understanding with explainable AI for trees. *Nature Machine Intelligence, 2*(1), 56–67.

Ma, D., Duan, S., Zhang, X., Xu, B., & Xu, Y. (2024). Spatiotemporal dynamic assessment of water resources carrying capacity and identification of obstacle factors in Yunnan province based on grey water footprint theory. *Water, 16*(24), 3651.

Madronich, S., & de Gruijl, F. (1993). Skin cancer and UV radiation. *Nature, 366*, 23–29.

Mahmood, F., Khokhar, M. F., & Mahmood, Z. (2020). Examining the relationship of tropospheric ozone and climate change on crop productivity using the multivariate panel data techniques. *Journal of Environmental Management, 272*, 111024.

Manisalidis, I., Stavropoulou, E., Stavropoulos, A., & Bezirtzoglou, E. (2020). Environmental and health impacts of air pollution: A review. *Frontiers in Public Health, 8*, 14.

Merz, B., Kuhlicke, C., Kunz, M., Pittore, M., Babeyko, A., Bresch, D. N., Domeisen, D. I. V., Feser, F., Koszalka, I., Kreibich, H., Pantillon, F., Parolai, S., Pinto, J. G., Punge, H. J., Rivalta, E., Schröter, K., Strehlow, K., Weisse, R., & Wurpts, A. (2020). Impact forecasting to support emergency management of natural hazards. *Reviews of Geophysics, 58*(4), e2020RG000704.

Min, K., Li, Y., Wang, D., Chen, B., Ma, M., Hu, L., et al. (2021). 3D printing-induced fine particle and volatile organic compound emission: An emerging health risk. *Environmental Science & Technology Letters, 8*(8), 616–625.

Mishra, R. K., Mohammad, N., & Roychoudhury, N. (2016). Soil pollution: Causes, effects and control. *Van Sangyan, 3*(1), 1–14.

Nawaz, M., Heitor, A., & Sivakumar, M. (2020). Geopolymers in construction - Recent developments. *Construction and Building Materials, 260*, 120472.

Nguyen, L. S. P., Chang, J. H. W., Griffith, S. M., Hien, T. T., Kong, S. S. K., Le, H. N., Huang, H. Y., Sheu, G. R., & Lin, N. H. (2022). Trans-boundary air pollution in a southeast Asian megacity: Case studies of the synoptic meteorological mechanisms and impacts on air quality. *Atmospheric Pollution Research, 13*(4), 101366.

Noel, C., Vanroelen, C., & Gadeyne, S. (2021). Qualitative research about public health risk perceptions on ambient air pollution. A review study. *SSM-Population Health, 15*, 100879.

Nowak, D. J., Greenfield, E. J., Ash, R. M., & Comendador, J. (2023). Assessing urban forest effects and values: The greater Kansas City region. *Forest Ecology and Management, 529*, 120667.

Nystrøm, V., Lurås, H., Moger, T., & Leonardsen, A. L. (2024). Patient experiences and clinical outcomes of admissions to municipal acute wards versus a hospital: a multicentre randomised controlled trial in Norway. *Scandinavian Journal of Primary Health Care, 42*(4), 659–667.

Park, S., & Kim, Y. (2022). A metaverse: Taxonomy, components, applications, and open challenges. *IEEE Access, 10*, 4209–4251.

Piracha, A., & Chaudhary, M. T. (2022). Urban air pollution, urban heat Island and human health: A review of the literature. *Sustainability, 14*(15), 9234.

Podda, A., Pisuttu, C., Hoshika, Y., Pellegrini, E., Carrari, E., Lorenzini, G., et al. (2019). Can nutrient fertilization mitigate the effects of ozone exposure on an ozone-sensitive poplar clone? *Science of the Total Environment, 657*, 340–350.

Pope, C. A., Burnett, R. T., Thun, M. J., Calle, E. E., Krewski, D., Ito, K., & Thurston, G. D. (2002). Lung cancer, cardiopulmonary mortality, and long-term exposure to fine particulate air pollution. *JAMA, 287*(9), 1132–1141.

Popescu, S. M., Mansoor, S., Wani, O. A., Kumar, S. S., Sharma, V., Sharma, A., Arya, V. M., Kirkham, M. B., Hou, D., Bolan, N., & Chung, Y. S. (2024). Artificial intelligence and IoT driven technologies for environmental pollution monitoring and management. *Frontiers in Environmental Science, 12*, 1336088.

Poterek, Q., Caretto, A., Braun, R., Clandillon, S., Huber, C., & Ceccato, P. (2025). INterpolated FLOod surface (INFLOS), a rapid and operational tool to estimate flood depths from earth observation data for emergency management. *Remote Sensing, 17*(2), 329.

Qi, Y., Li, Q., Karimian, H., & Liu, D. (2023). AirFormer: A transformer-based model for urban air quality prediction. *IEEE Transactions on Knowledge and Data Engineering, 35*(1), 123–135.

Rahman, M. M., McConnell, R., Schlaerth, H., Ko, J., Silva, S., Lurmann, F. W., Palinkas, L., Johnston, J., Hurlburt, M., Yin, H., & Ban-Weiss, G. (2022). The effects of coexposure to extremes of heat and particulate air pollution on mortality in California: Implications for climate change. *American Journal of Respiratory and Critical Care Medicine, 206*(9), 1117–1127.

Rao, N. D., Kiesewetter, G., Min, J., Pachauri, S., & Wagner, F. (2021). Household contributions to and impacts from air pollution in India. *Nature Sustainability, 4*(10), 859–867.

Rastogi, U., Kumar, S., Maurya, S. P., & Yadav, A. K. (2025). Air pollution forecasting using genetic algorithm and LSTM deep learning technique. *Aerosol Science and Engineering*. https://doi.org/10.1007/s41810-025-00343-y

Ryan, R. G., Marais, E. A., Balhatchet, C. J., & Eastham, S. D. (2022). Impact of rocket launch and space debris air pollutant emissions on stratospheric ozone and global climate. *Earth's Future, 10*(6), e2021EF002612.

Rentschler, J., & Leonova, N. (2023). Global air pollution exposure and poverty. *Nature Communications, 14*(1), 4432.

Requia, W. J., Adams, M. D., Arain, A., Papatheodorou, S., Koutrakis, P., & Mahmoud, M. (2020). Association between air pollution and coronary artery calcification within six metropolitan areas in the USA. *Environmental Research, 185*, 109435.

Rudin, C. (2019). Stop explaining black box machine learning models for high stakes decisions and use interpretable models instead. *Nature Machine Intelligence, 1*(5), 206–215.

Saber, A. T., Jacobsen, N. R., Jackson, P., Poulsen, S. S., Kyjovska, Z. O., Halappanavar, S., et al. (2014). Particle-induced pulmonary acute phase response may be the causal link between particle inhalation and cardiovascular disease. *Wiley Interdisciplinary Reviews: Nanomedicine and Nanobiotechnology, 6*(6), 517–531.

Sahani, J., Kumar, P., Debele, S., Spyrou, C., Loupis, M., Aragão, L., Porcù, F., Shah, M., & Di Sabatino, S. (2019). Hydro-meteorological risk assessment methods and management by nature-based solutions. *The Science of the Total Environment, 696*, 133936.

Sahu, S. K., Liu, S., Liu, M., & Shi, X. (2021). A random forest model for predicting PM2.5 concentration: Case study in Delhi. *Science of the Total Environment, 768*, 144522.

Saini, D. K., Garg, S. K., & Kumar, M. (2019). Major air pollutants and their effects on plant and human health: A review. *Plant Archives, 19*(2), 3273–3278.

Saxena, V. (2025). Water quality, air pollution, and climate change: Investigating the environmental impacts of industrialization and urbanization. *Water, Air, and Soil Pollution, 236*(2), 4.

Schöpp, W., Posch, M., Mylona, S., & Johansson, M. (2021). Long-term development of acid deposition (1880-2030) in sensitive freshwater regions in Europe. *Environmental Pollution, 268*, 115943.

Sharma, S., Chandra, M., & Kota, S. H. (2022). Four year long simulation of carbonaceous aerosols in India: Seasonality, sources and associated health effects. *Environmental Research, 213*, 113676.

Shayanmehr, M., Aarabi, S., Ghaemi, A., & Hemmati, A. (2025). A data driven machine learning approach for predicting and optimizing sulfur compound adsorption on metal organic frameworks. *Scientific Reports, 15*(1), 3138.

Sicard, P., Agathokleous, E., De Marco, A., Paoletti, E., & Calatayud, V. (2022). Urban population exposure to air pollution in Europe over the last decades. *Sustainable Cities and Society, 76*, 103438.

Singh, E., Tiwari, S., & Agrawal, M. (2009). Effects of elevated ozone on photosynthesis and stomatal conductance of two soybean varieties: a case study to assess impacts of one component of predicted global climate change. *Plant Biology, 11*(Suppl. 1), 101–108.

Skirienė, A. F., & Stasiškienė, Ž. (2021). COVID-19 and air pollution: Measuring pandemic impact to air quality in five European countries. *Atmosphere, 12*(3), 290.

Smith, A. H., Lingas, E. O., & Rahman, M. (2019). Contamination of drinking-water by arsenic in Bangladesh: A public health emergency. *Bulletin of the World Health Organization, 78*(9), 1093–1103.

Sokhi, R. S., Moussiopoulos, N., Baklanov, A., Bartzis, J., Coll, I., Finardi, S., Friedrich, R., Geels, C., Grönholm, T., Halenka, T., Ketzel, M., Maragkidou, A., Matthias, V., Moldanova, J., Ntziachristos, L., Schäfer, K., Suppan, P., Tsegas, G., Carmichael, G., & Kukkonen, J. (2022). Advances in air quality research—current and emerging challenges. *Atmospheric Chemistry and Physics, 22*(7), 4615–4703.

Somani, M., Srivastava, A. N., Gummadivalli, S. K., & Sharma, A. (2020). Indirect implications of COVID-19 towards sustainable environment: An investigation in Indian context. *Bioresource Technology Reports, 11*, 100491.

Stewart, C., Damby, D. E., Horwell, C. J., Elias, T., Ilyinskaya, E., Tomašek, I., Longo, B. M., Schmidt, A., Carlsen, H. K., Mason, E., & Baxter, P. J. (2022). Volcanic air pollution and human health: Recent advances and future directions. *Bulletin of Volcanology, 84*(1), 11.

Tainio, M., Andersen, Z. J., Nieuwenhuijsen, M. J., Hu, L., De Nazelle, A., An, R., Garcia, L. M., Goenka, S., Zapata-Diomedi, B., Bull, F., & de Sá, T. H. (2021). Air pollution, physical activity and health: A mapping review of the evidence. *Environment International, 147*, 105954.

Teramura, A. (2006). Effects of UV-B radiation on the growth and yield of crop plants. *Physiologia Plantarum, 58*, 415–427.

Tian, L., Li, J., Zhao, S., Tang, J., Li, J., Guo, H., et al. (2021). DDT, chlordane, and hexachlorobenzene in the air of the Pearl River Delta revisited: A tale of source, history, and monsoon. *Environmental Science & Technology, 55*(14), 9740–9749.

Toptaş, A. (2024). Innovative nanofibrous air filters: Advancing air quality and health protection. *Sürdürülebilir Çevre Dergisi, 4*(1), 29–40.

Torgersen, C. E., Pichon, C. L., Fullerton, A. H., Dugdale, S. J., Duda, J. J., Giovannini, F., Tales, É., Belliard, J., Branco, P., Bergeron, N. E., Roy, M. L., Tonolla, D., Lamouroux, N., Capra, H., & Baxter, C. V. (2021). Riverscape approaches in practice: perspectives and applications. *Biological Reviews/Biological Reviews of the Cambridge Philosophical Society, 97*(2), 481–504.

Tóth, G., Hermann, T., Szatmári, G., & Pásztor, L. (2016). Heavy metals in agricultural soils of the European Union with implications for food safety. *Environment International, 88*, 299–309.

Tran, H. M., Tsai, F. J., Lee, Y. L., Chang, J. H., Chang, L. T., Chang, T. Y., Chung, K. F., Kuo, H. P., Lee, K. Y., Chuang, K. J., & Chuang, H. C. (2023). The impact of air pollution on respiratory diseases in an era of climate change: A review of the current evidence. *Science of the Total Environment, 898*, 166340.

Ulrich, B. (1989). Effects of acidic precipitation on forest ecosystems in Europe. In D. C. Adriano & A. H. Johnson (Eds.), *Advances in environmental science. Acidic precipitation. volume 2: Biological and ecological effects* (pp. 189–262). Springer-Verlag.

United Nations Environment Programme. (2021). *Actions on air quality: A global summary of policies and programmes to reduce air pollution*. UNEP.

Veefkind, J. P., Aben, I., McMullan, K., Förster, H., de Vries, J., Otter, G., & Levelt, P. F. (2021). TROPOMI on the ESA Sentinel-5 precursor: A GMES mission for global observations of the atmospheric composition for climate, air quality and ozone layer applications. *Atmospheric Measurement Techniques, 14*(4), 2967–2995.

Vivan, E. L., Bashir, F. M., Eziashi, A. C., Gammoudi, T., & Dodo, Y. A. (2023). Ground water quality evaluation using hydrogeochemical characterization and novel machine learning in the Chikun local government area of Kaduna state, Nigeria. *Water Science & Technology, 88*(7), 1875–1892.

Wang, Y., Chen, H., Wu, J., & Zhang, Y. (2022). Graph neural networks for urban air pollution modeling: A comprehensive survey. *Nature Computational Science, 2*(3), 123–134.

Wang, Y., Wang, M., Zhang, R., Ghan, S. J., Lin, Y., Hu, J., & Jiang, J. H. (2019). Assessing the effects of anthropogenic aerosols on Pacific storm track using a multiscale global climate model. *Nature Communications, 10*(1), 1234.

Wilcke, W. (2007). Global patterns of polycyclic aromatic hydrocarbons (PAHs) in soil. *Global Biogeochemical Cycles, 21*(3), GB3013.

Wittwer, S. H., & Castilla, N. (1995). Protected cultivation of horticultural crops worldwide. *HortTechnology, 5*(1), 6–23.

Woolley, T. (2024). *Building materials, health and indoor air quality* (Vol. 2). Routledge.

World Health Organization. (2021). *WHO global air quality guidelines: Particulate matter ($PM_{2.5}$ and PM_{10}), ozone, nitrogen dioxide, sulfur dioxide and carbon monoxide*. World Health Organization.

World Health Organization. (2023). *WHO global air quality guidelines: Particulate matter ($PM_{2.5}$ and PM_{10}), ozone, nitrogen dioxide, sulfur dioxide and carbon monoxide*. World Health Organization.

Wuana, R. A., & Okieimen, F. E. (2011). Heavy metals in contaminated soils: A review of sources, chemistry, risks and best available strategies for remediation. *ISRN Ecology, 2011*, 1–20.

Xia, Y., Zhang, Y., & Tang, S. (2022). Machine learning for predicting asthma exacerbations from environmental exposures. *Nature Communications, 13*(1), 1–12.

Xu, X., Yang, H., & Li, C. (2022). Theoretical model and actual characteristics of air pollution affecting health cost: A review. *International Journal of Environmental Research and Public Health, 19*(6), 3532.

Yadav, A. K. (2022). Characterization and effective dose of gross alpha and gross beta activities in particulate matters ($PM_{2.5}$, PM_{10}, SPM) in India. *Aerosol Science and Engineering, 6*(4), 360–369.

Yadav, A. K., Gupta, K., Bodhale, R., & Phuleria, H. C. (2025). Metals and trace element associated health risk in size-segregated household dust of an industrial town in India. *Air Quality, Atmosphere & Health*. https://doi.org/10.1007/s11869-025-01794-5

Yadav, A. K., & Hopke, P. K. (2020). Characterization of radionuclide activity concentrations and lifetime cancer risk due to particulate matter in the Singrauli coalfield. *India. Environ Monit Assess, 192*, 680. https://doi.org/10.1007/s10661-020-08619-1

Yadav, A. K., Sahoo, S. K., Dubey, J. S., Kumar, A. V., Pandey, G., & Tripathi, R. M. (2019). Assessment of particulate matter, metals of toxicological concentration, and health risk around a mining area, Odisha, India. *Air Quality, Atmosphere & Health, 12*(7), 775–783.

Yadav, A. K., Sahoo, S. K., Kumar, A. V., Shirin, S., Jamal, A., Patra, A. C., Dubey, J. S., Thakur, V. K., Lenka, P., Singh, S., Jha, V., & Tripathi, R. M. (2023). Identifying size-segregated particulate matter ($PM_{2.5}$, PM_{10} and SPM) sources in an industrial town of India. *Aerosol Science and Engineering, 7*(4), 455–473.

Yadav, A. K., Sahoo, S. K., Patra, A. C., Dubey, J. S., Lenka, P., Sagar, D. V., Kumar, A. V., & Tripathi, R. M. (2014). Source identification of particulate matter and associated intake of elements through inhalation in an industrial area of Odisha, India. *Toxicological & Environmental Chemistry, 96*(3), 410–425.

Yadav, A. K., Shirin, S., Emmanouil, C., & Jamal, A. (2022). Effect of seasonal and meteorological variability of air pollution in Singrauli coalfield. *Aerosol Science and Engineering, 6*, 61–70.

Yang, Q., Liu, Y., Chen, T., & Tong, Y. (2023). Federated machine learning: Concept and applications. *ACM Transactions on Intelligent Systems and Technology, 10*(2), 1–19.

Yang, Y., Jiang, J., Shen, X., & Wu, J. (2023). Digital health and air pollution: Opportunities for intervention. *npj Digital Medicine, 6*(1), 45.

Yao, W., Zhao, Y., Chen, R., Wang, M., Song, W., & Yu, D. (2023). Emissions of toxic substances from biomass burning: A review of methods and technical influencing factors. *Processes, 11*(3), 853.

Zhan, Y., Luo, Y., Deng, X., Zhang, K., Zhang, M., Grieneisen, M. L., & Di, B. (2021). Satellite-based estimates of daily NO2 exposure in China using hybrid random forest and spatiotemporal kriging model. *Remote Sensing of Environment, 253*, 112203.

Zhang, C., Patras, P., & Haddadi, H. (2019). Deep learning in mobile and wireless networking: A survey. *IEEE Communications Surveys & Tutorials, 21*(3), 2224–2287.

Zhang, Q., Wang, Q., Zhu, J., Xu, L., Li, M., Rengel, Z., & He, N. (2021). Higher soil acidification risk in southeastern Tibetan plateau. *Science of the Total Environment, 755*, 143372.

Zhang, R., Zhang, Y., Lin, H., Feng, X., Fu, T.-M., & Wang, Y. (2023). Physics-informed machine learning for air quality predictions: A case study of Beijing. *Geoscientific Model Development, 16*(2), 567–580.

Zhang, X., Liu, K., Wang, Y., Liu, X., & Zhang, Y. (2023). Deep learning for soil pH mapping by combining spectral and terrain features using hyperspectral images. *Geoderma, 430*, 116312.

Zheng, S., Wang, Q., Yuan, Y., & Sun, W. (2020). Human health risk assessment of heavy metals in soil and food crops in the Pearl River Delta urban agglomeration of China. *Food Chemistry, 316*, 126213.

Zhou, Y., Li, X., Liu, Y., & Zhang, J. (2021). Hyperspectral sensing of heavy metals in soil and vegetation: Feasibility and challenges. *ISPRS Journal of Photogrammetry and Remote Sensing, 173*, 189–203.

Zhou, Y., et al. (2019). A Gaussian process mixture model-based hard-cut iterative learning algorithm for air quality prediction. *Applied Soft Computing, 85*, 105789.

Zhuoxian, W., Noor, N. A. M., & Mohamed, A. M. (2025). Impact of environmental concern, government role, and attitude on green patronage behaviour: Does green patronage intention matter? *Deleted Journal, 2025*(1), 1026–1037.

Zou, X., Yan, Y., Hao, X., Hu, Y., Wen, H., Liu, E., Zhang, J., Li, Y., Li, T., Zheng, Y., & Liang, Y. (2024). Deep learning for cross-domain data fusion in urban computing: Taxonomy, advances, and outlook. *Information Fusion, 113*, 102606.

Chapter 5
Air Pollution and Its Impact on Health and Environment in India

B. C. M. Patnaik, Ipseeta Satpathy, and Vishal Jain

5.1 Introduction

Air pollution constitutes a critical environmental challenge that profoundly influences human health, ecological systems, and climate dynamics. It refers to the existence of deleterious substances in the atmosphere, which include gases, particulates, and biological molecules that pose risks to living organisms and undermine the quality of existence (Kumar & Gupta, 2023). The predominant origins of air pollution include industrial discharges, vehicular emissions, agricultural practices, and deforestation, resulting in increased concentrations of pollutants such as carbon monoxide (CO), sulfur dioxide (SO_2), nitrogen oxides (NO_x), and particulate matter (PM) (Smith & Wang, 2022; Yadav, 2021; Yadav et al., 2023). In addition, natural phenomena such as volcanic eruptions and wildfires further intensify air pollution. Industrial operations exert a significant influence on the release of hazardous emissions into the atmosphere. Manufacturing facilities and power generation plants that use fossil fuels discharge substantial volumes of CO_2, SO_2, and various other pollutants, thus exacerbating the occurrence of acid precipitation and respiratory health disorders (Chen & Zhao, 2021). Emissions from vehicles represent another critical factor, particularly within metropolitan regions. Motor vehicles release NO_x, CO, and particulate matter, which contribute to the formation of smog and respiratory complications. Recent research suggests that the transportation sector accounts for nearly 25% of global air pollution, posing considerable health threats to urban inhabitants (Smith & Wang, 2022).

B. C. M. Patnaik (✉) · I. Satpathy
School of Management, KIIT Deemed to be University, Bhubaneswar, India

V. Jain
Sharda University, Greater Noida, India

A. K. Yadav (ed.), *Computational Approach for Air Pollution*,
https://doi.org/10.1007/978-3-032-09215-1_5

Agricultural practices also contribute greatly to air pollution. In many areas, burning the remains of open fields, which are common practices, produces high concentrations and CO_2 in the air (Chen & Zhao, 2021). In addition, the excessive use of fertilizers and pesticides leads to ammonia (NH_3), which reacts with other contaminants for the formation of secondary particles. The picture-like agriculture is further released by methane (CH_4), a powerful greenhouse gas that exacerbates global warming. Air pollution is still an important issue in developing countries. Supporting biomass fuel, such as wood, charcoal, and waste of agricultural crops for cooking and heating, leads to the discharge of dangerous contaminants such as $PM_{2.5}$ and black carbon (Patel & Dey, 2023). Long-term exposure to these contaminants increases the risk of respiratory diseases, cardiovascular disorders, and early mortality in women and children. This causes a decrease in the ecosystem and the acid rain, which causes the rain and the rain of water, becomes sour, which has a negative effect on plants and lifespan (Rajan & Thomas, 2022). Furthermore, due to the reaction of NO_x and volatile organic compounds (VOC) in the sun, the level of ozone of the earth damages vegetation and reduces the growth of agricultural productivity and forests. Pollutants such as black carbon and methane help accelerate the melting of the glacier and increase global temperature, which exacerbates climate change (Cohen et al., 2017). The impact of air pollution on health is amazing. The effects of high levels of pollutants are associated with respiratory diseases such as asthma, bronchitis, and chronic obstructive pulmonary disease (COPD). Cardiovascular disease, stroke, and lung cancer are common in areas with severe air pollution (Patel & Dey, 2023). Vulnerable population groups, including children, the elderly, and those with existing health, are more likely to cause complications due to contaminated air. To solve the air pollution problem, various approaches are required, including state policies, technical achievements, and community participation. Regulatory measures, such as strict discharge, support for clean production technology, and transition to renewable energy sources, are important for reducing industrial emissions. The use of electric vehicles, the improvement of public transportation, and the introduction of strict vehicle emissions can greatly reduce pollution in the transportation sector (Smith & Wang, 2022). In the agricultural sector, the adoption of stable practices such as the use of burns without burns, accurate agricultural and organic fertilizers, can soften emissions. The implementation of the air quality monitoring system and providing real-time data to the public can raise awareness and encourage behavioral changes (Chen & Zhao, 2021). In addition, international cooperation is important for the fight against air pollution. Contracts such as Paris Climate Contract and Regional Initiative, which are dedicated to aviation quality control, can contribute to joint efforts to reduce global emissions. Investment in pure energy technology and carbon capture and storage (CCS) can further maintain the purpose of reducing emissions (Kumar & Gupta, 2023).

Paramedical personnel are essential healthcare professionals that support physicians and nurses in the processes of diagnosis, treatment, and rehabilitation of patients. This group includes laboratory technicians, radiographers, physiotherapists, emergency medical technicians, and pharmacists. Their specialized knowledge in medical technology, therapeutic practices, and patient care plays a vital role in the effective delivery of healthcare services, facilitating precise diagnostics and comprehensive treatment assistance. Environmental activists advocate for the protection and conservation of the natural world by endorsing sustainable practices, increasing public awareness, and contesting regulations that negatively impact the environment. They participate in campaigns, demonstrations, and educational initiatives to address issues such as climate change, pollution, and deforestation, working toward a more sustainable planet for future generations through proactive measures and policy reform.

5.2 Factors Leading to Air Pollution: Issues and Challenges

Air pollution arises from a complex interplay of anthropogenic and natural factors, each contributing uniquely to the degradation of atmospheric quality. One of the main contributors is industrial activity. The combustion of fossil fuels such as coal and petroleum in energy generation plants, along with emissions from manufacturing and chemical industries, releases large amounts of sulfur dioxide (SO_2), nitrogen oxides (NO_x), and particulate matter into the atmosphere (World Health Organization, 2021a). These pollutants are detrimental both to the environment and human health.

Vehicle emissions constitute another major source of air pollution. Automobiles, buses, trucks, and motorcycles release carbon monoxide (CO), NO_x, and unburned hydrocarbons, particularly from diesel engines that produce substantial amounts of soot and fine particulate matter (Environmental Protection Agency, 2020a). Urban areas with dense traffic are particularly vulnerable to these emissions, which exacerbate respiratory diseases and degrade air quality.

Deforestation is also a significant factor. Trees naturally absorb carbon dioxide and filter pollutants; therefore, their removal leads to elevated greenhouse gas concentrations and increased particulate matter due to soil erosion and loss of vegetative cover (Intergovernmental Panel on Climate Change, 2021). Similarly, agricultural practices contribute to air pollution through the use of synthetic fertilizers and pesticides, which emit harmful gases, and through activities such as livestock farming and rice cultivation, which produce methane (CH_4), a potent greenhouse gas (Food and Agriculture Organization, 2020b). Burning crop residues further adds to the release of smoke and particulate matter.

Household practices, particularly in low-income regions, exacerbate indoor air pollution by burning wood, coal, or biomass for cooking and heating. In addition, the use of aerosol sprays, paints, and chemical cleaning agents contributes volatile organic compounds (VOC) to indoor environments, especially where ventilation is inadequate (World Health Organization, 2021b). Waste management issues, including open burning and the accumulation of organic material in landfills, release methane and toxic gases. Improper disposal of electronic waste also leads to the emission of heavy metals such as lead and mercury (United Nations Environment Programme, 2019b).

Construction and demolition activities emit dust, fine particulate matter, and pollutants from materials such as cement and asphalt. Heavy machinery operation during these processes also contributes to noise pollution and secondary emissions that deteriorate air quality (Environmental Protection Agency, 2020b). Natural causes such as volcanic eruptions, wildfires, and dust storms further exacerbate pollution levels. These events release massive quantities of ash, sulfur dioxide, carbon monoxide, and fine particulate matter over wide areas (Intergovernmental Panel on Climate Change, 2021).

Mining operations contribute significantly to localized air contamination through drilling, blasting, and the release of dust and hazardous gases. The use of explosives and exposure to heavy metals such as lead and mercury pose serious environmental and health risks (International Council on Mining and Metals, 2020). Lastly, climate change and weather conditions can influence air pollution patterns. Temperature inversions can trap pollutants near the earth's surface, while reduced rainfall limits the natural cleansing of the atmosphere. Additionally, higher temperatures can accelerate the formation of ground-level ozone, contributing to urban smog (Intergovernmental Panel on Climate Change, 2021).

5.3 Impact on Health

Air pollution poses a significant threat to human health, with wide-ranging effects on various physiological systems. One of the most immediate and evident consequences is the development of respiratory issues. Exposure to airborne pollutants, such as particulate matter ($PM_{2.5}$) and nitrogen dioxide (NO_2), can result in symptoms such as coughing, wheezing, and difficulty breathing, as well as exacerbation of asthma, chronic bronchitis, and susceptibility to lung infections (World Health Organization, 2021c). These respiratory conditions are especially severe among vulnerable populations, including children and the elderly (Yadav, 2015, 2021; Yadav et al., 2014, 2019).

Cardiovascular conditions are also closely related to air pollution. Pollutants contribute to systemic inflammation and oxidative stress, which can increase the risk of myocardial infarctions (heart attacks) and cerebrovascular accidents (strokes). Chronic exposure has been shown to induce hypertension and contribute to the development of atherosclerosis, or hardening of the arteries (Brook et al., 2010). Furthermore, prolonged exposure to toxic air components, such as ozone and fine particulate matter, has been associated with lung disorders, including chronic obstructive pulmonary disease (COPD), lung fibrosis, reduced lung function, and an increased risk of lung cancer (Pope et al., 2020).

Air pollution can compromise immune function, weakening the body's defense mechanisms and increasing vulnerability to infections (World Health Organization, 2021c). Furthermore, there are growing concerns regarding its impact on the nervous system. Exposure to air pollutants such as heavy metals and ultrafine particles has been linked to symptoms such as headaches, dizziness, and fatigue, as well as an increased risk of neurodegenerative conditions such as Alzheimer's and Parkinson's disease. In children, these exposures can affect neurodevelopment, leading to lower IQ scores, attention deficits, and learning difficulties (Calderón-Garcidueñas et al., 2016).

The effects of air pollution on children and pregnant women are particularly alarming. Exposure during pregnancy is associated with premature birth, low birth weight, and congenital anomalies. Infants and young children are more prone to respiratory infections due to their developing immune and respiratory systems (Perera et al., 2019). Additionally, common symptoms of exposure to air pollution include irritation of the eyes, nose, and throat—manifesting as redness, tearing, nasal congestion, and sore throat—which can increase the risk of sinus infections (Environmental Protection Agency, 2020b).

Long-term exposure to carcinogenic air pollutants, such as benzene, asbestos, and $PM_{2.5}$, significantly increases the risk of developing various cancers, including those of the lung, throat, and skin (Loomis et al., 2013). Mental health is also adversely affected, with studies linking poor air quality with increased incidences of depression, anxiety, cognitive decline, and sleep disturbances (Braithwaite et al., 2019). Ultimately, these health impacts can result in reduced life expectancy and elevated mortality rates, especially from cardiovascular and respiratory diseases (World Health Organization, 2021b).

5.4 Impact on Environment

Air pollution has far-reaching effects on the environment, contributing significantly to global ecological degradation. One of the most pressing issues is global warming and climate change, primarily driven by the accumulation of greenhouse gases such as carbon dioxide (CO_2), methane (CH_4), and nitrogen oxides (NO_x). These gases trap heat in the atmosphere of the Earth, leading to a rise in global temperatures. Consequently, there has been an increase in the frequency and intensity of extreme weather events such as hurricanes, droughts, and floods. Furthermore, glacial melting is accelerating, contributing to rising sea levels and threatening coastal communities (Intergovernmental Panel on Climate Change, 2021).

Acid rain is another severe environmental result of air pollution, formed when sulfur dioxide (SO_2) and NO_x react with atmospheric moisture. This precipitation is highly acidic and leads to acidification of soils and water bodies, negatively affecting plant life, aquatic ecosystems, and soil quality. Furthermore, acid rain contributes to the deterioration of infrastructure, including buildings, historical monuments, and transportation systems (United States Environmental Protection Agency, 2020).

The depletion of the ozone layer is largely due to the emission of chlorofluorocarbons (CFCs) and other ozone-depleting substances. The thinning of the ozone layer results in increased penetration of ultraviolet (UV) radiation, which can cause skin cancer, cataracts, and weakened immune systems in humans and animals (World Health Organization, 2021a). Similarly, the formation of smog, caused by a combination of ground-level ozone (O_3) and fine particulate matter, reduces visibility and air quality, particularly in urban areas. Smog not only impairs respiratory health but also hinders plant growth and agricultural productivity (Chen et al., 2017).

Water pollution is closely related to air pollution, as airborne toxins can deposit in water bodies such as lakes, rivers, and oceans. These pollutants can acidify aquatic environments and introduce heavy metals and other contaminants, degrading water quality and threatening marine biodiversity (United Nations Environment Programme, 2019a). Similarly, soil degradation results from the deposition of acidic and toxic compounds that reduce soil fertility and nutrient content, thus lowering crop yields and impacting agricultural sustainability (Food and Agriculture Organization, 2020b).

Airborne pollutants also inflict harm on wildlife and biodiversity. Many animal species suffer respiratory complications due to inhalation of toxic substances, while ecosystems and food chains are disrupted by habitat loss, often driven by climate change and deforestation. Pollution also impacts forests and vegetation by damaging leaf tissues, reducing photosynthetic efficiency, and increasing vulnerability to pests and diseases. In extreme cases, entire forest ecosystems suffer extinction due to the combined stresses of pollution and climate change (Intergovernmental Panel on Climate Change, 2021).

In marine environments, ocean acidification occurs when excess CO_2 is absorbed by seawater, reducing pH levels. This process harms shell-forming organisms such as corals, mollusks, and some plankton species, thus disrupting marine food webs and biodiversity (National Oceanic and Atmospheric Administration, 2020). Finally, air pollution contributes to the deterioration of built environments, causing metal corrosion and decay of construction materials. Historical monuments, buildings, and critical infrastructure are particularly vulnerable to long-term damage from acidic pollutants (Environmental Protection Agency, 2020b).

5.5 Research Methodology

This study is geographically confined to the capital region of Odisha, with particular emphasis on the cities of Bhubaneswar and Cuttack. The target population comprises paramedical personnel and environmental activists. Data were collected from multiple hospitals situated within these urban centers, as well as from environmental activists associated with nongovernmental organizations (NGOs) and the state's pollution control board. For ethical reasons and to maintain confidentiality, the names of the participating hospitals are not disclosed.

The research design adopted for this study was developed through a comprehensive review of the literature and core group discussions, which initially identified 37 variables. These included 13 variables related to factors leading to air pollution, 12 related to the impact of air pollution on health, and another 12 focused on environmental impact. However, after a pilot study involving 32 respondents, the number of variables was reduced to 10 for each parameter to improve the study's

Table 5.1 Sampling frame

Respondents	Gender	Frequency
Paramedical staff	Male (PM)	37
	Female (PF)	32
Environmental activists	Male (EM)	34
	Female (EF)	29
Total		132

Source: Primary data

manageability and focus of the study. The study used a combination of snowball and cluster sampling techniques to reach the desired sample size. Data were analyzed using a simple weighted average method, based on responses collected using a five-point Likert-type scale. The scale assigned weights ranging from 0 to 4, where 4 corresponded to "completely agree," 3 to "agree," 2 to "neutral," 1 to "disagree," and 0 to "completely disagree." The calculated mean weight of the scale was 2, determined by averaging the five response levels [(4 + 3 + 2 + 1 + 0)/5].

To gather qualitative insights, six focus group discussions were conducted with participants from both target groups: paramedical personnel and environmental activists. These discussions were held within the Bhubaneswar–Cuttack metropolitan area. The environmental activists included in the study were those working either with NGOs dedicated to environmental improvement or with the state's pollution control board.

The rationale for selecting the sample population was based on a pragmatic combination of cluster and snowball sampling approaches. This method enabled researchers to reach participants who were directly or indirectly connected to the primary population, facilitating efficient collection of relevant data within the defined geographic area. To determine the appropriate sample size, the study adhered to the guidelines established by Rummel's 1970 study and Schwab's 1980 research, which suggest that the number of attributes be multiplied by a ratio ranging from 1:4 to 1:10. Given the final total of 30 attributes, the recommended minimum and maximum sample sizes were 120 and 300, respectively. In alignment with these guidelines, the final sample size was determined to be 132, ensuring both methodological rigor and representativeness. In Table 5.1, 37 respondents are male paramedical staff, 32 are female paramedic staff, 34 are male environmental activists, and the remaining 29 are female environmental activists, out of a total sample size of 132.

5.5.1 Analysis of Data

In Table 5.2, the given dataset assesses different factors that lead to air pollution, evaluated by paramedical personnel (male and female) and environmental activists (male and female). Since all values are above the mean value (2.0), it indicates that all listed factors contribute significantly to air pollution.

Table 5.2 On the basis of factors leads to air pollution

Attributes	Code	Paramedical staff male (PM)	Paramedical staff female (PF)	Environmental activists male (EM)	Environmental activists female (EF)
Industrial activities	F1	2.68	2.56	2.76	2.79
Vehicle emissions	F2	2.68	2.78	3.00	2.79
Deforestation	F3	2.46	3.06	3.15	3.03
Agricultural practices	F4	2.78	2.88	3.06	2.86
Household practices	F5	3.00	2.94	3.09	2.97
Waste management practices	F6	2.59	2.88	2.91	2.86
Construction and demolition activities	F7	2.92	2.81	2.91	2.86
Natural causes	F8	2.43	2.81	2.94	2.97
Mining operations	F9	2.70	3.00	3.12	3.17
Climate change and weather conditions	F10	3.03	3.00	2.94	3.07

Source: Appendixes A, D, G, and J

Key Observations

Analysis of data collected from paramedical staff and environmental activists reveals that certain factors consistently emerge as the most significant contributors to air pollution, as indicated by their high weighted average scores across multiple response groups. Among these, *Climate Change and Weather Conditions* (F10) demonstrated the highest overall concern, with average scores of 3.03 from paramedical males (PM), 3.00 from paramedical females (PF), 2.94 from environmental males (EM), and 3.07 from environmental females (EF). This finding underscores the widespread acknowledgment of the severe environmental and atmospheric disturbances caused by climate variability. Similarly, *Mining Operations* (F9) also received strong concern in all groups, scoring 2.70 (PM), 3.00 (PF), 3.12 (EM), and 3.17 (EF), highlighting the detrimental environmental impact of extraction-related activities, including dust emission and toxic gas release.

Deforestation (F3) was another highly rated factor, particularly among environmental respondents, with PF (3.06), EM (3.15), and EF (3.03), reflecting concern about vegetation loss and subsequent rise in atmospheric carbon dioxide and particulate matter. *Household Practices* (F5), such as the use of biomass fuels and poor indoor air ventilation, was also highly rated, with average scores of 3.00 (PM), 2.94 (PF), 3.09 (EM), and 2.97 (EF), indicating recognition of the indoor contribution to ambient pollution levels. Collectively, these findings suggest a strong consensus that climate change, deforestation, mining, and household pollution are among the leading causes of air pollution in the region.

In terms of moderate contributors, factors such as *Agricultural Practices* (F4), *Vehicle Emissions* (F2), *Construction and Demolition Activities* (F7), and *Waste Management Practices* (F6) also showed significant, although comparatively lower, levels of concern. These factors were acknowledged for their considerable contribution to air quality degradation, particularly through the release of methane, particulate matter, and other pollutants. However, their average scores were marginally lower than those of the highest contributing factors, suggesting a slightly reduced level of urgency perceived by the respondents.

Lower contributing factors, while still rated above the overall mean value of 2.0, included *Industrial Activities* (F1) and *Natural Causes* (F8). Despite being commonly acknowledged sources of pollution, these were perceived as less critical than human-induced and climate-related factors. This suggests a public perception that more controllable and preventable human activities should be prioritized in mitigation efforts over natural-occurring events such as dust storms or volcanic eruptions.

In summary, all identified factors were rated above the mean value, indicating a general consensus that each play a role in contributing to air pollution. However, *climate change and weather conditions*, *mining operations*, and *deforestation* remained the most pressing concerns. Interestingly, the data also revealed that female environmental activists (EF) consistently reported the highest levels of concern across the categories, indicating increased environmental awareness and sensitivity within this subgroup. These findings reinforce the urgent need for targeted policy interventions and awareness programs focused on the most impactful sources of pollution.

In Table 5.3, the dataset evaluates the health impact of air pollution based on responses from paramedical personnel (male and female) and environmental activists (male and female). Since all values exceed the mean (2.0), all listed health condition is significantly affected by air pollution.

Key Observations

The study revealed that several health conditions were consistently rated as highly severe by paramedical personnel and environmental activists, indicating a shared concern about the long-term consequences of air pollution on human health. *Reduced Life Expectancy* (H10) emerged as the most critical health concern, receiving the highest average scores across all participant groups: 3.14 among paramedical males (PM), 3.06 among paramedical females (PF), 3.00 among environmental males (EM), and 3.03 among environmental females (EF). This suggests a strong perception that air pollution contributes significantly to premature mortality. Similarly, *Impacts on the Nervous System* (H5) were perceived as highly severe, with scores of 3.03 (PM), 3.06 (PF), 3.03 (EM), and 2.90 (EF), highlighting the potential for neurodegenerative diseases, cognitive impairments, and developmental disorders, particularly in vulnerable populations such as children.

Table 5.3 Impact of air pollution on health

Attributes	Code	Paramedical staff male (PM)	Paramedical staff female (PF)	Environmental activists male (EM)	Environmental activists female (EF)
Respiratory issues	H1	2.78	2.66	2.65	2.83
Cardiovascular conditions	H2	2.84	2.91	2.88	2.83
Pulmonary disorders	H3	2.68	3.13	3.09	3.07
Compromised immune function	H4	2.76	2.97	2.94	2.93
Impacts on the nervous system	H5	3.03	3.06	3.03	2.90
Effects on children and pregnant women	H6	2.81	2.97	3.03	2.93
Eye, nose, and throat irritation	H7	2.97	2.88	2.85	2.83
Increased risk of cancer	H8	2.51	2.88	3.00	3.00
Mental health effects	H9	2.68	3.06	3.12	3.14
Reduced life expectancy	H10	3.14	3.06	3.00	3.03

Source: Appendixes B, E, H, and K

Pulmonary Disorders (H3) also ranked among the most severe, with especially great concern from PF (3.13), EM (3.09), and EF (3.07), reflecting the perceived threat of chronic respiratory diseases, including chronic obstructive pulmonary disease (COPD), fibrosis, and lung cancer. Furthermore, *Mental Health Effects* (H9) were viewed as critical, particularly by EF (3.14), EM (3.12), and PF (3.06), suggesting a growing awareness of the psychological consequences of air pollution, such as anxiety, depression, cognitive decline, and sleep disturbances.

In terms of moderately severe health impacts, factors such as *Effects on Children and Pregnant Women* (H6), *Compromised Immune Function* (H4), and *Increased Risk of Cancer* (H8) also generated significant concern, although with slightly lower scores than the top-tier conditions. These findings indicate that pollution is perceived to pose substantial risks to vulnerable populations, including unborn children, infants, and individuals with compromised immune systems. Recognition of long-term developmental and immunological harm underlines the need for targeted protective interventions for at-risk groups.

Less severe but still impactful conditions included *Respiratory Problems* (H1), which received the lowest scores among all health impacts but remained above the mean value of 2.0. Similarly, *Eye, Nose, and Throat Irritation* (H7) scored relatively lower but still indicated the presence of short-term discomfort and irritation caused by pollutant exposure. These conditions, while not perceived immediately as life-threatening, remain important indicators of environmental stress and are particularly relevant in urban and industrial settings.

In summary, all identified health effects were rated above the mean threshold, indicating that air pollution is universally regarded a significant public health issue. The most critical health impacts were related to *reduced life expectancy*, *pulmonary and nervous system disorders*, and *mental health effects*. In addition, there was clear concern about the heightened risks faced by vulnerable populations such as children, pregnant women, and immunosuppressed individuals. Although short-term symptoms such as respiratory and sensory irritation were considered less severe, they still represent important early indicators of environmental health hazards.

According to Table 5.4 of the report, the dataset evaluates the environmental impact of air pollution based on responses from paramedical personnel (men and women) and environmental activists (males and females). Since all values exceed the mean (2.0), every listed environmental issue is significantly impacted by air pollution.

Key Observations

The findings of this study indicate that several environmental consequences of air pollution are regarded highly severe by both paramedical personnel and environmental activists. *Water Pollution* (E5) emerged as the most critical concern, receiving the highest ratings in all respondent groups: paramedical males (PM: 3.22),

Table 5.4 Impact of air pollution on environment

Attributes	Code	Paramedical staff male (PM)	Paramedical staff female (PF)	Environmental activists male (EM)	Environmental activists female (EF)
Global warming and climate change	E1	2.84	2.69	2.74	2.90
Acid rain	E2	2.97	2.94	2.91	2.83
Ozone layer depletion	E3	2.76	3.09	3.06	3.07
Smog formation	E4	2.92	3.00	3.03	2.93
Water pollution	E5	3.22	3.09	3.06	3.00
Soil degradation	E6	2.86	2.97	2.97	2.86
Harm to wildlife and biodiversity	E7	3.05	2.91	2.82	2.83
Impact on forests and vegetation	E8	2.65	2.94	2.88	2.90
Ocean acidification	E9	2.86	3.13	3.06	3.14
Deterioration of built environments	E10	2.95	3.00	3.00	3.00

Source: Appendixes C, F, I, and L

paramedical females (PF: 3.09), environmental males (EM: 3.06), and environmental females (EF: 3.00). This indicates widespread concern about the deposition of airborne pollutants in aquatic ecosystems, which can lead to the acidification of water bodies, bioaccumulation of toxins, and threats to marine biodiversity. Similarly, *Ocean Acidification* (E9) was another top-rated issue, with PF (3.13), EM (3.06), and EF (3.14) assigning it high scores, reflecting concerns about increased absorption of atmospheric CO_2 by oceans, which adversely affects coral reefs, shellfish, and planktonic organisms.

Ozone Layer Depletion (E3) was also rated among the most severe environmental impacts, particularly by EF (3.07), PF (3.09), and EM (3.06), signaling alarm over the release of ozone-depleting substances such as chlorofluorocarbons (CFC) and their role in increasing harmful ultraviolet radiation. *Smog Formation* (E4), characterized by elevated ground-level ozone and fine particulate matter, was also highly rated: PM (2.92), PF (3.00), EM (3.03), and EF (2.93), indicating concern about its harmful effects on visibility, respiratory health, and vegetation growth.

In the category of moderately severe environmental impacts, issues such as *Acid Rain* (E2), *Harm to Wildlife and Biodiversity* (E7), *Deterioration of Built Environments* (E10), and *Soil Degradation* (E6) were all perceived as significant, although they scored slightly lower than the impacts in the top tier. These factors underscore the diverse consequences of air pollution, including leaching of soil nutrients, corrosion of the infrastructure, and disruption of food chains due to ecosystem imbalance.

Lower-rated but still significant environmental impacts included *Impact on Forests and Vegetation* (E8), which received the lowest average scores but remained above the mean value of 2.0, suggesting that plant life is still considered an important, albeit less immediate, victim of pollution. *Global Warming and Climate Change* (E1) also received slightly lower ratings despite being a major environmental concern. This may reflect the perception that its effects are broader and longer-lasting compared to more localized and observable impacts such as smog and water pollution.

In summary, the results indicate that all environmental impacts considered in the study are strongly related to air pollution, each scoring above the calculated mean of 2.0. The most critical issues include *water pollution*, *ocean acidification*, *smog formation*, and *ozone layer depletion*. Although *biodiversity loss*, *acid rain*, and *soil degradation* are also important concerns, the effects of *climate change* and *forest degradation*, although serious, were slightly lower in perceived direct impact. In general, these findings underscore the widespread recognition of the multifaceted threats air pollution poses to ecological systems and infrastructure.

5.6 Strategies Based on Findings of the Study

Based on the findings of this study, a multipronged strategy is recommended to address the complex and multifaceted issue of air pollution. In the domain of policy and regulation, it is essential to implement and rigorously enforce comprehensive emission standards for industries, automobiles, and energy production units. Regulatory frameworks should prioritize the transition to renewable energy sources such as solar, wind, and hydropower, thus reducing dependency on fossil fuels. Policies that promote the adoption of electric vehicles (EV), the expansion of efficient public transport systems, and vehicle scrappage schemes can significantly reduce vehicular emissions. Industrial operations must be mandated to install pollution control technologies such as electrostatic precipitators and scrubbers. In addition, strict enforcement of bans on open burning of agricultural waste, biomass, and municipal refuse should be implemented to mitigate the release of particulate matter into the atmosphere.

Technological and innovative interventions also play a central role. Industries must be encouraged to adopt sustainable and energy-efficient technologies to minimize their environmental footprint. Establishing extensive real-time air quality surveillance systems can help monitor pollutant concentrations and trigger timely responses. Urban environmental health can be further improved through the implementation of eco-friendly infrastructure, such as urban greening projects, the development of green belts and vertical gardens, which function as natural air filtration systems.

Efforts to improve public awareness and community participation are equally vital. Educational initiatives should aim to inform citizens about the sources and detrimental effects of air pollution, along with practical strategies to reduce individual carbon footprints. Community participation should be actively encouraged through tree plantation drives and participatory waste management practices. From a healthcare perspective, medical systems should be equipped to manage pollution-induced health problems, including respiratory and neurological conditions. Regular health screenings and the distribution of protective gear such as masks, along with recommendations for the use of air purifiers in high-risk areas, are essential safety protocols.

Furthermore, sustainable urban planning and development practices should be integral to long-term air quality management. Cities must adopt integrated land use policies that prioritize environmentally friendly development, alleviate traffic congestion, and improve pedestrian and cycling infrastructure. Upgrading waste management systems to support segregation, recycling, and composting can minimize landfill emissions. Finally, increased investment in research and innovation is critical. Research should focus on the development of pollution control technologies and the discovery of low-emission or alternative materials that can replace conventional, high-polluting fuels and products.

Collectively, these strategies, rooted in regulatory reform, technological advancement, public participation, and sustainable planning, offer a comprehensive approach to mitigate the impacts of air pollution as revealed by the study findings.

5.7 Concluding Observations

The analysis reveals that all the identified factors play a significant role in contributing to air pollution, with climate change, mining, and deforestation identified as the most critical elements. On the contrary, industrial activities and natural causes, while less significant, still have a notable impact. Paramedical professionals and environmental activists recognize the seriousness of air pollution, and female environmental activists (EF) demonstrate the highest degree of concern. The health implications of air pollution are severe, as all health-related metrics exceed average values. Key health issues include reduced life expectancy, respiratory disorders, complications that affect the nervous system, and mental health challenges. Vulnerable populations, such as children, pregnant women, and those with weakened immune systems, face increased risks. Although short-term effects, such as respiratory problems and eye irritation, are acknowledged, they are considered less critical when compared to chronic health conditions.

The environmental impacts are closely related to air pollution. The most critical consequences include water pollution, ocean acidification, smog formation, and the depletion of the ozone layer. In addition, significant issues such as biodiversity loss, soil degradation, and acid rain warrant attention. Although climate change and forest degradation are important concerns, they are perceived to have a somewhat lesser impact than issues related to water and air quality. The overall findings highlight the pressing necessity for comprehensive and targeted strategies to reduce air pollution and its extensive effects on public health and the environment. The effective resolution of these urgent challenges requires collaborative efforts among policymakers, industries, and communities.

Acknowledgments The authors sincerely thank our prestigious institute and other people who are involved in this work.

Note: Annexure are available with the authors and will be provided on demand.

References

Braithwaite, I., Zhang, S., Kirkbride, J. B., Osborn, D. P. J., & Hayes, J. F. (2019). Air pollution (particulate matter) exposure and associations with depression, anxiety, bipolar, psychosis and suicide risk: A systematic review and meta-analysis. *Environmental Health Perspectives, 127*(12), 126002. https://doi.org/10.1289/EHP4595

Brook, R. D., Rajagopalan, S., Pope, C. A., Brook, J. R., Bhatnagar, A., Diez-Roux, A. V., & Kaufman, J. D. (2010). Particulate matter air pollution and cardiovascular disease. *Circulation, 121*(21), 2331–2378. https://doi.org/10.1161/CIR.0b013e3181dbece1

Calderón-Garcidueñas, L., Kavanaugh, M., Block, M., D'Angiulli, A., Delgado-Chávez, R., Torres-Jardón, R., & Perry, G. (2016). Neuroinflammation, hyperphosphorylated tau, diffuse amyloid plaques, and down-regulation of the cellular prion protein in air pollution exposed children and young adults. *Journal of Alzheimer's Disease, 28*(1), 93–107. https://doi.org/10.3233/JAD-2011-110715

Chen, R., Yin, P., Meng, X., Wang, L., Liu, C., Niu, Y., & Zhou, M. (2017). Fine particulate air pollution and daily mortality: A nationwide analysis in 272 Chinese cities. *American Journal of Respiratory and Critical Care Medicine, 196*(1), 73–81. https://doi.org/10.1164/rccm.201609-1862OC

Chen, Y., & Zhao, X. (2021). Agricultural practices and air quality: The influence of crop burning and fertilizer use. *Agricultural and Environmental Sciences, 29*(2), 230–255.

Cohen, A. J., Brauer, M., Burnett, R., Anderson, H. R., Frostad, J., Estep, K., & Forouzanfar, M. H. (2017). Estimates and 25-year trends of the global burden of disease attributable to ambient air pollution: An analysis of data from the global burden of disease study 2015. *The Lancet, 389*(10082), 1907–1918.

Dockery, D. W., & Pope, C. A. (2022). Air pollution and its ecological consequences: Understanding atmospheric deposition and ecosystem damage. *Environmental Research Journal, 45*(3), 112–130.

Environmental Protection Agency. (2020a). *Health and environmental effects of particulate matter (PM)*. https://www.epa.gov/pm-pollution

Environmental Protection Agency. (2020b). *Integrated science assessment (ISA) for particulate matter*. https://www.epa.gov/

Food and Agriculture Organization. (2020a). *Livestock and climate change*. https://www.fao.org/

Food and Agriculture Organization. (2020b). *Soil pollution: A hidden reality*. https://www.fao.org/

Intergovernmental Panel on Climate Change. (2021). *Climate change 2021: The physical science basis*. https://www.ipcc.ch/

International Council on Mining and Metals. (2020). *Air quality management in mining*. https://www.icmm.com/

Kumar, P., & Gupta, R. (2023). Industrial emissions and their impact on urban air quality: A case study. *Environmental Pollution Research, 45*(3), 123–145.

Kumar, R., & Singh, P. (2022). The role of air pollutants in climate change: Atmospheric chemistry and environmental feedbacks. *Climate and Atmospheric Science, 15*(2), 198–215.

Landrigan, P. J., Fuller, R., Acosta, N. J. R., Adeyi, O., Arnold, R., Basu, N., & Zhong, M. (2018). The lancet commission on pollution and health. *The Lancet, 391*(10119), 462–512.

Lee, J., & Chen, R. (2020). Air pollution and its influence on soil and water quality: A regional assessment. *Environmental Monitoring and Assessment, 192*(7), 556–574.

Lelieveld, J., Evans, J. S., Fnais, M., Giannadaki, D., & Pozzer, A. (2015). The contribution of outdoor air pollution sources to premature mortality on a global scale. *Nature, 525*(7569), 367–371.

Loomis, D., Grosse, Y., Lauby-Secretan, B., Ghissassi, F. E., Bouvard, V., Benbrahim-Tallaa, L., & Straif, K. (2013). The carcinogenicity of outdoor air pollution. *The Lancet Oncology, 14*(13), 1262–1263. https://doi.org/10.1016/S1470-2045(13)70487-X

National Oceanic and Atmospheric Administration. (2020). *Ocean acidification*. https://www.noaa.gov/education/resource-collections/ocean-coasts/ocean-acidification

Patel, S. K., & Dey, M. (2023). Household air pollution from biomass fuel consumption: Health impacts and mitigation strategies. *Journal of Clean Energy Solutions, 18*(1), 78–99.

Perera, F. P., Tang, W. Y., Herbstman, J., Tang, D., Levin, L., Miller, R. L., & Rauh, V. (2019). Relation of DNA methylation of 5'-CpG Island of ACSL3 to transplacental exposure to airborne polycyclic aromatic hydrocarbons and childhood asthma. *PLoS One, 4*(2), e4488. https://doi.org/10.1371/journal.pone.0004488

Pope, C. A., Turner, M. C., Burnett, R. T., Jerrett, M., Gapstur, S. M., Diver, W. R., & Krewski, D. (2020). Relationships between fine particulate air pollution, cardiometabolic disorders, and cardiovascular mortality. *Circulation Research, 126*(10), 1259–1275. https://doi.org/10.1161/CIRCRESAHA.120.316831

Rajan, R., & Thomas, L. (2022). The role of deforestation and land-use changes in atmospheric pollution. *Global Environmental Challenges, 11*(5), 301–328.

Schraufnagel, D. E., Balmes, J. R., Cowl, C. T., De Matteis, S., Jung, S. H., Mortimer, K., & Thurston, G. D. (2019). Air pollution and noncommunicable diseases: A review by the forum

of international respiratory societies' environmental committee, part 1: The damaging effects of air pollution. *Chest, 155*(2), 409–416.
Smith, J. L., & Wang, H. (2022). Vehicular emissions and their role in air pollution across metropolitan cities. *Journal of Environmental Studies, 37*(4), 567–590.
Smith, K. R., & Desai, M. A. (2021). The environmental effects of particulate matter and gaseous pollutants on ecosystems. *Journal of Environmental Management, 40*(4), 345–360.
Thurston, G. D., Kipen, H., Annesi-Maesano, I., Balmes, J., Brook, R. D., Cowl, C. T., & Szema, A. M. (2017). A joint ERS/ATS policy statement: What constitutes an adverse health effect of air pollution? An analytical framework. *European Respiratory Journal, 49*(1), 1600419.
United Nations Environment Programme. (2019a). *Air pollution and water quality*. https://www.unep.org/
United Nations Environment Programme. (2019b). *Waste management outlook for Asia and the Pacific*. https://www.unep.org/
United States Environmental Protection Agency. (2020). *Acid rain*. https://www.epa.gov/acidrain
World Health Organization. (2021a). *Air pollution*. https://www.who.int/
World Health Organization. (2021b). *Air pollution and health*. https://www.who.int/health-topics/air-pollution
World Health Organization. (2021c). *Ozone depletion and human health*. https://www.who.int/
Yadav, A. K. (2015). Elemental composition and source apportionment of suspended particulate matters and health risk assessment in mining and nonmining areas of Odisha, India. *Journal of Hazardous, Toxic, and Radioactive Waste, 19*(3), 04014037.
Yadav, A. K. (2021). Human health risk assessment in opencast coal mines and coal-fired thermal power plants surrounding area due to inhalation. *Environmental Challenges, 3*, 100074.
Yadav, A. K., Sahoo, S. K., Dubey, J. S., Kumar, A. V., Pandey, G., & Tripathi, R. M. (2019). Assessment of particulate matter, metals of toxicological concentration, and health risk around a mining area, Odisha, India. *Air Quality, Atmosphere & Health, 12*(7), 775–783.
Yadav, A. K., Sahoo, S. K., Kumar, A. V., Shirin, S., Jamal, A., Patra, A. C., Dubey, J. S., Thakur, V. K., Lenka, P., Singh, S., Jha, V., & Tripathi, R. M. (2023). Identifying size-segregated particulate matter ($PM_{2.5}$, PM_{10} and SPM) sources in an industrial town of India. *Aerosol Science and Engineering, 7*(4), 455–473.
Yadav, A. K., Sahoo, S. K., Patra, A. C., Dubey, J. S., Lenka, P., Sagar, D. V., Kumar, A. V., & Tripathi, R. M. (2014). Source identification of particulate matter and associated intake of elements through inhalation in an industrial area of Odisha, India. *Toxicological & Environmental Chemistry, 96*(3), 410–425.
Yadav, A. K., Shirin, S., Emmanouil, C., & Jamal, A. (2022). Effect of seasonal and meteorological variability of air pollution in Singrauli coalfield. *Aerosol Science and Engineering, 6*, 61–70.
Zhang, Y., & Wang, H. (2023). Acid rain and its long-term impact on forest biodiversity: A case study in East Asia. *Ecological Indicators, 68*(5), 234–251.

Chapter 6
Analyzing Air Pollution Sources Effect and Mitigation Strategies

Rajrupa Ghosh, Sonali Paul, and Susmita Mukherjee

6.1 Introduction

A major environmental problem is air pollution because of its complexity, causes, and effects on the environment and human health, and it is a worldwide issue that needs immediate action. Many people are concerned about the impacts of the declining quality of air on human beings, damage on environment, and climate. There are different sources of air pollutants, including events of nature such as volcanic eruptions and wildfires, industrial emissions, vehicle exhaust, and agricultural practices. Sources mainly involved are major primary pollutants (SO_x, NO_x, CO, methane and non-methane VOCs, particulate matters, CFCs), secondary pollutants (PAN, POPs, smog, acid rain), and so on (World Health Organization, 2021). The regressing impacts of polluted air are very much extensive and affect not only the atmosphere but also the biodiversity and human health, such as throat and nose irritations, bronchoconstriction, which are the results of long-time exposure to SO_x, NO_x, or, to some extent, heavy metals such as arsenic (As), nickel (Ni), cadmium (Cd), vanadium (V), and so on (Meo et al., 2024).

Climate, terrain, and population density are only a few of the variables that affect air pollution levels, which differ significantly between nations and areas. Because of the concentration of man-made sources, including industry, electricity generation, and traffic, air pollution is persistently worse in metropolitan areas. A research carried out in Beijing, China, found that the burning of fossil fuels, such as coal, oil, and natural gas, was the main source of the city's air pollution, amounting to annual average of 70 $\mu g/m^3$ for PM_{10} and 35 $\mu g/m^3$ for $PM_{2.5}$ and 12.3 billion tons of CO_2, which is almost equivalent to more over 27% of global greenhouse gas emission.

R. Ghosh · S. Paul · S. Mukherjee (✉)
Department of Biotechnology, Institute of Engineering and Management, University of Engineering and Management, Kolkata, India
e-mail: susmita.mukherjee@uem.edu.in

A. K. Yadav (ed.), *Computational Approach for Air Pollution*,
https://doi.org/10.1007/978-3-032-09215-1_6

This may be connected to the population density and technology growth of the country (Patel et al., 2014).

This study aims to discuss the basic principles of air pollution, such as various kinds of pollutants, where they come from, how they affect the environment and human health, and what can be done to lessen these negative consequences. Understanding these principles is essential for putting into practice practical ways to diminish the pollution effect of air and make the environment healthier and more sustainable for coming generations.

6.2 The Sources

Anthropogenic (man-made) and natural causes both contribute to air pollution. Because of urbanization, industrialization, and the widespread use of motor vehicles, the contribution from human-made activities has grown dramatically in recent decades.

6.2.1 Natural Sources

Although most of the air pollutions are caused by human activity, natural sources also contribute significantly. Large volumes of particulate matter (PM), various oxides of sulfur, carbon monoxide, and other gases can be released into the atmosphere by forest fires. Change in climate has contributed to an increase in the occurrence of wildfires, which has raised air pollution levels. Black carbon, a super pollutant that increases heat waves, changes weather patterns, and speeds up the melting of ice and snow, is mostly produced by wildfires. This contributes to the cycle of climate change and intense wildfires (Kampa & Castanas, 2008).

Particularly in adjacent areas, the volcanic activities can drastically alter the quality of the air or atmosphere by releasing sulfur dioxide (SO_2), different ash particles, and other chemicals into the environment. The gases (CO_2, SO_2, H_2S, CO, HF, and HCl), silicate ash, volatile metal vapors, and sulfate aerosol, which is created when SO_2 gas is converted to particles, are among the various combinations of airborne volcanic emissions. Large numbers of respirable-sized particles (less than 4 μm in diameter) that can enter the lungs can be found in ash, which can be produced during a number of eruptive events (Horwell, 2007). Ash's physical and chemical characteristics can differ greatly between eruptions and across distance (Oppenheimer et al., 2003). Volcanic gases may condense into particles or adhere to the surfaces of ash when they cool and react in the atmosphere. Long-distance ash clouds can contribute to bad air quality over wide regions (Kaur et al., 2007).

Airborne particles can also originate from natural phenomena such as pollen and desert dust dispersal. Even though these pollutants are typically less dangerous than industrial pollutants, they can nevertheless have an impact on air quality, especially for people who have asthma or other respiratory disorders (Sofiev et al., 2013).

6.2.2 Anthropogenic Sources of Air Pollution

Due to urbanization, population expansion, and industry, human activity has taken over as the primary cause of air pollution. Among the most important source is industrial emissions. One of the main causes of being polluted air is emissions from the factories and different power plants. They emit a lot of dangerous pollutants into the environment, including CO, NO_x, SO_2, and volatile organic compounds (VOCs). Particularly, polluting industries include the production of chemicals, cement, and refineries. Refineries are industrial facilities that convert raw materials, such as natural gas and crude oil, into a variety of fuels, chemicals, and other components that are necessary for daily living. Refineries release a variety of airborne pollutants that have an impact on human health and air quality, such as $PM_{2.5}$, which can lead to cardiovascular and respiratory diseases; sulfur dioxide and nitrogen oxides, which cause smog, acid rain, and respiratory problems; volatile organic compounds, which cause ground-level ozone formation and other health effects; carbon monoxide, which can hinder the body's ability to deliver oxygen; benzene, a known carcinogen; and other dangerous air pollutants such as toluene, xylene, and formaldehyde, which are known to pose serious health risks (Munsif et al., 2021).

Oxides of nitrogen or NO_x, particulate matter or PM, CO, and hydrocarbons are among pollutants released by vehicles that run on gasoline, diesel, and other fuels (Bhandarkar, 2013). One of the biggest reasons for air pollution in cities is the transportation industry. In metropolitan areas, road transportation has emerged as the primary cause of both traffic congestion and environmental pollution. The need for a unified regulatory framework for the control of traffic, air quality, and emissions at the urban, regional, and national levels has increased as a result of the ongoing traffic growth, which has sparked worries about the effects of traffic emissions on human health and urban environmental quality (Ban-Weiss et al., 2008; Yadav, 2015; Yadav et al., 2014, 2023, 2025). Agricultural practices is also an important cause of air pollution. Ammonia (NH_3), methane (CH_4), and other substances are released into the atmosphere by agriculture. Volatile organic compounds (VOCs) are released when pesticides and fertilizers are used. CH_4 is a powerful major greenhouse gas that is released in enormous quantities during the production of livestock, especially when cattle are raised (Pawar et al., 2024). Both interior and outdoor air pollution are greatly increased by the domestic burning of coal, wood, and other fuels, especially for cooking and heating. Burning biomass fuels is a familiar source of polluted air in low-income areas.

6.3 Types of Air Pollutants

Depending on their development in the atmosphere, air pollutants can be roughly divided into primary and secondary pollutants.

6.3.1 *Primary Pollutants*

Substances that are released into the atmosphere directly from a source are known as primary pollutants. Among the examples are:

- Particulate matter: A mixture of droplets of liquid and microscopic particles of solid floating within atmospheric content is known as particulate matter (Kim et al., 2015). It falls into two primary categories:
  - PM_{10}: particles that have basically diameters of 10 μm or less. Numerous health problems can result from breathing in these particles into the respiratory system (Kim et al., 2015).
  - $PM_{2.5}$: particles that have at least the diameters of 2.5 μm. These particles pose serious health hazards because of their size, which allows them to enter the circulation and travel deep into the lungs (Barmpadimos et al., 2012).
- Nitrogen oxides: Fossil fuel combustion, which occurs in power plants and automobiles, produces nitrogen oxides. These exclusive pollutants are associated with the production of acid rain, smog, and ground-level ozone. Nitric oxide (NO) and nitrogen dioxide (NO_2) are the two primary oxides of nitrogen. The troposphere is home to these two significant types. Nitrous oxide (N_2O), a greenhouse gas, is an additional significant gas. It is the primary form of nitrogen oxide and is tasteless and colorless. Because nitric oxide has a short residence period in nature, it is very difficult to quantify. As a result, it is easily transformed into nitrogen dioxide, which is hazardous in nature when it reacts chemically with ozone. NO_2 is a reddish-brown to yellowish-orange gas that has a potent oxidizing capability and a strong, unpleasant smell (Patel et al., 2014). The other and mostly extremely important oxides are NO_2 and N_2O_5, but they are found in smaller concentrations. The majority of NO_2 is transformed to NO and ammonium salts. The removal of nitrate aerosol and sulfate aerosol requires the same technique, which is wet or dry deposition.
- Sulfur dioxide: Burning sulfur, which basically possess fossil fuels such as coal and oil, is the ultimate main reason of source of sulfur dioxide emissions. It is a precursor to acid rain, which can harm flora and aquatic environments (Craig, 2019). Sulfur trioxide, or SO_3, is released into the environment directly or is created from SO_2, which is readily transformed into H_2SO_4.
- Carbon monoxide: This gas, which is significant as odorless and colorless, is created when fuels burn incompletely or in the absence of oxygen. CO_2 is directly soluble in water and easily produces carbonic acid or H_2CO_3. High quantities can

be dangerous because they disrupt the bloodstream's ability to carry oxygen, especially in confined areas (Penney, 2007).

6.3.2 Secondary Pollutants

Within the atmospheric substances, primary basic pollutants react chemically to generate secondary pollutants, which are not released directly from the sources. Among the examples are:

- Ozone (O_3): When NO_x and VOCs combine with ground-level ozone in the presence of sunlight, it creates basically the photochemical smog. Smog means smoke in the combination of fog. One of the main ingredients of smog is ground-level ozone, which can harm crops and create respiratory problems. PAN, or peroxyacetyl nitrate, is another element of photochemical smog that can irritate eyes and create respiratory problems. Pollution is most likely to become a problem in large cities with high levels of industry and traffic, particularly those located in basins, such as Mexico City and Los Angeles, where the topography can trap the pollution and keep it from being dispersed by the wind (Neal, 2005).
- Acid rain: Whenever sulfuric acid or H_2SO_4 and nitric acid or HNO_3 are produced as a result of the reactions of SO_2 and NO_x along with the atmospheric vapor of water, they form acid rain. These acids harm aquatic life, infrastructure, and ecosystems when they descend to the earth as rain (Afifa et al., 2024).

6.4 Health Impacts of Air Pollution

Air pollution has caused minor irritations to serious illnesses and, in certain situations, early death, with a significant impact on human health.

6.4.1 Respiratory Diseases

Respiratory illnesses are largely caused by air pollution, especially ozone and fine particulate matter ($PM_{2.5}$). Pollutants and allergens in the air can cause asthma episodes and worsen symptoms in people who already have the illness. Chronic obstructive pulmonary disease (COPD) is most importantly associated with chronic bronchitis as well as emphysema, which can develop after prolonged exposure to polluted air (Cazzola et al., 2007). Exposure through a long period to the polluted air can lead to COPD, i.e., a condition of such emphysema and chronic bronchitis. These effects, which fall into two categories—short-term and long-term—can include worsening preexisting symptoms, lung function impairment, and higher rates of hospitalization and death. The prevalence of COPD may also rise with prolonged exposure to air that contains high levels of pollution. In the future, the combined effects of several contaminants might become increasingly complicated (Ling & van Eeden, 2009). An elevated risk of lung cancer is familiar to be associated with exposure to airborne pollutants such as diesel exhaust, tobacco smoke, and other carcinogens (Xue et al., 2022).

6.4.2 Cardiovascular Diseases

Studies have demonstrated that air pollutions, specifically NO_x and fine particulate matter or $PM_{2.5}$, can be threaten to health issues of cardiovascular conditions such as hypertension, heart attacks, and strokes (Kolli, 2013). The risk of cardiovascular disease from particle pollution is lower for an individual than the risk from several of the other known risk factors mentioned above. Short- and long-term exposure, however, has been demonstrated to raise hospitalizations for major cardiovascular events, including coronary syndrome, arrhythmia, heart failure, stroke, and sudden cardiac death, especially in individuals with preexisting heart disease, for the general population.

Inflammation and oxidative stress in the cardiovascular system brought on by continuous exposure to such pollutants can aid in the development of atherosclerosis or the hardening of the arteries (Fiordelisi et al., 2017).

6.4.3 Premature Deaths

As per the WHO studies, air pollution causes millions of early deaths globally. Indeed, it is among the most important environmental risk factors for death. Among the most common illnesses that are extremely associated to air pollution are heart disease, stroke, and respiratory infections.

6.4.4 Vulnerable Populations

Some demographics are particularly susceptible to the regressive consequences of air pollution, such as children. Due to their developing respiratory systems, children are particularly fragile to the detrimental impact of polluted air, which can culminate in impaired lung growth and asthma (Tzivian, 2011). Health issues associated with air pollution are more likely to affect older persons, especially those who already have medical disorders. Living in places with more industrial activity and traffic makes low-income people more likely to be exposed to pollution, especially in urban areas.

6.5 Environmental Impact of Air Pollution

In addition to endangering human health, air pollution has detrimental consequences on the environment as a whole. Among these effects are climate change, acid rain, and ecosystem degradation.

6.5.1 Climate Change

Strong greenhouse gases that contribute to global warming include nitrous oxide or N_2O, methane or CH_4, and black carbon (soot). The greenhouse effect is exacerbated and climate change is accelerated by these pollutants' ability to trap heat in the atmosphere (Pasha et al., 2016). Further affecting climate patterns, air pollution can also change the water cycle, precipitation patterns, and cloud formation.

6.5.2 Acid Rain

Acid rain is observed mainly when the SO_2 and NO_x undergoes emission from industrial sources. Acid rain harms ecosystems in a number of ways. Acid rain may reduce the alkalinity of the soil, which disturbs the availability of key nutrients for plants and harms soil organisms. Rivers, lakes, and streams may be transformed to acidic ones after acid rain, which alters and destroys the aquatic life and upsets ecosystems. By removing nutrients from the soil and causing direct harm to plant tissues, acid rain harms crops and trees.

6.5.3 Ecosystem Degradation

There are several ways that air pollution can upset ecosystems, including biodiversity loss. Plant species can be harmed by air pollution, which lowers biodiversity in impacted areas. For instance, although nitrogen deposition can change the makeup of plant communities, elevated ozone levels can harm crops, forests, and other plants. Pollutants that harm plants can have an effect on herbivores and carnivores that rely on them, as well as those higher up the food chain.

6.6 Economic Impact of Air Pollution

Air pollution has significant financial impacts all around the world. These expenses are caused by a number of things, such as medical bills, lost production, and environmental rehabilitation projects.

6.6.1 Health Care Costs

It takes a large amount of public and private funding to treat illnesses brought on by air pollution. Every year, billions of dollars are spent on hospital stays, prescription drugs, and long-term care for people with cardiovascular and pulmonary conditions.

6.6.2 Loss of Productivity

People with air pollution-related illnesses are less able to work, which might result in a loss in labor output. The impacts of air pollution also significantly affect outdoor workers, resulting in lost working hours and lower economic production.

6.6.3 Damage to Agriculture

Crops, animals, and agricultural production are all negatively impacted by air pollution. For instance, the genesis of ground-level ozone has the potential to drastically lower crop output, which has both direct and indirect economic effects.

6.7 Mitigation Strategies and Solutions

Regulations, technological advancements, and individual acts are all necessary to address the problem of air pollution. The following tactics are crucial.

6.7.1 Regulatory Policies

Stricter laws governing automobile exhaust, industrial emissions, and other significant sources of air pollution must be put into place and enforced by the government. Among these rules ought to be:

- Establishing emissions guidelines for automobiles and industry.
- Promoting the use of greener fuels and technology.
- Developing real-time data-gathering air quality monitoring devices.

6.7.2 Technological Innovation

In order to reduce air pollution, technological breakthroughs are essential. Among the solutions are using renewable energy. Emissions from different power plants or thermal plants can be immensely decreased by switching to renewable energy sources such as hydropower, wind, and solar energy. Besides, the transportation sector's emissions of carbon monoxide, nitrogen oxides, and particulate matter can be decreased by the widespread use of electric vehicles. In some cases the hazardous emissions can be greatly decreased by installing scrubbers, filters, and catalytic converters in automobiles and industrial facilities.

6.7.3 Public Awareness and Behavioral Change

Mitigating the air pollution requires public awareness about the negative health effects of air pollution and promoting eco-friendly practices, which includes cutting back on car use, using less energy, and using cleaner technologies.

6.7.4 International Cooperation

International cooperation is required to manage transboundary contamination since air pollution transcends national boundaries. A framework for lowering emissions and addressing global environmental challenges is provided by agreements such as the Paris Climate Agreement (COP21).

Climate and Clean Air Coalition (CCAC): The CCAC is a voluntary collaboration between governments, NGOs, and intergovernmental organizations that aims to reduce short-lived climate pollutants (SLCPs) such as hydrofluorocarbons (HFCs), black carbon, and methane.

UNEP's Global Air Quality Cooperation Network: Created in accordance with UNEA resolution 6/10, this network seeks to improve access to funding and assistance, increase the adoption of solutions, and boost international cooperation on air pollution.

The Air Quality Management Exchange Platform (AQMx): As the focal point of the Global Air Quality Cooperation Network, AQMx promotes collaboration and information exchange between governments and interested parties.

6.8 Conclusion

One of the biggest environmental hazards that society is currently confronting is air pollution. Its impacts are perceived globally and have a substantial impact on the economy, ecology, and human health. Implementing practical solutions to lower emissions, advance cleaner technology, and increase public awareness is essential as urbanization, industry, and vehicle activity all continue to increase. By working together, we can lessen the negative consequences of air pollution and make progress toward a better, cleaner future for everybody.

Acknowledgments The authors gratefully acknowledge all the members of the Department of Biotechnology and Centre of Excellence for Interdisciplinary Research on Environmental Monitoring and Management for completion of this work.

References

Afifa, K. A., Hussain, N., Ashraf, M. H., & Saleem, M. Z. (2024). Air pollution and climate change as grand challenges to sustainability. *Science of the Total Environment, 928*, 172370.

Ban-Weiss, G. A., McLaughlin, J. P., Harley, R. A., Lunden, M. M., Kirchstetter, T. W., Kean, A. J., Strawa, A. W., Stevenson, E. D., & Kendall, G. R. (2008). Long-term changes in emissions of nitrogen oxides and particulate matter from on-road gasoline and diesel vehicles. *Atmospheric Environment, 42*(2), 220–232.

Barmpadimos, I., Keller, J., Oderbolz, D., Hueglin, C., & Prévôt, A. S. (2012). One decade of parallel fine (PM 2.5) and coarse (PM 10–PM 2.5) particulate matter measurements in Europe: Trends and variability. *Atmospheric Chemistry and Physics, 12*(7), 3189–3203.

Bhandarkar, S. (2013). Vehicular pollution, their effect on human health and mitigation measures. *Vehicle System Dynamics, 1*(2), 33–40.

Cazzola, M., Donner, C. F., & Hanania, N. A. (2007). One hundred years of chronic obstructive pulmonary disease (COPD). *Respiratory Medicine, 101*(6), 1049–1065.

Craig, K. (2019). A review of the chemistry, pesticide use, and environmental fate of sulfur dioxide, as used in California. *Reviews of Environmental Contamination and Toxicology, 246*, 33–64.

Fiordelisi, A., Piscitelli, P., Trimarco, B., Coscioni, E., Iaccarino, G., & Sorriento, D. (2017). The mechanisms of air pollution and particulate matter in cardiovascular diseases. *Heart Failure Reviews, 22*, 337–347.

Horwell, C. J. (2007). Grain-size analysis of volcanic ash for the rapid assessment of respiratory health hazard. *Journal of Environmental Monitoring, 9*, 1107–1115.

Kampa, M., & Castanas, E. (2008). Human health effects of air pollution. *Environmental Pollution, 151*(2), 362–367.

Kaur, S., Nieuwenhuijsen, M. J., & Colvile, R. N. (2007). Fine particulate matter and carbon monoxide exposure concentrations in urban street transport microenvironments. *Atmospheric Environment, 41*(23), 4781–4810.

Kim, K. H., Kabir, E., & Kabir, S. (2015). A review on the human health impact of airborne particulate matter. *Environment International, 74*, 136–143.

Kolli H. (2013). Examining the relationship between particulate matter, nitrogen oxide, carbon oxides, Sulfur dioxide and hypertension in urban areas in India: A review. *VCU Scholars Compass, AUCTUS: The Journal of Undergraduate Research and Creative Scholarship.*

Ling, S. H., & van Eeden, S. F. (2009). Particulate matter air pollution exposure: role in the development and exacerbation of chronic obstructive pulmonary disease. *International Journal of Chronic Obstructive Pulmonary Disease, 4*, 233–243.

Meo, S. A., Salih, M. A., Alkhalifah, J. M., Alsomali, A. H., & Almushawah, A. A. (2024). Environmental pollutants particulate matter (PM2. 5, PM10), Carbon Monoxide (CO), Nitrogen dioxide (NO2), Sulfur dioxide (SO2), and Ozone (O3) impact on lung functions. *Journal of King Saud University Science, 36*(7), 103280.

Munsif, R., Zubair, M., Aziz, A., & Zafar, M. N. (2021). *Industrial air emission pollution: Potential sources and sustainable mitigation, environmental emissions*. IntechOpen.

Neal, P. (2005). Air pollution—with special reference to acid rain, the greenhouse effect and ozone layer depletion. In *Environment in question* (pp. 125–143). Routledge.

Oppenheimer, C., Pyle, D. M., & Barclay, J. (2003). *Volcanic degassing* (p. 213). Geological Society of London.

Pasha, N., Reddy, P. S., Sagar, T. V., Lingaiah, N., & Prasad, P. S. (2016). Combating global warming due to nitrous oxide. *Science Spectrum, 1*, 410–423.

Patel, A. K., Chaudhary, H. H., Patel, K. S., & Sen, D. J. (2014). Air pollutants all are chemical compounds hazardous to ecosystem. *World Journal of Pharmaceutical Sciences, 1*, 729–744.

Pawar, S., Ikram, M., & Khan, S. (2024). Chapter–6, health and environmental impacts of agricultural pollution. In *A comprehensive exploration of soil, water, and air pollution in agriculture* (Vol. 103). BFC Publication.

Penney, D. G. (2007). *Carbon monoxide poisoning*. CRC Press.

Sofiev, M., Belmonte, J., Gehrig, R., Izquierdo, R., Smith, M., Dahl, Å., & Siljamo, P. (2013). Airborne pollen transport. In *Allergenic pollen: A review of the production, release, distribution and health impacts* (pp. 127–159). Springer.

Tzivian, L. (2011). Outdoor air pollution and asthma in children. *Journal of Asthma, 48*(5), 470–481.

World Health Organization. (2021). *WHO global air quality guidelines: particulate matter ($PM_{2.5}$ and PM_{10}), ozone, nitrogen dioxide, sulfur dioxide and carbon monoxide*. World Health Organization.

Xue, Y., Wang, L., Zhang, Y., Zhao, Y., & Liu, Y. (2022). Air pollution: A culprit of lung cancer. *Journal of Hazardous Materials, 434*, 128937.

Yadav, A. K. (2015). Elemental composition and source apportionment of suspended particulate matters and health risk assessment in mining and nonmining areas of Odisha, India. *Journal of Hazardous, Toxic, and Radioactive Waste, 19*(3), 04014037.

Yadav, A. K., Gupta, K., Bodhale, R., & Phuleria, H. C. (2025). Metals and trace element associated health risk in size-segregated household dust of an industrial town in India. *Air Quality, Atmosphere & Health*. https://doi.org/10.1007/s11869-025-01794-5

Yadav, A. K., Sahoo, S. K., Kumar, A. V., Shirin, S., Jamal, A., Patra, A. C., Dubey, J. S., Thakur, V. K., Lenka, P., Singh, S., Jha, V., & Tripathi, R. M. (2023). Identifying size-segregated particulate matter ($PM_{2.5}$, PM_{10} and SPM) sources in an industrial town of India. *Aerosol Science and Engineering, 7*(4), 455–473.

Yadav, A. K., Sahoo, S. K., Patra, A. C., Dubey, J. S., Lenka, P., Sagar, D. V., Kumar, A. V., & Tripathi, R. M. (2014). Source identification of particulate matter and associated intake of elements through inhalation in an industrial area of Odisha, India. *Toxicological & Environmental Chemistry, 96*(3), 410–425.

Part II
Computational and Data-Driven Approaches

Chapter 7
Toward a Data-Driven Air Pollution Control Strategy

Trisha Paul and Sankar Narayan Sinha

7.1 Introduction

Air pollution represents an urgent global environmental problem that endangers human health together with ecosystems and economic systems. According to the World Health Organization (WHO, 2021), air pollution results in seven million premature deaths yearly, which mostly affects low- and middle-income countries (LMICs). The combination of nitrogen oxides (NO_x), volatile organic compounds (VOCs), ground-level ozone, sulfur oxides (SO_x), and fine particulate matter ($PM_{2.5}$) leads to respiratory and cardiovascular and neurological diseases (Cohen et al., 2017; Yadav et al., 2022). The various sources of air pollution include emissions from vehicles along with industrial fossil fuel combustion, residential heating, and agricultural burning (Yadav, 2015; Yadav et al., 2014, 2019, 2025). Natural occurrences such as wildfires together with dust storms increase pollution levels, but human activities remain the main cause of air pollution in urban areas. The situation worsens because of rapid urban development combined with uncontrolled industrial operations and inadequate enforcement of emission control regulations. Numerous regulatory and technological efforts including electrostatic precipitators, catalytic converters, and fabric filters have not successfully reduced air pollution in various regions. Conventional methods operate independently because they focus exclusively on final treatment solutions. These methods do not tackle the fundamental elements of pollutant movement and chemical reactions and atmospheric movement patterns. The methods lack predictive and adaptive features that restrict their ability to function in changing urban environments. The implementation of computational methods has revolutionized air pollution control by introducing advanced tools,

---

T. Paul · S. N. Sinha (✉)
Environmental Microbiology Research Laboratory, Department of Botany, University of Kalyani, Kalyani, India

A. K. Yadav (ed.), *Computational Approach for Air Pollution*,
https://doi.org/10.1007/978-3-032-09215-1_7

including mathematical modeling along with simulation techniques and artificial intelligence (AI) and data analytics and geospatial information systems (GIS). Methods support comprehensive air quality management through integrated and predictive practices. These computational tools show the spread of emissions and detect pollution centers while evaluating mitigation strategies through multiple space–time frameworks. Atmospheric dispersion modeling stands out as an important tool. The Gaussian plume model, Eulerian grid-based models, and Lagrangian particle tracking models predict pollutant behavior through different weather patterns (Holmes & Morawska, 2006). The combination of computational fluid dynamics (CFD) with these tools delivers detailed information about pollution behavior in confined areas such as street canyons on the streets and industrial zones. Spatial mapping of emission sources and affected populations through GIS enables better zoning and infrastructure planning (Briggs et al., 1997). Real-time air quality prediction along with pollution source detection and emission control strategy optimization use artificial intelligence and machine learning systems. These technologies process IoT sensor information together with satellite data, which allows for immediate decision-making. Through computational methods, scientists can connect scientific findings directly with policy development processes. Decision support systems (DSS) together with integrated assessment models (IAMs) allow policymakers to evaluate various scenarios and develop strategic plans for effective implementation (Amann et al., 2011). Open-source platforms, together with cloud-based tools and affordable low-cost sensors, help LMICs establish smart air quality management systems. Computational methods now reshape air pollution management through their ability to create integrated solutions that offer predictive power and adaptability. The chapter provides a detailed examination of the ways computational methods are transforming the field of air pollution control. The integration of environmental science with computational technology and policy frameworks enables these methods to create sustainable urban development along with climate resilience and cleaner air.

7.2 Fundamentals of Air Pollution and Control: A Contemporary Scientific Perspective

Air pollution control requires knowledge of both traditional engineering methods and contemporary computational approaches. Pollutants exist in two distinct categories: primary and secondary types. Sources that produce primary emissions include fossil fuel combustion and industrial releases and vehicle exhaust and farming activities. The generation of secondary contaminants such as ground-level ozone (O_3) and secondary particulate matter occurs through complex nonlinear atmospheric reactions, which are triggered by precursors such as nitrogen oxides (NO_x) and volatile organic compounds (VOCs) in the presence of sunlight. The major air pollutants of concern consist of nitrogen oxides (NO_x), sulfur dioxide (SO_2), ozone (O_3), carbon monoxide (CO), VOCs and suspended particulate matter ($PM_{2.5}$ and

PM_{10}), and trace toxicants, including heavy metals (e.g., Pb, Hg, As). The pollutants create multiple adverse effects on the environment. The small size of fine particulate matter ($PM_{2.5}$) allows it to reach deep into human respiratory alveolar regions, where it causes chronic pulmonary and cardiovascular diseases. The oxidative properties of ozone make it a powerful agent that worsens asthma and COPD and bronchitis conditions. Research has shown that sustained exposure to SO_2, NO_x, and CO results in higher rates of illness together with neurological problems and systemic inflammation (Chadalavada et al., 2025). Different control strategies have been developed for ambient air pollutants such as particulate matter (PM_{10}, $PM_{2.5}$), nitrogen oxides (NO_x), sulfur dioxide (SO_2), volatile organic compounds (VOC), carbon monoxide (CO), and ozone (O_3) based on their source characteristics and atmospheric movement patterns. The main tools for pollution control have traditionally been conventional air pollution control techniques. These include filtration systems such as fabric filters and electrostatic precipitators (ESP), which mechanically and electrostatically remove suspended particulates from emission streams. Similarly, scrubbing technologies, particularly wet and dry scrubbers, are widely deployed to chemically absorb SO_2 and NO_x from flue gases. In the transportation sector, catalytic converters are used to convert toxic gases such as CO and NO_x to less harmful CO_2 and N_2. Adsorption systems, particularly activated carbon filters, are also utilized in industrial applications to capture VOCs and odorous compounds. Additionally, regulatory instruments such as emission standards, fuel regulations, and zoning policies have complemented these technological approaches to managing air quality. However, these traditional methods have several critical limitations. Most systems are sector-specific and lack holistic integration across spatial and temporal scales. They operate reactively rather than proactively, often failing to adapt in real time to changing pollution dynamics or meteorological conditions. Traditional pollution control systems focus mainly on primary pollutants but do not address secondary pollutants such as ozone and PM, which develop through intricate photochemical processes. The scalability of such methods remains limited in low- and middle-income countries due to high operational costs and restricted technological infrastructure, according to El-Harbawi (2013).

Integration of advanced computational methods has become prominent in addressing these limitations. Through Gaussian plume and Eulerian grid-based models, researchers can obtain spatial and temporal information about how pollutants move and spread throughout the atmosphere. Photochemical models (e.g., CAMx, CMAQ, WRF-Chem) simulate secondary pollutant formation by integrating atmospheric chemistry mechanisms (El-Harbawi, 2013).

Deep learning (DL) and machine learning (ML) methods have transformed pollution forecasting operations by delivering precise predictions about air quality levels and pinpointing areas with high pollution concentrations. Random Forest along with LSTM and XGBoost algorithms shows exceptional accuracy (up to 98.2%) in air quality forecasting according to Chadalavada et al. (2025). The implementation of computational intelligence and real-time modeling frameworks has become essential for effective management of advanced pollution scenarios, which traditional methods fail to address.

7.3 Computational Methods for Air Pollution and Control

The development of computational methods has become essential tools for air pollution modeling as well as control strategies over the last several decades. Through these approaches, researchers can create detailed simulations of atmospheric systems to determine pollutant movements and develop evidence-driven solutions for reduction. The traditional approach to air quality modeling employed deterministic methods that included Gaussian plume models to make basic predictions about steady-state pollutant concentrations. The current requirement for precise spatial and temporal data has forced scientists to use advanced Eulerian and Lagrangian models, which solve advection–diffusion equations across dynamic grids that couple with meteorological data (El-Harbawi, 2013).

The three-dimensional grid-based system of Eulerian models in CMAQ and CAMx allows for dynamic simulations of pollutant advection, diffusion, and chemical reactions and deposition processes that occur under changing meteorological conditions. Models operate with chemical mechanisms such as CB05 and SAPRC07 that simulate the formation of photochemical smog together with secondary organic aerosol production and atmospheric species relationships (Antuña et al., 2023). The tracking of pollutant parcels through time and space using Lagrangian models CALPUFF provides excellent results for non-steady-state conditions and long-range transport, especially in complex terrain.

AERMOD uses AERMAP for terrain preprocessing and AERMET for meteorological dynamics within its framework. CALIOPE operates as a modular framework that joins HERMES (emission inventory) and WRF-ARW (meteorology) with CMAQ (chemistry) for conducting high-resolution forecasting. The regulatory modeling field heavily depends on two specialized tools named COPERT for road transport emissions and SMOKE as an emissions processing system. Each model differs in its spatial resolution capacity together with the complexity of the input data and pollutant-specificity and visualization tools of pollutants. Open-source models enable both accessibility and flexibility, but proprietary tools such as ANSYS Fluent using computational fluid dynamics (CFD) provide detailed microscale analysis through numerical solutions of Navier–Stokes equations (Antuña et al., 2023). The development of pollution control strategies experienced a revolution due to recent advances in machine learning (ML) and artificial intelligence (AI). The use of Random Forest algorithms with LSTM networks, Support Vector Machines (SVM), and XGBoost enables AI-based systems to accurately predict pollution hotspots and source locations while performing real-time air quality predictions (Adhikary & Batra, 2020). Large datasets from sensors and meteorological and behavioral sources enable these data-driven models to detect nonlinear patterns and temporal shifts that traditional methods would overlook.

Wearable sensors are currently being developed to establish hyperlocal exposure assessment through their sensing capabilities. Several prototypes with integrated $PM_{2.5}$ sensors, VOC detectors, and CO_2 and temperature sensors are undergoing field tests for instant monitoring of workers at construction sites and industrial kitchens. The devices employ LoRa and BLE communication systems with solar power supply and microcontroller interfaces for low-energy data recording and communication. These advancements in exposure monitoring advancements show individual exposure monitoring progress through smart masks that measure lung function by analyzing MEMS microphones and spirometry analogs (Adhikary & Batra, 2020).

Eco-feedback mechanisms together with computational visualization tools serve as important tools to improve both awareness and behavioral change. OpenAir and QGIS together with browser-based interactive dashboards enable users to view pollutant concentration maps while also performing source attribution analysis and conducting comparative analytics. The platforms deliver essential data-driven services to researchers and policymakers, as well as community stakeholders, which enables them to adopt a participatory approach to managing air pollution. The foundation of modern air quality management consists of computational approaches that transform environmental science into adaptive predictive citizen-focused practices.

7.4 Integrative Methods for Air Pollution and Control

The complex and dynamic behavior of pollutants demands a multidimensional approach that integrates technological systems with computational frameworks and policymaking strategies for effective control of air pollution. The exclusive use of end-of-pipe technologies as a reactive approach no longer meets modern requirements. Modern integrative methods combine predictive modeling with health impact assessment and AI-driven cost optimization and adaptive solutions to develop a resilient holistic air quality management system.

A successful integration system requires the examination of various pollution-causing elements that create the problem. Research by Rautela and Goyal (2024) demonstrate that Indian regions experience pollution from multiple sources, including vehicles along with power plants and industrial facilities, and biomass combustion and environmental changes. The study revealed that the concentration of $PM_{2.5}$ showed strong spatiotemporal patterns due to the presence of organic carbon (OC) and sea salt and black carbon (BC) and dust and sulfates. Using satellite-based surface mass concentration data and AI-enabled convolutional autoencoder models, they predicted the $PM_{2.5}$ distribution with remarkable precision. Their model scoring parameters, including a Structural Similarity Index (SSIM >0.60) and the peak signal-to-noise ratio (28–30 dB), confirmed the efficacy of deep learning systems for air quality forecasting. These predictive tools enable the development of early warning systems along with dynamic urban planning and policy-driven emission reduction strategies, particularly in the Indo-Gangetic Plain due to its high pollution risk zone.

The integrative approach simultaneously links policy development with modeling methodologies. Qiu et al. (2022) introduced a two-objective optimization model aimed at regional $PM_{2.5}$ control. The system combines the results of the health impact assessment results from meta-analysis with economic cost optimization techniques to address the dual challenges of environmental pollution and health impacts from air pollution. The first objective function aims to decrease mortality rates from exposure to $PM_{2.5}$ using exposure–response coefficients obtained through meta-analysis. The second goal is to reduce the overall control system cost over multiple time frames. Policymakers achieve efficient control pathways through dual-objective optimization by weighing population health benefits against economic costs. This methodological framework was applied in the Nanshan District of Shenzhen, China, offering a transferable model for urban air quality governance. Advanced machine learning systems, together with intelligent hybrid methods, play a vital role in integrated air pollution control strategies. Multiple machine learning models including Random Forest (RF), Gaussian Process (GP), Adaptive Neuro-Fuzzy Inference System (ANFIS), and Multilayer Perceptron (MLP) were studied by Mottahedin et al. (2024) for their capability to forecast pollutant concentrations in Semnan, Iran. The ANFIS model produced the best results by demonstrating high reliability when processing seasonal datasets. Hybrid learning approaches used historical weather patterns and pollution records to achieve exact $PM_{2.5}$ forecasts. The models showed superior performance compared to conventional statistical models when detecting complex pollutant patterns, as well as short-term changes in pollutant levels. A major transformation in air pollution control has been achieved through the combination of artificial intelligence models with health-based optimization and environmental sensing technologies. AI algorithms receive real-time data from Continuous Ambient Air Quality Monitoring Systems (CAAQMS) that operate as sensor networks to create automated feedback systems. These systems enable government institutions to enforce regulatory actions through emission restrictions combined with traffic route adjustments and factory operation planning. Visualization tools consisting of geospatial dashboards and eco-feedback platforms let communities become involved in environmental protection efforts as this important yet frequently overlooked element of integrated control approaches.

Despite technological advances, challenges still exist. Rautela and Goyal (2024) stress the need for well-established institutional frameworks that need to be enforced consistently to enhance the effectiveness of technological advancements. Qiu et al. (2022) show that economic costs must be properly balanced with health advantages during model implementation. Mottahedin et al. (2024) emphasize that AI-driven systems require better data infrastructure and financial investments in impoverished areas to reach their maximum potential. Air pollution control systems integrate computational intelligence with health impact evaluation and economic optimization and legislative frameworks through cohesive methods. The integration of these methods creates a unified system that transforms how future cities manage their air quality through synchronized instruments. These methods will advance as long as developers prioritize scalability together with inclusivity and cross-sectorial collaboration to achieve sustainable and fair air pollution reduction.

7.5 Decision Support Systems and Policy Modeling

Modern air pollution control practices have moved past traditional engineering methods to implement integrated data-based policies that support sustainable air quality management. The development of air quality management decisions is heavily dependent on two crucial instruments known as decision support systems (DSS) and policy modeling tools. The tools merge environmental data with simulation models and health risk assessments and socioeconomic evaluations into one framework that boosts the performance and speed of pollution mitigation approaches. DSS platforms deliver systematic instructions for the evaluation of complex air pollution scenario through the combination of real-time data analysis with forecasting models and optimization algorithms. Yerramilli et al. (2011) demonstrated that GIS-based DSS systems use atmospheric dispersion models and meteorological data and emission inventories and land use parameters to determine spatial air quality risks. The systems enable authorities to predict pollution hotspots through scenario-based decision-making while assessing different mitigation strategy outcomes.

The development of AirNet by Rahman et al. (2024) involved building a machine learning-driven DSS with Support Vector Machines, Random Forest, and Decision Trees implemented through a web-based Django interface. The system uses environmental data from 23,000 cities around the world to generate fast air quality forecasts that allow local officials and public users to initiate immediate actions.

Specific DSS models developed for industrial purposes concentrate on sectors that produce high emissions. The intelligent DSS framework for coal-fired power plants enables predictive emission management along with real-time stack monitoring and compliance tracking. These systems combine combustion parameters with operational efficiencies and environmental benchmarks to improve decision-making accuracy. The Iranian oil and gas industry adopted a DSS-based sustainability prioritization tool as described by Tayebi et al. (2022). The system assesses air pollution control technologies through a combination of efficiency factors including pollutant reduction, energy usage, economic viability, and carbon emission reduction. The evaluation of air quality objectives by industries is highly dependent on these prioritization models for achieving carbon management objectives. Urban areas need hybrid DSS policy modeling systems because of their dense population and complex origins of pollution. The development of a DSS to evaluate PM_{10} and $PM_{2.5}$ particulate matter control strategies demonstrated significant success in Thessaloniki Greece (Vlachokostas et al., 2009). The system analyzed vehicle emissions along with factory and residential emissions to identify the most affordable control methods. Local authorities used this method to rank control alternatives according to their combined environmental impact and social cost.

Rautela and Goyal (2024) demonstrated the power of integrating AI and machine learning (ML) with DSS frameworks in India. The $PM_{2.5}$ prediction model based on convolutional autoencoders produced accurate results through its analysis of aerosol components such as black carbon and dust and sea salt and sulfates. The approach received enhanced evaluation through the application of SSIM and PSNR spatial metrics. The authors demonstrated how the integration of AI-DSS offers potential for extensive air quality prediction throughout the Indo-Gangetic Plain region, which faces severe pollution problems. The dual-objective DSS model presented by Qiu et al. (2022) combines policy modeling techniques with health-risk optimization processes and economic cost optimization. This model, applied in Shenzhen, China, used exposure–response coefficients derived from meta-analysis to calculate population mortality risks associated with $PM_{2.5}$. The second objective of the model addressed the total system cost of implementing various emission reduction measures. This dual-objective modeling enables decision-makers to align public health goals with budgetary realities, ensuring an equitable and economically sustainable air quality strategy. Integrative DSS and policy modeling approaches are no longer optional but essential instruments in the global response to air pollution. The amalgamation of domain-specific models, AI-driven analysis, health-economic optimization, and regulatory scenario simulation within DSS architecture offers unparalleled precision, adaptability, and stakeholder inclusiveness. The future lies in the development of cloud-based, interoperable DSS platforms, embedding real-time air quality monitoring, AI prediction algorithms, and interactive policy dashboards—all contributing to the solution to resilient, data-informed environmental governance.

7.6 Case Studies and Applications

Real-world applications in different sectors and geographies demonstrate the practical implementation of computational methods for air pollution control strategies. The combination of CFD, GIS, ML, and integrated modeling systems enables pollution control approaches that are both data-driven and cost-effective and adaptive to different environments.

7.6.1 Urban Air Quality Management Using Integrated Models

Cities undergoing rapid urbanization need integrated solutions because their sources of pollution include emissions from vehicles, industries, construction activities, and household fuel consumption. The cities of Beijing, Delhi, and Los Angeles have established air quality forecasting models with GIS mapping and machine learning prediction tools to create low-emission zones and implement public health alerts and traffic rerouting plans. The Urban Emission Model (SIM-air) provides Southeast Asia and India with a system that links emissions data to dispersion models and health impact assessments to help local governments make effective decisions. These tools support local governments in creating action plans through initiatives such as NCAP (National Clean Air Program) in India for each city.

7.6.2 Industrial Emission Control Planning Using CFD and AI

The industrial areas in the urban and peri-urban regions generate most of the air pollution in these areas. CFD modeling helps industrial facilities improve pollutant distribution patterns from chimney cooling towers and stack emissions to maximize their plant design efficiency and emission reduction systems. A Vietnamese cement plant conducted CFD simulations that led to stack height and placement changes resulting in a 23% decrease in PM concentrations in localized areas. Applied artificial intelligence optimization algorithms are implemented in Middle Eastern oil refineries through Genetic Algorithms (GA) and Particle Swarm Optimization (PSO), which reduce emissions by controlling fuel usage and implementing process adjustments.

7.6.3 Regional Pollution Transport Modeling and Transboundary Analysis

Air pollutants move between different administrative areas and countries, so regional modeling systems become essential for analysis. The pollution transport modeling system HYSPLIT (Hybrid Single-Particle Lagrangian Integrated Trajectory Model) and WRF-Chem (Weather Research and Forecasting Model coupled with Chemistry) are used extensively for modeling pollutant transport across regions. Transboundary haze episodes caused by Indonesian biomass burning in Southeast Asia have been studied using HYSPLIT modeling and satellite data research by Promsiri et al. (2022). The models offered essential proof that supported ASEAN dialogues and joint mitigation protocols on border crossing warning systems and fire management teamwork.

7.6.4 Use of AI and ML in Low- and Middle-Income Countries

The affordability and scalability of AI and ML tools have shown promise in low- and middle-income countries, where conventional monitoring infrastructure is often limited. Countries such as Bangladesh, Nigeria, and Nepal have adopted low-cost sensor networks combined with machine learning algorithms to map urban pollution patterns and conduct exposure risk assessments. Hasan et al. (2024) selected the most effective forecasting models for PM_{10} and $PM_{2.5}$ concentrations in Dhaka, Bangladesh, through machine learning and statistical analysis that supports active air quality management and policy development (Table 7.1).

Table 7.1 Case studies on integrative computational approaches

Set of information	Model/ algorithm	Pollutants studied	Outcome	References
6 years of Indian city data	Gaussian naive Bayes, XGBoost	$PM_{2.5}$, NO_2, SO_2	XGBoost achieved the best AQI prediction accuracy among tested models	Kumar and Pande (2022)
Google street view & EDF	Ridge regression, gradient boosting	NO_2, black carbon	Predicted pollutant concentrations in urban neighborhoods with high precision	Gopalakrishnan (2021)
Taiwan EPA and weather bureau	AdaBoost	PM_{10}, $PM_{2.5}$, CO	AdaBoost outperformed other models for seasonal AQI forecasting	Liang et al. (2020)
US EPA	Support vector regression	CO, NO_2, SO_2	Established an innovative method for hourly atmospheric pollution modeling	Castelli et al. (2020)
Taiwan air quality monitoring network	Random Forest, XGBoost	$PM_{2.5}$	Identified close alignment between predicted and actual pollutant levels	Doreswamy et al. (2020)

7.7 Future Advancing AI Integration for Air Quality Prediction

Future research must concentrate on integrating AI with heterogeneous data sources, for instance satellite imagery, meteorological conditions, social media feeds, and citizen-stated pollution observations. This will enable a better comprehensive understanding of air pollution dynamics. Machine learning models, particularly deep learning architectures, need to be enhanced to process. The patterns of air pollution exist under the combined effects of climate change together with urbanization processes and industrial activities. The fixed nature of AI models prevents them from successfully tracking modifications in pollution levels. Future research should focus on developing AI models with learning capabilities that adapt to new data points in real time. AI agents can develop through reinforcement learning to offer real-time pollution control strategies (Chadalavada et al., 2025). Multimodal data streams efficiently, allowing improved spatial and temporal determination in air pollution predictions (Chadalavada et al., 2025).

7.7.1 *Explainable AI and Uncertainty Quantification*

Although AI models have demonstrated high predictive accuracy, their "black box" nature restricts their applicability in policy decision-making. Future research should explore explainable AI techniques such as Local Interpretable Model-agnostic

Explanations (LIME) and SHAP (SHapley Additive exPlanations) to recognize key predictors and sources of pollution. Additionally, Bayesian deep learning approaches must be refined to quantify prediction uncertainties, enhancing the reliability of AI-driven air quality assessments (Chadalavada et al., 2025).

7.7.2 *Hybrid Modeling Approaches*

Computational air quality models should integrate physics-based approaches (such as chemical transport models) with ML-driven predictive techniques. Hybrid models that combine Gaussian plume dispersion modeling with AI-driven parameter optimization can amplify real-time monitoring capabilities. This will aid in linking the gap among theoretical simulations and actual world applications (Badawy, 2025).

7.7.3 *Incorporation of Emerging Pollutants*

Current air pollution models primarily focus on classical pollutants such as NO_x, $PM_{2.5}$, and SO_2. Future research must incorporate emerging pollutants, including nanoparticles and persistent organic pollutants, into computational models. Advanced sensor technology and data assimilation techniques will be critical in detecting and quantifying these underrepresented contaminants (Badawy, 2025).

7.7.4 *Dynamic and Adaptive AI Models*

Air pollution patterns are influenced by climate change, urbanization, and industrial activities. Static AI models struggle to capture evolving pollution trends. Therefore, future research should prioritize self-learning AI models that continuously modify new data. Reinforcement learning techniques can be employed to develop AI agents that suggest optimal pollution control strategies in real-time (Chadalavada et al., 2025).

7.7.5 *Blockchain for Decentralized Air Quality Monitoring*

Data reliability and transparency remain significant challenges in air pollution research. Future studies should explore the integration of blockchain technology to generate decentralized air quality monitoring networks. This approach could improve data authenticity, facilitate cross border pollution tracking, and enhance collaborative environmental policymaking. (Chadalavada et al., 2025).

7.7.6 Enhancing GIS-Based Decision Support Systems

Geographic information systems (GIS)-based decision support systems (DSS) should be further developed to include real-time air quality data. These systems can be integrated with traffic control measures, industrial emission regulations, and land-use planning to optimize urban air quality management strategies (Yerramilli et al., 2011).

7.7.7 Policy-Oriented Computational Models

To facilitate evidence-based policymaking, computational air quality models should integrate socioeconomic and health impact assessments. Future models should include public health information combined with economic loss assessments from pollution and mitigation expense assessments for making legislative decisions (Badawy, 2025). Computational methods for air pollution control will evolve toward better policy relevance and interpretability through these research advancements, which will produce more effective global air quality management strategies.

7.8 Conclusion

The incorporation of computational methods into air pollution modeling and control represents a major transformative step from traditional reactive approaches to proactive data-driven adaptive frameworks. The combination of machine learning, artificial intelligence, chemical transport models, decision support systems, and geographic information systems in this interdisciplinary methodology produces deeper pollution understanding and actionable interventions.

These computational strategies possess the potential to transform current systems through their precise execution and efficient processing and scalable capabilities. The combination of real-time sensor networks and cloud computing and multiscale simulations and blockchain and quantum computing enables policymakers to develop specific mitigation strategies that match contexts. Additionally, it enables better public health results by adding health information and offering solid risk evaluations for at-risk groups. The path forward includes addressing transparency requirements for AI models along with dataset accessibility and cost-effective computational solutions. The resolution of these barriers depends on interdisciplinary teamwork and initiatives to build capacities together with international partnerships. Open-source platforms together with participatory frameworks serve as essential tools for resource-limited regions to gain access to advanced technologies. The integration of computation serves as a fundamental requirement for achieving sustainable and effective air quality management. The approaches connect scientific discoveries with real-time monitoring systems to policy execution to create methods for global air quality improvement and community wellness. The achievement of this vision demands sustained financial backing for computational research together with stakeholder engagement and dedication to environmental sustainability and equity.

Acknowledgments The authors sincerely thank the Department of Botany at University of Kalyani, India, for offering important resources alongside complete backing and an academic space that produced this chapter.

References

Adhikary, R., & Batra, N. (2020). Computational tools for understanding air pollution. In *Adjunct proceedings of the 2020 ACM international joint conference on pervasive and ubiquitous computing and proceedings of the 2020 ACM international symposium on wearable computers*. https://doi.org/10.1145/3410530.3414327

Amann, M., Bertok, I., Borken-Kleefeld, J., Cofala, J., Heyes, C., Höglund-Isaksson, L., Klimont, Z., Binh, N., Posch, M., Rafaj, P., Sander, R., Schoepp, W., Wagner, F., & Winiwarter, W. (2011). Cost-effective control of air quality and greenhouse gases in Europe: Modeling and policy applications. *Environmental Modelling and Software, 26*, 1489–1501. https://doi.org/10.1016/j.envsoft.2011.07.012

Antuña, Y. E., Pacheco, F., Manuel, V., Álvarez, E., Carús-Candás, J., López, S., & José, M. (2023). A CFD-based methodology for impact assessment of industrial emissions and evaluation of mitigation measures for regulatory purposes. *Process, 11*, 2039. https://doi.org/10.3390/pr11072039

Badawy, H. (2025). Recent trends in spatial modeling studies of air quality (2010–2023). *Earth Science Informatics, 18*. https://doi.org/10.1007/s12145-024-01667-y

Briggs, D., Collins, S., Elliott, P., Fischer, P., Kingham, S., Lebret, E., Pryl, K., Reeuwijk, H., Smallbone, K., & Veen, A. (1997). Mapping urban air pollution using GIS: A regression-based approach. *International Journal of Geographical Information Science, 11*, 699–718. https://doi.org/10.1080/136588197242158

Castelli, M., Clemente, F., Popovič, A., Silva, S., & Vanneschi, L. (2020). A machine learning approach to predict air quality in California. *Complexity, 2020*, 1–23. https://doi.org/10.1155/2020/8049504

Chadalavada, S., Faust, O., Salvi, M., Seoni, S., Raj, N., Raghavendra, U., Gudigar, A., Barua, P., Molinari, F., & Acharya, R. (2025). Application of artificial intelligence in air pollution monitoring and forecasting: A systematic review. *Environmental Modelling and Software, 185*, 106312. https://doi.org/10.1016/j.envsoft.2024.106312

Cohen, A., Brauer, M., Burnett, R., Anderson, H., Frostad, J., Estep, K., Balakrishnan, K., Brunekreef, B., Dandona, R., Feigin, V., Freedman, G., Hubbell, B., Jobling, A., Kan, H., Knibbs, L., Liu, Y., Martin, R., Morawska, L., & Forouzanfar, M. (2017). Estimates and 25-year trends of the global burden of disease attributable to ambient air pollution: An analysis of data from the global burden of diseases study 2015. *The Lancet, 389*. https://doi.org/10.1016/S0140-6736(17)30505-6

Doreswamy, H., Harishkumar, K. S., Km, Y., & Gad, I. (2020). Forecasting air pollution particulate matter (PM2.5) using machine learning regression models. *Procedia Computer Science, 171*, 2057–2066. https://doi.org/10.1016/j.procs.2020.04.221

El-Harbawi, M. (2013). Air quality modelling, simulation, and computational methods: a review. *Environmental Reviews, 21*, 149–179. https://doi.org/10.1139/er-2012-0056

Gopalakrishnan V. (2021). Hyperlocal air quality prediction using machine learning. *Towards data science*. https://towardsdatascience.com/hyperlocalairqualitypredictionusingmachine-learning-ed3a661b9a71

Hasan, K., Rahman, M., Akhter, M., Mohinuzzaman, M., Kayes, I., & Rahman, S. (2024). A new dynamic approach using data-driven and machine learning models for particulate matter forecasting in Dhaka megacity. *Environmental Pollution and Management, 1*, 235–247. https://doi.org/10.1016/j.epm.2024.11.005

Holmes, N., & Morawska, L. (2006). A review of dispersion modelling and its application to the dispersion of particles: An overview of different dispersion models available. *Atmospheric Environment, 40*, 5902–5928. https://doi.org/10.1016/j.atmosenv.2006.06.003

Kumar, K., & Pande, B. (2022). Air pollution prediction with machine learning: A case study of Indian cities. *International Journal of Environmental Science and Technology, 20*. https://doi.org/10.1007/s13762-022-04241-5

Liang, Y.-C., Maimury, Y., Chen, A., & Juarez, J. (2020). Machine learning-based prediction of air quality. *Applied Sciences, 10*, 9151. https://doi.org/10.3390/app10249151

Mottahedin, P., Chahkandi, B., Moezzi, R., Fathollahi-Fard, A. M., Ghandali, M., & Gheibi, M. (2024). Air quality prediction and control systems using machine learning and adaptive neuro-fuzzy inference system. *Heliyon, 10*, e39783. https://doi.org/10.1016/j.heliyon.2024.e39783

Promsiri, P., Tekasakul, S., Thongyen, T., Suwattiga, P., Morris, J., Latif, M. T., Tekasakul, P., & Dejchanchaiwong, R. (2022). Transboundary haze from peatland fires and local source-derived PM2.5 in southern Thailand. *Atmospheric Environment, 294*, 119512. https://doi.org/10.1016/j.atmosenv.2022.119512

Qiu, B., Zhou, M., Qiu, Y., Ma, Y., Ma, C., Tu, J., & Li, S. (2022). An integration method for regional PM2.5 pollution control optimization based on meta-analysis and systematic review. *International Journal of Environmental Research and Public Health, 19*(344). https://doi.org/10.3390/ijerph19010344

Rahman, M., Nayeem, M., Ahmed, M., Tanha, K., Sakib, S., Uddin, M., Mohammad, K., & Hasan Babu, H. (2024). AirNet: Predictive machine learning model for air quality forecasting using web interface. *Environmental Systems Research, 13*. https://doi.org/10.1186/s40068-024-00378-z

Rautela, K. S., & Goyal, M. K. (2024). Transforming air pollution management in India with AI and machine learning technologies. *Scientific Reports, 14*, 20412. https://doi.org/10.1038/s41598-024-71269-7

Tayebi, M., Bemani, A., Fetanat, A., & Fehresti-Sani, M. (2022). A decision support system for sustainability prioritization of air pollution control technologies in energy and carbon management: Oil & gas industry of Iran. *Journal of Natural Gas Science and Engineering, 99*, 104416. https://doi.org/10.1016/j.jngse.2022.104416

Vlachokostas, C., Achillas, C., Moussiopoulos, N., Hourdakis, E., Tsilingiridis, G., Ntziachristos, L., Georgios, B., Stavrakakis, N., & Sidiropoulos, C. (2009). Decision support system for the evaluation of urban air pollution control options: Application for particulate pollution in Thessaloniki, Greece. *The Science of the Total Environment, 407*, 5937–5948. https://doi.org/10.1016/j.scitotenv.2009.07.040

WHO. (2021). *Air pollution*. World Health Organization. https://www.who.int/news-room/fact-sheets/detail/ambient-(outdoor)-air-quality-and-health

Yadav, A. K. (2015). Elemental composition and source apportionment of suspended particulate matters and health risk assessment in mining and nonmining areas of Odisha, India. *Journal of Hazardous, Toxic, and Radioactive Waste, 19*(3), 04014037.

Yadav, A. K., Gupta, K., Bodhale, R., & Phuleria, H. C. (2025). Metals and trace element associated health risk in size-segregated household dust of an industrial town in India. *Air Quality, Atmosphere & Health*. https://doi.org/10.1007/s11869-025-01794-5

Yadav, A. K., Sahoo, S. K., Dubey, J. S., Kumar, A. V., Pandey, G., & Tripathi, R. M. (2019). Assessment of particulate matter, metals of toxicological concentration, and health risk around a mining area, Odisha, India. *Air Quality, Atmosphere & Health, 12*(7), 775–783.

Yadav, A. K., Sahoo, S. K., Patra, A. C., Dubey, J. S., Lenka, P., Sagar, D. V., Kumar, A. V., & Tripathi, R. M. (2014). Source identification of particulate matter and associated intake of elements through inhalation in an industrial area of Odisha, India. *Toxicological & Environmental Chemistry, 96*(3), 410–425.

Yadav, A. K., Shirin, S., Emmanouil, C., & Jamal, A. (2022). Effect of seasonal and meteorological variability of air pollution in Singrauli coalfield. *Aerosol Science and Engineering, 6*, 61–70.

Yerramilli, A., Dodla, B. R., & Yerramilli, S. (2011). *Air pollution, modeling and GIS based decision support systems for air quality risk assessment*. IntechOpen. https://doi.org/10.5772/22055

Chapter 8
Air Pollution Modeling Using Computational Approaches

Sanjeev Kumar Pandey and Riddhi Mohan Bora

8.1 Introduction

8.1.1 Background

Air pollution is a substantial global concern as a result of its detrimental impacts on human health, ecosystems, and climate change (Kampa and Castanas 2008). Research indicates that long-term exposure to air pollutants can contribute to respiratory diseases, cardiovascular problems, and even premature death (Bernstein et al., 2004; Brunekreef & Holgate, 2002; Vallero, 2014). Furthermore, contaminants such as particulate matter (PM), nitrogen oxides (NO_x), sulfur dioxide (SO_2), and volatile organic compounds (VOCs) contribute substantially to environmental deterioration, adversely impacting biodiversity and agricultural production. Climate change is another major consequence of air pollution, as greenhouse gases (GHGs) and aerosols alter atmospheric composition, leading to an increase in global temperature and unpredictable weather patterns.

One major factor contributing to air pollution is the rapid growth of industrial activities, urbanization, and transportation. Primary sources of air pollution can be broadly classified into five major sectors: transportation, industrial emissions, agriculture, residential sources, and natural sources, as illustrated in Fig. 8.1. Transportation accounts for approximately 29% of total emissions, mainly due to vehicle exhaust containing carbon monoxide (CO), nitrogen oxides (NO_x), and particulate matter. Industrial emissions are the most significant contributor, accounting for 32% of total pollution, as manufacturing processes and fossil fuel combustion

S. K. Pandey (✉) · R. M. Bora
Department of Electrical Engineering, Indian Institute of Technology Delhi, New Delhi, India
e-mail: sanjeev@ee.iitd.ac.in

A. K. Yadav (ed.), *Computational Approach for Air Pollution*,
https://doi.org/10.1007/978-3-032-09215-1_8

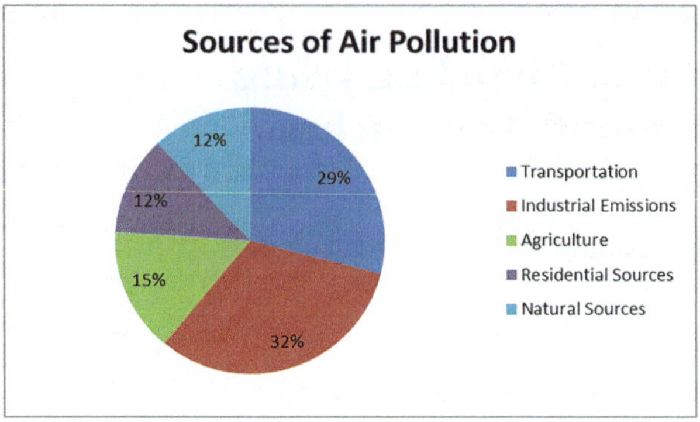

Fig. 8.1 Breaking down air pollution: a visual insight into its major sources

release large amounts of pollutants into the atmosphere. Agricultural activities contribute 15%, primarily through ammonia (NH_3) emissions from fertilizers and methane (CH_4) from livestock. Residential sources, such as household heating and cooking, account for 12%, while natural sources, including wildfires and volcanic eruptions, also contribute 12% (Boubel et al., 2013; Mayer, 1999; Zannetti, 2013).

Computational models play a vital role in air pollution research, as they allow scientists and policymakers to assess air quality, detect pollution hotspots, and design targeted mitigation strategies. These models take advantage of mathematical formulations to describe how pollutants disperse in the atmosphere, accounting for various physical and chemical interactions. Unlike traditional analytical methods, which struggle to capture the nonlinear and dynamic nature of pollutant transport, computational models provide a more robust and accurate framework.

The dispersion of pollutants is affected by various factors, including meteorological conditions (wind velocity, outside temperature, and humidity levels), topographical features (mountains, valleys, and urban structures), and chemical transformations (reactions among pollutants in the atmosphere) (Baklanov, 2000; Yadav, 2015; Yadav et al., 2014, 2025; Zlatev, 1995). Table 8.1 provides a structured comparison of these numerical methods, highlighting their respective strengths and weaknesses. Although traditional numerical methods remain valuable due to their well-established theoretical foundations, deep learning-based approaches offer a promising avenue for improving the predictive capabilities of air pollution models, particularly in handling large-scale and complex atmospheric interactions. Numerical simulation techniques have been widely adopted to overcome the limitations of conventional analytical approaches. These techniques involve solving complex governing equations, such as the Navier–Stokes equations for fluid dynamics and the advection–diffusion equations for pollutant transport, using advanced computational algorithms (Dimov et al., 2004; Rastogi et al., 2025; Sanín & Montero, 2007; Wu et al., 2011; Zlatev & Berkowicz, 1988). Researchers can generate

Table 8.1 Comparison of numerical methods for solving advection diffusion equations

Finite difference method	Finite element method	Alternating direction implicit	Deep learning *n*/*n*	References
Structured grids only	Unstructured grids supported	Structured grids only	Mesh-free (no grid required)	Olaiju et al. (2018), Salman et al. (2022)
First second order accuracy	2nd higher order accuracy	2nd order accuracy	High data-driven accuracy	Szymkiewicz (2010), Zhang (2009)
Conditional (CFL compliance)	Unconditional	Unconditional	Unconditional	Salman et al. (2022)
Low computational cost	Moderate computational cost	Moderate (requires matrix inversion)	High (training cost dominates)	Mazumder (2015), Özişik et al. (2017)
Limited to structured grids	Excellent for irregular domains	Limited to structured grids	Excellent	Thalib and Bakar (2021)
Not directly addressed	Supported (weak formulation)	Not directly addressed	Excellent (learns nonlinear patterns)	Gill et al. (2025)
Simple discretization	Moderate (basis functions, weak form)	Moderate (splitting schemes)	High (requires training data and optimization)	Pepper and Heinrich (2005), Wen (2024)
Moderate scalability	Moderate scalability	Moderate scalability	High scalability	Özişik et al. (2017)
Simple, efficient for structured grids	Flexible for irregular domains	Unconditionally stable	High accuracy, handles coarse grids well	Salman et al. (2022), Olaiju et al. (2018)
Mesh dependency, numerical diffusion	Computationally expensive for large grids	Requires structured grids	Training data-dependent, high computational cost	Mazumder (2015), Thongmoon et al., (2006).

NS, not specified, *NDA* not directly addressed, *NDC* not directly considered

high-resolution pollution maps and forecast air quality under various conditions discussed in Table 8.1 by applying numerical methods such as finite element analysis, finite volume methods, and machine learning-based predictive models. This advancement in computational modeling has significantly enhanced our ability to formulate data-driven policies and implement effective air pollution control measures. Thus, atmospheric science, computer modeling, and data analytics are needed to combat air pollution. Researchers can better understand pollutant behavior, forecast air quality trends, and support evidence-based decision-making to reduce the negative effects of air pollution on human health and the environment by using advanced numerical approaches and predictive models.

8.1.2 Motivation

Analyzing and studying air pollution is complex in its own right due to the interplay of different atmospheric processes. These include advection, diffusion, chemical interactions, and turbulence. All of these processes affect how pollution is distributed and deposited in ways that are too complex for analytical models to be accurate. Historically and modern approaches to evaluating air and environmental concerns have been based on the collection and analysis of quantitative data. However, by observation, analyzing these processes through observation yields nonlinear and dynamic results. There is no capture of the essence of pollutant transportation, which is why greater computational capabilities are required.

Advanced numerical methods have become indispensable tools for air pollution. These methods transform complex governing equations, such as the Navier–Stokes equations for fluid motion and advection–diffusion equations for pollutant transport, into discrete formulations that can be solved iteratively. By discretizing continuous equations, numerical techniques enable researchers to obtain approximations of pollutant concentrations across spatial and temporal domains. This computational approach is instrumental in simulating several types of environmental scenarios, including examining the effect of meteorological conditions on the amount of pollution, the spread of contaminants in urban environments, and the effectiveness of emission control measures (Salman et al., 2022; Yadav et al., 2022).

Air pollution modeling has been transformed by recent developments in machine learning and high-performance computing (HPC). Due to the high dimensionality and complexity of actual pollution problems, traditional numerical methods frequently encounter computational limitations. However, the inclusion of HPC methods has dramatically increased the efficiency of these models, allowing high-resolution simulations to be hitherto unfeasible. Furthermore, machine learning methods have been progressively used to improve prediction ability with large datasets and intricate links between pollution sources and atmospheric conditions. In particular, deep learning models have shown a remarkable ability for predicting air quality and pollution control strategies (Salman et al., 2022).

Since computational techniques are crucial for air quality, this chapter seeks to give an in-depth examination of the primary numerical techniques applied in air pollution modeling. The finite difference method (FDM), finite element method (FEM), Monte Carlo simulations, and computational fluid dynamics (CFD) (Baklanov, 2000; Sanín & Montero, 2007) are discussed. Each method is discussed in terms of its analytical foundation, computational implementation, and real-world applications. The chapter also highlights the advantages and limitations of each approach, providing information on its suitability for different air pollution modeling scenarios.

This chapter aims to bridge the distance between theoretical atmospheric science and useful environmental management by offering a complete study of computational methods. By thorough awareness of these numerical techniques, researchers, legislators, and engineers can create more efficient air pollution mitigating systems. This contributes to better public health and air quality.

8.1.3 Focus Areas

The field of air pollution modeling encompasses a wide range of computational techniques, each tailored to address specific challenges in simulating pollutant behavior. Different numerical approaches have special benefits in terms of computing efficiency, accuracy, and adaptation to real-world scenarios, considering the complexity of atmospheric processes and the varied character of pollution transport mechanisms. This chapter focuses on five key computational approaches that are widely used in air pollution modeling, each of which plays a crucial role in advancing our understanding of air quality dynamics.

- Finite difference method: A widely used numerical technique for solving partial differential equations (PDEs) in structured domains.
- Finite element method: A versatile approach for handling irregular geometries and unstructured domains.
- Monte Carlo simulation: A probabilistic method for uncertainty quantification and risk assessment.
- Computational fluid dynamics: A high-resolution modeling technique for simulating airflow and pollutant dispersion in urban environments.
- Pollutant dispersion and atmospheric modeling: Uses of computer techniques to comprehend and forecasting environmental pollution behavior.

8.2 Finite Difference Method

8.2.1 Overview

The finite difference method is one of the oldest and most direct approaches to discretizing partial differential equations (PDEs) (Johari et al., 2018; Salman et al., 2022; Sanín & Montero, 2007). It is particularly suited for solving PDEs in bounded domains and has been extensively used in air pollution modeling to simulate pollutant transport governed by the advection–diffusion equation. One of the first and most straightforward approaches to discretize partial differential equations (Johari et al., 2018; Salman et al., 2022; Sanín & Montero, 2007) is the finite difference method. It has been extensively applied in air pollution modeling to replicate pollution transport controlled by the advection–diffusion equation and is especially appropriate for solving PDEs in limited domains. By means of finite differences, FDM transforms the continuous derivatives in PDEs into discrete approximations, hence producing algebraic equations with numerical solvable solutions. For structured grids, FDM performs effectively, is computationally efficient, and is easy to implement. Still, it calls for great attention to stability conditions, including diffusion number restrictions and the Courant–Friedrichs–Lewy (CFL) condition.

8.2.2 Mathematical Formulation

The advection–diffusion equation describes the transport of pollutants in the atmosphere, given by

$$\frac{\partial c}{\partial t} + u\frac{\partial c}{\partial x} = D\frac{\partial^2 c}{\partial x^2},$$

where $c(x, t)$ is the pollutant concentration, u is velocity of the airflow, D is diffusion coefficient, and t and x represent time and space, respectively.

In FDM, the derivatives are approximated using finite differences:

- Time derivative of the forward difference:

$$\frac{\partial c}{\partial t} \approx \frac{\left(c_i^{n+1} - c_i^n\right)}{\Delta t}.$$

- Space derivative (advection term) of the upwind scheme:

$$\frac{\partial c}{\partial x} \approx \frac{\left(c_i^n - c_{i-1}^n\right)}{\Delta x}.$$

- Second space derivative (diffusion term) of the central difference:

$$\frac{\partial^2 c}{\partial x^2} \approx \frac{\left(c_{i+1}^n - 2c_i^n + c_{i-1}^n\right)}{\Delta x^2}.$$

Substituting these approximations into the governing equation:

$$\frac{c_i^{n+1} - c_i^n}{\Delta t} + u\frac{c_i^n - c_{i-1}^n}{\Delta x} = D\frac{c_{i+1}^n - 2c_i^n + c_{i-1}^n}{\Delta x^2}.$$

Rearranging for c_i^{n+1}:

$$c_i^{n+1} = c_i^n - u\,\Delta t\frac{c_i^n - c_{i-1}^n}{\Delta x} + D\Delta t\frac{c_{i+1}^n - 2c_i^n + c_{i-1}^n}{\Delta x^2}.$$

8.2.3 Example: Numerical Simulation of Pollutant Transport

To demonstrate the application of FDM, we solve the 1D advection–diffusion equation for a pollutant released at $x = 0$ and transported downstream due to advection and diffusion. The parameters are:

Domain length: $L = 10$ m,

Total simulation time: $T = 5$ s
Velocity: $u = 1$ m/s,
Diffusion coefficient: $D = 0.1$ m²/s.
The initial and boundary conditions are the following:

$$c(x,0) = 0, \quad c(0,t) = 1, \quad c(L,t) = 0.$$

8.2.3.1 Numerical Implementation

The spatial and temporal domains are discretized with $\Delta x = 0.1$ m and $\Delta t = 0.01$ s. The stability criteria are checked and given by

$$C = \frac{u\Delta t}{\Delta x} = 0.1 \le 1, \; D_n = \frac{D\Delta t}{\Delta x^2} = 0.1 \le 0.5.$$

8.2.3.2 Results and Discussion

The numerical results obtained are compared with the analytical solution:

$$c(x,t) = \exp\left(-\frac{(x-ut)^2}{4Dt}\right)$$

The numerical results of MATLAB illustrate the pollutant concentration profile over a 10-m domain length (L) after a total simulation time of 5 s (T). The spatial grid is discretized into 100 points (N_x), and the temporal domain is divided into 500-time steps (N_t). The airflow velocity is set to $u = 1$ m/s, while the diffusion coefficient is $D = 0.1$ m²/s. These parameters are used to calculate the spatial step size ($\Delta x = L/N_x$) and the time step size ($\Delta t = T/N_t$). To ensure numerical stability, the Courant number $\left(CFL = u \cdot \frac{\Delta t}{\Delta x}\right)$ and the Peclet number $\left(Pe = u \cdot \frac{\Delta x}{D}\right)$ are computed. The CFL condition ensures that the information does not propagate further than one point in the grid per time step, while the Pe condition checks the balance between advection and diffusion. The code would terminate with an error message indicating unstable parameters, suggesting that Δt should be reduced or Δx increased, if $CFL > 1$ or $Pe > 2$. However, in this case, the chosen parameters satisfy the stability criteria. Using the finite difference scheme, the time-stepping loop updates the concentration at each grid point. Specifically, the concentration at any point i is updated based on its previous value, the upstream concentration, and the concentrations of neighboring points. This scheme accounts for both advection and diffusion effects. At each time step, the boundary conditions are enforced: a constant source of pollutants is maintained at $x = 0$, and zero concentration is assumed at $x = L$.

The resulting graph shows a smooth decay in pollutant concentration from the source at $x = 0$ toward $x = L$. Initially, the concentration is high near the source as a result of the continuous release of pollutants. The concentration decreases exponentially as we progress downstream, a phenomenon that is a result of the combined effects of advection, which transports pollutants away from the source, and diffusion, which disperses them. The effective transport and dispersion of pollutants are suggested by the abrupt decrease in concentration near the midpoint of the domain. Beyond approximately 7 m, the concentration drops to nearly zero, consistent with the boundary condition at $x = L$.

In summary, the numerical solution accurately captures the expected behavior of pollutant transport governed by the advection–diffusion equation:

$$\frac{\partial c}{\partial x} + u\frac{\partial c}{\partial x} = D\frac{\partial^2 c}{\partial x^2}.$$

The concentration profile (such as physical occurrences) exhibits a gradual decline from the source, confirming the efficacy of the finite difference method in modeling. Stability criteria were met, guaranteeing precise and precise results. Figure 8.2 illustrates the profile of the pollutant concentration at various time intervals. The bridge between numerical and analytical solutions is excellent, validating the accuracy of the FDM implementation. The results demonstrate how the pollutant spreads downstream as a result of advection and diffusion.

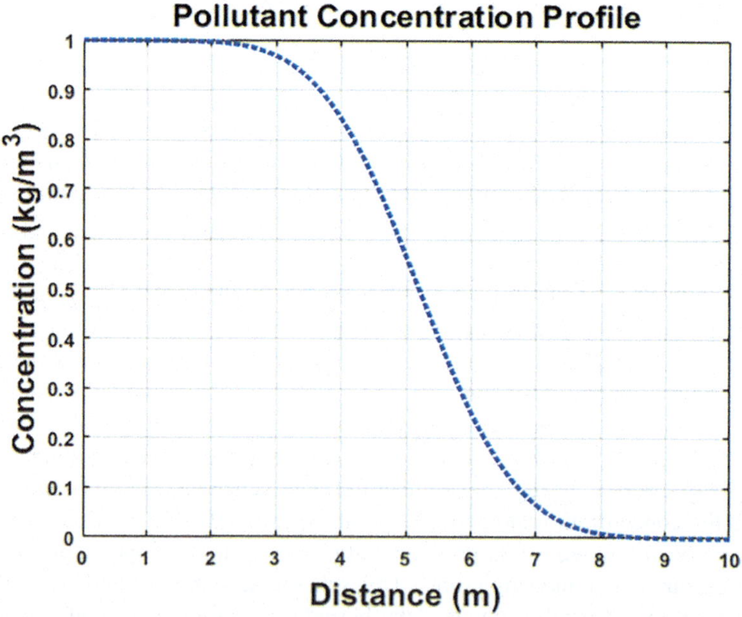

Fig. 8.2 Pollutant concentration profiles obtained using FDM at different time steps

8.2.3.3 Air Pollution Modeling Applications

FDM has been frequently used to simulate pollutant transport in a variety of scenarios, such as:

- Continuous pollutant release: Predicting the spatial and temporal distribution of pollutants emitted frequently from sources such as factories or vehicles.
- Fixed period release: Investigate the behavior of pollutants released over a specific period.
- Urban air quality assessment: Modeling the dispersion in urban environments to identify pollution hotspots.

Simplicity and computational efficiency make it suitable for structured grids and regular domains. However, its limitations include challenges in handling irregular geometries and the need for small time steps to ensure stability.

8.3 Finite Element Method

8.3.1 Overview

The finite element method (FEM) is a potent numerical technique for resolving partial differential equations (PDEs) (Albani et al., 2015), particularly in unstructured domains and complex geometries. Unlike the finite difference method (FDM), which relies on structured grids, FEM divides the domain into smaller elements and approximates the solution within each element using basis functions. This makes FEM highly versatile for modeling irregular geometries, such as those encountered in air pollution studies involving urban environments or topographically complex regions. FEM is widely used to model engineering and environmental systems due to its simplicity and ability to handle complex boundary conditions and nonlinearities. It is particularly suited for problems where the domain geometry is irregular or where high accuracy is required in localized regions.

8.3.2 Mathematical Formulation

The advection–diffusion equation governing pollutant transport is given as

$$\frac{\partial c}{\partial t} + u.\nabla c = D\nabla^2 c,$$

In this equation, c represents the concentration of the pollutant at a particular location x and time t; it indicates how much pollutant is present at that point in space and time. The symbol u denotes the velocity field of the surrounding medium, such as air or water, and describes the direction and speed with which the pollutant is transported. The parameter D is the diffusion coefficient, which quantifies the rate at which the pollutant spreads from higher concentration to lower concentration regions due to molecular diffusion. In FEM, the weak form of the governing equation is derived using variational principles. The solution is approximated as a linear combination of basis functions:

$$c(x,t) \approx \sum_{i=1}^{n} c_i(t)\phi_i(x),$$

where $c_i(t)$ are the unknown coefficients and $\phi_i(x)$ are the basis functions defined over the finite elements, t represents time, and x denotes the spatial position, which may be one-, two-, or three-dimensional depending on the specific context of the problem. The weak form of the advection–diffusion equation is obtained by multiplying the governing equation by a test function $v(x)$ and integrating it over the domain:

$$\int_\Omega v\frac{\partial c}{\partial t}\, d\Omega + \int_\Omega v u.\nabla c\, d\Omega = \int_\Omega v D\nabla^2 c\, d\Omega.$$

Using integration by parts, second-order derivatives are reduced to first-order terms, resulting in

$$\int_\Omega v\frac{\partial c}{\partial t} d\Omega - \int_\Omega \nabla v.(uc)d\Omega + \int_\Omega \nabla v.(D\nabla c)d\Omega = 0.$$

This weak form is then discretized using the finite element basis functions, leading to a system of algebraic equations:

$$\mathbf{M}\frac{d\mathbf{c}}{dt} + \mathbf{Kc} = \mathbf{f},$$

In this discretized form of the advection–diffusion equation, $\boldsymbol{M}$ represents the mass matrix, which arises from the time-dependent part of the weak form and accounts for how the concentration is distributed over the elements. The matrix $\boldsymbol{K}$ is the stiffness matrix, which comes from the spatial derivative terms and reflects the influence of diffusion and advection across the elements. Vector $\mathbf{c}$ contains the nodal concentrations, representing unknown pollutant concentrations at each node in the finite element mesh. Finally, $\mathbf{f}$ is the forcing vector, including contributions from any source terms or boundary conditions applied to the system.

8.3.3 Example: Numerical Simulation of Pollutant Transport

To demonstrate the application of FEM, we solve the 1D advection–diffusion equation for a pollutant released at $x = 0$ and transported downstream due to advection and diffusion. The parameters are:

Domain length: $L = 10$ m,
Total simulation time: $T = 5$ s,
Velocity: $u = 1$ m/s,
Diffusion coefficient: $D = 0.1$ m^2/s.
The initial and boundary conditions are

$$c(x,0) = 0,\ c(0,t) = 1,\ c(L,t) = 0.$$

8.3.3.1 Numerical Implementation

The domain is divided into finite elements, and the solution is approximated using piecewise linear basis functions. The weak form of the advection–diffusion equation is discretized, and the resulting system of equations is solved using a time-stepping scheme, such as the implicit Euler method. The stability of the solution is ensured by selecting appropriate time steps and element sizes. Unlike FDM, FEM does not have strict stability constraints like the CFL condition, but care must be taken to avoid numerical oscillations in advection-dominated problems.

8.3.3.2 Results and Discussion

The numerical results are compared with the analytical solution

$$c(x,t) = \exp\left(-\frac{(x-ut)^2}{4Dt}\right)$$

The numerical results obtained from the provided MATLAB code illustrate the pollutant concentration profile over a 10-m domain length (L) after a total simulation time of 5 s (T). The spatial grid is discretized into 100 points (N_x), and the temporal domain is divided into 500 time steps (N_t). The airflow velocity is set to $u = 1$ m/s, while the diffusion coefficient is $D = 0.1$ m^2/s. These parameters are used to calculate the spatial step size $\left(\Delta x = \frac{L}{N_x}\right)$ and the time step size $\left(\Delta t = \frac{T}{N_t}\right)$. To ensure numerical stability, the Courant number $\left(CFL = u \cdot \frac{\Delta t}{\Delta x}\right)$ and the Peclet number $\left(Pe = u \cdot \frac{\Delta x}{D}\right)$ are computed. The CFL condition ensures that the information does not propagate further than one point in the grid per time step, while the *Pe* condition checks the balance between advection and diffusion. If either $CFL > 1$ or $Pe > 2$, the code would terminate with an error message indicating unstable parameters, suggesting that Δt should be reduced or Δx increased. However, in this case, the parameters chosen to satisfy these stability criteria.

The time-stepping loop updates the concentration at each point on the grid using the finite element scheme. Specifically, the concentration at any point i is updated based on its previous value, the upstream concentration, and the concentrations of neighboring points. This scheme accounts for both advection and diffusion effects. At each time step, the boundary conditions are enforced: a constant source of pollutants is maintained at $x = 0$, and zero concentration is assumed at $x = L$. The resulting graph shows a smooth decay in pollutant concentration from the source at $x = 0$ toward $x = L$. Initially, the concentration is high near the source as a result of the continuous release of pollutants. As we move downstream, the concentration decreases exponentially, reflecting the combined effects of advection, which transports the pollutants away from the source, and diffusion, which spreads them out. The sharp decline in concentration around the middle of the domain indicates the effective transport and dispersion of pollutants. Beyond approximately 7 m, the concentration drops to nearly zero, consistent with the boundary condition at $x = L$.

In summary, the numerical solution accurately captures the expected behavior of pollutant transport governed by the advection–diffusion equation:

$$\frac{\partial c}{\partial t} + u\frac{\partial c}{\partial x} = D\frac{\partial^2 c}{\partial x^2}.$$

The concentration profile exhibits a gradual decline from the source, which confirms the efficacy of the finite element method in modeling such physical occurrences. The stability standards have been met by guaranteeing precise and precise results. Figure 8.3 illustrates the pollutant concentration profile at various time intervals. The concordance between the numerical and analytical responses is exceptional, which affirms the precision of the FEM implementation. The results demonstrate how the pollutant spreads downstream as a result of advection and diffusion.

In summary, the numerical solution accurately captures the expected behavior of pollutant transport governed by the advection–diffusion equation:

$$\frac{\partial c}{\partial t} + u\frac{\partial c}{\partial x} = D\frac{\partial^2 c}{\partial x^2}$$

The concentration profile demonstrates a smooth decay from the source, validating the effectiveness of the finite element method in simulating such physical phenomena.

8.3.3.3 Air Pollution Modeling Applications

The FEM has been widely applied to simulate pollutant transport in a variety of scenarios. In the assessment of urban air quality, FEM is used to model pollutant dispersion of pollutants within complex urban environments characterized by

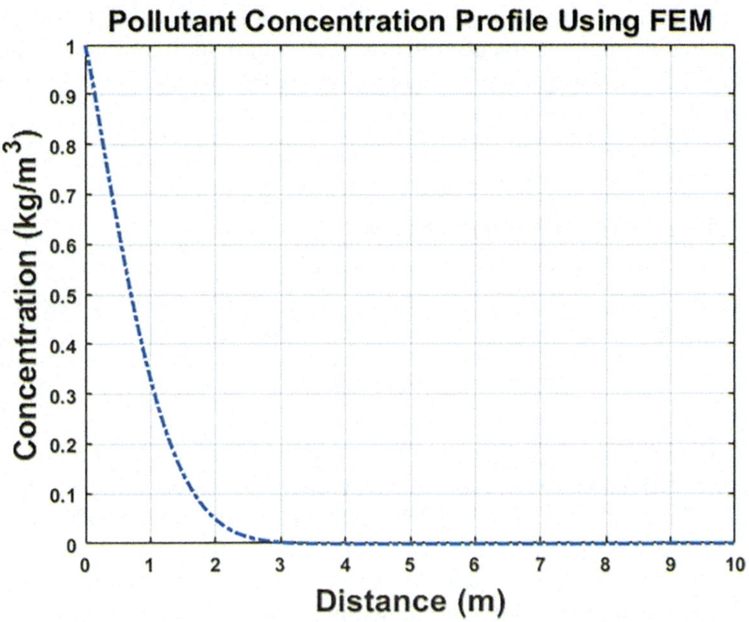

Fig. 8.3 Pollutant concentration profiles obtained using FEM at different time steps

irregular geometries, such as buildings and streets. It is also used in topographically difficult areas, such as coastal zones or mountainous regions, where geography greatly affects pollution transport. FEM also helps replicate the movement of pollutants across subsurface water systems with different soil and rock qualities, thus modeling groundwater pollution. The versatility and ability to handle irregular geometries make it suitable for a wide range of environmental modeling applications. However, its limitations include higher computational costs compared to FDM and the need for expertise in mesh generation and numerical implementation.

8.4 Monte Carlo Simulations for Uncertainty Quantification

8.4.1 Overview

Monte Carlo simulations are a probabilistic numerical approach used for uncertainty quantification and risk assessment in air pollution modeling (Cheng & Sandu, 2009; Lo et al., 2005). Monte Carlo simulations generate random samples from probability distributions, therefore accounting for uncertainty in input parameters unlike deterministic techniques such as the finite difference method. These simulations give a probabilistic understanding of model outputs, therefore supporting strong decision-making in uncertain circumstances. Monte Carlo simulation (MCS) is particularly well suited for complex systems that contain numerous sources of uncertainty, including meteorological data, emission rates, and diffusion coefficients. Although computationally intensive, MCS provides substantial benefits in capturing the sensitivity and variability of pollutant transport models (Gao et al., 2025).

8.4.2 Mathematical Formulation

The advection–diffusion equation governing pollutant transport is given as

$$\frac{\partial c}{\partial t} + u.\nabla c = D\nabla^2 c.$$

The concentration of pollutant at a specific point in space x and time t is denoted by $c(x, t)$, which represents the spatial and temporal variation in the amount of pollutant. The direction and rate at which the pollutant is advected or conveyed through the domain are determined by the velocity field of the fluid medium, such as air or water, denoted by u. The diffusion coefficient, denoted by the parameter D, quantifies the rate at which the pollutant disseminates due to molecular diffusion. The temporal and spatial dimensions in which pollutant transport is being analyzed are represented by the variables t and x.

In Monte Carlo simulations, the input parameters (u, D, and initial/boundary conditions) are treated as random variables with a class of specified probability distributions. For example,

$$u \sim \mathcal{N}\left(\mu_u, \sigma_u^2\right),\ D \sim \text{Uniform}\left(D_{\min}, D_{\max}\right)$$

Here $\mathcal{N}$ signifies the normal distribution and Uniform implies a uniform distribution. The Monte Carlo procedure consists of numerous vital phases. To reflect the uncertainty of the input of the model, the first input parameters are taken from their respective probability distributions. The advection–diffusion equation is then solved for every sampled set of values using an appropriate numerical technique, the finite difference method or the finite element method. Finally, the aggregated results of all simulations are used to calculate statistical measures such as mean, variance, and confidence intervals, thereby revealing information on expected behavior and system variability under uncertainty. The output concentration $c(x, t)$ is expressed as a probability distribution:

$$c\left(x,t\right) \sim P\left(c|,u|,D|,\text{boundary / initial conditions}\right),$$

where $P(c)$ represents the likelihood of different concentration values given the uncertainties in input parameters.

8.4.3 Example: Numerical Simulation of Pollutant Transport

To demonstrate the application of Monte Carlo simulations, we solve the 1D advection–diffusion equation for a pollutant released at $x = 0$ and transported downstream due to advection and diffusion. The parameters are:

Domain length: $L = 10$ m,
Total simulation time: $T = 5$ s,
Velocity: $u \sim \mathcal{N}(1{,}0.12)\,\mathrm{m/s}$,
Diffusion coefficient: $D \sim$ Uniform(0.05, 0.15) $\mathrm{m^2/s}$.
The initial and boundary conditions are

$$c(x{,}0) = 0,\ c(0{,}t) = 1,\ c(L{,}t) = 0.$$

8.4.3.1 Numerical Implementation

Spatial and temporal domains are discretized into $N_x = 100$ grid points and $N_t = 500$ time steps, respectively. The Monte Carlo process involves the following steps:

1. Generate $N_{samples} = 1000$ random samples for u and D from their respective distributions.
2. For each sample, solve the advection–diffusion equation using the finite difference method (FDM) with the explicit scheme:

$$c_i^{n+1} = c_i^n - u\Delta t\frac{c_i^n - c_{i-1}^n}{\Delta x} + D\Delta t\frac{\left(c_{i+1}^n - 2c_i^n + c_{i-1}^n\right)}{\Delta x^2}$$

3. Compute statistical metrics (mean, variance, and confidence intervals) for the concentration profiles across all samples

Stability criteria such as the Courant–Friedrichs–Lewy (CFL) condition and diffusion number constraints are checked for each sample to ensure numerical stability.

8.4.3.2 Results and Discussion

Figure 8.4 shows the mean and 95% confidence interval of the pollutant concentration profile in different time steps. The results demonstrate the variability in concentration due to uncertainties in the velocity (u) and diffusion coefficient (D). Specifically

- The mean concentration profile exhibits a smooth decay from the source ($x = 0$) to the far-end boundary ($x = L$).
- The confidence intervals widen as x increases, reflecting the growing uncertainty in pollutant transport due to cumulative effects of parameter variability.

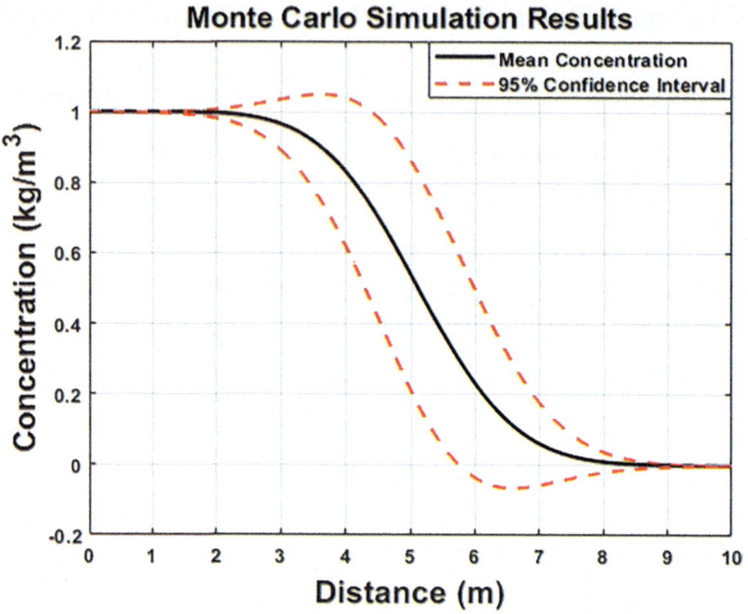

Fig. 8.4 Mean and 95% confidence intervals of pollutant concentration profiles obtained using Monte Carlo simulations

The probabilistic characteristics of pollution transport are shown by the numerical findings of the Monte Carlo simulations. The temporal domain is split into 500-time steps (N_t) while the spatial grid is discretized into 100 points (N_x). By means of random sampling of input parameters, the solution guarantees to reflect the intrinsic system variability. For each sample, we compute the Courant number ($CFL = u \cdot \Delta t/\Delta x$) and the Peclet number ($Pe = u \cdot \Delta x/D$) to guarantee numerical stability. The corresponding sample is discarded, and a new sample is generated if either $CFL > 1$ or $Pe > 2$.

The resulting graphical representations clearly demonstrate the efficacy of Monte Carlo simulations in quantifying the uncertainty of pollutant transport. The confidence intervals delineate the range of potential outcomes, but the mean concentration profile aligns effectively with deterministic solutions. The numerical solution accurately represents the anticipated dynamics of pollution transfer governed by the advection–diffusion equation.

$$\frac{\partial c}{\partial t} + u\frac{\partial c}{\partial x} = D\frac{\partial^2 c}{\partial x^2}.$$

The concentration profile indicates a consistent decrease from the source, illustrating the capacity of Monte Carlo simulations to emulate such actual phenomena. The probabilistic method provides a valuable study of the sensitivity and variability.

8.4.3.3 Applications in Air Pollution Modeling

The Monte Carlo simulations method has been generally applied to investigate air pollution employing simulated pollution movements under several conditions, including:

- Evaluating the effects of uncertainty in meteorological data, emission rates, and diffusion coefficients is uncertain quantification.
- Examining the likelihood of exceeding air quality criteria and spotting high-risk regions helps to assess (Risk Assessment).
- Key characteristics affecting pollution dispersion should be found using sensitivity analysis, which will help prioritize data-collecting initiatives.

This approach (Monte Carlo simulations) is appropriate for complicated environmental modeling uses since it can manage uncertainties or stochasticity. Meanwhile, its constraints include extraordinary computational expenses and the precise definition of input probability distributions.

8.5 Computational Fluid Dynamics for High-Resolution Dispersion Modeling

8.5.1 Overview

Computational fluid dynamics (CFD) is a robust numerical method employed to resolve fluid flow and transport phenomena dictated by partial differential equations (PDEs) (Reiminger et al., 2025). It is widely used in air pollution modeling to replicate pollutant dispersion at high resolutions, especially in intricate geometries and turbulent flows (Karim & Nolan, 2011). CFD is appropriate for high-fidelity simulations because, unlike simpler techniques such as the finite difference method (FDM), it captures complex flow dynamics using advanced discretization schemes and turbulence models.

Although computationally demanding, CFD provides unmatched accuracy in resolving temporal and spatial fluctuations in pollution concentrations (Chu et al., 2005). Applications needing thorough depictions of irregular domains, including those involving industrial facilities or urban settings, especially benefit from it. Its application does, however, indicate close attention to grid resolution, boundary conditions, and turbulence modeling.

8.5.2 Mathematical Formulation

The governing equation for pollutant transport in CFD is the advection–diffusion equation:

$$\frac{\partial c}{\partial t}+\nabla.(\mathbf{u}c)=\nabla.(D\nabla c)$$

In this context, $c(x,t)$ denotes the concentration of the pollution as a function of space x and time t, thereby expressing the distribution of the pollutant over the domain over time. $\boldsymbol{u}$ is the velocity field of the fluid medium, such as air or water, which controls the direction and speed of pollution movement via advection. Described by the diffusion coefficient, D is the rate of spread of the pollution from areas of greater concentration to areas of lower concentration resulting from molecular diffusion. Within which the pollution transport is being modeled, the variables t and x correspond according to the temporal and spatial dimensions accordingly. The advection–diffusion equation is solved in CFD using numerical methods including finite-volume method (FVM), therefore guaranteeing conservation of mass, momentum, and energy. The discretization of the derivatives follows:

- Time derivative—first-order implicit scheme:

$$\partial c/\partial t\approx\frac{\left(c^{n+1}-c^{n}\right)}{\Delta t}$$

- Advection term—second-order upwind scheme:

$$\nabla.(\mathbf{u}c)\approx\frac{1}{\Delta x}\left[F_{i+\frac{1}{2}}-F_{i-\frac{1}{2}}\right]$$

- where $F_{i+\frac{1}{2}}$ and $F_{i-\frac{1}{2}}$ are the fluxes at cell interfaces.
- Diffusion term—central difference scheme:

$$\nabla.(D\nabla c)\approx\frac{1}{\Delta x^{2}}\left[D_{i+\frac{1}{2}}\left(c_{i+1}-c_{i}\right)-D_{i-\frac{1}{2}}\left(c_{i}-c_{i-1}\right)\right]$$

Substituting these approximations into the governing equation yields:

$$\frac{c_i^{n+1}-c_i^{n}}{\Delta t}+\frac{1}{\Delta x}\left[F_{i+\frac{1}{2}}-F_{i-\frac{1}{2}}\right]=\frac{1}{\Delta x^{2}}\left[D_{i+\frac{1}{2}}\left(c_{i+1}-c_{i}\right)-D_{i-\frac{1}{2}}\left(c_{i}-c_{i-1}\right)\right]$$

Rearranging for c_i^{n+1} :

$$c_i^{n+1}=c_i^{n}-\Delta t\frac{1}{\Delta x}\left[F_{i+\frac{1}{2}}-F_{i-\frac{1}{2}}\right]+\Delta t\frac{1}{\Delta x^{2}}\left[D_{i+\frac{1}{2}}\left(c_{i+1}-c_{i}\right)-D_{i-\frac{1}{2}}\left(c_{i}-c_{i-1}\right)\right].$$

8.5.3 Example: Numerical Simulation of Pollutant Transport

To demonstrate the application of CFD, we solve the 3D advection–diffusion equation for a pollutant released at a point source and transported downstream due to advection and diffusion. The parameters are:

- Domain dimensions: $L_x = 100$ m, $L_y = 50$ m, $L_z = 20$ m,
- Total simulation time: $T = 60$ s,
- Velocity field: $u = (2, 0, 0)$ m/s,

Diffusion coefficient: $D = 0.5$ m^2/s.
The initial and boundary conditions are

$$c(x,y,z,0) = 0, c(0,y,z,t) = 1, c(L_x,y,z,t) = 0.$$

8.5.3.1 Numerical Implementation

The computational domain is discretized using a structured grid with $\Delta x = 1$ m, $\Delta y = 1$ m, and $\Delta z = 1$ m. The temporal domain is divided into $\Delta t = 0.1$ s. The stability criteria are checked:

$$CFL = \max\left(\frac{u\Delta t}{\Delta x}, \frac{v\Delta t}{\Delta y}, \frac{w\Delta t}{\Delta z}\right) \leq 1,$$

$$Pe = \max\left(\frac{u\Delta x}{D}, \frac{v\Delta y}{D}, \frac{w\Delta z}{D}\right) \leq 2$$

A second-order upwind scheme is used for the advection term, and the SIMPLE (Semi-Implicit Method for Pressure-Linked Equations) algorithm is employed to couple the velocity and pressure fields. Turbulence is modeled using the $k - \varepsilon$ model, which accounts for turbulent mixing and diffusion.

8.5.3.2 Results and Discussion

Comparing the numerical findings with the analytical response helps, one evaluate the dependability and correctness of the computational method. Figures 8.5a show the pollution concentration profiles derived from both numerical and analytical approaches. Consistent with expected advection–diffusion behavior, the data show a gradual decline in the concentration of pollution from the source at $x = 0$ toward the far-end boundary $x = L_x$. The concentration gradient of the numerical and analytical solutions is clearly defined, and the pollution spreads gradually downstream. The consistency among these approaches shows how well the computational model captures critical transport pathways.

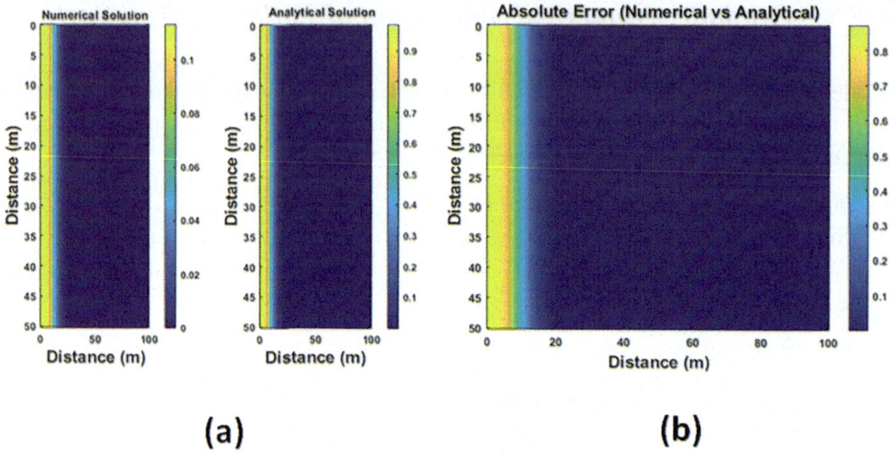

Fig. 8.5 (**a**) Comparison of pollutant concentration profiles obtained using numerical (left) and analytical (right) methods. (**b**) Absolute error between numerical and analytical solutions

The total variance (absolute error) between the numerical and analytical results is shown in Fig. 8.5b to examine the accuracy of the numerical technique even more. With few deviations close to the source region, the error distribution over the computing domain remains relatively low. Localized numerical diffusion effects and small changes in implementing initial conditions can help explain these differences. However, beyond the first region, the numerical model nearly corresponds to the analytical predictions, verifying the accuracy of the CFD application.

With minor errors that have no significant impact on the broader properties of pollution dispersion, the results typically demonstrate that the numerical method reasonably approximates the analytical solution. Reliable for simulating the dispersion in such flow conditions, the computational model captures the basic transport processes well. These results underline the accuracy of the CFD method and its possible use in practical environmental modeling conditions.

8.5.3.3 Air Pollution Modeling Applications

Understanding the dispersion, transformation, and deposition of contaminants in different ecosystems is highly dependent on air pollution modeling. With a thorough understanding of air quality dynamics made possible by computational fluid dynamics (CFD) and other numerical techniques, researchers and legislators can create sensible mitigating plans. These modeling approaches are used in many fields, from large-scale metropolitan assessments to localized indoor air quality investigations.

Urban Air Quality Assessment

Urban areas are one of the most essential uses of air pollution modeling, since industrial activity and high population density help to create complicated pollution dispersion patterns. Analyzing the movement of pollutants in cities using CFD simulations that incorporate the impact of buildings, transportation networks, and weather variables is somewhat common. Through the modeling of airflow patterns and pollution dispersion, researchers can pinpoint pollution hotspots, evaluate the success of green infrastructure, and improve urban planning techniques, thus improving air quality. Furthermore, these models help authorities to establish early warning systems for air pollution incidents, allowing for prompt actions.

Industrial Emissions and Source Apportionment

When particulate matter, nitrogen oxides, sulfur dioxide, and volatile organic compounds are emitted into the atmosphere, industrial activity seems to be the primary contributor to air pollution. The dispersion of pollutants from industrial stacks and point sources is simulated using CFD and other numerical methods considering stack height, emission velocity, and atmospheric stability. These simulations enable industries to comply with environmental rules, maximize pollution control systems, evaluate and assess possible effects of emissions on nearby populations. Furthermore, source apportionment studies help to establish the relative contributions of various pollution sources by employing numerical models, therefore supporting focused mitigating strategies.

Traffic-Related Air Pollution Modeling

Producing carbon monoxide (CO), hydrocarbons (HC), and delicate particulate matter ($PM_{2.5}$), vehicle emissions are the leading cause of urban air pollution. The dispersion of contaminants along roadways, tunnels, and highways is examined using numerical models such as Gaussian plume models and CFD. These models forecast air quality levels at various times of the day including traffic density, wind patterns, and temperature intervals. This information helps designers in cities establish low-emission zones, maximize traffic control plans, and assess how well hybrid and electric cars help lower pollution levels.

Indoor Air Quality and Ventilation Studies

In particular, in residences, companies, hospitals, and places of employment, the quality of indoor air is becoming more prominent. Cooking, heating systems, tobacco smoke, and volatile organic compounds (VOCs) from building materials constitute sources of indoor air pollution. CFD simulations are extensively used to investigate the circulation of airborne contaminants in enclosed spaces, giving valuable information on the effectiveness of ventilation, how effective it is in pollutant elimination, and the influence of air filtration systems. Through the modeling of several ventilation scenarios, researchers can optimize the HVAC (heating, ventilation, and air conditioning) designed system to improve indoor air quality and reduce health hazards associated with continuous exposure to pollution.

Impact of Air Pollution on Climate and Ecosystems

In addition to influencing human health, air pollution has wider consequences on the environment. In particular, with regard to aerosol–cloud interactions, radiative forcing, and greenhouse gas (GHG) emissions, numerical models help to examine the connection between air pollutants and climate systems. These models allow one to forecast how air pollution can affect the decline of biodiversity, agricultural production, and climate patterns. Chemical transport models help one to investigate, for instance, the production of acid rain resulting from nitrogen oxides (NO_x) and sulfur dioxide (SO_2) emissions in relation to soil degradation, water bodies, and forest ecosystems.

Health Risk Assessment and Epidemiological Studies

Computational models are applied to determine human exposure to hazardous chemicals and their associated health consequences in view of growing concern over air pollution-related health hazards. To evaluate the correlation between contaminants and respiratory disorders, cardiovascular problems, and other health effects, these models combine epidemiological data with atmospheric dispersion data. These evaluations support the development of public health policies, air quality regulations, and mitigating strategies to safeguard the sensitive groups including children, seniors, and people with preexisting medical illnesses.

8.6 Conclusion

By offering exact and consistent techniques to simulate and forecasting pollution dispersion under challenging circumstances, computational models have transformed air pollution research. Although this is beneficial, conventional analytical techniques fail to adequately capture the complex nonlinear dynamics of pollution movement, which emphasizes the requirement of sophisticated numerical approaches. Key computational techniques, such as the finite difference method (FDM), the finite element method (FEM), Monte Carlo simulations, and computational fluid dynamics (CFD), each with special advantages suited to particular difficulties in air quality assessment, have been discussed in this chapter.

Especially useful for simulating pollution distribution over time and space, FDM offers a computationally efficient framework for solving discretized numerical solutions on structured grids. FEM is fairly good at managing irregular geometries, which helps to create accurate models in metropolitan areas and rugged terrain. Strong tools for estimating uncertainty provided by Monte Carlo simulations allow probabilistic risk assessments essential for decision-making. CFD facilitates the identification of pollution hotspots by means of high-resolution modeling of air flow and pollutant dispersion, therefore providing a comprehensive understanding of urban air quality dynamics.

When implemented together, these computerized techniques create a complete toolkit for handling the various challenges of air pollution modeling. Using their individual uses, academics and legislators can improve the accuracy of forecasting pollution levels, evaluate environmental effects, and create workable mitigating policies. These models will grow even more powerful as developments in high-performance computing and machine learning keep changing, opening the path for better air quality control and sustainable urban design. Ultimately, computer modeling is the pillar of contemporary air pollution research since it provides a strong basis for knowledge and solutions to address this crucial worldwide issue.

Acknowledgments I would like to thank the Control and Automation Group at IIT Delhi for their support and guidance. I am also grateful to the Editor and the Technical Review Committee for their valuable comments and suggestions, which helped to improve this manuscript.

References

Albani, R. A., Duda, F. P., & Pimentel, L. C. G. (2015). On the modeling of atmospheric pollutant dispersion during a diurnal cycle: A finite element study. *Atmospheric Environment, 118*, 19–27. https://doi.org/10.1016/j.atmosenv.2015.07.048

Baklanov, A. (2000). Application of CFD methods for modelling in air pollution problems: Possibilities and gaps. In *Urban air quality: Measurement, modelling and management, proceedings of the second international conference on urban air quality* (pp. 181–189). Springer.

Bernstein, J. A., Alexis, N., Barnes, C., Bernstein, I. L., Nel, A., Peden, D., Diaz-Sanchez, D., Tarlo, S. M., & Williams, P. B. (2004). Health effects of air pollution. *Journal of Allergy and Clinical Immunology, 114*(5), 1116–1123. https://doi.org/10.1016/j.jaci.2004.08.030

Boubel, R. W., Vallero, D., Fox, D. L., Turner, B., & Stern, A. C. (2013). *Fundamentals of air pollution* (5th ed.). Elsevier.

Brunekreef, B., & Holgate, S. T. (2002). Air pollution and health. *The Lancet, 360*(9341), 1233–1242. https://doi.org/10.1016/S0140-6736(02)11274-8

Cheng, H., & Sandu, A. (2009). Uncertainty quantification and apportionment in air quality models using the polynomial chaos method. *Environmental Modelling & Software, 24*(8), 917–925. https://doi.org/10.1016/j.envsoft.2009.01.003

Chu, A., Kwok, R. C.-W., & Yu, K. (2005). Study of pollution dispersion in urban areas using computational fluid dynamics (CFD) and geographic information system (GIS). *Environmental Modelling & Software, 20*(3), 273–277. https://doi.org/10.1016/j.envsoft.2003.12.008

Dimov, I., Georgiev, K., Ostromsky, T., & Zlatev, Z. (2004). Computational challenges in the numerical treatment of large air pollution models. *Ecological Modelling, 179*(2), 187–203. https://doi.org/10.1016/j.ecolmodel.2004.05.011

Gao, Z., Mei, E. J., He, X., Hopke, P. K., Ebelt, S., Rich, D. Q., & Russell, A. G. (2025). Multicity accountability and uncertainty assessment of the impacts of regulations on air quality in Atlanta, New York City, and Southern California. *Atmospheric Environment, 342*, 120947. https://doi.org/10.1016/j.atmosenv.2022.120947

Gill, E. Z., Mastromatteo, M., Cangelmi, L., Cellini, P., Cardone, D., & Amelio, A. (2025). A comparative analysis of artificial intelligencemethods for air quality prediction on construction sites. Proceedings ofthe 48th MIPRO ICT and Electronics Convention, Opatija, Croatia, 2127–2132. https://doi.org/10.1109/MIPRO65660.2025.11131894

Johari, H., Rusli, N., & Yahya, Z. (2018). Finite difference formulation for prediction of water pollution. *IOP Conference Series: Materials Science and Engineering, 318*, 012005. https://doi.org/10.1088/1757-899X/318/1/012005

Kampa, M., & Castanas, E. (2008). Human health effects of air pollution. *Environmental Pollution, 151*(2), 362–367. https://doi.org/10.1016/j.envpol.2007.06.012

Karim, A., & Nolan, P. (2011). Modelling reacting localized air pollution using computational fluid dynamics (CFD). *Atmospheric Environment, 45*(4), 889–895. https://doi.org/10.1016/j.atmosenv.2010.10.013

Lo, S.-C., Ma, H.-W., & Lo, S.-L. (2005). Quantifying and reducing uncertainty in life cycle assessment using the Bayesian Monte Carlo method. *Science of the Total Environment, 340*(1–3), 23–33. https://doi.org/10.1016/j.scitotenv.2004.08.014

Mayer, H. (1999). Air pollution in cities. *Atmospheric Environment, 33*(24–25), 4029–4037. https://doi.org/10.1016/S1352-2310(99)00144-2

Mazumder, S. (2015). Numerical methods for partial differential equations: finite difference and finite volume methods. Academic Press.

Olaiju, O. A., Hoe, Y. S., & Ogunbode, E. B. (2018). Finite Element and Finite Difference Numerical Simulation Comparison for Air Pollution Emission Control to Attain Cleaner Environment. *Chemical Engineering Transactions*, 63, 679–684.

Özişik, M. N., Oran, E. S., & Tzou, D. Y. (2017). Finite difference methods in heat transfer (2nd ed.). CRC Press.

Pepper, D. W., & Heinrich, J. C. (2005). The finite element method: basic concepts and applications. Taylor & Francis.

Rastogi, U., Kumar, S., Maurya, S. P., & Yadav, A. K. (2025). Air pollution forecasting using genetic algorithm and LSTM deep learning technique. *Aerosol Science and Engineering*. https://doi.org/10.1007/s41810-025-00343-y

Reiminger, N., Jurado, X., Maurer, L., Vazquez, J., & Wemmert, C. (2025). Advancing urban air quality modeling with solar radiation-included computational fluid dynamics simulations. *Atmospheric Pollution Research, 16*(2), 102383. https://doi.org/10.1016/j.apr.2024.102383

Salman, A. K., Pouyaei, A., Choi, Y., Lops, Y., & Sayeed, A. (2022). Deep learning solver for solving advection–diffusion equation in comparison to finite difference methods. *Communications in Nonlinear Science and Numerical Simulation, 115*, 106780. https://doi.org/10.1016/j.cnsns.2022.106780

Sanín, N., & Montero, G. (2007). A finite difference model for air pollution simulation. *Advances in Engineering Software, 38*(6), 358–365. https://doi.org/10.1016/j.advengsoft.2006.09.004

Szymkiewicz, R. (2010). Numerical Solution of the Advection-Diffusion Equation. In: Numerical Modeling in Open Channel Hydraulics. Water Science and Technology Library, vol 83. Springer, Dordrecht. https://doi.org/10.1007/978-90-481-3674-2_7

Thalib, R., Bakar, M. A., & Ibrahim, N. F. (2021). Application of support vector regression in Krylov solvers. Annals of Emerging Technologies in Computing (AETiC), 5(5), 178–186.

Thongmoon, M., & McKibbin, R. (2006). A comparison of some numerical methods for the advection-diffusion equation.

Vallero, D. A. (2014). *Fundamentals of air pollution* (5th ed.). Academic Press.

Wen, Y., Zhang, S., Wang, Y., Yang, J., He, L., Wu, Y., & Hao, J.(2024). Dynamic traffic data in machine-learning air quality mappingimproves environmental justice assessment. *Environmental Science & Technology*, 58(7), 3118–3128

Wu, Q., Wang, Z., Gbaguidi, A., Gao, C., Li, L., & Wang, W. (2011). A numerical study of contributions to air pollution in Beijing during CAREBeijing-2006. *Atmospheric Chemistry and Physics, 11*(12), 5997–6011. https://doi.org/10.5194/acp-11-5997-2011

Yadav, A. K. (2015). Elemental composition and source apportionment of suspended particulate matters and health risk assessment in mining and nonmining areas of Odisha, India. *Journal of Hazardous, Toxic, and Radioactive Waste, 19*(3), 04014037.

Yadav, A. K., Gupta, K., Bodhale, R., & Phuleria, H. C. (2025). Metals and trace element associated health risk in size-segregated household dust of an industrial town in India. *Air Quality, Atmosphere & Health*. https://doi.org/10.1007/s11869-025-01794-5

Yadav, A. K., Sahoo, S. K., Patra, A. C., Dubey, J. S., Lenka, P., Sagar, D. V., Kumar, A. V., & Tripathi, R. M. (2014). Source identification of particulate matter and associated intake of elements through inhalation in an industrial area of Odisha, India. *Toxicological & Environmental Chemistry, 96*(3), 410–425.

Yadav, A. K., Shirin, S., Emmanouil, C., & Jamal, A. (2022). Effect of seasonal and meteorological variability of air pollution in Singrauli coalfield. *Aerosol Science and Engineering, 6*, 61–70.

Zannetti, P. (2013). *Air pollution modeling: Theories, computational methods and available software*. Springer Science & Business Media.

Zhang, Y. (2009). A finite difference method for fractional partial differential equation. Applied Mathematics and Computation, 215(2), 524–529. https://doi.org/10.1016/j.amc.2009.05.060

Zlatev, Z. (1995). Numerical treatment of large air pollution models. In *Computer treatment of large air pollution models* (pp. 69–109). Springer.

Zlatev, Z., & Berkowicz, R. (1988). Numerical treatment of large-scale air pollution models. *Computers & Mathematics with Applications, 16*(1–2), 93–109. https://doi.org/10.1016/0898-1221(88)90241-7

Chapter 9
Machine Learning in Air Quality Prediction and Control

Muhammad Saad Waseem, Zahid Ali, and Saeed Ahmed

9.1 Introduction

The world is developing faster and faster with each passing day. We can see, industries are being made around us everywhere, which is undoubtedly causing a change in the environment due to this massive growth. What was once a visible environmental situation is now diagnosed as a complicated hazard with deep roots in social, financial, and ecological systems. Far from being a localized problem, air pollutants have emerged as a major worldwide hazard, which endangers the environment and poses critical dangers to human health and financial stability (Mason et al., 2019; Sueyoshi & Yuan, 2015). Problems such as acid rain, eutrophication, and concrete smog directly result from this developing menace. Human value is staggering. According to the World Health Organization (WHO), air pollutants are responsible for about 13 deaths each minute. Economically, the harm is similarly alarming because it hinders worldwide development, strains healthcare systems, and forces industries to undertake transformative modifications to reduce their emissions (Buekers et al., 2011; Hering & Poncet, 2014; Landrigan et al., 2018).

M. S. Waseem
Department of Chemistry, University of Chakwal, Chakwal, Pakistan
e-mail: saeedahmedlaghari@chemist.com

Z. Ali
School of Chemistry and Chemical Engineering, Shandong University, Jinan, China

S. Ahmed (✉)
Department of Biological & Chemical Sciences, University of Rasul, Mandi Bahauddin, Pakistan
e-mail: saeedahmedlaghari@chemist.com

A. K. Yadav (ed.), *Computational Approach for Air Pollution*,
https://doi.org/10.1007/978-3-032-09215-1_9

9.1.1 Background

In response to those escalating concerns, the WHO launched its first air quality recommendations in 2010, laying the basis for first-class indoor air quality standards. This change was accompanied by developing global attention: in 2014, the inaugural United Nations Environment Assembly (UNEA) formally classified air pollutants as an essential environmental issue (Chen et al., 2017; Shahbazi et al., 2019). By 2016, the second UNEA offered principal reviews: Air Quality and The Process of Air Pollution Control in Beijing: 1998–2013. These reviews provided a worldwide evaluation of air quality from 1998 to 2015 and mentioned key techniques Beijing used to manipulate its air pollutants over 115 years (Hering & Poncet, 2014; Yang et al., 2020).

1/3 UNEA in 2017 similarly highlighted the pressing need for worldwide action. In 2018, the WHO and UNEA, among others, hosted the global conference on air pollution and health, intending to reduce pollutant-associated deaths by 2030 (Kaldellis et al., 2007). Building on that momentum, UNEA released the Air Pollution Series: Regulating Air Quality in 2021, comparing federal and institutional efforts in 194 nations and the EU. Given their enormous and interconnected results, air pollutants have become one of the most urgent environmental and public health crises of our time (Lin et al., 2018; Yadav, 2021; Zhang et al., 2021).

For more than 70 years, the effects of pollutants on human health have been drastically studied, with studies specializing in measurable effects, including hospitalizations and mortality. Economists have played a key role in enhancing causal inference in outdoor laboratory settings. However, health effects constitute just a small fraction of the total impact. Pollution also influences much less-seen, however, critical, "non-health" effects, along with productivity, cognitive function, overall college performance, decision-making, or potentially criminal activity. Although each health and non-health effect comes from nontraditional physiological mechanisms, they vary in how they are observed. Health-associated studies generally focus on intense effects such as respiration or cardiovascular disease, while non-health studies explore subtler, subclinical results that could impact conduct and competencies without immediate, observable symptoms (Deschênes et al., 2017).

Understanding those non-health results is critical because their subtlety complicates efforts to avoid pollution-related publicity. Many humans may also enjoy reduced productivity, decreased cognitive abilities, or impaired overall performance without determining that pollutants are the underlying cause. As a result, proactive steps to avoid pollutants, including staying indoors on days with excessive pollutants, are much less likely. However, the effects of non health results, which include decreased taking a look at ratings or reduced painting efficiency, regularly cause reactive remedial measures. These compensatory actions, such as tutoring or extra task training, incur monetary expenses and affect average well-being, even though the cause (pollutants) remains unidentified.

For instance, if public publicity reduces pupils' ratings, colleges and universities may also take corrective steps, including tutoring, even without understanding that

Air Great is the cause behind the decline. Although this remediation may enhance ratings, it no longer eliminates the monetary expenses imposed by using pollutants.

The in-between pollutants and non health effects are much less truthful than the well-established pathways connecting pollutants to diseases such as bronchial allergies or cardiovascular diseases. Toxicological research maps clean relationships between publicity and unique health issues. However, the mechanisms for non health effects regularly depend upon oblique or speculative connections. Some studies involve animal research to recognize how pollutants affect human cognition and conduct. Although the absence of agreed-upon mechanisms may also increase concerns about capability biases, the exploratory nature of this study enables manual destiny investigations. It promotes collaboration between environmental economics and organic sciences, growing a virtuous cycle.

Another distinction is the various timings of non-health results. Some effects, including impaired awareness due to excessive pollutants, may also arise immediately but fade as soon as air quality improves. Others, however, may also have long-lasting effects. For example, exposure to pollutants in early adolescence can affect learning, hinder instructional development, and decrease future earnings. This dynamic is consistent with the "fetal origin hypothesis" which shows that youth exposures have long-term effects, and theories emphasizing the cumulative nature of human capital development.

Finally, the dose–reaction relationship for non-health effects differs from that for health effects. While intense health problems such as breathing failure tend to occur at excessive pollutant levels and, by and large, affect susceptible populations, non-health consequences may also emerge at much-decreased pollutant levels and affect healthy individuals in any other case. This broader applicability will increase the societal impact of pollutants' unfavorable consequences. Understanding those subtler but significant influences is essential for growing potent pollutants, manipulating regulations, and mitigating hidden monetary and human capital expenses related to air pollutants.

Gathering biological, ecological, and climatic facts is critical to analyzing complex systems, and deploying large-scale sensor networks will become a key environmental tracking device (Polastre et al., 2004; Werner-Allen et al., 2006). The National Science Foundation (NSF) facilitates several cyber-infrastructure tasks for large-scale data collection and data assessment based on large-scale self-reliant sensor networks, along with the National Ecological Observatory Network (NEON) and the Long-Term Ecological Research Network (LTERN).

Designing a large-scale sensor network may encounter computational challenges (e.g., network architecture, walking gadgets, programming environments, facts collection, assessment, synthesis, and inference) (Akyildiz et al., 2007). For example, while sensor networks are used to show spatial phenomena, selecting the high-quality placement of sensors to maximize report advantage while minimizing communication costs is a complex problem that requires new techniques (Krause & Guestrin, 2009).

Environmental problems such as air pollutants, extreme weather, deforestation, and water shortages threaten human health, ecosystems, and economies. Traditional strategies for analyzing and mitigating these issues regularly fail to address their complexity, scale, and dynamic nature. More and more, this hole has been filled through computational strategies pushed through improvements in statistics science, artificial intelligence (AI), and extreme performance computing. These strategies provide revolutionary answers and deeper insights into environmental processes. Computational strategies embody numerous gears and technologies, including machine learning (ML), computational fluid dynamics (CFD), chemical transport models (CTM), and remote sensing. These use massive satellite, sensors, and climate station datasets to examine, expect, and visualize environmental phenomena in unparalleled detail.

One of the most transformative packages of computational strategies is in air pollutant tracking and forecasting. Accurate air high-satisfactory predictions are crucial for public health, especially in city regions where pollutant levels differ due to visitors, commercial activities, and climate. Machines gaining knowledge of fashion systems' real-time statistics from the Internet of Things (IoT) enabled air high-satisfactory sensors and satellite TV for PC imagery to offer brief- and long-term period pollutants forecasts. For example, superior models, which include Convolutional Neural Networks (CNNs) and Recurrent Neural Networks (RNNs), have verified excessive precision in forecasting particulate matter ($PM_{2.5}$) concentrations (Rastogi et al., 2025; Zhang & Woo, 2019). Pollutant dispersion modeling is another vital application. Computational fluid dynamics (CFD) simulations examine how pollution develops in urban environments, taking into account wind patterns, building structures, and visitor density. It will allow city planners to optimize visitors to go with the flow and layout of inexperienced areas to lessen pollutant hotspots (Tominaga & Stathopoulos, 2013).

Chemical transport models (CTMs) provide a broader perspective by simulating the sources, movement, and chemical transformation of air pollution on nearby and international scales. These models combine meteorological statistics, emissions inventories, and chemical reactions to show pollution over time. CTMs were instrumental in evaluating high sulfur air policies and those geared toward lowering sulfur dioxide (SO_2) emissions to combat acid rain within the United States (Napelenok et al., 2006). Computational strategies expand past air pollutants to cope with different environmental challenges. For instance, AI algorithms stumble on deforestation from satellite TV for PC images, optimize water utilization in agriculture, and forecast the extreme effects of weather on biodiversity. These numerous packages highlight the flexibility of computational gear in fixing complicated environmental issues. Despite their advantages, computational strategies face particular challenges. Data, version uncertainties, and excessive computational expenses can affect their effectiveness. Ethical concerns, statistical privacy, and algorithmic bias must also be addressed to ensure the responsible and equitable use of this technology.

Looking to the future, rising technologies such as quantum computing, part AI, and virtual twins are poised to enhance the strength and scalability of computational strategies in environmental research. These improvements may allow faster, more accurate analyses and contribute to more significant, sustainable, and resilient ecological answers. The computational approach is gearing up to address environmental degradation. Scientists, policymakers, and groups can recognize and mitigate the complicated environment by harnessing the strength of statistics and superior algorithms.

9.2 Computational Methods in Air Pollution

Computational strategies have become vital equipment in air pollutant research, imparting superior talents for tracking, modeling, and predicting pollutant patterns. These techniques are based on various record sources, including satellite TV for PC records, IoT-enabled air quality sensors, and ground-based monitoring stations, which give real-time and large-scale information on pollutant levels. Machine learning (ML) and artificial intelligence (AI) are critical in studying these records, allowing predictive modeling to forecast air quality and clustering or classification algorithms to perceive pollutant sources. Computational fluid dynamics (CFD) is another practical approach to how pollution disperses in complicated city environments, considering elements such as wind patterns, visitor emissions, and construction layouts. Additionally, chemical transport models (CTMs) simulate the movement, chemical transformation, and shipping of air pollution on nearby and worldwide scales, helping researchers in verifying pollutant developments and comparing the effects of emission management policies. Together, these computational processes offer a complete framework for information and mitigation of air pollutants on public health and the environment.

9.3 Efficient Data Gathering

First-class air tracking has historically relied on deploying a confined, wide variety of constant tracking stations throughout city regions (Zou et al., 2009). Although this technique offers a trendy assessment of pollutant stages, it has predominant limitations. Most notably, it gives a coarse-grained attitude with low spatial resolution, meaning that localized pollutants are frequently undetected. In addition, constant stations are inflexible and cannot easily adapt to modifications in city infrastructure, weather conditions, or land use, making some places outdated as towns develop and evolve. To triumph over those challenges, researchers have added the idea of Vehicular Sensor Networks (VSNs), (Wang, 2014), which contain equipping cars along with cars, buses, and motorcycles with cell sensors. Vehicle powered sensors encompass fuel line detectors, GPS units, and Wi-Fi verbal exchange interfaces such as Wi-Fi or LTE-A. As those cars pass through the city, they collect first-class real-time air statistics from formerly unreachable regions, imparting a much finer spatial and temporal resolution than conventional constant stations. The accumulated statistics are then transmitted to an essential server for aggregation and analysis. This cell-sensing approach allows for especially dynamic air first-class tracking that highlights the variety of city pollutant patterns.

In addition, drivers participating in these networks can be financially incentivized primarily on the amount and quality of the data they provide. However, the sampling price, the frequency with which the car collects statistics, is managed centrally through the server, no longer via the means of drivers. This is completed to ensure stability among statistics, first-class verbal exchange, and monetary costs. Achieving this stability requires dynamic sampling price modifications primarily on factors such as site visitors density and pollutant concentration levels (Wang et al., 2009). Additionally, the gadget avoids amassing redundant statistics by stopping cars from sampling close together because pollutant levels in nearby places tend to be especially correlated.

Another vital component is the ability to estimate air quality in regions in which real-time statistics cannot be accumulated, along with places obstructed by means of homes or otherwise inaccessible. The server uses the statistics acquired from nearby positions to infer first-class air values in those unsampled regions statistically. These estimations, without delay, impact the accuracy of general tracking. To enhance the device, gadget learning algorithms may be hired to understand traits and predict future pollutant levels, primarily based on historical and environmental data. Altogether, VSNs provide a flexible, scalable, and cost-effective answer for complete city air first-class tracking, addressing the shortcomings of conventional constant tracking infrastructures.

9.4 Air-Pollutants Evaluation Model: AQI

The Air Quality Index (AQI), installed through the US Environmental Protection Agency (EPA), is a standardized device to assess and report how clean or polluted the air is on a given day. It is designed to assist the general public in recognizing the fitness effects of outdoor air pollutants. The AQI scale is split into six categories, each similar to a particular variety of values and represented through a unique shade to convey the severity of air quality.

When the AQI is 50 or less, air quality is considered “Good” (green), indicating that outdoor air poses little threat to fitness. If the AQI falls between 50 and 100, it is labeled “Moderate” (yellow). In this variety, the air quality is generally acceptable. However, folks who are strangely sensitive to air pollutants may also forget proscribing extended or extreme outdoor sports. An AQI from 101 to 150 falls below the category “Unhealthy for Sensitive Groups” (orange). In this manner, those more vulnerable, including children, adults aged, outdoor workers, and people with respiratory diseases such as asthma, have to reduce extended or heavy outdoor exertion. When the AQI rises to the level (red, 151–200), those touchy companies are recommended to stay away from prolonged outdoor exertion altogether, even as the general population is advised to restrict such sports.

In the case that AQI reaches the "Very Unhealthy" category (purple, 201–300), outdoor sports become risky for most people who have to keep away from outdoor exercise, even the healthy individuals have to decrease their outdoor exposure. At the best level, "hazardous" (maroon, AQI $\geq$ 301), air quality represents a critical threat to fitness for everyone, and all outdoor activities should be avoided irrespective of health status. The EPA determines the AQI primarily based on six key air pollutants: ozone (O_3), particulate matter ($PM_{2.5}$ and PM_{10}), carbon monoxide (CO), sulfur dioxide (SO_2), and nitrogen dioxide (NO_2). A particular subindex value (Ik) is calculated for every pollutant using a formula described in Eq. (9.1). The maximum of those subindex values turns into the general AQI, ensuring that the most risky pollutant at any given time dictates the highest air quality rating. This gadget allows people and governments to respond accurately to various levels of pollutants, with the aim of effectively protecting public health.

$$I_k = \left(\frac{I_{\text{high}} - I_{\text{low}}}{B_{\text{high}} - B_{\text{low}}} \right) \times \left(C_k - B_{\text{low}} \right) + I_{\text{low}} \tag{9.1}$$

where B_{high} is a breakpoint C_k, Blow is a breakpoint C_k, and $I_{\text{high}}/I_{\text{low}}$ denotes the AQI cost for $B_{\text{high}}/B_{\text{low}}$ (Wang & Chen, 2017).

9.5 Metropolitan Traffic Model: SUMO

The Metropolitan Traffic Model using SUMO (Simulation of Urban Mobility) effectively simulates visitor flows in city environments. SUMO is an open-source visitor simulation bundle that allows for modeling numerous visitor scenarios, along with multimodal transportation structures and the effect of various visitor control strategies. Using SUMO, city planners and researchers can examine visitor patterns, compare the outcomes of infrastructure changes, and optimize visitor sign timings. The version can include real-global information from GPS, visitor counts, and public shipping schedules, bearing in mind a more accurate illustration of metropolitan visitors' dynamics. By simulating visitors under numerous conditions, stakeholders could make informed selections to lower congestion, improve safety, and improve ordinary mobility in metropolitan areas. In addition, SUMO gives functions such as path planning, automobile conduct modeling, and emission calculations, making it a complete device for visitor analysis. SUMO is a powerful open-source simulation tool designed to model actual global avenue networks and vehicular site visitors with high precision and realism (Bedogni et al., 2015). It gives a microscopic degree of modeling site visitors, considering several distinct parameters: speed limits, street widths, lane hierarchies (e.g., principal roads vs. minor lanes), site visitors' mild configurations, and more. Each automobile within the simulation is, for my part, described with a unique identifier and is assigned various attributes such as departure time, pace, and route. This fine-grained modeling permits distinctly correct automobile conduct simulations under exceptional conditions.

SUMO successfully simulates how motors interact with the site visitors' infrastructure and everything else. For example, it could be a version of the motors slowing down the vehicle motors regularly while approaching a red light and accelerating as the light turns green. The motion and conduct of every vehicle are dynamically controlled, allowing for the practical simulation of visitors to the city site. Additionally, every automobile can be assigned emission profiles for pollution or noise, making SUMO a beneficial device for environmental effect tests associated with city site visitors. One of SUMO's superb capabilities is its ability to remember the impact of nearby motors. For instance, while motoring at a crossroad without traffic signals, SUMO guarantees that one vehicle yields until the opposite has passed, mimicking real-world traffic etiquette. In the case of collisions, SUMO carries a postponement to simulate the disruption due to site visitors' injuries earlier than removing the affected motors, thereby improving the realism of the simulation. SUMO consists of the number one module: the Street Builder (SB) and the Traffic Signal Timing (TS). The SB module uses geographical street information acquired from systems such as OpenStreetMap (Fonte et al., 2017) to recreate metropolis maps with actual global accuracy. Additionally, it determines the position and timing of the site visitors' lights. Using the SUMO street community importer device called netconvert, customers can robotically assemble street layouts, including streets and lane structures, from the map information. This complete setup allows researchers and concrete planners to simulate, analyze, and optimize site visitors' control techniques successfully in a digital environment before implementing them in the real world.

9.6 Pollutant Dispersion Model: ISC3

The ISC3 (Industrial Source Complex Version 3) is a steady-state Gaussian plume version designed to simulate the diffusion, dispersion, and sedimentation of pollution inside the air. It is broadly used to control air pollutants from commercial assets and bills for various things that affect pollutant behavior, such as more than one type of supply, pollutant settling, and interactions with terrain and climate conditions (Fonte et al., 2017).

9.6.1 Key Features of ISC3

The five key features of the ISC3 are given as follows:

1. *Source types*: ISC3 can simulate distinct pollutant supply configurations, such as:
   (a) Point reasserts (e.g., smokestacks),
   (b) Line reasserts (e.g., highways),
   (c) Area reasserts (e.g., commercial zones),
   (d) Volume reasserts (e.g., pollutant clouds).

2. *Point source division*: ISC3 can handle instances in which emissions are divided into more than one asset of a minor factor.
3. *Dry deposition and settlement*: The version consists of the consequences of particle deposition and settling due to gravity, which can be specifically applicable to particulate pollution.
4. *Downwash*: It bills for plume conduct stimulated by nearby structures, which cause pollution to be deflected downward, affecting floor-level concentrations.
5. *Terrain adjustment*: ISC3 restricts the adjustment of various terrains, allowing it to version pollutant dispersion in regions with slight elevation changes.

The mathematical formulation for the awareness can be calculated using Eq. (9.2). The $C(x, y, z)$ at any 3D factor (x, y, z) is calculated using the subsequent Eq. (9.2):

$$C(x,y,z) = \frac{Q \cdot W \cdot V \cdot D}{2\pi \cdot u \cdot \sigma_y \cdot \sigma_z} \exp\left[-\frac{1}{2}\left(\frac{y}{\sigma_y}\right)^2\right] \tag{9.2}$$

where Q pollutant discharge rate (in g/s), W conversion aspect to explicit awareness in suitable units (e.g., μg/m³), V reflection aspect accounting for the vertical pollutant-mirrored image (because of the floor and environment layers), D disintegration coefficient (to represent pollution with a finite half-life, e.g., sulfur dioxide, SO_2), Vertical reflection factor, V To account for the mirrored image of pollution from the floor and atmospheric inversion layers, V is calculated using the subsequent Eq. (9.3):

$$V = \exp\left[-\frac{1}{2}\left(\frac{z-H}{\sigma_z}\right)^2\right] + \exp\left[-\frac{1}{2}\left(\frac{z+H}{\sigma_z}\right)^2\right] \tag{9.3}$$

where H effective top of the pollutant supply (e.g., the peak of a smokestack), z height above the floor on the statement factor, *tau_z* vertical diffusion coefficient.

This method guarantees that the version captures pollutant dispersion above and beneath the supply top, accounting for atmospheric mixing.

9.7 Machine Learning and Artificial Intelligence

Air quality forecasting techniques are advancing unexpectedly due to the developing demand for accurate pollutant measurement. Conventional models, including HYSPLIT-4 and Kalman Filters (KF), are typically hired to simulate the behavior of air pollutants, in particular how they collect and are used within the atmosphere (Pun et al., 2014). Similarly, hidden Markov Models (HMMs) have been implemented to estimate pollutant concentrations primarily based on historical statistical trends (Yetilmezsoy & Abdul-Wahab, 2012). While these conventional strategies had been valuable, they came with numerous limitations. Chief among them are their excessive computational demands, pushed via means of the complicated nature of atmospheric transport mechanisms. Additionally, their inflexible version systems make it difficult to evaluate performance even if large education datasets are available. This restricts their adaptability and effectiveness in dealing with dynamic and nonlinear environmental conditions. As a result, there is a shift in the direction of extra-flexible, statistics-driven forecasting techniques.

Compared to standard models, deep mastering techniques, including Recurrent Neural Networks (RNNs) and Long Short-Term Memory (LSTM) networks, have confirmed a drastic step forward in overall performance in coping with complicated and unstructured data. These models excel at mastering temporal patterns, making them well applicable for meteorological forecasting and air pollutants prediction, in which pollutant tiers rely closely on time-based variables such as wind, temperature, and emissions history. By taking pictures of long-term dependencies and nonlinear relationships within the data, LSTMs, in particular, outperformed conventional statistical and machine learning models in numerous predictive tasks. Their capacity to examine without delay uncooked sequential inputs eliminates the need for extensive function engineering, providing an extra scalable and correct solution. In air quality forecasting, this results in higher predictions of pollutant concentrations and faster variation in converting environmental conditions (Deng et al., 2009). These models excel because of their capacity to seize temporal patterns, outperforming conventional systems studying models. However, deep studying calls for large-scale datasets to optimize overall performance, as visible in different domains, such as computer vision, in which datasets such as ImageNet have been extensively used in research.

Despite progress, the best forecasting sector lacks a complete dataset for education and evaluation. To address this gap, researchers have evolved and launched the AirNet dataset to help improve version and set up a standardized evaluation framework. Experimental validations on AirNet establish an overall performance baseline for air pollutant prediction, facilitating similar research, improving prediction accuracy, and inspiring innovation in air-best prediction technologies. Predicting air quality is a much more complex undertaking than forecasting precipitation due to the complexity of things that influence the awareness and unfolding of air pollution. These elements are dynamic and rely on historical pollutant levels and dynamic climate situations, consisting of wind patterns, temperature shifts, humidity, and atmospheric pressure. This interaction between pollution and meteorological variables makes dealing with air first-class forecasting vital as a sequential prediction that mixes the facts beyond expected.

To address this, the awareness of the pollutant at any given time step, t, denoted as x_t, and the meteorological situations at that identical time, b_t, are modeled based on Eq. (9.4):

$$P(xt) = P(xt \quad xt-1,;,,,xt-2,;,,,,\ldots,;,,,x0,;,,,bt,;,,,bt-1,;,,,,\ldots,;,,,\mathrm{b0}) \tag{9.4}$$

where *xtxt* the concentration of air pollutants at time *t*, *btbt* the weather conditions at time *t*, *xt* − 1,*xt* − 2,…,*x*0*xt* − 1,*xt* − 2,…,*x*0: historical pollutant concentration values, *bt* − 1,*bt* − 2,…,*b*0*bt* − 1,*bt* − 2,…,*b*0: historical weather conditions.

Essentially, this means forecasting the degree of pollutants at times that rely on each beyond pollutant concentration and a mixture of historical and expected climate factors. Climate plays a vital role in how pollution disperses or accumulates, and the destiny of meteorological situations inside the prediction procedure notably improves the model's accuracy. A numerical forecasting technique can be applied, producing facts on key elements, such as wind speed, temperature, and humidity. This fact is then entered into the version to task pollutants, grade by grade, over time, making allowance for forecasts that extend into the future. Large datasets are vital for the education of those models. For instance, the AirNet dataset incorporates over ten million samples from 1498 tracking stations, masking pollution such as $PM_{2.5}$, PM_{10}, NO_2, CO, O_3, SO_2, and AQI between April 2015 and September 2017. Smaller datasets, such as Beijing PM25 and Five-Cities PM25, provide facts at extra localized scales.

To compare prediction performance, metrics such as hit (effectively predicting excessive pollutants), miss (failing to predict a pollutant spike), and false alarm (predicting excessive pollutants while they stay low) are regularly used. Combining those datasets and probabilistic modeling can lead to extra accurate, real-time air quality forecasting. As artificial intelligence (AI) evolves, machine learning (ML) is becoming increasingly essential to cope with big datasets in atmospheric technology, mainly in air pollutant studies. ML algorithms, including Artificial Neural Networks (ANN), Random Forest (RF), and Support Vector Machines (SVM), are typically hired for category and regression responsibilities, assisting in modeling complicated relationships among facts. These ML models are compelling in nonlinear situations, presenting higher prediction accuracy and robustness by decreasing the effect of outliers. This capacity to create direct relationships among facts ends in a forward version match, and a smaller root suggests a rectangular error (RMSE), as validated via various studies.

Furthermore, ML models are of high quality because of their potential to process various fact types, including integers and strings, and their relative simplicity, low cost, and time performance compared to standard numerical models. These blessings make ML strategies mainly appropriate for atmospheric technology, where they may be implemented in responsibilities such as short-term forecasting, chemical evaluation of pollutants, and effect assessments. In addition to their sensible packages, ML models contribute to advancing atmospheric technology studies by allowing more accurate and environmentally friendly analyses of air pollutants. Bibliometric evaluation, a device used to evaluate study developments, performs a crucial function in mapping the improvement of ML in this field. This approach no longer best tracks study topics but also illustrates the interrelations between countries, institutions, and authors, providing a complete view of the field's evolution. A bibliometric evaluation of ML packages in air pollutants studies, specializing in courses from the Web of Science (WoS) center database between 1990 and 2021. The facts series conducted a preliminary search for the usage of popular key phrases, including "system learning," which yielded 1834 files. Due to the exclusion of

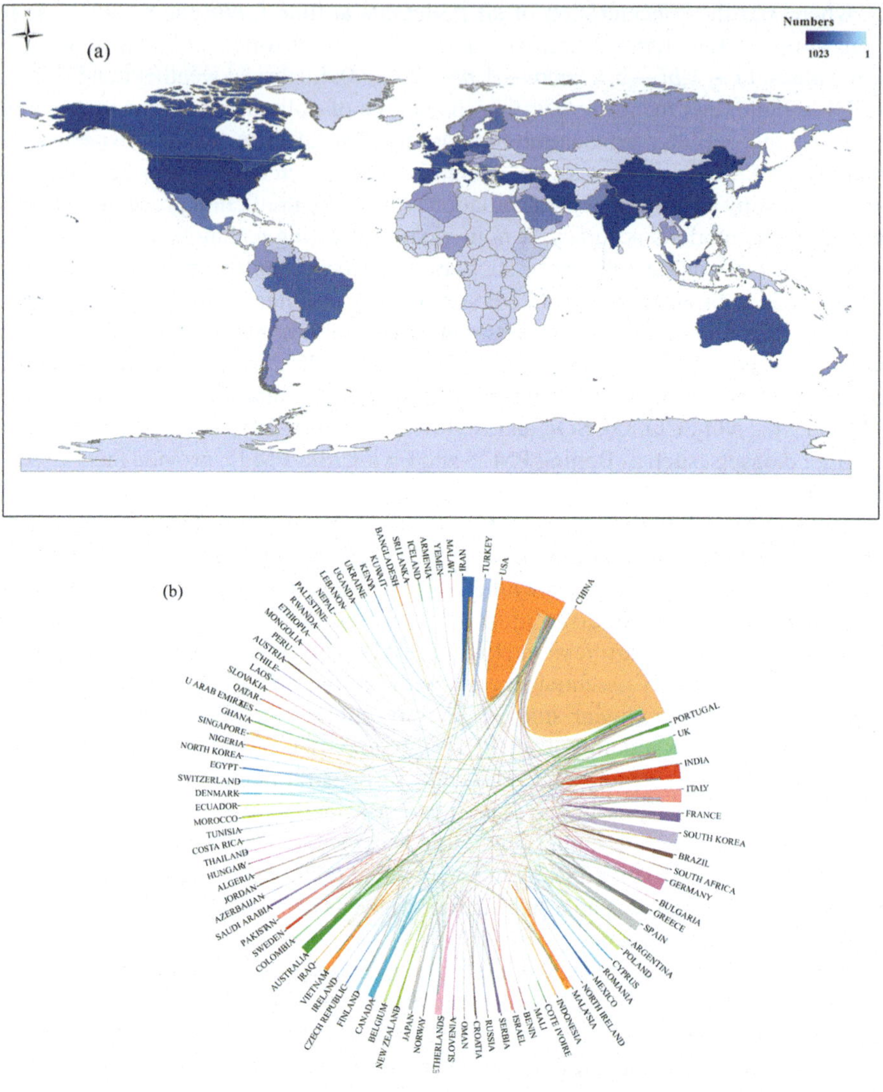

Fig. 9.1 (**a**) Division of publications by different countries; (**b**) collaboration network of countries (Li et al., 2023)

papers, using a precise set of rule names such as ANN, RF, and SVM, a second series with extra specified key phrases introduced a wide variety of files to 3930. After eliminating duplicates and inappropriate papers, 2962 courses have been selected for additional evaluation. These files provided insight into the improvement and developments in the utility of ML in air pollutants studies, presenting valuable guidance for destiny studies and guidelines in atmospheric technology.

Research on air pollutants, gadget usage, and knowledge of ML has been carried out using establishments in 101 countries. As illustrated in Fig. 9.1, China leads in electronic book production with 1023 articles, followed by the United States (376) and India (186). The collaboration patterns, illustrated in Fig. 9.1b, show that the most common partnerships came about between China and the United States (and 80 joint publications), observed with the aid of the United Kingdom (59) and the United Kingdom and the United States (39). Despite these numbers, the multiple country publication (MCP) fee for the three international locations remains below 30%, indicating that most research has been conducted regionally with limited worldwide cooperation.

Since 1990, the utility of ML in air quality studies has gradually increased, with a significant increase after 2017. This exponential increase shows that the sector is transitioning from an early improvement stage to a more mature stage. China and the United States dominate the landscape, contributing more than one-half of all publications. However, study establishments remain widely dispersed, and robust worldwide networks or large-scale collaborative hubs are lacking.

Recent research has increasingly focused on improving pollutant detection systems, improving short-term air quality forecasting, carrying out chemical composition evaluation of numerous pollutants, and optimizing techniques for emission control. Moreover, there is a developing hobby in integrating satellite TV for PC data, sensor networks, and advanced ML algorithms, including ensemble techniques and deep knowledge of fashions to enhance predictive accuracy and spatial coverage. These traits replicate a larger push closer to real-time, data-driven answers in environmental management, specifically in city regions that face chronic pollutant challenges.

Publications that follow ML to air pollutants studies have accelerated rapidly in 30 years and grown exponentially after 2017. The dominant studies business was primarily based in China and the United States, collectively accounting for over 50% of all articles. Research firms and establishments had been highly scattered, and massive studies and collaborations had no longer been formed. However, ML software for air pollutants studies has reached maturity, with a strong recognition of the chemical characteristics of pollutants, short-term pollutant forecasts, enhancing pollutant detection performance, and optimizing layout for emission reduction.

9.8 Computational Fluid Dynamics Analysis of Pollutants

This examines how pollutant dispersion around high upward pushes home modifications while surrounding systems are present. Using wind tunnel experiments and large eddy simulation (LES), primarily based on computational fluid dynamics (CFD), researchers analyzed pollutants unfolding around a standalone high-rise construction and a primary high-upward push surrounded by low-upward push homes. Pollutants had been launched on the pedestrian peak at the leeward aspect of the high-upward push. The findings confirmed that the surrounding homes extensively impact the dispersion of pollutants, causing pollutants inside the wake location to move vertically upward closer to the high-upward push's roof. At two-thirds of the construction peak, there were five instances of pollutant concentrations better inside the surrounded case than inside the remote scenario. Consequently, citizens on higher floors might enjoy better air quality despite being far above the street-level source.

Additionally, turbulent diffusive flux and implied convective flux inside the streamwise path were found to be similar but oppositely directed. The look underscores the danger of underestimating the stages of pollutant around high-rise buildings if surrounding systems are not adequately considered. The computational fluid dynamics (CFD) simulation has numerous benefits over bodily measurements and empirical statistical models, specifically (1) it can offer complete flow and dispersion area statistics and can be accomplished at a complete scale, and (2) similarity requirements can no longer afflict it. Consequently, it has been diagnosed as an effective device for pollutant dispersion problems.

Previous research on pollutant dispersion around city homes through computational fluid dynamics (CFD) modeling primarily focused on simplified scenarios: remote homes and avenue canyons. These idealized setups have allowed researchers to recognize the turbulent delivery mechanisms more clearly and determine the overall performance of CFD models, overcoming the complexities of real city environments (Ioannidis et al., 2024; Meroney, 2016; Schatzmann et al., 1997). However, global city regions contain more elaborate interactions among homes, atmospheric flows, and pollutant dispersion styles. Although some studies have focused on grids of homes or constructing blocks, studies on how high-upward push homes affect pollutant dispersion in neighborhoods remain exceptionally scarce. High-upward push homes, not unusual places in big cities, can considerably adjust airflows and pollutant motion because of their interplay with the atmospheric boundary layer (ABL).

High-upward push houses generate complicated flow features, consisting of downwash nook streams, which affect the pollutant dispersion field. Additionally, city avenue intersections, which regularly improve automobile emissions, similarly complicate the way with drift and pollutant dynamics. The go-drift styles at those intersections decide how pollution unfolds into adjoining streets and upward pushes into the atmosphere. Although some case studies have tested pollutant dispersion in actual city configurations, abnormal construction geometries make it hard to analyze the specified mechanisms of dispersion prompted through high-rise homes. Notable studies have tested how tall homes affect pollutant behavior. For example, they explored the impact of a tall tower on airflows and pollutant dispersion using a version of a city community using wind tunnel experiments and CFD simulations. Their findings discovered that the tower laterally drawn pollution toward it, more substantial vertical dispersion in its wake, and decreased pollutant residence time close to the construction. Similarly, pollutant diffusion around a high upward push in Hong Kong was investigated, and pollution migrated downward because of a downwash in windward emissions, while leeward emissions ended in turbulent mixing and upwind migration of pollution. It highlighted the formation of “useless zones” and “pollutant hotspots” close to high-rise homes, similarly emphasizing the effect of such systems on dispersion styles. Despite these findings, the specified dispersion mechanisms specified in actual city environments are not entirely understood because of the unique configurations utilized in most research. The existing examination followed a usual organizational configuration, such as a high-upward push with a 1:1:3 shape. The pollutants had been released from the lowest of the leeward aspects of the high-upward push. This reasonable summary setup bridges the information of remote homes, avenue canyons, and constructed arrays from previous studies, offering broader information on city pollutant dispersion.

9.9 Numerical Model

Research on air pollutants and machine learning (ML) knowledge has been performed through establishments throughout one-zero and other nations. As illustrated in Fig. 9.1, China leads in e-book output with 1023 articles, followed by the United States (376) and India (186). The collaboration patterns, illustrated in Fig. 9.1b, show that the maximum of common partnerships between China and the United States (180 joint publications), accompanied by China and the United Kingdom (59) and the United Kingdom and the United States (39). Despite these numbers, the multiple country publication (MCP) price for the top three nations stays below 30%, indicating that most research has been performed locally with limited global cooperation. Since 1990, the number of ML in air-best studies has regularly increased, with a substantial surge after 2017. This exponential boom shows that the sphere is transitioning from an early improvement to an advanced stage. China and the United States dominate the landscape, contributing more than half of all publications. However, study establishments remain widely dispersed and robust worldwide networks or large-scale collaborative hubs are lacking. Recent research has been increasingly aimed at improving the accuracy and performance of pollutant detection systems, refining short-term air-best forecasting, performing chemical composition evaluation of diverse pollutants, and optimizing techniques for emission control. Moreover, there is a developing hobby in integrating satellite TV for PC data, sensor networks, and advanced ML algorithms, including ensemble strategies and deep learning knowledge of fashions to enhance predictive accuracy and spatial coverage. These tendencies reflect a broader push toward real-time, data-driven answers in environmental management, especially in city regions facing chronic pollution challenges:

$$\partial C \partial t + \partial (uj^{-}C)\ \partial xj - \partial \partial x\ j\ (Deff\, \partial C \partial x\ j) = 0 \tag{9.5}$$

where D represents the robust diffusion coefficient, the sum of the turbulent diffusion coefficient Dt and the molecular diffusion coefficient D_m. However, because of the negligible impact of D_m on pollutant dispersion in low-speed winds, it was excluded from the version. This exclusion aligns with the findings of Bonifacio et al. (2014) and Bedogni et al. (2015), who have proven that D_m may not be noted because of its notably minor significance relative to D_t. Additionally, Holmes and

Morawska (2006) and Tominaga and Stathopoulos (2013) concluded that modeling particle dispersion in road environments should disregard coagulation phenomena over brief timescales. The study has modeled nitrogen oxides (NO_x) as nonreactive pollutants, given the quick simulation timescales and impartial balance in city microscale environments. The turbulent Schmidt number *S*, used to compute the turbulent diffusion term Dt, was transformed into a set of 0.7, primarily based on Eq. (9.6):

$$Sct = vtdt \tag{9.6}$$

where v_t is the turbulent viscosity, which characterizes airflow inside the simulation. The air kinematic viscosity was set at 1 in five instances, 10^{-5} m^2/s at the area boundaries, reflecting atmospheric conditions.

9.10 Turbulence and Velocity Profiles

The RANS simulations used the same old k-ε turbulence version, usually used in pollutant dispersion studies. Pantusheva et al. (2022) confirmed that the same old k-ε version was used in 38% of the reviewed city pollutant dispersion cases, which was the most appropriate for such applications. The atmospheric speed profile on the inlets can be calculated using Eq. (9.7):

$$(z) = u * (z + z_0 z) \tag{9.7}$$

Here, *u* represents the friction velocity, *κ* (kappa) is the Von Karman constant (0.41), and z_0 is the length of aerodynamic roughness, which is about 0.25 m for hard urban surfaces (Pun et al., 2014). This logarithmic wind profile captures the conduct of the atmospheric boundary layer and is vital in enhancing the accuracy of pollutant dispersion modeling in city environments (Ioannidis et al., 2024). Meteorological facts are crucial for air quality modeling because they offer key atmospheric conditions for the duration of the simulation. For this look-at, wind velocity and route facts had been received from a sensor positioned 15 km southeast of the case look-at site, outside the city material, and far from capability disruptions resulting from surrounding buildings. The tracking device was placed at a height of 3 m and shaped as a necessary issue of the Smart AQ Net community in Augsburg, Germany. This modern air fine tracking machine is a hybrid framework, combining high-precision government-established stations with several mid- and low-priced sensors operated collaboratively by citizens and research groups. The community tracks more than 30 meteorological and aerosol-associated parameters, including wind speed and direction, temperature, atmospheric pressure, humidity, and precipitation. Smart AQ Net covers a total region of about 16 × 16 km^2, with the best sensor density focused in a 6 × 4 km^2 area that encompasses the maximum of Augsburg's urban core of Augsburg.

During the simulation period, evaluation of wind situations discovered dominant wind guidelines from the northeast and west. Wind speeds have typically been low, ranging from 2 to 1.2 m, in line with the second (m/s). A closer breakdown confirmed that 21% of wind speeds fell below 20.4 m/s, 29% ranged from 0 to 4.6 m/s, and every other 13% has been lightly distributed between 0–0.8 m/s and 0.8–1.0 m/s. The final 24% passed off inside the 1.0–1.2 m/s bracket. In total, 76% of recorded wind situations have been categorized as calm, probably inspired by the fantastically low setup height of the sensor unit. To enhance the accuracy of pollutant dispersion modeling, those recorded wind styles have been integrated into the atmospheric boundary layer profile carried out on the CFD version of boundary conditions. Incorporation enabled the version to simulate fundamental international airflow dynamics, which might be critical for correct pollutant delivery predictions in city settings. By integrating real low-wind situations into the center of the city of Augsburg, the version more reliably captured the conduct of atmospheric flows and the dispersion under solid situations. This step was essential to ensure that the simulated pollutant dispersion fields reflected actual environmental dynamics, boosting the predictive fidelity (Ioannidis et al., 2024).

9.11 Simulation of the Chemical Transport Model

Machine learning (ML) has been around for more than 70 years. However, its utility in environmental technology gained momentum in the late 1990s. Early studies explored various ML strategies for predicting and classifying atmospheric states, with neural networks (NNs) rising as one of the most famous methods. Initial research confirmed the ability of NNs to predict destiny pollutant concentrations primarily based on air quality and climate data, with fantastic examples from Stockholm and Helsinki. Over time, NNs have evolved and are seen as an up-and-coming tool for applications in environmental and weather technology. Neural networks are especially valued for their flexibility and effectiveness. According to the universality theorem, with high accuracy, NNs can approximate any complicated, nonstop function, including the evolution of atmospheric chemical states. However, NNs are based on a "black-box" approach. The developers outline the structure of the NNs, which includes defining the type and range of neuron layers. However, they no longer explicitly specify the mathematical relationships among these layers. Instead, a set of schooling rules determines those connections by optimizing the connection between the input and output data. This technique eliminates the need to manually version complicated processes, reducing the probability of introducing errors.

In practice, the schooling technique is not perfect. NNs use iterative algorithms that shape the community by becoming its output to the schooling data, which may fit nicely for positive styles of trouble. However, it can fail when applied to others. NN builders often aim to improve performance by designing architectures with problem-specific knowledge. This includes prescribing unique connections or systems within the NN and allowing manual schooling techniques toward applicable solutions. Another approach to decorating NN schooling is the usage of techniques-associated consequences. For instance, they educated NNs on using consequences that represent primary bodily principles. During schooling, the NN adjusted itself to reduce this penalty, which helped it produce bodily regular solutions. Achieving this requires a structure that accommodates such constraints, which could improve accuracy and reliability. However, deciding on the ultimate structure is not straightforward. NN builders must solve diverse and sensible questions to ensure that the selected structure fits the unique problem (Vlasenko et al., 2021).

Dispersion and chemical transport models (CTMs) are critical for comparing ambient air pollutant exposure, typically used in epidemiological studies and health impact assessments (HIA). Dispersion models simulate pollutant shipping, dispersion, and deposition, offering predictions at various spatial and temporal scales. CTMs amplify these abilities by incorporating atmospheric chemistry and aerosol dynamics, providing a complete evaluation of air pollutants' behavior. A most important benefit of CTMs is their ability to supply specific pollutant components, allowing policymakers to assess the influences of emission discount techniques on air quality. Such techniques can also include decreasing on-street car emissions on the street through public transportation promotion, adopting low- or zero-emission fuels, and helping electric vehicles use by enhancing the power sector. However, CTMs are particularly complicated and require specialized expertise to operate. Policymakers often depend upon air quality modeling professionals for behavior simulations, interpreting results, and dealing with uncertainties while assessing the costs and benefits of pollutant-discount interventions. Despite improvements in computational performance, the complexity of CTMs remains a barrier, particularly for studies that require several versions of runs, including HIAs, regarding a couple of intervention scenarios.

Surrogate models have been advanced as green options to full-scale chemical transport models (CTMs), presenting quicker approximations of atmospheric processes under simplified assumptions. These fashions combine environmental records over prolonged periods and across extensive geographic areas to reduce computational complexity. A not-unusual place technique entails the usage of reduced complexity models (RCMs), which can be designed to estimate fitness effects from precise pollutant sources, along with coal-fired energy plants, without the need for in-depth simulations. For example, the TM5-FASST version uses a source–receptor matrix from the worldwide TM5 CTM outputs to evaluate how emissions from one location affect air quality and fitness effects in others. While those surrogate fashions beautify computational performance and are particularly beneficial for large-scale coverage tests or state-of-the-art analyses, they come with trade-offs in accuracy. Their reliance on simplifications frequently ends in omitting vital atmospheric chemistry procedures or nuances in pollutant transport. As a result, they will no longer seize short-term fluctuations, nonlinear interactions, or localized environmental dynamics, which may be necessary for particular forecasting and intervention planning. Despite those limitations, surrogate fashions remain treasured within the broader modeling ecosystem, specifically when rapid, large-scale reviews are required. Furthermore, those simplified models are commonly aware of large-scale local analyses and the absence of precision for intracity air quality assessments. The Community Multiscale Air Quality (CMAQ) system incorporates specialized variations such as CMAQ-DDM-3D and ISAM. An extra state-of-the-art answer is available. These variations allow for sensitivity analyses and supply attribution calculations within a single simulation, decreasing computational demand. Although they offer progressive state-of-the-art evaluation abilities, they continue to be complex and resource-intensive. A CTM provides designated insights into pollutant dispersion and supply influences, but its computational needs pose challenges, particularly for multistate-of-affairs HIAs. Surrogate models and specialized gear, such as CMAQ-DDM-3D and ISAM, are intended to stabilize computational performance and accuracy; nevertheless, they require a substantial understanding to be enforced effectively.

9.12 Gaussian Process Emulation

Gaussian process emulation (GPE) is a statistical method that replicates computationally intensive models' outputs, including those utilized in atmospheric chemistry and air quality prediction. It permits fast scenario evaluation through emulating the outputs Y of a complicated model such as the CCAM-CTM, using a set of input emission parameters $X = \{X1,X2,\ldots,Xd\}$ defined over a d-dimensional space. For any scenario S, a set of selected parameters x is entered to estimate the output Y = f(x), which usually requires the complete model execution. Instead, the emulator approximates the output with $Y = f(x)$, where f is the surrogate function built through GPE, trained on a dataset of previous simulation output.

GPE operates within a Bayesian framework, where the prior distribution of *fff* is updated using training data to produce a posterior predictive distribution. The prior mean is modeled as:

$$m(x) = h(x)T\beta.$$

where $h(x)$ is a known function of x and β is estimated by regression on the training outputs. The prior covariance for two parameter points x and x' is given by:

$$\mathrm{cov}(x,x') = \sigma 2c(x,x') \tag{9.8}$$

where $c(x,x')$ is a Gaussian correlation function dependent on the distance between the inputs and the smoothness parameters δi, which quantify the sensitivity of the output to each input variable and are inferred from the training data.

The resulting emulator f defines a posterior distribution with a mean function: $m*(x) = h(x)T\beta + t(x)TA - 1(y - H\beta)$.

and posterior covariance:

$$\mathrm{cov}*\left(x,x'\right)=\sigma 2\left[\begin{array}{l} c\left(x,x'\right)-t\left(x\right)TA-1t\left(x'\right)+\left(h\left(x\right)-t\left(x\right)TA-1H\right) \\ \left(HTA-1H\right)-1\left(h\left(x'\right)-t\left(x'\right)TA-1H\right)T \end{array}\right].$$

Here, the output vector yT = (f(x1),…,f(xn))yT = (f(x1), the design matrix HT = (h(x1),…,h(xn)) and the matrix A consists of the correlations c(xi,xj). The vector t(x)T = (c(x,x1),…,c(x,xn))t(x)T = (c(x,x1),…,c(x,xn)) represents the correlation of the new input x with training data points. The coefficients β and the variance estimate σ^2 are computed as:

$$\begin{array}{l} \beta=\left(HTA-1H\right)-1HTA-1y,\sigma 2=n-q-21yT \\ \left[A-1-A-1H\left(HTA-1H\right)-1HTA-1\right]y \end{array} \tag{9.9}$$

where n is the number of training data points and q is the number of parameters plus one. These formulations, originally presented by O'Hagan and extended by Johnson et al. (2022), form the core mathematical framework of GPE, highlighting the interaction between emulator fidelity and training data structure.

9.13 Conclusion

Computational techniques have revolutionized the sector of air pollution control, imparting innovative, scalable, and particular answers to address this urgent environmental challenge. These strategies permit identifying and tracking emission sources, simulation of pollutant dispersion, and excellent assessment of real-time air, contributing to information-driven decision-making. Technologies that include chemical transport models (CTMs), computational fluid dynamics (CFD), and machine learning (ML) algorithms offer more detailed information about pollutant behaviors across various spatial and temporal scales. In addition, integration of far-away sensing and GIS technology complements the functionality of music and controls pollutants in real time, equipping policymakers and researchers with actionable insights. Moreover, massive information analytics helps process and interpret considerable datasets, reworking excellent raw air statistics into predictive models and actionable outcomes. Computational equipment additionally empowers policymakers by comparing the effectiveness of regulatory measures, ensuring a systematic foundation for environmental regulation and intervention strategies. Through the synergy of these techniques, it is feasible to design city and commercial infrastructures that prioritize sustainability while minimizing air pollutants. By providing cost-effective, efficient, and scalable responses, computational strategies bridge the distance between medical studies and sensible solutions in air pollutant control. They empower communities, industries, and governments to collaborate to combat the detrimental influences of air pollutants on public health and the environment. As the sector advances, the mixing of artificial intelligence, superior sensors, and interdisciplinary studies will also enhance the ability of computational techniques to attain purified air and a more sustainable future.

Acknowledgments The authors would like to express their sincere gratitude to the University of Chakwal, and University of Rasul, Mandi Baha Uddin, Pakistan, for providing the facilities and resources necessary for the completion of this chapter. The authors are also deeply thankful to Dr. Zahoor Ahmed and Dr. Muhammad Naeem Ashiq for their insightful feedback and valuable suggestions, which significantly improved the quality of this work. Special appreciation is extended to our families and friends for their patience, encouragement, and unwavering support throughout this project.

References

Akyildiz, I. F., Melodia, T., & Chowdhury, K. R. (2007). A survey on wireless multimedia sensor networks. *Computer Networks, 51*(4), 921–960. https://doi.org/10.1016/j.comnet.2006.10.002

Bedogni, L., Gramaglia, M., Vesco, A., Fiore, M., Härri, J., & Ferrero, F. (2015). The Bologna Ringway dataset: Improving road network conversion in SUMO and validating urban mobility via navigation services. *IEEE Transactions on Vehicular Technology, 64*(12), 5464–5476. https://doi.org/10.1109/TVT.2015.2475608

Bonifacio, S. H., de Oliveira, P. S., & Loureiro, F. S. (2014). "A simplified methodology for the simulation of the dispersion of pollutants in the atmosphere." International Journal of Environmental Science and Technology, 11(6), pp. 1655–1662.

Buekers, J., Stassen, K., Panis, L. I., Hendrickx, K., & Torfs, R. (2011). Ten years of research and policy on particulate matter air pollution in hot spot Flanders. *Environmental Science & Policy, 14*(3), 347–355. https://doi.org/10.1016/j.envsci.2010.10.012

Chen, L., Shi, M., Li, S., Gao, S., Zhang, H., Sun, Y., & Zhou, J. (2017). Quantifying public health benefits of environmental strategy of PM2.5 air quality management in Beijing–Tianjin–Hebei region, China. *Journal of Environmental Sciences, 57*, 33–40. https://doi.org/10.1016/j.jes.2016.11.014

Deng, J., Dong, W., Socher, R., Li, L. J., Kai, L., & Li, F. F. (2009). ImageNet: A large-scale hierarchical image database. In *Paper presented at the 2009 IEEE Conference on Computer Vision and Pattern Recognition*.

Deschênes, O., Greenstone, M., & Shapiro, J. S. (2017). Defensive investments and the demand for air quality: Evidence from the NOx budget program. *The American Economic Review, 107*(10), 2958–2989. https://doi.org/10.1257/aer.20131002

Fonte, C. C., Minghini, M., Patriarca, J., Antoniou, V., See, L., & Skopeliti, A. (2017). Generating Up-to-Date and Detailed Land Use and Land Cover Maps Using OpenStreetMap and GlobeLand30. *Geo Information, 6*(4), 125.

Hering, L., & Poncet, S. (2014). Environmental policy and exports: Evidence from Chinese cities. *Journal of Environmental Economics and Management, 68*(2), 296–318. https://doi.org/10.1016/j.jeem.2014.06.005

Holmes, N. S., & Morawska, L. (2006). "A review of dispersion modelling and its application to the dispersion of particles from road, near-road and vehicular sources." *Atmospheric Environment*, 40(26), pp. 4909–4938.

Ioannidis, G., Li, C., Tremper, P., Riedel, T., & Ntziachristos, L. (2024). Application of CFD modelling for pollutant dispersion at an urban traffic hotspot. *Atmosphere, 15*(1). https://doi.org/10.3390/atmos15010113

Johnson, J. J., et al. (2022). "Machine Learning Applications for Air Quality Prediction: A Comprehensive Review." *Journal of Cleaner Production*, 345, 131154.

Kaldellis, J. K., Chalvatzis, K. J., & Spyropoulos, G. C. (2007). Transboundary air pollution balance in the new integrated European environment. *Environmental Science & Policy, 10*(7), 725–733. https://doi.org/10.1016/j.envsci.2007.06.002

Krause, A., & Guestrin, C. (2009). Optimizing sensing: From water to the web. *Computer, 42*(8), 38–45. https://doi.org/10.1109/MC.2009.265

Landrigan, P. J., Fuller, R., Acosta, N. J. R., Adeyi, O., Arnold, R., Basu, N., & Zhong, M. (2018). The lancet commission on pollution and health. *The Lancet, 391*(10119), 462–512. https://doi.org/10.1016/S0140-6736(17)32345-0

Li, Y., Sha, Z., Tang, A., Goulding, K., & Liu, X. (2023). The application of machine learning to air pollution research: A bibliometric analysis. *Ecotoxicology and Environmental Safety, 257*, 114911. https://doi.org/10.1016/j.ecoenv.2023.114911

Lin, C. Q., Liu, G., Lau, A. K. H., Li, Y., Li, C. C., Fung, J. C. H., & Lao, X. Q. (2018). High-resolution satellite remote sensing of provincial PM2.5 trends in China from 2001 to 2015. *Atmospheric Environment, 180*, 110–116. https://doi.org/10.1016/j.atmosenv.2018.02.045

Mason, T. G., Schooling, C. M., Chan, K. P., & Tian, L. (2019). An evaluation of the air quality health index program on respiratory diseases in Hong Kong: An interrupted time series analysis. *Atmospheric Environment, 211*, 151–158. https://doi.org/10.1016/j.atmosenv.2019.05.013

Meroney, R. N. (2016). Ten questions concerning hybrid computational/physical model simulation of wind flow in the built environment. *Building and Environment, 96*, 12–21. https://doi.org/10.1016/j.buildenv.2015.11.005

Napelenok, S. L., Cohan, D. S., Hu, Y., & Russell, A. G. (2006). Decoupled direct 3D sensitivity analysis for particulate matter (DDM-3D/PM). *Atmospheric Environment, 40*(32), 6112–6121. https://doi.org/10.1016/j.atmosenv.2006.05.039

Pantusheva, M. G., Gadzhev, & Ganev, K. (2022). "A Review of Numerical Simulations of the Urban Air Quality." Atmosphere, 13(10), 1618.

Polastre, J., Szewczyk, R., Mainwaring, A., Culler, D., & Anderson, J. (2004). Analysis of wireless sensor networks for habitat monitoring. In C. S. Raghavendra, K. M. Sivalingam, & T. Znati (Eds.), *Wireless sensor networks* (pp. 399–423). Springer.

Pun, V. C., Yu, I. T.-S., Qiu, H., Ho, K.-F., Sun, Z., Louie, P. K. K., & Tian, L. (2014). Short-term associations of cause-specific emergency hospitalizations and particulate matter chemical components in Hong Kong. *American Journal of Epidemiology, 179*(9), 1086–1095. https://doi.org/10.1093/aje/kwu026

Rastogi, U., Kumar, S., Maurya, S. P., & Yadav, A. K. (2025). Air pollution forecasting using genetic algorithm and LSTM deep learning technique. *Aerosol Science and Engineering*. https://doi.org/10.1007/s41810-025-00343-y

Schatzmann, M., Rafailidis, S., & Pavageau, M. (1997). Some remarks on the validation of small-scale dispersion models with field and laboratory data. *Journal of Wind Engineering and Industrial Aerodynamics, 67-68*, 885–893. https://doi.org/10.1016/S0167-6105(97)00126-8

Shahbazi, H., Hassani, A., & Hosseini, V. (2019). Evaluation of Tehran clean air action plan using emission inventory approach. *Urban Climate, 27*, 446–456. https://doi.org/10.1016/j.uclim.2019.01.002

Sueyoshi, T., & Yuan, Y. (2015). China's regional sustainability and diversified resource allocation: DEA environmental assessment on economic development and air pollution. *Energy Economics, 49*, 239–256. https://doi.org/10.1016/j.eneco.2015.01.024

Tominaga, Y., & Stathopoulos, T. (2013). CFD simulation of near-field pollutant dispersion in the urban environment: A review of current modeling techniques. *Atmospheric Environment, 79*, 716–730. https://doi.org/10.1016/j.atmosenv.2013.07.028

Vlasenko, A., et al. (2021). "SmartAQnet—A Novel Approach for High-Resolution Air Quality Monitoring in Urban Areas." Sensors, 21(23), 7856.

Vlasenko, A., Matthias, V., & Callies, U. (2021). Simulation of chemical transport model estimates by means of a neural network using meteorological data. *Atmospheric Environment, 254*, 118236. https://doi.org/10.1016/j.atmosenv.2021.118236

Wang, Y. C., & Chen, G. W. (2017). Efficient data gathering and estimation for metropolitan air quality monitoring by using vehicular sensor networks. *IEEE Transactions on Vehicular Technology, 66*(8), 7234–7248. https://doi.org/10.1109/TVT.2017.2655084

Wang, Y. C., Hsieh, Y. Y., & Tseng, Y. C. (2009). Multiresolution spatial and temporal coding in a wireless sensor network for long-term monitoring applications. *IEEE Transactions on Computers, 58*(6), 827–838. https://doi.org/10.1109/TC.2009.20

Wang, Y.-C. (2014). Mobile sensor networks: System hardware and dispatch software. *ACM Computing Surveys, 12*, 47. https://doi.org/10.1145/2617662

Werner-Allen, G., Lorincz, K., Ruiz, M., Marcillo, O., Johnson, J., Lees, J., & Welsh, M. (2006). Deploying a wireless sensor network on an active volcano. *IEEE Internet Computing, 10*(2), 18–25. https://doi.org/10.1109/MIC.2006.26

Yadav, A. K. (2021). Human health risk assessment in opencast coal mines and coal-fired thermal power plants surrounding area due to inhalation. *Environmental Challenges, 3*, 100074.

Yang, H.-H., Dhital, N. B., Wang, Y.-F., Huang, S.-C., & Zhang, H.-Y. (2020). Effects of short-duration vehicular traffic control on volatile organic compounds in roadside atmosphere. *Atmospheric Pollution Research, 11*(2), 419–428. https://doi.org/10.1016/j.apr.2019.11.016

Yetilmezsoy, K., & Abdul-Wahab, S. A. (2012). A prognostic approach based on fuzzy-logic methodology to forecast PM10 levels in Khaldiya residential area, Kuwait. *Aerosol and Air Quality Research, 12*(6), 1217–1236. https://doi.org/10.4209/aaqr.2012.07.0163

Zhang, D., & Woo, S. S. (2019). Predicting air quality using moving sensors (poster). In *Paper presented at the proceedings of the 17th annual international conference on Mobile systems, applications, and services, Seoul, Republic of Korea*. https://doi.org/10.1145/3307334.3328647

Zhang, G., Jia, Y., Su, B., & Xiu, J. (2021). Environmental regulation, economic development and air pollution in the cities of China: Spatial econometric analysis based on policy scoring and satellite data. *Journal of Cleaner Production, 328*, 129496. https://doi.org/10.1016/j.jclepro.2021.129496

Zou, B., Wilson, J. G., Zhan, F. B., & Zeng, Y. (2009). Air pollution exposure assessment methods utilized in epidemiological studies. *Journal of Environmental Monitoring, 11*(3), 475–490. https://doi.org/10.1039/B813889C

Chapter 10
Transforming Air Quality Forecasting with Machine Learning Techniques

Vaishnav Bartaria, Ashok Jangid, Siddharth Agarwal, and Ranjit Kumar

10.1 Introduction

Air pollution caused by toxic substances, including particulate matter ($PM_{2.5}$, PM_{10}), originates mainly from anthropogenic sources such as vehicle exhaust, industrial emissions, and agricultural activities (Gupta et al., 2019). The World Health Organization (World Health Organization, 2021) estimates that 99% of the world's population breathes air above recommended pollution levels, causing almost seven million yearly premature deaths. Pollutants aggravate asthma, heart disease, and other major medical conditions. Environmentally, air pollution speeds up climatic feedback loops that increase global warming, biodiversity loss, soil, and water pollution. With losses from air pollution as infrastructure deterioration, decreased productivity, and health costs that amount to approximately US $8 trillion per year, the economic impact at 6.1% of global GDP is striking (World Bank Group, 2016). With these impacts being so great, accurate air quality modeling is acutely needed for real-time public health warnings and providing early warning for vulnerable groups such as asthma sufferers. It ensures compliance with air quality criteria, most importantly the EPA's National Ambient Air Quality Standards (NAAQS), therefore enabling legislators to achieve emission reductions. Air quality models greatly contribute to slowing climate change by tracking pollution trends, simulating radiative forcing effects, and implementing targeted actions.

V. Bartaria · R. Kumar (✉)
Faculty of Science, Department of Chemistry, Dayalbagh Educational Institute, Agra, India

A. Jangid
Faculty of Science, Department of Physics and Computer Science, Dayalbagh Educational Institute, Agra, India

S. Agarwal
Department of Emergency Medicine, Government Medical College, Firozabad, India

A. K. Yadav (ed.), *Computational Approach for Air Pollution*,
https://doi.org/10.1007/978-3-032-09215-1_10

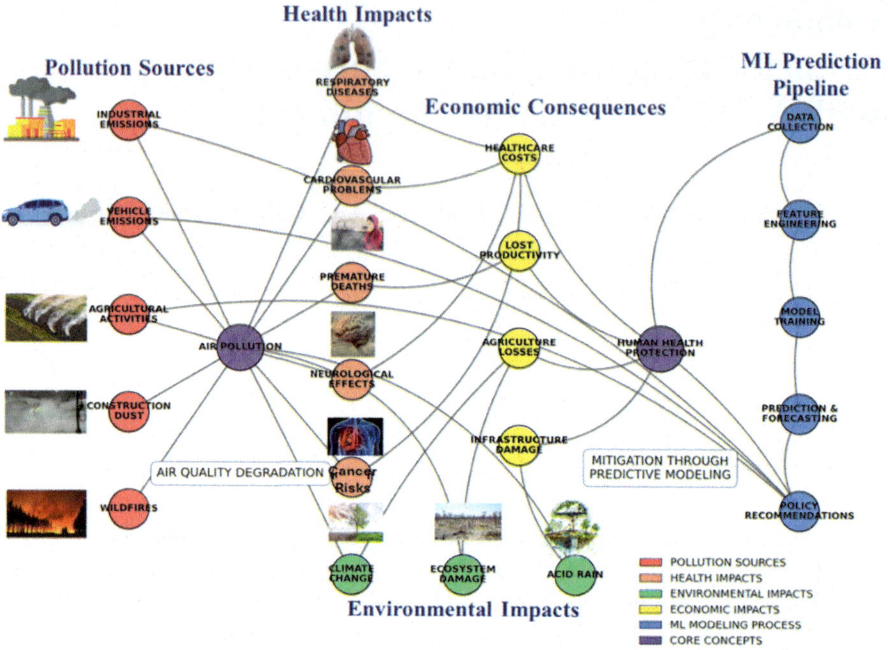

Fig. 10.1 Machine learning for air pollution prediction: protecting human health

Pollution dispersion and chemical transformation simulation have made extensive use of traditional chemical transport models (CTMs) such as CMAQ and WRF-Chem. However, quantifying nonlinearity in pollution interactions, high processing demands, and unavailability of data in some regions (Gao & Zhou, 2024; Wang et al., 2024; Yadav, 2015) pose challenges. Enhanced methodologies are necessary for precise air quality modeling under these constraints. Figure 10.1 demonstrate the interconnection of air pollution to human health and how mitigation is possible through predictive modeling.

Machine learning (ML) has emerged as a potent alternative to improve predictive accuracy and scalability by leveraging various data sources, including satellite imaging, Internet of Things (IoT) sensors, and meteorological indicators. Although hybrid methods combining machine learning (ML) with physics-based computational traffic models (CTMs) maintain interpretability while increasing predictive effectiveness, models containing LSTM networks facilitate real-time prediction. Data-based machine learning technologies have replaced deterministic mathematical forecasts in the prediction of air pollution. Machine learning algorithms analyze complex pollution interactions and provide scalable, real-time monitoring systems for proactive air quality control. Air pollution harms human health, the environment, and the economy; hence, machine learning (ML) in air quality prediction is essential for public health advice, regulatory compliance, and climate change mitigation. Figure 10.2 shows flow chart of evolution from traditional to ML models with time.

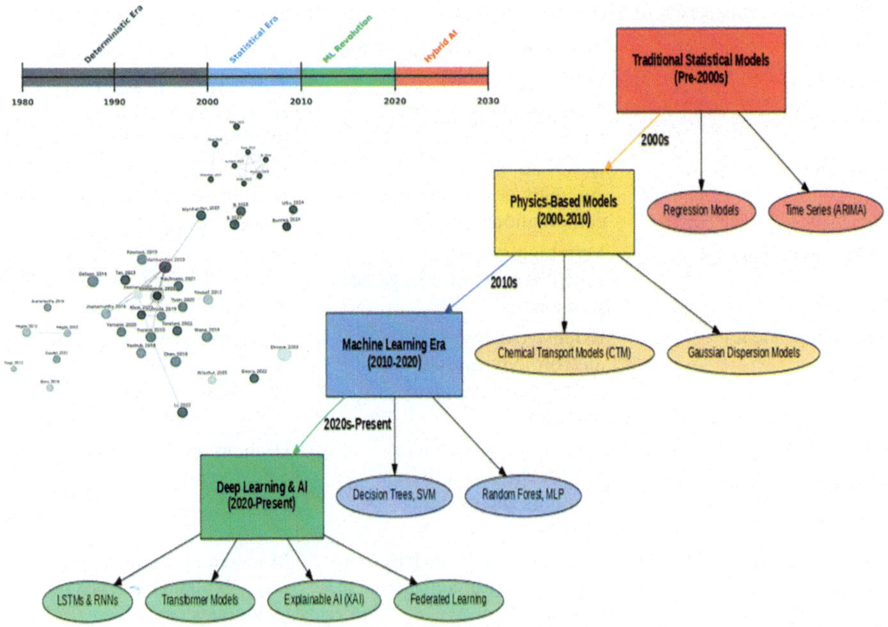

Fig. 10.2 Flow chart of evolution from traditional to ML models

10.2 Conventional Air Quality Models

Air quality models are important in terms of tracking and predicting levels of pollution, assessing compliance with regulations, and informing public health interventions. Models tend to fall into either the deterministic or statistical category. Deterministic models emulate the processes of pollution using physical and chemical principles, while statistical models provide links between the pollutants and driving factors. Despite their widespread usage, both approaches have shortcomings, particularly in dealing with complex atmospheric interactions, the availability of data, and scalability. A comparison table of traditional and machine learning models with different features is shown in Table 10.1.

10.2.1 Deterministic Models

Mathematical representations of atmospheric phenomena facilitate deterministic models, specifically chemical transport models (CTMs) and Gaussian dispersion models, in simulating air pollution. These models are relevant for policy formulation, source attribution, and regulatory objectives.

Table 10.1 Comparison of traditional and machine learning models

Feature	Traditional models (CTMs and statistical models)	Machine learning (ML) models
Accuracy	Moderate, depends on parameterization and input data	High learns complex relationships from large datasets
Scalability	Low- and high-resolution simulations require extensive computational power	High, easily scalable with cloud computing and parallel processing
Computational cost	Very high, requiring supercomputers for large-scale simulations	Moderate, can be optimized for efficiency with GPUs and TPUs
Data dependency	Extensive datasets on meteorology, emission, and atmospheric chemistry are required	Flexible, can work with diverse datasets (satellite data, IoT, crowdsourced data)
Adaptability	Low, as models require recalibration for different locations and evolving conditions	High, as models dynamically adjust to new data sources
Interpretability	High, as models are based on known physical and chemical equations	Moderate, but can be improved using explainable AI (XAI) techniques
Real-time prediction	Limited due to high computational requirements	High, can generate real-time air quality predictions
Handling of nonlinear interactions	Limited, as deterministic models struggle with complex, nonlinear pollutant interactions	Excellent, ML algorithms excel at recognizing nonlinear patterns
Handling of extreme events (e.g., wildfires, dust storms, industrial leaks)	Poor, as traditional models may not react well to unexpected or rare pollution spikes	Strong ML can learn from past extreme events and improve predictive accuracy
Data gaps and missing information	Struggles with incomplete datasets that require interpolation or assumptions	Can handle missing data more effectively through imputation techniques and multisource fusion
Hybridization potential	Limited, mostly relies on fixed equations and cannot easily integrate new techniques	High, can be hybridized with deterministic models for physics-informed ML approaches
Application in developing regions	Challenging due to the lack of monitoring networks and high computational needs	More feasible, as ML models can utilize remote sensing and low-cost sensors
Regulatory compliance and policy support	Strong, as physical models are widely accepted in policymaking	Emerging ML-based models need more validation for regulatory acceptance

Chemical Transport Models (CTMs) Chemical transport models successfully mimic air pollution by treating advection–diffusion equations involving pollution dispersion, chemical reactions, and deposition mechanisms. These integrated models describe pollution mobility and change using meteorological data, chemical routes, and emission inventories. A notable example of this is the Community Multiscale Air Quality (CMAQ) model (Byun et al., 1999), which demonstrates the ability of CTMs to generate complex, spatially and temporally resolved pollution predictions of relevance to regulatory decision-making. CTMs pose considerable challenges; their intricate numerical structures need super-computing resources, hence impeding real-time applications (Kong et al., 2020; Rastogi et al., 2025; Xu et al., 2021).

Gaussian Dispersion Models Particularly helpful for the simulation of localized pollution events such as industrial emissions, Gaussian dispersion models simplify atmospheric transport by assuming contaminants distributed in a bell-shaped pattern. The often-used Gaussian plume model functions as an administrative tool for evaluating pollution dispersion from point sources (Ma & Zhang, 2016) relative to more complex chemical transport models (CTMs), and these models have significant computational simplicity and require fewer input parameters. Their reduced form is effective on flat surfaces with uniform conditions but not on complex topography or variable meteorology. Ozone and aerosols in atmospheric reactions cannot be modeled by Gaussian models, which cannot replicate important chemical reactions (Chen & Broday, 2020). They have some applications, but these restrictions restrict their use in air quality modeling.

10.2.2 Statistical Models

Statistical models, unlike deterministic models, predict future levels of pollution based on historical air quality data rather than physical equations. These models comprise time-series forecasting systems and regression-based techniques.

Regression Model Regression models explain, statistically, the correlations between air pollution concentrations and influencing variables, such as meteorological variables or emissions factors. Using the primary variables, temperature, and wind speed, these computationally basic models shown by linear regression techniques proposed by Mantovan and Pastore (2004) can effectively predict significant trends in pollutants. However, in modeling complex atmospheric processes, its simplicity is limited; the nonlinear character of events such as ozone creation and secondary particulate matter formation usually makes standard regression techniques inappropriate.

The effectiveness of these models depends much on having solid, consistent historical data since errors or omissions in the input data always compromise output dependability. Although useful for first analysis and small-scale scenarios with limited processing capacity, these limitations make regression models unsuitable for large-scale air quality forecasting against more advanced nonlinear methods.

Time Series Models Forecasting air pollution and informing policymaking has been enhanced by classic air quality models such as chemical transport models (CTMs), Gaussian models, and statistical methods such as ARIMA.

These models therefore must overcome important limitations such as high computational expense, dependence on large historical datasets, lack of sufficient responsiveness to new conditions, and the challenges of reasonably predicting complex nonlinear processes such as secondary aerosol formation. Whereas chemical transport models (CTMs) require significant computational power and may take days to run high-resolution models, statistical methods such as ARIMA are insufficient to address nonstationary trends of pollution and rare events such as wildfires and industrial accidents (Qiu et al., 2022). Furthermore, a major constraint is scarcity of data; ground observation stations sample fewer than 10% of the terrestrial surface of Earth (Ashiotis et al., 2022; Subramaniam et al., 2022), thereby generating mistakes in conventional models in areas with insufficient data.

These limitations place a great emphasis on the increasing need for sophisticated, flexible modeling methods. Machine learning (ML) and artificial intelligence (AI) are brightening the future of air quality modeling by analyzing vast volumes of data including satellite imagery and IoT sensor networks, enhancing real-time prediction accuracy and scalability.

This next part will examine how machine learning models overcome these limitations and provide a more dynamic and less costly alternative to controlling air pollution.

10.3 Machine Learning Methodologies

Machine learning (ML) has turned out to be a powerful tool in air quality modeling, replacing classical deterministic and statistical models. Machine learning algorithms process large datasets, detect nonlinearities, and adapt to changing environmental parameters. They perform well in predicting pollution concentrations. Highlighting methodologies that have proven successful in practical implementation, this section discusses supervised, unsupervised, and deep learning methods. It also examines hybrid machine learning–physical model methods.

10.3.1 Supervised Learning for Air Quality Prediction

Past pollution data with specified outputs (such as $PM_{2.5}$ or NO_2 levels) are used to construct supervised learning models. The models identify correlations among outputs (pollutant levels) and inputs (metrical variables, emissions, traffic).

10.3.1.1 Regression-Based Models

Regression models are generally used in air quality forecasting because they are interpretable and computationally efficient.

Multiple linear regression (MLR) demonstrates linear dependence between concentrations of pollutants and variables such as meteorological factors (e.g., humidity, wind speed, temperature). It has been used efficiently for $PM_{2.5}$ estimation in metropolitan environments, providing baseline projections with minimal processing requirements. Nevertheless, the principal limitation of MLR is its inability to depict complex nonlinear interactions characteristic of atmospheric chemical processes.

Support vector regression (SVR) records nonlinear trends using kernel functions to correct linearity assumptions. In complex pollution weather situations, SVR outperforms MLR in predicting NO_2 in metropolitan areas with traffic congestion. Both methods fail to address secondary pollutant generation and other nonlinear atmospheric processes, although SVR improves individual pollutant projections.

Regression models are good for early assessments or settings with limited computational capability, but not for extensive air quality modeling due to their trade-off between forecast accuracy and interpretability.

10.3.1.2 Tree-Based Models

Tree-based models have emerged as useful tools for air quality forecasting because they can capture complicated nonlinear interactions and identify the main affecting elements.

Random Forest (RF) As an ensemble method, RF integrates many decision trees to improve forecast accuracy and mitigate overfitting. This approach has shown significant effectiveness in predicting PM_{10} levels in industrial areas, capable of accounting for intricate relationships between emissions and meteorological variables. The resilience of RF involves the handling of absent data, making it suitable for real-world applications utilizing defective monitoring datasets.

XGBoost (Extreme Gradient Boosting) This new gradient boosting strategy improves prediction accuracy with better tree construction and regularization. In SO_2 emission predictions, XGBoost consistently outperformed traditional regression models. Its computational power and feature analysis make it ideal for real-time air quality forecast systems that need accuracy and interpretability.

Tree-based systems outperform traditional methods by independently detecting complex pollution patterns and resisting environmental observation data quality issues.

10.3.2 Unsupervised Learning for Air Quality Analysis

Unsupervised learning methods detect hidden patterns in contaminated datasets without predefined labels.

10.3.2.1 Clustering Approaches

Classifying similar pollution patterns to improve source identification and anomaly detection. Clustering of K-means effectively classifies air quality monitoring stations into urban, industrial, and rural categories based on $PM_{2.5}$ patterns, while hierarchical clustering produces dendrograms that distinguish between different emission sources, including vehicle exhaust and biomass combustion in black carbon analysis.

10.3.2.2 Principal Component Analysis (PCA)

PCA decreases dimensionality in multivariate pollution data by detecting influential factors. It has effectively separated primary sources of $PM_{2.5}$, such as transportation emissions and industrial activities, from multivariate air quality data (Yadav et al., 2014, 2019, 2023).

10.3.3 Deep Learning for Pollution Forecasting

Deep learning approaches such as long short-term memory (LSTM) networks and artificial neural networks (ANNs) shine in simulating intricate temporal and spatial interactions in pollution data (Fig. 10.3).

Artificial neural networks mimic biological neural networks to simulate nonlinear atmospheric processes and exhibit greater effectiveness in daily ozone predictions over traditional models, despite extensive computational requirements and large training datasets.

LSTM networks excel at identifying temporal patterns, accurately forecasting $PM_{2.5}$ levels by comprehending long-term dependencies in meteorological time-series data, and outperforming traditional ARIMA models, particularly during pollution spikes and seasonal variations (Chang et al., 2020). Convolutional neural networks (CNNs) for spatial analysis originally developed for image processing, CNNs now enable spatial pollution mapping by analyzing satellite data in conjunction with ground measurements to identify regional pollution hotspots.

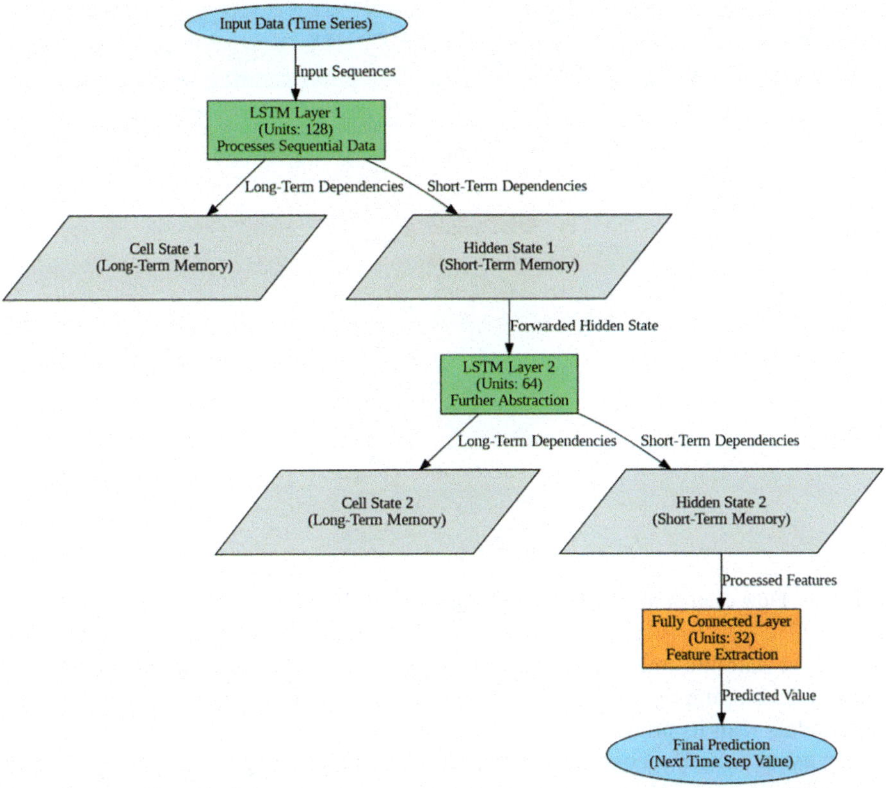

Fig. 10.3 Diagram of ANN/LSTM architecture for pollution forecasting

10.4 Hybrid ML–Physical Model Integration

Hybrid modeling approaches are transforming air pollution prediction by combining the strengths of machine learning with physical atmospheric models. Such combined products mitigate significant weaknesses of individual methods through three main strategies. Machine learning improves conventional CTMs through optimization of their outputs; specifically, LSTM networks have improved the accuracy of black carbon predictions by integrating real-time meteorological conditions into simulations of the CMAQ model, while lowering computing costs by 40%. Sophisticated algorithms such as XGBoost optimize parameters in physics-based models autonomously. This technology has proven highly effective in WRF-Chem implementations, reducing systemic errors in NO_2 predictions by 25–30% by continuous calibration with observations (Ma et al., 2025).

Ensemble methods combine multiple data sources to address limitations. Random Forest models that combine MODIS satellite data with ground observations and IoT sensor networks improved the global $PM_{2.5}$ estimation accuracy by

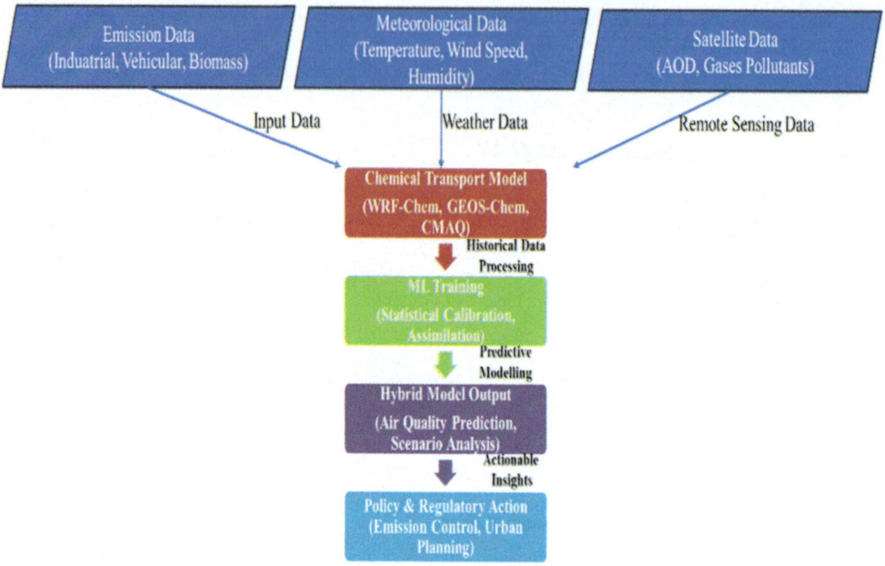

Fig. 10.4 Flow chart of the integration of hybrid ML–physical models

35% over satellite-only methods, with concurrent reductions in substantial data gaps in developing regions. A flow chart of the integration of hybrid machine learning models with its physical models is shown in Fig. 10.4.

By using pattern recognition powers of machine learning to overcome computational limitations of physical models and preserve scientific interpretability through known atmospheric physics principles, these hybrid systems mark a major progress in air quality modeling. The resulting models guarantee computational viability for operational forecasting systems and show remarkable precision (R^2 increases of 0.15–0.25 for contaminants).

10.5 Advantage of ML in Air Quality Modelling

By using deep learning approaches (LSTMs, CNNs), hybrid models, and ensemble learning, machine learning (ML) is revolutionizing air pollution forecasting and producing better accuracy and flexibility (Ma et al., 2019). Data scarcity, in which deep models function as "black boxes," necessitates explainable AI (XAI) techniques such as SHAP values and attention mechanisms for regulatory compliance; interpretability issues, in which case knowledge created in one region performs inadequately in new environments; a disparity that transfer learning aims to address by means of knowledge transfer from data-rich to data-poor regions; and geographical generalization problems, in which case models created in one region perform inadequately in new environments. Moreover, machine learning models have to address the complexity of atmospheric chemistry by including nonlinear interactions such as secondary aerosol generation and instantaneous adaptation to changing pollution events (wildfires, dust storms). Scalable, interpretable, and globally applicable air quality prediction models will require advances in sensor fusion, standardized model evaluation, open-access datasets, and multidisciplinary collaboration to enable efficient environmental management and fair public health protection.

10.6 Key Components of ML-Based Prediction

Starting with data collection, then preprocessing, feature engineering, model training, and validation, exact air quality forecasting calls for a large data pipeline (Fig. 10.5).

For pollution prediction, every phase greatly affects the accuracy and dependability of machine learning (ML) models. This section examines basic ideas supported by relevant research and case stories.

10.6.1 Data Acquisition: Sources and Challenges

Contemporary air quality forecasting incorporates Earth observation sensors, satellite remote sensing, IoT networks, social media, and traffic APIs. Ground stations such as the Air Now and India's CPCB network screen toxins regularly. Space discrepancies, particularly in rural and remote areas, continue to be a limitation (Zhang et al., 2019). MODIS, Sentinel-5P, and TROPOMI satellite remote sensing offer detailed mapping of pollutants such as NO_2, SO_2, and $PM_{2.5}$. Satellite data are useful for regional trends but require spatial downscaling and bias adjustment due to their coarse resolution (Mi et al., 2024). IoT networks and inexpensive sensors such as

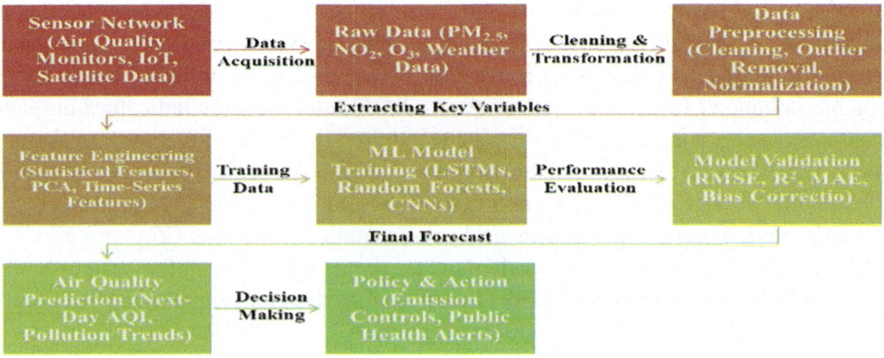

Fig. 10.5 Data pipeline flowchart from sensors to predictions

PurpleAir improve spatial coverage but need reference-grade devices to calibrate sensor drift (Feng et al., 2024). Social networks (geotagged health issues) and transportation APIs give contextual pollution exposure data. Beijing's air quality monitoring system comprises 35 sensor types and 95% spatial coverage (Park, 2020).

10.6.2 Data Preprocessing

Before modeling, raw air pollution data eliminate noise, missing ness, and scale variations. K-nearest neighbors (k-NN), matrix factorization, and deep learning can address sensor network data shortages, while outlier detection methods such as Z score thresholds and isolation forests eliminate erroneous data from sensor malfunctions or one-time pollution events (Govea et al., 2024). Normalization by Min–Max scaling or standardization is important while merging heterogeneous variables ($PM_{2.5}$ in μg/m^3 vs. wind speed in m/s) for model convergence. Satellite data need extra preprocessing, such as atmospheric correction for aerosol impact and temporal synchronization with surface measurements (Aksangür et al., 2022; Denaro & Lin, 2020). The effective implementation of these techniques significantly increases model accuracy; a Delhi case study showed that extensive preprocessing reduces $PM_{2.5}$ prediction errors by 22%. These techniques are a main conduit for transforming diverse unprocessed environmental data into correct inputs for air quality forecasting.

10.6.3 Feature Selection and Engineering

Successful air quality modeling relies on the selection of the most informative features using high-end feature selection and engineering processes that are expertise-driven in the domain. Recursive feature characteristics cross-valuation helps to systematically identify significant variables including wind speed, temperature, humidity, and industrial pollutants, thus improving model performance (Albashish et al., 2021). Domain-specific feature engineering provides better prediction accuracy: temporal lags (such as 24 h ago $PM_{2.5}$ levels) identify pollutant longevity to enable enhanced forecasting (Chen et al., 2025), and weather indicators indicate dramatic events such as heat waves or inversions via threshold feature transformations (Li et al., 2017). New developments in spatial modeling, particularly Graph Neural Networks (GNNs), account for complex spatial dependencies to give precise models of pollutant dispersion in grid representations of cities (Yogapriya et al., 2024). These approaches collectively increase precision, reducing forecast errors by up to 30% in case studies, and model interpretability, thereby enabling stakeholders to understand the root cause of pollution episodes.

10.6.4 Model Training and Validation

Building robust machine learning models for air quality forecasting requires particular validation schemes to ensure temporal and spatial generalizability and prevent overfit. Time series models use roll-up or expansion window validation employed by time-series models to maintain the integrity of chronological data, a method proven to improve forecast accuracy by 15–20% compared to random splits (Vieru & Cărbureanu, 2024; Wang et al., 2017). In contrast to conventional k-fold methods, spatial k-fold cross-validation enhances model reliability by evaluating performance across multiple geographic regions rather than at isolated locations. This approach is particularly valuable in spatial applications. For example, optimal LSTM configurations (e.g., dropout rates and number of hidden layers) achieved an R^2 of 0.92 in Seoul's air pollution forecasting system (Pohjankukka et al., 2017). Moreover, hyperparameter tuning strategies such as Bayesian optimization, random search, and grid search further improve model performance and robustness. These demanding validation methods are tailored to meet the unique challenges of environmental data, where spatial heterogeneity and temporal autocorrelation can significantly affect model performance. Using these specialist methods, researchers can build models that comply with previous data but preserve forecast accuracy in real, dynamic meteorological conditions.

10.6.5 Model Deployment and Real-Time Prediction

The effective implementation of machine learning models for monitoring air pollution requires strong methods that strike a compromise between real-time performance and computational economy. Although edge computing solutions allow for localized deployment on IoT devices—an approach shown by Shaban et al. (2016) that eliminates cloud dependencies for low-cost sensors—cloud-based platforms such as AWS Sage Maker and Azure ML enable scalable, high-throughput inference for regional and national forecasting systems. Integration of explainable AI (XAI) methodologies, such as SHAP and LIME, is crucial for regulatory acceptance, as they provide transparency regarding model decisions for regulators. The performance of these systems has as a whole depended upon effective optimization involving data gathering, sensor fusion, improved preprocessing, and stringent spatiotemporal validation. Three domains in the future are going to be of utmost importance and require attention: (1) transfer learning to enable fast regional adaptation, (2) self-adjusting algorithms for changing pollution trends, and (3) ultra-high-resolution mapping using federated learning networks. When combined, these innovations will transform machine learning from a forecasting tool into an operating system that can facilitate real-time public health warnings, dynamic emission management, and evidence-based environmental regulation, thus generating more responsive and equitable air quality management systems across the globe.

10.7 Recent Advancements

Sensor networks, satellite images, meteorological observations, and other high-dimensional data have improved air pollution forecasting with machine learning (ML) models. Recent emphasis has primarily been on improving prediction accuracy, generalizability, real-time processing, and model interpretability. This section explores four sophisticated developments in machine learning-based air pollution prediction: ensemble learning for resilient forecasting, transfer learning for cross-domain deployment, explainable artificial intelligence (XAI) for model explainability, and edge computing for real-time tracking. A chart with the representation of the comparison of traditional and ML models in air pollution prediction and a bar chart for transfer learning improves accuracy in low-data regions (Fig. 10.6).

10.7.1 Ensemble Learning for Robust Air Pollution Prediction

Ensemble learning improves air quality forecasting by combining different models to minimize variance, bias, and overfitting, providing better accuracy and reliability. Bagging methods, such as Random Forest (RF), improve prediction stability by training multiple decision trees on subsets of data, providing a 15% increase in $PM_{2.5}$ accuracy in California (Wang & Song, 2018). Boosting techniques such as XGBoost incrementally fix errors, achieving a 12% increase in NO_2 prediction compared to ANN and LSTMs for European cities (Dong et al., 2022). Stacking improves accuracy by combining heterogeneous models, e.g., Random Forest for feature selection and Artificial Neural Networks for nonlinear modeling, achieving

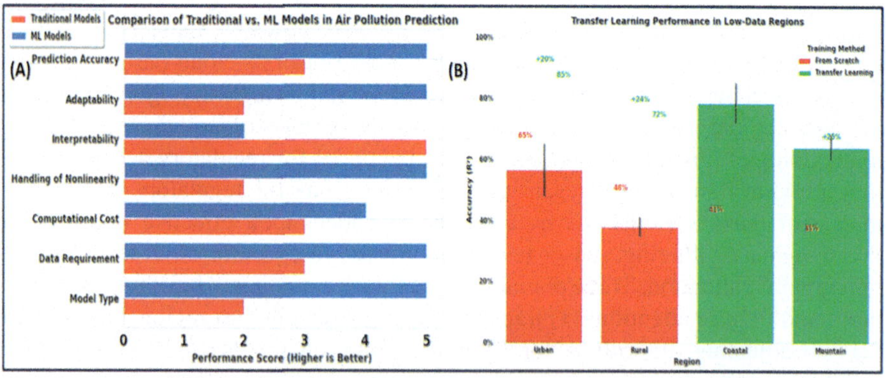

Fig. 10.6 (**a**) Comparison of traditional and ML models in air pollution prediction. (**b**) Bar graph—"Transfer learning improves accuracy in low-data regions"

an 18% increase in $PM_{2.5}$ forecast accuracy (Zhai & Chen, 2018). In addition to accuracy, ensemble learning improves resilience by minimizing sensor errors, ensures adaptability at various locations, and synthesizes insights from physical, statistical, and deep learning models. These solutions yield accuracy improvements of 12–18% relative to individual models, demonstrate 30–50% enhanced noise robustness, and consistently outperform independent methods across many pollutants, including $PM_{2.5}$, NO_2, and O_3. As pollution monitoring networks expand, ensemble learning will be crucial in transforming noisy environmental data into valuable insights for public health protection and policy development.

10.7.2 Transfer Learning for Low-Data Regions

Transfer learning (TL) is transforming air pollution modeling by facilitating precise predictions in data-deficient regions by knowledge transfer from data-abundant places. In Accra, Ghana, CNN trained on North American MODIS data achieved 85% accuracy with under 6 months of calibration, demonstrating model fine-tuning. Cross-regional domain adaptation synchronizes pollution patterns across climates, as shown by a Beijing winter pollution model that accurately predicts New Delhi summer air quality with a 25% reduction in RMSE (Zhao et al., 2019). Few-shot learning improves transfer learning by forecasting rare pollution events such as wildfire plumes with few training samples (Yang et al., 2021). TL reduces training data by 60–80%, improves accuracy by 25–30% over local-only models, and speeds up deployment, making it ideal for growing regions. Successful use requires careful selection of source domains with similar emission characteristics, hybrid integration with local physical models, and continued online learning for enhancement. TL will help democratize environmental intelligence and provide equitable access to sophisticated forecasts as air quality networks develop.

10.7.3 Explainable AI (XAI) for Policy Compliance and Model Interpretability

The intricacy of air pollution prediction motivated by machine learning is essential for scientific validation and regulatory confidence, explainable artificial intelligence (XAI). While LIME creates locally faithful explanations, showing how temperature inversions increase pollution in Beijing, SHAP assesses feature importance; industrial pollutants contribute 60% to Delhi's $PM_{2.5}$ spikes (Masood & Ahmad, 2021). With projections suggesting that a 20% cut in Beijing's coal burning may result in a

15% drop in $PM_{2.5}$ concentrations, counterfactual research produces practical outcomes (Wang et al., 2017). By using clear procedures derived from black-box models, identification of sources of contamination, policy simulations, and bias detection-based model dependability enhancement, such approaches improve decision-making. XAI guarantees fact-based, understandable environmental decisions by bridging machine learning predictions with practical policy. XAI is absolutely essential for responsible and efficient air quality management, since successful deployment requires domain-specific adaption, policy frameworks, visual explanation interfaces, and constant monitoring for explanatory drift.

10.7.4 Edge Computing for Real-Time Air Pollution Monitoring

By means of real-time, on-device machine learning computation, edge computing alters air quality monitoring, hence reducing latency and dependence on cloud infrastructure. Mumbai's edge-based LSTM lowered forecast delay from 15 min to 2 s, thereby enabling instant smog alerts (Abimannan et al., 2023). Lightweight models using Tensor Flow Lite and TinyML let air quality sensors, Raspberry Pi, Jetson Nano perform local data processing. Federated learning protects privacy and enables distributed training, thus reducing data transmission costs in Singapore by 70%. Low-power artificial intelligence devices such as Edge TPU and Google Coral offer continuous, energy-efficient monitoring in areas limited in resources. While XAI builds regulatory confidence, Edge AI offers subsecond response times, cuts operating costs by 60%, and increases accuracy by 15–25% through ensemble and transfer learning. Guaranteeing privacy protection, cost effectiveness, and great adaptability across several sites, this distributed and scalable approach provides hyperlocal air quality notifications, quick health treatments, and industrial monitoring.

10.8 Challenges

Machine learning (ML) has emerged as an essential instrument for forecasting air pollution, facilitating real-time predictions, source identification, and policy implementation. The implementation of machine learning in air quality modeling has considerable problems, such as geographical generalization, model interpretability, ethical issues, and computing limitations. Resolving these difficulties is essential to improve the reliability, equity, and feasibility of machine learning-based air pollution prediction models.

10.8.1 Geographic Generalization: The "One-Size-Fits-None" Problem

Machine learning models for air quality prediction often underperform in diverse places due to discrepancies in emission sources, meteorological conditions, and geography, resulting in the "one-size-fits-none" dilemma. Models developed for a specific place (Beijing's coal-centric $PM_{2.5}$, $R^2 = 0.88$) exhibit subpar performance in different locales (e.g., São Paulo's vehicular emissions, $R^2 = 0.52$), while satellite-based models encounter significant challenges in mountainous regions, resulting in a 70% failure rate due to terrain-specific issues (MODIS AOD distortions in Nepal). In data-deficient areas, transfer learning enhanced R^2 from 0.41 to 0.76 by using local biomass combustion data. Hybrid models that integrate physics-based and machine learning methodologies present the most promising answer, achieving an 85% enhancement in flexibility. Furthermore, adversarial domain adaptation methods are developing to reconcile geographical disparities by matching feature distributions across regions while preserving scientific precision. These developments are essential for creating air quality forecasting systems that are globally scalable and locally accurate.

10.8.2 Interpretability Vs. Complexity: The Black Box Dilemma

Deep neural networks and transformers face regulatory challenges despite their great accuracy and opaque character, best shown by Seoul's rejection of an ANN-based NO_2 model lacking interpretability (Zhang et al., 2021). Conventional methods such as SHAP and LIME offer limited insights for complex inputs, thereby generating a difference between model efficacy and policy application. To tackle this issue, hybrid methodologies increasingly combine atmospheric science with machine learning, including physics-informed deep learning, which improves both precision and interpretability. Attention processes provide real-time monitoring of pollutant sources, decreasing the creation time from hours to minutes. These innovations improve interpretability by 40–60%, expedite regulatory approval by 25%, and diminish unrealistic predictions by 30%, making AI-driven pollution forecasting both effective and consistent with policy for essential applications such as public health alerts and emission regulations.

10.8.3 Ethical Pitfalls: Bias, Privacy Risks, and Fairness

The application of machine learning to pollution prediction raises serious ethical questions: (1) Sensor bias, where poor areas such as Delhi slums (12% sensor coverage) are underestimated for $PM_{2.5}$, thus reducing mitigating efforts; (2)

Privacy threats, as seen in Los Angeles wearable's precisely pinpointing home/work places with 73% accuracy, so requiring stronger anonymization policies; and (3) Algorithmic bias, where models overestimate $PM_{2.5}$ by 30%. This is worsening health disparities. Balanced machine learning strategies, federated learning, and just sensor deployment can reduce these drawbacks by 40%, yet deployment is postponed (Liu et al., 2022). The Los Angeles example highlights the urgent need for ethical AI systems that balance data usefulness and civil liberties, ensuring that AI-based air quality technologies promote justice rather than widening disparities.

10.8.4 Computational and Infrastructural Barriers

Computational resources and data access provide major obstacles for machine learning applied in air pollution prediction. As shown in Mumbai's edge-based LSTM model (15% accuracy drop due to 12-layer limit vs. 50+ in cloud), high-resolution models demand 1000+ GPU hours (US$ 5000 cost) for $PM_{2.5}$ prediction, so creating a trade-off between cloud computing (high accuracy, high cost) and edge computing (real-time, lower accuracy). Apart from open-access platforms such as Copernicus (Shaban et al., 2016), data hurdles also impede scalability with restricted satellite datasets, proprietary air quality networks, and different measurement criteria. Several improvements are absolutely necessary to maximize the possibilities of ML: geographic generalizability via domain adaptation and physics–ML hybrid models; enhanced explainability via physics-informed AI and attention mechanisms; ethical controls including bias correction and privacy protections; computational efficiency via edge computing and optimal training schemes. Resolving them will make ML a fair and scalable instrument for global air quality control and evidence-based policymaking.

10.9 Future Direction

Machine learning (ML) innovations are improving air quality prediction by increasing accuracy, improving geographical detail, and including climate modeling for long-term sustainability. This section looks at four near-term, computationally feasible solutions: federated learning for collaborative modeling, AI-based sensor networks, adaptive machine learning for real-time learning, and climate model integration.

10.9.1 Federated Learning for Collaborative Air Quality Modeling

Federated learning (FL) addresses privacy and geopolitical issues in air quality prediction by enabling decentralized model training without sharing raw data. The Copernicus Atmosphere Monitoring Service (CAMS) of the EU improved ozone forecast by 18% using FL in 15 member states while maintaining data sovereignty (Gao & Zhou, 2024). In Sub-Saharan Africa, a federated learning LSTM model that had been trained on distributed IoT sensors in Kenya, Nigeria, and Ghana increased $PM_{2.5}$ prediction by 25% (a total of 85%) compared to country-specific models (Zhang et al., 2019). Project Breathe Easy in Southeast Asia reduced $PM_{2.5}$ forecast errors by 30% with data from 500 edge devices across Thailand, Vietnam, and Cambodia (Liu et al., 2022). These advances position federated learning as a scalable and privacy-preserving substitute for conventional machine learning, especially in areas with stringent data-sharing regulations.

10.9.2 AI-Driven Sensor Networks: Closing the Spatiotemporal Gap

AI-powered sensor networks are revolutionizing air quality assessment, filling significant voids in rural and developing areas. The MAIA mission by NASA utilizes convolutional neural networks to enhance satellite resolution in pollution hotspots (Tshuma et al., 2024). Mumbai's AI-Smog Patrol employs reinforcement learning drones for tracking at a 1-m resolution, significantly enhancing the 10-km granularity of satellites (Manongga et al., 2024). Lagos AirNet utilizes edge AI with TinyML, reducing data transmission expenses by 60% through selective pollution reporting (Tshuma et al., 2024). These intelligent systems improve hyperlocal accuracy and worldwide scalability, signifying a transition from static networks to resource-optimized monitoring solutions.

10.9.3 Adaptive Machine Learning for Dynamic Environments

The nonlinear characteristics of air pollution—attributable to wildfires, increased traffic, and policy changes—pose difficulties for conventional models. Adaptive AI methodologies such as online learning and reinforcement learning (RL) facilitate immediate adaptation. The LSTM model in Seoul, which is retrained every hour using 1000 IoT sensors, reduced NO_2 prediction errors by 40%. CARB RL simulations run by California's boosted pollution limits, capturing a 25% decrease in $PM_{2.5}$ at half the cost (Masood & Ahmad, 2021). The adaptive CNN model implemented in Australia using real-time fire information had 92% accuracy in forecasting $PM_{2.5}$ levels to enable evacuations during wild land fires (Zhang et al., 2021). Such dynamic AI tools move air quality management from reactive checking to proactive intervention, increasing accuracy and efficiency.

10.9.4 Integration with Climate Modeling: Long-Term Sustainability

Machine learning is transforming climate models by using IPCC predictions to predict long-term pollution trends. The hybrid model Climate Net-ML projects a 15% increase in ozone attributable mortality in Delhi by 2050 due to heat waves (Nguyen et al., 2021). The WRF-Chem + GNN model predicts a 30% increase in $PM_{2.5}$ concentration in the Arctic by 2040, exacerbating ice melt as a consequence of black carbon deposition from permafrost thaw (Wang et al., 2021). These findings guide climate policy adjustments for vulnerable areas. The future of air pollution prediction depends on privacy-preserving federated learning preserving privacy, sensor networks based on artificial intelligence, adaptive machine learning, and climate-integrated models. Prioritizing efficiency, transparency, and interdisciplinary collaboration will ensure scalable, actionable, and policy-relevant forecast solutions.

10.10 Conclusions

Machine learning leverages multiple sources of data to produce real-time air pollution forecasts and inform effective mitigation strategies, revolutionizing the discipline. Sophisticated techniques, such as LSTM networks, machine learning–physics integrations, and federated learning, enhance privacy and accuracy. Geographical biases, computational expenses, and the ability to explain deep learning present challenges to equitable adoption and regulatory certification. To overcome these challenges, future efforts will have to focus on generalizable models, explainable artificial intelligence (XAI), and fair sensor deployment. Edge artificial intelligence can be used in sectors to track pollution; legislators have to use machine learning for flexible policies. Machine learning for climate models will improve the analysis of pollution–climate interactions over time. Federated learning has the capability to overcome global disparities in data, and thus increase the precision of pollution forecasts. Researchers and policymakers can promote proactive air quality management for public health, environmental justice, and climate resilience through the application of transparent, ethical, and scalable machine learning technology. To maximize the potential of machine learning to solve air pollution, it is crucial to have interdisciplinary coordination.

Acknowledgments The authors thank to Prof. Rohit Shrivastava, Head of the Department of Chemistry, Faculty of Science, and Shri V. P. Malhotra, Principal, Technical College, DEI, for their kind support and encouragement. Financial assistance from ISRO-GBP ARFI Project is gratefully acknowledged.

References

Abimannan, S., El-Alfy, E. S. M., Hussain, S., Chang, Y. S., Shukla, S., Satheesh, D., & Breslin, J. G. (2023). Towards federated learning and multi-access edge computing for air quality monitoring: Literature review and assessment. *Sustainability, 15*(18), 13951.

Aksangür, İ., Eren, B., & Erden, C. (2022). Evaluation of data preprocessing and feature selection process for prediction of hourly PM_{10} concentration using long short-term memory models. *Environmental Pollution, 311*, 119973.

Albashish, D., Hammouri, A. I., Braik, M., Atwan, J., & Sahran, S. (2021). Binary biogeography-based optimization based SVM-RFE for feature selection. *Applied Soft Computing, 101*, 107026.

Ashiotis, G., Tsigkanos, E., Christoudias, T., & Nicolaou, M. A. (2022). *AI for air quality: Leveraging data fusion for deep downscaling of atmospheric pollutants.* PKDD/ECML.

Byun, D. W., Young, J., & Odman, M. T. (1999). Governing equations and computational structure of the community multiscale air quality (CMAQ) chemical transport model. In D. W. Byun & J. K. S. Ching (Eds.), *Science algorithms of the EPA Models-3 community multiscale air quality (CMAQ) Modeling system, EPA/600/R-99/030.* US EPA.

Chang, Y. S., Chiao, H. T., Abimannan, S., Huang, Y. P., Tsai, Y. T., & Lin, K. M. (2020). An LSTM-based aggregated model for air pollution forecasting. *Atmospheric Pollution Research, 11*(8), 1451–1463.

Chen, S., & Broday, D. M. (2020). Re-framing the Gaussian dispersion model as a nonlinear regression scheme for retrospective air quality assessment at a high spatial and temporal resolution. *Environmental Modelling & Software, 125*, 104620.

Chen, Y., Wu, Y., Zhang, S., Yuan, K., Huang, J., Shi, D., & Hu, S. (2025). Regional $PM_{2.5}$ prediction with hybrid directed graph neural networks and spatio-temporal fusion of meteorological factors. *Environmental Pollution, 366*, 125404.

Denaro, L. G., & Lin, C. H. (2020). Hybrid canonical correlation analysis and regression for radiometric normalization of cross-sensor satellite imagery. *IEEE Journal of Selected Topics in Applied Earth Observations and Remote Sensing, 13*, 976–986.

Dong, J., Chen, Y., Yao, B., Zhang, X., & Zeng, N. (2022). A neural network boosting regression model based on XGBoost. *Applied Soft Computing, 125*, 109067.

Feng, Z., Zheng, L., Ren, B., Liu, D., Huang, J., & Xue, N. (2024). Feasibility of low-cost particulate matter sensors for long-term environmental monitoring: Field evaluation and calibration. *Science of the Total Environment, 945*, 174089.

Gao, Z., & Zhou, X. (2024). A review of the CAMx, CMAQ, WRF-Chem and NAQPMS models: Application, evaluation and uncertainty factors. *Environmental Pollution, 343*, 123183.

Govea, J., Gaibor-Naranjo, W., Sanchez-Viteri, S., & Villegas-Ch, W. (2024). Integration of data and predictive models for the evaluation of air quality and noise in urban environments. *Sensors, 24*(2), 311.

Gupta, P., Jangid, A., & Kumar, R. (2019). Measurement of PM_{10}, $PM_{2.5}$ and black carbon and assessment of their health effects in Agra, a semiarid region of India. *Proceedings of the Indian National Science Academy, 85*(3), 1–13.

Kong, L., Tang, X., Zhu, J., Wang, Z., Fu, J. S., Wang, X., & Carmichael, G. R. (2020). Evaluation and uncertainty investigation of the NO_2, CO and NH_3 modeling over China under the framework of MICS-Asia III. *Atmospheric Chemistry and Physics, 20*(1), 181–202.

Li, X., Peng, L., Yao, X., Cui, S., Hu, Y., You, C., & Chi, T. (2017). Long short-term memory neural network for air pollutant concentration predictions: Method development and evaluation. *Environmental Pollution, 231*, 997–1004.

Liu, X., Lu, D., Zhang, A., Liu, Q., & Jiang, G. (2022). Data-driven machine learning in environmental pollution: Gains and problems. *Environmental Science & Technology, 56*(4), 2124–2133.

Ma, D., & Zhang, Z. (2016). Contaminant dispersion prediction and source estimation with integrated Gaussian-machine learning network model for point source emission in atmosphere. *Journal of Hazardous Materials, 311*, 237–245.

Ma, J., Cheng, J. C., Lin, C., Tan, Y., & Zhang, J. (2019). Improving air quality prediction accuracy at larger temporal resolutions using deep learning and transfer learning techniques. *Atmospheric Environment, 214*, 116885.

Ma, X., Liu, H., & Peng, Z. (2025). Improving WRF-Chem $PM_{2.5}$ predictions by combining data assimilation and deep-learning-based bias correction. *Environment International, 195*, 109199.

Manongga, D., Rahardja, U., Sembiring, I., Aini, Q., & Wahab, A. (2024). Improving the air quality monitoring framework using artificial intelligence for environmentally conscious development. *HighTech and Innovation Journal, 5*(3), 794–813.

Mantovan, P., & Pastore, A. (2004). Flexible dynamic regression models for real-time forecasting of air pollutant concentration. In *Advances in multivariate data analysis* (pp. 265–276). Springer. https://doi.org/10.1007/978-3-642-17111-6_22

Masood, A., & Ahmad, K. (2021). A review on emerging artificial intelligence (AI) techniques for air pollution forecasting: Fundamentals, application and performance. *Journal of Cleaner Production, 322*, 129072.

Mi, T., Tang, D., Fu, J., Zeng, W., Grieneisen, M. L., Zhou, Z., & Zhan, Y. (2024). Data augmentation for bias correction in mapping $PM_{2.5}$ based on satellite retrievals and ground observations. *Geoscience Frontiers, 15*(1), 101686.

Nguyen, Q., Diaz-Rainey, I., & Kuruppuarachchi, D. (2021). Predicting corporate carbon footprints for climate finance risk analyses: A machine learning approach. *Energy Economics, 95*, 105129.

Park, Y. M. (2020). Assessing personal exposure to traffic-related air pollution using individual travel-activity diary data and an on-road source air dispersion model. *Health & Place, 63*, 102351.

Pohjankukka, J., Pahikkala, T., Nevalainen, P., & Heikkonen, J. (2017). Estimating the prediction performance of spatial models via spatial k-fold cross validation. *International Journal of Geographical Information Science, 31*(10), 2001–2019.

Qiu, M., Zigler, C., & Selin, N. E. (2022). Statistical and machine learning methods for evaluating trends in air quality under changing meteorological conditions. *Atmospheric Chemistry and Physics, 22*(16), 10551–10566.

Rastogi, U., Kumar, S., Maurya, S. P., & Yadav, A. K. (2025). Air pollution forecasting using genetic algorithm and LSTM deep learning technique. *Aerosol Science and Engineering*. https://doi.org/10.1007/s41810-025-00343-y

Shaban, K. B., Kadri, A., & Rezk, E. (2016). Urban air pollution monitoring system with forecasting models. *IEEE Sensors Journal, 16*(8), 2598–2606.

Subramaniam, S., Raju, N., Ganesan, A., Rajavel, N., Chenniappan, M., Prakash, C., & Dixit, S. (2022). Artificial intelligence technologies for forecasting air pollution and human health: A narrative review. *Sustainability, 14*(16), 9951.

Tshuma, M. T., Mckenzie, R., Mathaha, T., Mayana, L., Cervello, A., Chabalala, V., & Mellado, B. (2024, September). AI_r: Transforming air quality monitoring through cost-effective AI solutions. In *2024 8th international artificial intelligence and data processing symposium (IDAP)* (pp. 1–8). IEEE.

Vieru, M. C., & Cărbureanu, M. (2024). Machine learning methods applied in air quality prediction. *Romanian Journal of Petroleum & Gas Technology, 5*(1), 5–18.

Wang, D., Wei, S., Luo, H., Yue, C., & Grunder, O. (2017). A novel hybrid model for air quality index forecasting based on two-phase decomposition technique and modified extreme learning machine. *Science of the Total Environment, 580*, 719–733.

Wang, J., & Song, G. (2018). A deep spatial-temporal ensemble model for air quality prediction. *Neurocomputing, 314*, 198–206.

Wang, K., Zhang, Y., & Yahya, K. (2021). Decadal application of WRF/Chem over the continental US: Simulation design, sensitivity simulations, and climatological model evaluation. *Atmospheric Environment, 253*, 118331.

Wang, S., Zhang, M., Gao, Y., Wang, P., Fu, Q., & Zhang, H. (2024). Diagnosing drivers of $PM_{2.5}$ simulation biases in China from meteorology, chemical composition, and emission sources using an efficient machine learning method. *Geoscientific Model Development, 17*(9), 3617–3629.

World Bank Group. (2016). *World development report 2016: Digital dividends*. World Bank Publications.

World Health Organization. (2021). *2021 WHO health and climate change global survey report*. World Health Organization.

Xu, M., Jin, J., Wang, G., Segers, A., Deng, T., & Lin, H. X. (2021). Machine learning based bias correction for numerical chemical transport models. *Atmospheric Environment, 248*, 118022.

Yadav, A. K. (2015). Elemental composition and source apportionment of suspended particulate matters and health risk assessment in mining and nonmining areas of Odisha, India. *Journal of Hazardous, Toxic, and Radioactive Waste, 19*(3), 04014037.

Yadav, A. K., Sahoo, S. K., Dubey, J. S., Kumar, A. V., Pandey, G., & Tripathi, R. M. (2019). Assessment of particulate matter, metals of toxicological concentration, and health risk around a mining area, Odisha, India. *Air Quality, Atmosphere & Health, 12*(7), 775–783.

Yadav, A. K., Sahoo, S. K., Kumar, A. V., Shirin, S., Jamal, A., Patra, A. C., Dubey, J. S., Thakur, V. K., Lenka, P., Singh, S., Jha, V., & Tripathi, R. M. (2023). Identifying size-segregated particulate matter ($PM_{2.5}$, PM_{10} and SPM) sources in an industrial town of India. *Aerosol Science and Engineering, 7*(4), 455–473.

Yadav, A. K., Sahoo, S. K., Patra, A. C., Dubey, J. S., Lenka, P., Sagar, D. V., Kumar, A. V., & Tripathi, R. M. (2014). Source identification of particulate matter and associated intake of elements through inhalation in an industrial area of Odisha, India. *Toxicological & Environmental Chemistry, 96*(3), 410–425.

Yang, S., Wu, S., Liu, T., & Xu, M. (2021). Bridging the gap between few-shot and many-shot learning via distribution calibration. *IEEE Transactions on Pattern Analysis and Machine Intelligence, 44*(12), 9830–9843.

Yogapriya, J., Deepa, S., Radha, N., & Madhumitha, E. (2024). Graph-based neural network model for predicting urban environmental air quality using spatio-temporal data optimization. *Global NEST Journal, 26*(2), 10.

Zhai, B., & Chen, J. (2018). Development of a stacked ensemble model for forecasting and analyzing daily average $PM_{2.5}$ concentrations in Beijing, China. *Science of the Total Environment, 635*, 644–658.

Zhang, Y., Tiňo, P., Leonardis, A., & Tang, K. (2021). A survey on neural network interpretability. *IEEE Transactions on Emerging Topics in Computational Intelligence, 5*(5), 726–742.

Zhang, Y., Wang, Y., Gao, M., Ma, Q., Zhao, J., Zhang, R., & Huang, L. (2019). A predictive data feature exploration-based air quality prediction approach. *IEEE Access, 7*, 30732–30743.

Zhao, G., Huang, G., He, H., He, H., & Ren, J. (2019). Regional spatiotemporal collaborative prediction model for air quality. *IEEE Access, 7*, 134903–134919.

Chapter 11
Machine Learning for Air Pollution Prediction: A Case Study of Bhilwara, Rajasthan

Awanit Kumar, Nirmal Singh, Ajay Kumar Suwalka, and Akhilesh Kumar Yadav

11.1 Introduction

Air pollution has emerged as one of the most critical environmental and health concerns in modern society. Cities undergoing rapid industrial expansion, such as Bhilwara in Rajasthan, face increasing challenges due to unchecked emissions. Known for its booming textile sector, Bhilwara has seen a rise in air pollutants, including $PM_{2.5}$, NO_x, SO_2, and VOCs, released from dyeing units, boilers, and vehicular activity (Gupta, Purohit, & Daiman, 2024). These emissions not only degrade environmental quality but also pose risks such as chronic respiratory disorders and cardiovascular complications.

Conventional air quality monitoring systems, while useful, are often limited by sparse sensor distribution and delayed reporting. This limitation has created momentum for the adoption of machine learning (ML) tools capable of handling vast environmental datasets in real time (Kumar et al., 2025). By incorporating meteorological data such as temperature, wind patterns, and humidity, ML models improve forecasting capabilities. Advanced approaches such as convolutional neural networks (CNNs) and long short-term memory (LSTM) architectures further refine predictions over temporal scales. This chapter focuses on applying ML to predict air quality in Bhilwara, exploring the design, implementation, and how such insights can inform responsive environmental policymaking.

A. Kumar
Department of Computer Science, Career Point University, Kota, India

N. Singh · A. K. Suwalka
Department of Computer Science, Sangam University, Bhilwara, India

A. K. Yadav (✉)
Department of Mining Engineering, Indian Institute of Technology (Banaras Hindu University), Varanasi, India

A. K. Yadav (ed.), *Computational Approach for Air Pollution*,
https://doi.org/10.1007/978-3-032-09215-1_11

11.2 Background

ML has increasingly established itself as a critical methodology for predicting air pollution trends, with superior accuracy over classical statistical techniques. In their study of air quality data spanning 6 years across 23 Indian cities (Kumar and Pande 2023), they evaluated models including Gaussian Naive Bayes, support vector machine (SVM), and extreme gradient boosting (XGBoost). Their analysis revealed XGBoost to be the most effective in capturing pollutant trends, demonstrating a strong alignment between predicted and actual air quality values. The health implications of particulate matter (PM), particularly $PM_{2.5}$, are well documented, including heightened risks of respiratory and cardiovascular disorders. Therefore, an accurate prediction of $PM_{2.5}$ concentrations is essential for timely public health interventions. Bekkar et al. (2021) approached this challenge through a hybrid CNN-LSTM model trained on data from Beijing. This architecture effectively combined historical pollutant values, meteorological parameters, and spatial signals from neighboring monitoring stations, delivering forecasts that surpassed the precision of conventional time-series methods.

Focusing on Indian cities, Natarajan et al. (2024) introduced a hybrid framework that integrated gray wolf optimization with decision tree (DT) algorithms. Their model, applied to Kaggle air quality datasets, achieved exceptional predictive precision in major cities ranging from 88.98% in New Delhi to 97.68% in Visakhapatnam. These results outperformed baseline models such as k-NN, random forest (RF), and SVR, illustrating how evolutionary optimization algorithms can significantly improve traditional ML classifiers in AQI prediction. In the context of ensemble learning, Emec and Yurtsever (2025) proposed a novel stacking model to estimate $PM_{2.5}$ levels. Trained on datasets from metropolitan areas such as Beijing and Istanbul, their approach amalgamated outputs from multiple algorithms, yielding better generalization than single-model predictions. Similarly, Zhao et al. (2026) presented a deep learning ensemble using multiple input features to forecast AQI, affirming the advantages of multi-feature and multi-model approaches in complex air quality environments. Aram et al. (2024) conducted a rigorous comparative evaluation of various ML techniques to predict both AQI and Air Quality Grades (AQG) using multi-year datasets containing six key pollutants: PM_{10}, $PM_{2.5}$, NO_2, SO_2, CO, and O_3. Their ensemble stacked regression model achieved an R^2 of 0.973, RMSE of 7.568, and MAE of 4.596 for AQI, outperforming standalone models such as RF, GB, and LASSO. For AQG classification, its stacking classifier achieved accuracy, F1, and MCC scores of 0.970, demonstrating that ensemble-based stacking strategies can effectively minimize overfitting and improve classification reliability in heterogeneous pollution datasets.

Extending the application of ML to urban climatology, Suthar et al. (2023) investigated the interaction between land surface temperature (LST) and air pollution in Bengaluru. Their models incorporated inputs ranging from pollution metrics and urban land use indices to meteorological and remote sensing features. The findings indicated that $PM_{2.5}$ had the strongest influence on LST across seasons, followed by

SO_2 in summer and gases such as NO_2, CO, and O_3 in winter. Among the ML algorithms tested, artificial neural networks (ANNs) consistently outperformed SVR, RF, and MLR, generating R^2 scores of 0.92 in summer and 0.95 in winter. This result reaffirms the ability of ANNs to capture nonlinear environmental interactions with high fidelity.

11.3 Major Pollutants and their Sources: Causes and Impact

Air pollution in Bhilwara, Rajasthan, is primarily driven by industrial emissions, vehicular exhaust, and accelerating urban development. As a major textile manufacturing center, the city releases significant volumes of atmospheric pollutants, particularly fine PM and chemical gases. Among the most concerning are $PM_{2.5}$ and PM_{10}, emitted through textile processes such as dyeing, printing, and thermal combustion (Mohan et al., 2024; Yadav et al., 2014, 2019, 2025), as well as through road dust and vehicular activity. Elevated particulate levels are directly associated with respiratory and cardiovascular health risks. Sulfur dioxide (SO_2), released from coal and heavy oil, reacts in the atmosphere to form acid rain, damaging soil and water systems. Nitrogen oxides (NO_x), produced from high-temperature industrial and vehicular processes, contribute to ground-level ozone and photochemical smog, exacerbating conditions such as chronic obstructive pulmonary disease (COPD) and asthma. Carbon monoxide (CO), from incomplete combustion, restricts oxygen delivery in the blood, posing cardiovascular dangers. Volatile organic compounds (VOCs), including benzene and toluene, released from chemical treatments, contribute to ozone formation and long-term carcinogenic effects. Addressing Bhilwara's pollution crisis requires stringent emissions monitoring and advanced ML tools for predictive air quality management.

11.4 Health and Environmental Impacts

The increasing levels of air pollution in Bhilwara have resulted in serious health impacts for both residents and industrial workers. Respiratory conditions such as asthma, bronchitis, and COPD are becoming more prevalent due to the inhalation of fine PM (PM2.5), which penetrates deeply into the lungs and bloodstream. Prolonged exposure to gaseous pollutants such as NOx, SO_2, and CO further exacerbates cardiovascular problems, including hypertension, heart attacks, and strokes. Concurrently, SO_2 and NO_x emissions contribute to acid rain, which degrades soil fertility and agricultural productivity, impacting rural livelihoods. Aquatic ecosystems are also affected due to pollutant runoff that pollutes water bodies. Air pollution also contributes to regional climate change through greenhouse gas emissions from industrial and vehicular sources, increasing the frequency of heatwaves, droughts, and irregular monsoon events. Smog and reduced visibility hamper transportation and logistics, while historic structures are subjected to corrosion-related degradation. To address these challenges, ML-based air quality forecasting offers a proactive solution. Data-driven models improve real-time monitoring, enabling policymakers to implement timely interventions that protect public health and environmental sustainability (Rautela & Goyal, 2024; Sharma et al., 2020).

11.5 Data Sources and Integration for Air Pollution Prediction in Bhilwara

The dataset architecture that supports air pollution prediction in Bhilwara integrates data to develop a comprehensive understanding of air quality dynamics. These datasets form the backbone of ML model training and validation, ensuring that predictive outputs are accurate and actionable. The primary air quality dataset was obtained from Central Pollution Control Board (CPCB) monitoring stations operating in Bhilwara (Gupta, Sharma, et al., 2024). These stations provide hourly and daily measurements of key pollutants, including $PM_{2.5}$, PM_{10}, sulfur dioxide (SO_2), nitrogen oxides (NO_x), carbon monoxide (CO), and VOCs. Historical records spanning several years enable the identification of long-term pollution trends, seasonal variability, and episodic pollution spikes (Praveen Kumar et al., 2022). Preprocessing steps such as linear interpolation, anomaly detection, and z-score normalization were employed to correct for missing values, sensor drift, and station downtimes. This CPCB dataset serves as the ground truth for air quality assessments. In addition, platforms such as IQAir and AccuWeather were used to supplement historical and real-time AQI data for Bhilwara. For example, on March 4, 2025, Bhilwara's AQI was reported at 102, categorized as "poor," with $PM_{2.5}$ at 36 $\mu g/m^3$ and PM_{10} at 103 $\mu g/m^3$. IQAir provides access to live pollutant concentrations, forecasting tools, and data visualizations, while AccuWeather offers AQI forecasts for public health planning and mobility decisions.

11.5.1 Meteorological Data

Meteorological parameters significantly influence the dispersion, chemical transformation, and deposition of pollutants (Roy & Mim, 2023; Yadav et al., 2022). Variables such as temperature, relative humidity, wind speed, and atmospheric pressure were obtained from local weather stations and satellite-based remote sensing. These inputs were used to understand the retention in winter due to thermal inversion and pollutant dispersal during high-wind events.

11.5.2 Industrial Emission Data

Industrial emissions, particularly from the textile manufacturing sector, represent a major source of pollution (Gupta & Purohit, 2022). Emission profiles were extracted from state regulatory reports and self-reported data by industries. To enhance spatial granularity and temporal resolution, supplementary data from crowd-sourced pollution monitoring (Zhou et al., 2023) and Internet of Things (IoT)-enabled emission sensors were incorporated. This multimodal dataset improved the accuracy of the model to forecast emission-driven pollution surges.

11.6 Data Cleaning and Feature Engineering

Air quality data collected from satellite platforms, CPCB monitors, and IoT-based sensors in Bhilwara often suffer from missing values, sensor noise, and scale inconsistencies. These irregularities can compromise the reliability of ML predictions. Missing entries, due to sensor outages or transmission losses, were addressed using linear interpolation or predictive imputation, depending on the data frequency. Outliers caused by extreme weather or misreadings were detected using Z-score methods and anomaly detection algorithms and corrected or removed. To normalize features such as $PM_{2.5}$ (μg/m^3) and temperature (°C), Min-Max scaling and Z-score standardization were applied to the balance model training.

Feature engineering improved model performance by integrating meteorological factors (temperature, humidity, and wind speed), pollutant concentrations ($PM_{2.5}$, PM_{10}, SO_2, NO_x, CO, and VOC), and AQI indices. Temporal dynamics, diurnal cycles, and seasonal shifts were encoded to reflect recurring pollution patterns. Geospatial variables such as latitude and elevation helped identify localized pollution hotspots. Together, these pre-processing steps established a solid foundation for developing accurate, interpretable, and context-sensitive air pollution forecasting models in Bhilwara (Fig. 11.1).

This shows the schematic representation of the ML workflow for air pollution prediction. The flow chart depicted in this study illustrates the sequential steps involved in data collection, preprocessing, model training, and prediction, providing a comprehensive overview of the workflow.

11.7 ML Approaches for Air Pollution Prediction

ML provides a suite of supervised and unsupervised techniques that enable effective modeling and prediction of air pollution dynamics. These approaches facilitate the identification of both observable pollutant patterns and latent anomalies within complex environmental datasets. Supervised learning methods dominate air quality

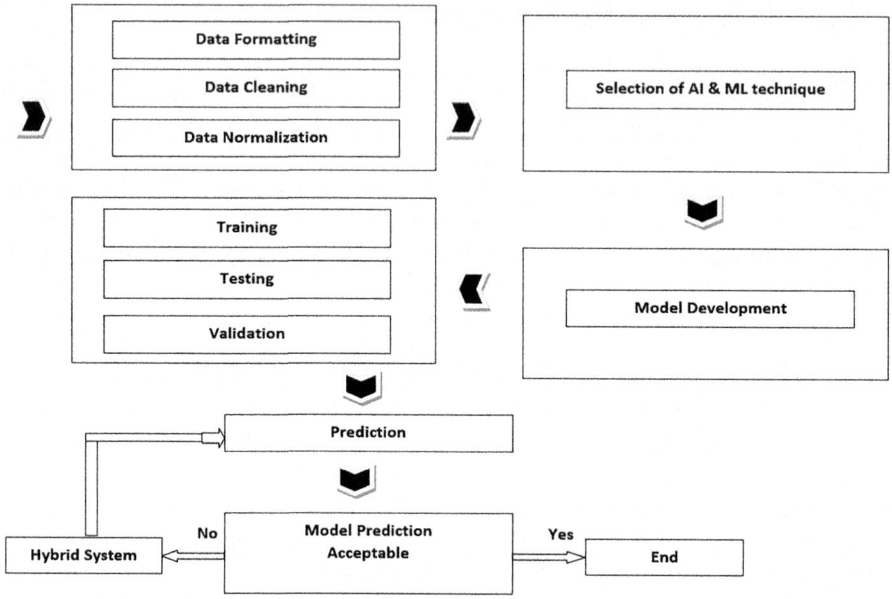

Fig. 11.1 ML algorithms and process flow

prediction due to their ability to associate pollutant concentrations with the influence of meteorological and anthropogenic variables. Linear and polynomial regression models serve as foundational techniques for estimating pollutant levels using features such as wind speed, ambient temperature, and industrial activity. Polynomial regression is particularly well-suited for modeling nonlinear environmental interactions. SVMs enhance classification by mapping feature sets into higher-dimensional spaces where pollution categories are separable via hyperplanes. Kernel functions allow SVMs to manage nonlinear dependencies while minimizing overfitting. DTs, although interpretable, are prone to overfitting. Ensemble methods like RFs overcome this limitation by aggregating multiple DTs, thereby improving generalization and robustness, especially when modeling pollutant–meteorology interactions. Unsupervised learning is essential in scenarios that lack labeled datasets. Techniques such as K-means clustering classify air quality data into meaningful groups, enabling the detection of pollution hotspots and temporal trends. In Bhilwara, K-means has proven effective in segmenting areas by emission intensity and tracking seasonal pollution variability. However, model performance is contingent on high-quality input data and careful selection of cluster parameters. Collectively, these ML techniques provide scalable, real-time, and interpretable solutions to manage and mitigate air pollution, offering a strong foundation for integration into urban planning and environmental health frameworks (Table 11.2).

11.8 Autoencoders for Anomaly Detection in Sensor Data

Autoencoders, a subset of unsupervised neural networks, have become increasingly valuable for detecting anomalies in air pollution datasets, particularly those collected from IoT-enabled sensors. By compressing high-dimensional environmental data into a latent space and reconstructing it, autoencoders highlight deviations when the reconstruction error exceeds a defined threshold, signaling unusual pollution events or sensor malfunctions. In complex urban environments such as Bhilwara, where industrial emissions and vehicular pollutants intersect, autoencoders effectively distinguish genuine air quality spikes from data inconsistencies. For example, a sudden surge in $PM_{2.5}$ may indicate industrial discharge or faulty readings; trained on historical norms, autoencoders learn to differentiate such events. Unlike rigid threshold-based detection methods, they dynamically adapt to evolving data patterns, improving real-time responsiveness. However, their reliability depends on careful model architecture and tuning, especially when handling multivariable, high-resolution data. When paired with clustering algorithms like K-means, autoencoders serve as robust tools for early detection, spatial analysis, and informed environmental policy planning.

11.9 Model Evaluation and Interpretive Insights

To assess the performance of ML models applied to air pollution forecasting in Bhilwara, a variety of regression and classification metrics were used (Tables 11.1 and 11.2). These included the root mean square error (RMSE) and the mean absolute error (MAE), which quantify the prediction accuracy by measuring the deviation between the observed and predicted values. Lower values of RMSE and MAE indicate better model precision. The coefficient of determination (R^2) was used to evaluate how well the input variables explained the variations in pollutant levels. For classification models, such as those predicting health risk categories, confusion matrices were analyzed to assess precision, recall, and overall classification accuracy. Among all models tested, LSTM networks achieved the highest accuracy for time-series predictions of $PM_{2.5}$, with an RMSE of 5.98 μg/m^3, MAE of 3.67 μg/m^3, and R^2 score of 0.93. RF also performed well, providing a balanced trade-off

Table 11.1 Performance metrics of air pollution prediction models

Model	RMSE ↓	MAE ↓	R^2 score ↑	Observations
Linear regression	10.456	7.120	0.72	Poor for complex pollution patterns
SVR	9.821	6.749	0.78	Moderate performance, limited interpretability
DT	8.432	5.230	0.85	Overfitting risk, decent accuracy
RF	6.823	4.190	0.91	Strong performance with good explainability
LSTM	5.982	3.675	0.93	Best for time-series forecasting

Table 11.2 Feature importance in air pollution prediction

Feature	Influence	Key observations
PM_{10}	High	Strongest correlation with $PM_{2.5}$
Humidity	High	Influences pollutant retention and dispersion
NOx	Medium	Key factor in smog and secondary pollution
Temperature	Medium	Affects the chemical transformation of pollutants
Wind speed	Low	Helps in dilution and dispersion

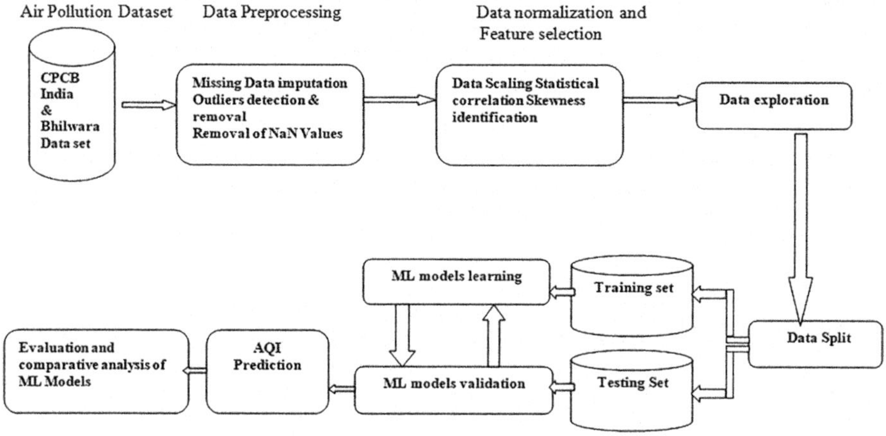

Fig. 11.2 Workflow for air pollution prediction

between interpretability and performance (RMSE: 6.82 µg/m^3, R^2: 0.91). In contrast, linear models such as linear regression and SVMs showed limitations in modeling non-linear pollution patterns prevalent in industrial cities like Bhilwara.

To identify the variables that most influenced the $PM_{2.5}$ predictions, a feature importance analysis was conducted. PM_{10} and humidity emerged as the dominant predictors, followed by nitrogen oxides (NO_x) and ambient temperature. This finding is consistent with previous studies that highlight the role of humidity in trapping fine PM, thus increasing ground-level pollution exposure (Fig. 11.2).

To visualize the dynamics of pollution, various techniques were applied, and we have combined them with model insights to offer valuable guidance for air quality management. The integration of ML tools such as RF and LSTM with diverse datasets, from CPCB monitors to industrial emission logs, demonstrates the power of AI to support real-time forecasting, policy planning, and health-centric mitigation strategies for cities like Bhilwara (Fig. 11.3).

A heatmap was generated to analyze the correlations between different pollutants and meteorological factors. The heatmap analysis confirmed a strong positive correlation between PM_{10} and $PM_{2.5}$, reinforcing the role of industrial emissions in the degradation of air quality (Fig. 11.4).

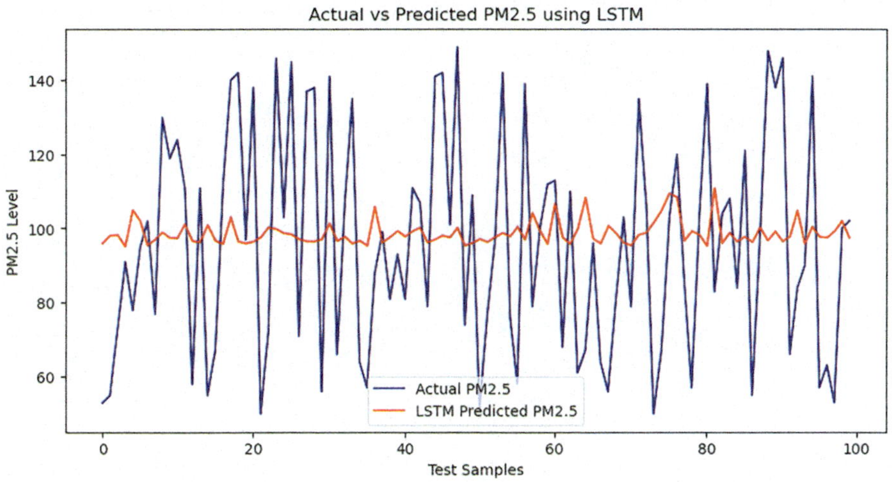

Fig. 11.3 Correlation heatmap of air pollution features

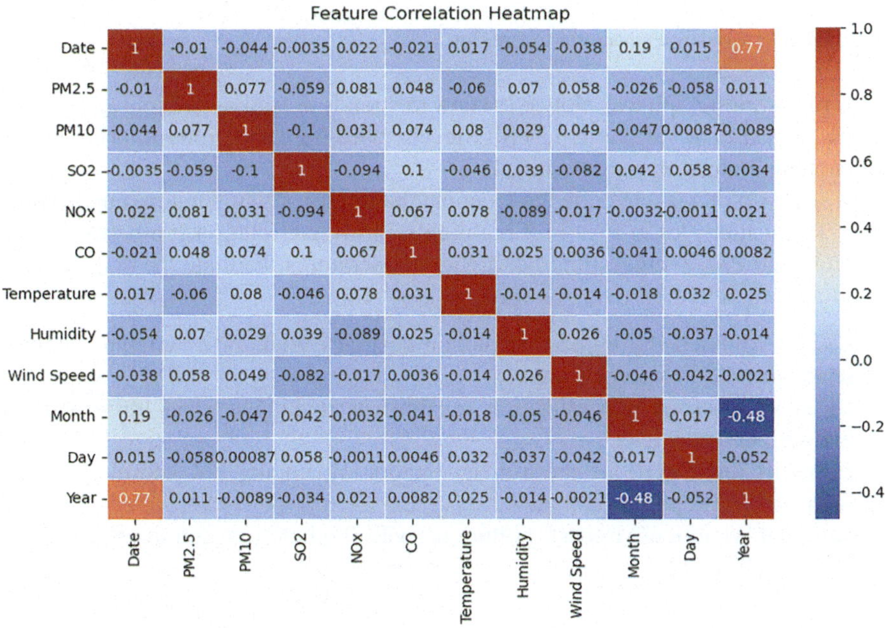

Fig. 11.4 $PM_{2.5}$ levels over time

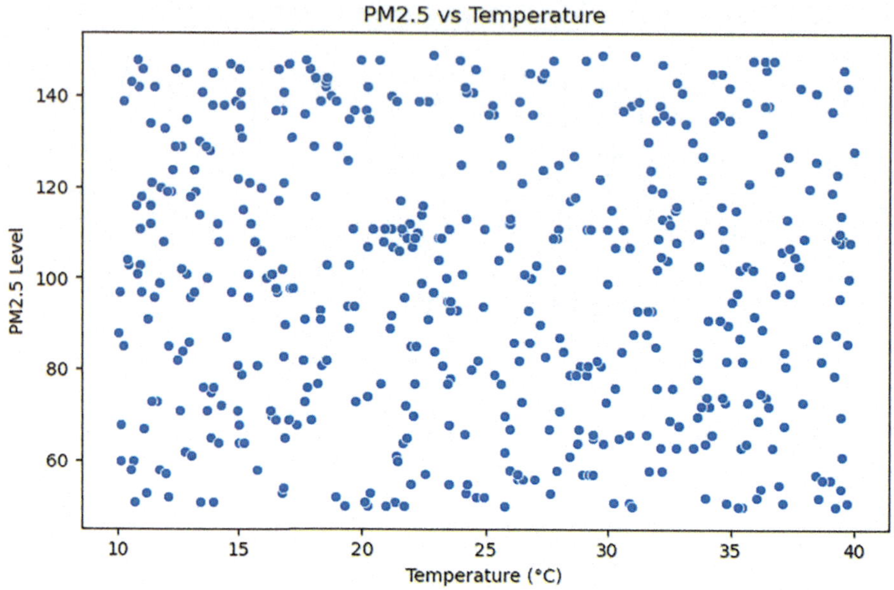

Fig. 11.5 $PM_{2.5}$ versus temperature scatter plot

A time series graph was plotted to show the trends in Bhilwara. Figure 11.2 illustrates the seasonal variations in $PM_{2.5}$ levels, with higher pollution during winter due to temperature inversions (Fig. 11.5).

A scatter plot was generated to examine the impact of temperature on air pollution. The scatter plot in Fig. 11.3 highlights an inverse relationship between temperature and $PM_{2.5}$, suggesting that higher temperatures facilitate the dispersion of pollutants, thereby improving air quality (Fig. 11.6).

- LSTM exhibited the highest prediction accuracy, making it suitable for time-series air pollution forecasting.
- RF showed strong generalization capabilities, making it ideal for structured, high-dimensional data.
- Humidity and PM10 are the primary factors influencing $PM_{2.5}$ levels, aligning with real-world pollution trends.
- Seasonal variations impact pollution, with higher concentrations of $PM_{2.5}$ observed in winter.

11.10 Conclusion and Future Prospect

ML has emerged as a transformative tool in air pollution prediction, offering data-driven insights with a level of accuracy and scalability that surpasses traditional statistical approaches. This chapter has explored the application of various

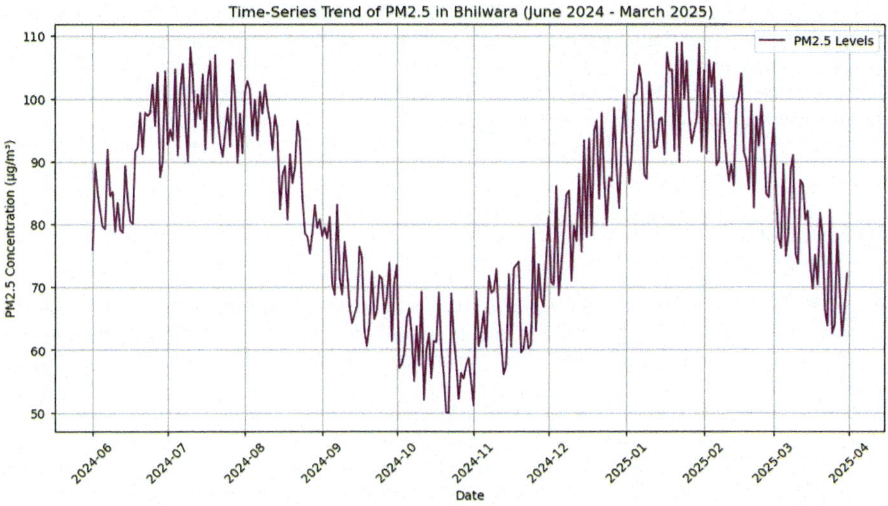

Fig. 11.6 Time series of Bhilwara ($PM_{2.5}$)

supervised and unsupervised ML techniques in forecasting air quality in Bhilwara, Rajasthan, an urban industrial hub with a complex pollution profile. By integrating meteorological data, pollutant concentrations, and geospatial inputs, ML models such as RFs and deep neural networks have demonstrated high predictive accuracy and operational feasibility.

The deployment of IoT-enabled sensors further strengthens real-time monitoring capabilities, enabling early warnings and data-informed interventions. Although challenges persist, such as data sparsity, sensor anomalies, and computational overhead, the integration of preprocessing pipelines and explainable AI methods (e.g., SHAP and LIME) enhances model robustness and transparency. Looking ahead, future research should focus on hybrid modeling approaches that combine the pattern recognition strengths of deep learning (e.g., LSTM and CNN) with the interpretability of ensemble methods and regression models. These models can address trade-offs between performance and explainability, especially in regulatory contexts.

Beyond prediction, ML holds promise in shaping environmental policy. By analyzing large-scale pollution trends, ML-powered decision support systems can guide policymakers in formulating dynamic regulations, setting adaptive emission thresholds, and deploying smart city interventions. Cloud-based ML platforms can also allow governments to automate air quality alerts and optimize resource deployment in high-risk zones. Ultimately, the convergence of ML, IoT, and cloud infrastructure offers a pathway toward fully automated real-time environmental monitoring systems. With continued advancements, ML will play a central role in safeguarding public health, guiding urban sustainability, and mitigating the long-term impacts of industrialization.

Acknowledgments The authors thank the IIT(BHU), Varanasi, India and Department of Computer Science and Engineering, Career Point University, Kota, Rajasthan, India, for providing facilities and support to carry out this research work. The authors also thank the Department of Computer Science and Engineering, Sangam University, Bhilwara, Rajasthan, India, for access to the characterization facility.

References

Aram, S. A., Nketiah, E. A., Saalidong, B. M., et al. (2024). Machine learning-based prediction of air quality index and air quality grade: A comparative analysis. *International Journal of Environmental Science and Technology, 21*, 1345–1360. https://doi.org/10.1007/s13762-023-05016-2

Bekkar, A., Hssina, B., Douzi, S., et al. (2021). Air-pollution prediction in smart city, deep learning approach. *Journal of Big Data, 8*, 161. https://doi.org/10.1186/s40537-021-00548-1

Emec, M., & Yurtsever, M. (2025). A novel ensemble machine learning method for accurate air quality prediction. *International Journal of Environmental Science and Technology, 22*, 459–476. https://doi.org/10.1007/s13762-024-05671-z

Gupta, A., Sharma, S. K., Mishra, A., et al. (2024). Effect of mine-derived pollution on photosynthetic pigments of plants in Rajasthan. *Vegetos, 37*, 1833–1842. https://doi.org/10.1007/s42535-024-00992-3

Gupta, N., Purohit, M. R., & Daiman, A. (2024). Assessing urban transformation and green infrastructure in Bhilwara through machine learning and earth observation. *Arabian Journal of Geosciences, 17*, 269. https://doi.org/10.1007/s12517-024-12076-2

Gupta, N., & Purohit, M. R. (2022). An analysis of the factors affecting the development of urban green spaces: a case study of Bhilwara City, Rajasthan. *Research Reinforcement A Peer Reviewed International Refereed Journal, 10*(1), 78–82.

Kumar, A., Somani, V., Krishna, P., & Amarawat, G. (2025). Comprehensive study of computational methodologies for air pollution control. In A. K. Yadav (Ed.), *Computational techniques in environmental engineering*. CRC Press. https://doi.org/10.1201/9781003605553-3

Kumar, K., Pande, B. P. (2023). Air pollution prediction with machine learning: a case study of Indian cities. *Int. J. Environ. Sci. Technol, 20*, 5333–5348. https://doi.org/10.1007/s13762-022-04241-5

Mohan, V., Mishra, R. K., & Soni, V. K. (2024). Air quality analysis in desert region in the northern state of India: GIS based approach. *Journal of the Indian Society of Remote Sensing*. https://doi.org/10.1007/s12524-024-02073-z

Natarajan, S. K., Shanmurthy, P., Arockiam, D., et al. (2024). Optimized machine learning model for air quality index prediction in major cities in India. *Scientific Reports, 14*, 6795. https://doi.org/10.1038/s41598-024-54807-1

Praveen Kumar, R., Samuel, C., Raju, S. R., et al. (2022). Air pollution in five Indian megacities during the Christmas and new year celebration amidst COVID-19 pandemic. *Stochastic Environmental Research and Risk Assessment, 36*, 3653–3683. https://doi.org/10.1007/s00477-022-02214-1

Rautela, K. S., & Goyal, M. K. (2024). Transforming air pollution management in India with AI and machine learning technologies. *Scientific Reports, 14*, 20412. https://doi.org/10.1038/s41598-024-71269-7

Roy, A., & Mim, M. R. (2023). An approach to detect air quality using machine learning and data analysis tools. In *2023 14th international conference on computing communication and networking technologies (ICCCNT)* (pp. 1–7). IEEE.

Sharma, M., Jain, S., & Lamba, B. Y. (2020). Epigrammatic study on the effect of lockdown amid Covid-19 pandemic on air quality of most polluted cities of Rajasthan (India). *Air Quality, Atmosphere and Health, 13*, 1157–1165. https://doi.org/10.1007/s11869-020-00879-7

Suthar, G., Kaul, N., Khandelwal, S., & Singh, S. (2023). Predicting land surface temperature and examining its relationship with air pollution and urban parameters in Bengaluru: A machine learning approach. *Urban Climate, 53*, 101830. https://doi.org/10.1016/j.uclim.2024.101830

Yadav, A. K., Gupta, K., Bodhale, R., & Phuleria, H. C. (2025). Metals and trace element associated health risk in size-segregated household dust of an industrial town in India. *Air Quality, Atmosphere & Health*. https://doi.org/10.1007/s11869-025-01794-5

Yadav, A. K., Sahoo, S. K., Dubey, J. S., Kumar, A. V., Pandey, G., & Tripathi, R. M. (2019). Assessment of particulate matter, metals of toxicological concentration, and health risk around a mining area, Odisha, India. *Air Quality, Atmosphere & Health, 12*(7), 775–783.

Yadav, A. K., Sahoo, S. K., Patra, A. C., Dubey, J. S., Lenka, P., Sagar, D. V., Kumar, A. V., & Tripathi, R. M. (2014). Source identification of particulate matter and associated intake of elements through inhalation in an industrial area of Odisha, India. *Toxicological & Environmental Chemistry, 96*(3), 410–425.

Yadav, A. K., Shirin, S., Emmanouil, C., & Jamal, A. (2022). Effect of seasonal and meteorological variability of air pollution in Singrauli coalfield. *Aerosol Science and Engineering, 6*, 61–70.

Zhao, Z., Qin, J., Hu, A., Zhang, N., Xie, J., & Sun, Y. (2026). WaveNet-LSTM: A deep learning air quality prediction model based on both air pollutants and meteorological factors. *Information Sciences*, 730, 122894. https://doi.org/10.1016/j.ins.2025.122894

Zhou, Z., Qiu, C., & Zhang, Y. (2023). A comparative analysis of linear regression, neural networks and random forest regression for predicting air ozone employing soft sensor models. *Scientific Reports, 13*, 22420. https://doi.org/10.1038/s41598-023-49899-0

Chapter 12
Emission Inventories to Artificial Intelligence-Based Prediction for Air Pollutants

Karigowda, S. K. Pandey, and G. Habib

12.1 Introduction

The environment is a mixture of several natural components, including water, air, and soil, and the presence of living entities depends on the state of the environment. Human involvement in making life pleasant has changed the ratios of natural resources used or exploited, compromising the environment. Attempts have been made to track and replicate the effects of the compromise, which is complicated because of the nesting of difficulties (Yadav et al., 2024). Usually, modeling an environmental system requires a compromise among social, financial, technical, and ecological factors. However, when identifying problems and developing multilevel systems, social and ecological factors interact regularly through feedback loops and interchangeability (Pu & Qiu, 2016). Complicated and dynamic interdependencies between social and ecological subsystems (Liu et al., 2015) are less than fully known and defined. However, ecosystems and sustainable resource management depend on awareness of complicated interconnections among environmental systems. Ignorance of such interdependencies and dynamics results in even more intricate environmental issues and development difficulties, such as degraded natural resources, biodiversity loss, loss of natural habitat, and climate change.

Karigowda (✉) · G. Habib
Department of Civil and Environmental Engineering, Indian Institute of Technology Delhi, New Delhi, India
e-mail: cez238123@iitd.ac.in

S. K. Pandey
Department of Electrical Engineering, Indian Institute of Technology Delhi, New Delhi, India

A. K. Yadav (ed.), *Computational Approach for Air Pollution*,
https://doi.org/10.1007/978-3-032-09215-1_12

12.2 Data Sources

The quality of the input data determines the accuracy of the assessment and prediction. Ground-based monitoring networks, satellite observations, emission inventories, and meteorological reanalysis data must be used to develop a reliable computational framework for air quality analysis. Especially when merging datasets with varied spatial and temporal resolutions, proper data preparation is crucial to ensure consistency, accuracy, and compatibility between several data sources.

12.2.1 Ground-Based Monitoring

Ground-based monitoring, satellite-based remote sensing, emission inventories, and meteorological data define four broad categories of primary data sources for air pollution modeling. Although they offer different insights into air pollution patterns and trends, each data source also brings particular difficulties in coverage, accuracy, and resolution.

Ground-based monitoring stations directly assess air pollutants, including particle matter ($PM_{2.5}$ and PM_{10}), nitrogen oxides (NO_x), sulfur dioxide (SO_2), carbon monoxide (CO), ozone (O_3), and volatile organic compounds (VOCs). Usually run by national and regional organizations such as the European Environment Agency, the Environmental Protection Agency (EPA) in the United States, and the Central Pollution Control Board (CPCB) in India, these stations are run because they keep large-scale monitoring systems with high frequency, usually at hourly or daily intervals. For the National Air Quality Monitoring Program, the CPCB operates more than 400 sites in India. By measuring important air pollutants and meteorological data, these stations generate a strong dataset for air quality research (CPCB, 2022). Comparably, the United States EPA maintains the Air Quality System, which offers long-term records of air pollution levels and meteorological data throughout the country (EPA, 2023). Even with their accuracy, ground-based monitoring stations are sometimes limited in spatial coverage, particularly in rural and distant places.

12.2.2 Satellite-Based Remote Sensing

Using statistical and machine learning (ML) methods, the Moderate Resolution Imaging Spectroradiometer offers aerosol optical depth data to estimate $PM_{2.5}$ concentration (Gupta et al., 2013). With great precision, the Ozone Monitoring Instrument detects NO_2 and SO_2 concentrations, helping to identify pollution hotspots and long-term trends (Krotkov et al., 2016; Veefkind et al., 2012). Suitable for urban-scale pollution research, the Tropospheric Monitoring Instrument provides daily global data with a high spatial resolution (up to 7×3.5 km). Combining ground-based and satellite-based data improves spatial and temporal coverage of air pollution estimations, improving model projection accuracy (Xue et al., 2020). Although satellite data offer broad coverage, they are sometimes constrained by cloud cover, atmospheric interference, and ground-based calibration needs. Combining satellite and ground-based data improves the accuracy and dependability.

12.2.3 Emission Inventories

Emission inventories give approximations of pollution from natural and man-made sources. Numerical modeling and source apportionment investigations contain comprehensive data on source categories, fuel types, and activity levels, all of which are vital. These inventories are based on designing focused mitigation plans and knowing the contribution of several sources to the general levels of pollution. Important global and regional emissions inventories include the Emission Database for Global Atmospheric Research, which offers $0.1° \times 0.1°$ resolution global emission estimates for air pollutants and greenhouse gases (Crippa et al., 2018). High-resolution estimates of anthropogenic emissions in Asia (Kurokawa & Ohara, 2020) are supplied by the Regional Emission Inventory in Asia. Sector-specific emissions for black carbon, methane, and other short-lived climate pollutants (Klimont et al., 2017) are produced by the Evaluating the Climate and Air Quality Impacts of Short-Lived Pollutants inventory. Under the National Clean Air Program, the National Emission Inventory assembled in India offers comprehensive sector-wise emissions for important metropolitan areas (Guttikunda & Calori, 2013). However, inadequate data, unpredictability in emission components, and activity data errors sometimes link emission inventories with uncertainty.

12.2.4 Meteorological Data

Models of pollution transport, dispersion, and chemical reactions depend critically on meteorological data, including wind speed (m/s), wind direction (°), temperature (°C), relative humidity (%), and boundary layer height (m). Reanalysis (ERA5) of the European Center for Medium-Range Weather Forecasts offers hourly worldwide meteorological data at 31 km spatial resolution (Hersbach et al., 2020). Long-term worldwide atmospheric data for historical and predictive modeling is supplied by the National Centers for Environmental Prediction (NCEP)/National Center for Atmospheric Research (NCAR) reanalysis (Kalnay et al., 2018). The Indian Meteorological Department offers real-time meteorological data and forecasts, including those for monsoons and severe storms in India. Improved pollution transport and chemical transformation simulations are made possible by merging meteorological data with emission inventories and pollution concentration data.

12.3 Statistical and Numerical Models

Predicting the distribution, transformation, and removal of pollutants in the atmosphere depends on air pollution models, which are crucial instruments. These models use mathematical equations to replicate the movement of pollution based on meteorological, spatial, and emission data. As Fig. 12.1 shows, the choice of a suitable model depends on elements, including the size of the analysis, the behavior of pollution, and data availability (Seinfeld & Pandis, 2016). Gaussian dispersion models are fundamental for calculating pollution concentrations from point sources among the extensively used models, including American Meteorological Society/ Environmental Protection Agency Regulatory Model (AERMOD) and California Puff Model (CALPUFF).

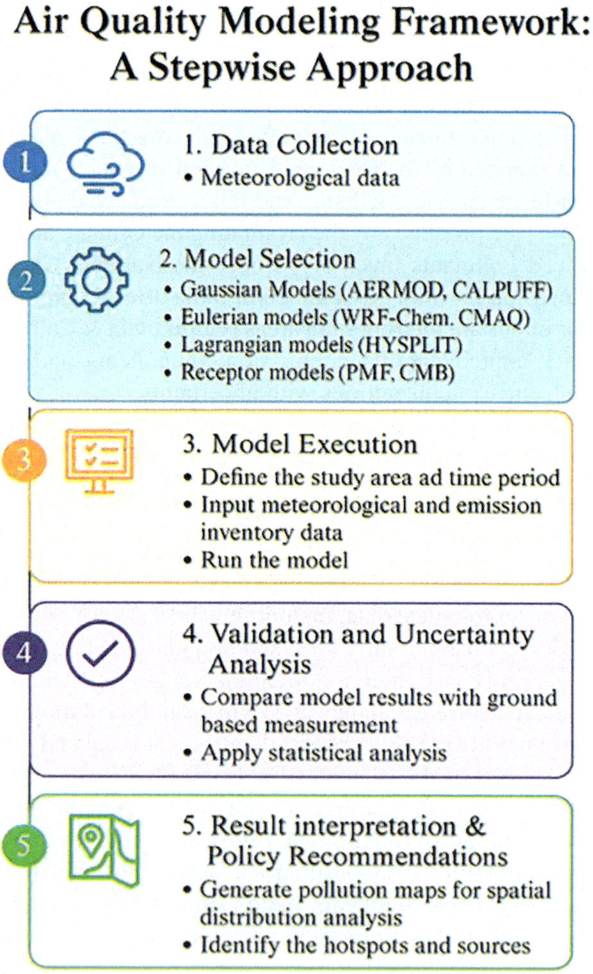

Fig. 12.1 Selection of models for air quality prediction

12.3.1 Gaussian Diffusion Models

Gaussian dispersion models show that air turbulence causes contaminants to spread in a bell-shaped (Gaussian) pattern. These models are especially successful in predicting the dispersal of contaminants from limited sources, including industrial stacks (Holmes & Morawska, 2006). Designed for short-range dispersion (up to 50 km), AERMOD combines meteorological and topographical influences to enhance concentration forecasts (Cimorelli et al., 2005). CALPUFF, on the other hand, is a sophisticated model fit for regional-scale research that allows long-range transportation, complex wind patterns, and chemical changes (Scire et al., 2000).

Emission data define source location (latitude, longitude), emission rate (g/s-m^2), and pollution type (SO_2, NO_x, and $PM_{2.5}$). Several factors influence initial dispersion dynamics, including stack height, exit velocity, exit temperature, and source type (point, line, or area) (Holmes & Morawska, 2006). Meteorological data govern pollution movement. Wind speed and direction determine dispersion patterns. Vertical mixing is affected by temperature, atmospheric stability (Pasquill-Gifford classification), and the height of the boundary layer. Further influencing chemical reactions and secondary pollutant generation are turbulence parameters, relative humidity, and sun radiation (Scire et al., 2000). The shape behavior, topography height, affects airflow, and land usage determines surface interactions. Turbulence, wind flow, and pollution removal rates (US EPA, 2023) are influenced by surface roughness, topographical features, and deposition parameters.

Defining emission sources and setting the simulation domain call for model setup. These covers defining point, line, or area sources, setting boundary conditions, and choosing suitable dispersion parameters (Nieuwstadt, 1991). Gaussian dispersion models assume steady-state conditions; hence, accuracy depends on the appropriate initiation of the data using real-time or historical weather. To enhance model predictions, the setup process also combines chemical transformation mechanisms and turbulence (Turner et al., 1994). To determine the concentrations of pollution, the computations consist of solving the Gaussian dispersion equation. The model computes horizontal and vertical dispersion using meteorological inputs and emission properties, producing spatial and temporal pollution distributions (Arya, 1999). Algorithms guarantee accurate simulations of pollution behavior in different atmospheric conditions by including turbulence, atmospheric mixing, and chemical transformations, thereby refining concentration forecasts (Scire et al., 2000). The results generate time-series data for analysis and pollution concentration maps. These outputs track pollution dispersion over time, help evaluate air quality levels, and show areas for regulatory intervention's hotspot development. Sensitivity analysis and validation of observational data improve model dependability, guaranteeing consistency between simulated and measured air pollution levels (Zhang et al., 2023).

12.3.2 *Eulerian Models (CMAQ, GEOS-Chem, and WRF-Chem)*

Taking into account emissions, chemical interactions, transportation, and removal techniques, Eulerian models split the atmosphere into a three-dimensional (3D) grid and calculate pollution concentrations inside each grid cell. These models offer a complete method for reproducing the dynamics of local, regional, and worldwide air pollution. The Weather Research and Forecasting (WRF)-Chem-linked climate prediction and forecasting model linked with Chemistry (WRF-Chem) generates real-time air quality forecasts by combining meteorological forecasting with atmospheric chemistry. Incorporating secondary pollutant generation and chemical interactions, the Community Multiscale Air Quality (CMAQ) model specializes in regional-scale air quality modeling. Globally tracking long-term pollution trends and transcontinental pollution movement, the Goddard Earth Observing System with Chemistry (GEOS-Chem) runs.

To guarantee reliable simulations, Eulerian models depend on a large dataset spanning emissions, meteorology, chemistry, and boundary conditions. Industrial, vehicle, and natural emissions are broken down by sector and source type in the gridded emission inventory. Included are primary pollutants (e.g., carbon monoxide (CO), nitrogen oxides (NO_x), sulfur dioxide (SO_2), particulate matter (PM), VOCs, and secondary aerosol precursors). Meteorological fields include temperature, pressure, humidity, cloud cover, planetary boundary layer height, precipitation, solar radiation, and wind speed and direction. These values define the dispersion and movement of pollutants. Detailed oxidation routes, gas- and aqueous-phase chemistry, gas-to-particle conversion, and heterogeneous reactions constitute chemical reaction mechanisms. Models of atmospheric changes commonly used on chemical systems include the carbon bond mechanism Z, the regional atmospheric chemistry mechanism, and the Statewide Air Pollution Research Center, transport of pollutants from outside the modeling domain, including background pollutant concentrations, stratospheric influx of ozone, and transboundary pollution contributions.

Eulerian models replicate pollution distribution, transformation, and removal using a methodical approach: Set up the model domain as a 3D grid where every cell denotes an atmospheric volume. The research scale determines the grid resolution; it ranges from local (1–5 km) to regional (10–50 km) to worldwide (100–500 km). Numerical weather prediction models, such as the WRF model, help replicate meteorological conditions to generate dynamic atmospheric input fields for pollution transfer. Differential equations controlling advection, diffusion, turbulent mixing, dry and wet deposition, and chemical modification of contaminants within every cell of the grid are solved in the Chemical Transport Simulation. Model performance and accuracy are evaluated using real-world air quality measurements from ground-based monitoring networks (e.g., Air Quality Index data, satellite retrievals, and lidar observations).

12.3.3 Lagrangian Models (HYSPLIT)

Lagrangian models follow individual pollutants through the atmosphere. These models follow air parcels across time rather than grid-like Eulerian models do. Studying pollution movement, source identification, and long-range dispersion extensively uses the hybrid single-particle Lagrangian integrated trajectory (HYSPLIT) model. Designed by the Air Resources Laboratory (NOAA ARL) of the National Oceanic and Atmospheric Administration, it finds use in operational air quality applications and research.

Lagrangian models call for important meteorological information and emission source characteristics. Typically derived from numerical weather prediction models such as the Global Forecast System and the North American Mesoscale model, meteorological inputs include wind fields (m/s), temperature (K or °C), atmospheric pressure (hPa), relative humidity (%), and precipitation (mm). Essential for describing pollution behavior in the atmosphere are also emission source factors such as location (latitude and longitude), release height (m), pollutant type (e.g., CO, NO_2, and $PM_{2.5}$), emission rate (kg/s), and length of release (hours or days).

The HYSPLIT model simulates the dispersion by a methodical approach. Specifying its location (latitude, longitude), height, and release timing first defines the emission source. Then, meteorological data replicate the movement of pollution using wind fields and atmospheric factors to forecast transport and dispersion. The model then tracks the pollution pathways, computes dispersion patterns, chemical changes, and finally deposition. HYSPLIT also produces outputs that help depict pollution paths, deposition zones, and concentration distributions, supporting impact analysis and air quality evaluations.

12.3.4 Positive Matrix Factorization, Chemical Mass Balance, Receptor Models

Using chemical compositions of the observed $PM_{2.5}$ concentrations at receptor locations, receptor models are widely applied to identify the sources of air pollution. Receptor models depend on statistical methods to allocate observed pollution to its sources, unlike dispersion models, which replicate the movement and modification. Receptor models often utilize chemical mass balance (CMB) and positive matrix factorization (PMF). PMF statistically breaks down data on ambient air quality into elements reflecting several sources of emission patterns (Paatero & Hopke, 2003). On the contrary, CMB projects source contributions using specified chemical signatures of contaminants from different sources (Bisht et al., 2016).

Receptor models use a disciplined approach to project pollutant sources. Data collection for air quality comes from state environmental authorities and national and regional monitoring networks, including the CPCB. Source Profile Development, laboratory investigations, and emission inventory data help to compile the chemical compositions of many pollution sources. PMF or CMB models are used in statistical processing to distribute the observed pollution levels to various sources. While CMB needs known source signatures, PMF detects elements without prior source knowledge. The projected source contributions are investigated to find seasonal trends, primary pollution sources, and the effect of several emission sources on air quality.

12.3.5 ML- and AI-Based Approaches

Policy planning, environmental protection, and public health depend on accurate prediction of air pollution. Estimating the dispersion of pollution, chemical transformations, and meteorological factors requires physical equations in conventional air quality models. However, these models can be limited due to complicated atmospheric chemistry, climatic projections, and the uncertainty of the emission inventory. By understanding patterns from past data to improve forecast accuracy, artificial intelligence (AI) and ML offer a substitute method. This part investigates several ML and AI models applied in air pollution forecasting, their data needs, technological workflow, and how they lower uncertainty than conventional approaches.

Trained on past air pollution data, supervised learning algorithms seek trends and project future pollution levels. These models need labeled datasets, in which target outputs (pollution concentrations) are linked to input features (e.g., traffic emissions). Random forest (RF) is an ensemble learning technique that creates several decision trees, and their predictions are averaged to improve accuracy while reducing overfitting. It ranks the relevance of different sources of pollution and manages missing data well. The support vector machine is a classification model that defines limits between several air quality categories (e.g., low, moderate, and high pollution levels) to recognize trends in complicated datasets. Artificial neural networks are collections of linked artificial neural layers that process data like the human brain. It fits non-linear data patterns since it can identify intricate interactions between pollution variables and outside variables. Originally designed for image analysis, CNNs can also examine spatial data, including satellite pictures or pollution maps, to identify emission hotspots and follow pollution dispersion patterns.

Deep learning (DL) models collect high-level features from vast datasets using several layers of neural networks, hence augmenting conventional ML methods. These models are especially successful in time series forecasting and long-term air quality trend analysis. Designed to manage sequential data, long short-term memory (LSTM) is a specialized variation of a recurrent neural network (RNN). It is appropriate to project future air quality based on historical observations, since it recalls prior pollution levels and patterns of pollution. Originally designed for language processing, these cutting-edge AI algorithms are now applied in air quality prediction. Transformers capture intricate relationships between meteorological circumstances, pollution sources, and air transport mechanisms by simultaneously analyzing many factors.

Information gathering: Get high-resolution spatiotemporal pollution data from satellite observations, traffic sensors, meteorological networks, and air quality monitoring stations. Land use trends, vehicle emission inventories, and industrial source emissions data should be included to improve prediction accuracy. Data preprocessing: Apply statistical and ML-based approaches to outlier discovery and eradication. Apply a missing value imputation based on DL, interpolation, or regression model-based missing value imputation. Standardize and normalize data to guarantee consistency across many scales and measuring units. Selection and feature engineering: Obtain and build pertinent features, including indices of air stability, chemical compositions, and transport patterns. You can maximize model input and lower dimensionality by using feature selection methods such as principal component analysis, SHAP values, or recursive feature elimination. Model training and optimization: With historical air quality data, train ML or DL models, including RFs, gradient boosting machines, LSTMs, or transformer-based architectures. Use cross-validation, grid search, or Bayesian optimization to optimize hyperparameters for the best prediction performance. Combining physics-informed ML or hybrid modeling techniques helps you incorporate domain knowledge. Model verification, cross-validation methods, and independent test datasets help you assess model performance. To guarantee strength, use several evaluation measures such as root mean square error (RMSE), mean absolute error (MAE), R^2, and skill scores. Spatial and temporal validation is used to evaluate the generalizability of the model over several sites and time periods.

12.3.6 Computational Fluid Dynamics for Air Pollution

Particularly at urban and industrial scales, computational fluid dynamics (CFD) is a potent tool for simulating the dispersion of pollutants in complicated systems. To replicate how contaminants spread under different air conditions, CFD methods solve the governing equations of fluid flow, including the Navier-Stokes equations. CFD offers high-resolution views of microscale turbulence, pollution transport, and interactions with buildings and terrain features, unlike conventional air quality models (Blocken, 2015). This approach is constructive in crowded cities because the pollution concentration distributions (Tominaga & Stathopoulos, 2013) are highly influenced by street canyons, traffic emissions, and wind flow patterns.

Numerical methods are fundamental tools used in CFD-based air pollution models to solve equations controlling fluid dynamics and pollution transport. Among these, the continuity equation guarantees mass conservation of air and contaminants. Describe how buoyancy, turbulence, and wind flow cause pollutants to migrate. Blocken (2015) considers the advection, diffusion, and chemical modification of contaminants in a transportation equation. Resolving small-scale fluctuations in pollution dispersion makes CFD one of the main benefits of air pollution modeling, appropriate for analysis of near-source effects, including emissions from industrial stacks, vehicle pollution in street canyons, and ventilation patterns in built environments (Gousseau et al., 2011; Yadav, 2015; Yadav et al., 2019, 2025).

12.3.7 LES, RANS, and DNS Approaches

Using several turbulence modeling techniques, each with differing degrees of complexity and computing expense, CFD simulations: By means of statistical averaging, Reynolds-averaged Navier-Stokes (RANS) solves the mean flow equations and represents turbulence. Although it may oversimplify transitory turbulence effects, it is computationally efficient and is widely applied in air quality studies (Blocken, 2015). Large Eddy simulation (LES) models smaller sizes by directly resolving significant turbulent eddies. It needs more computer resources but offers better accuracy in simulating pollution dispersal in complicated metropolitan settings (Tominaga & Stathopoulos, 2013). Direct numerical simulation (DNS) produces the most accurate results by resolving all turbulence scales without modeling. However, its significant computing cost usually limits it to basic research rather than functional air quality modeling (Lateb et al. 2016).

CFD simulations allow analysis of pollution dispersion in street canyons, thus evaluating the effects of building construction, traffic emissions, and natural ventilation on pollution concentrations (Gousseau et al., 2011). For example, research in Indian cities has evaluated the dispersion of $PM_{2.5}$ using CFD in relation to urban design and vehicle traffic. The evaluation of pollution dispersion from refineries, thermal power plants, and industrial stacks using CFD is common. Studies based on CFD-based models have been applied in India to maximize chimney height and emission control measures in industrial zones (Kumar et al., 2017). CFD enables the analysis of airflow patterns in buildings, optimizing ventilation systems to lower indoor pollution from cooking, smoking, or industrial activity (Kumar et al., 2017).

12.3.8 Emission Inventory Development and Simulation

An emission inventory is an all-encompassing database that covers a given geographic area and period of time and estimates pollutants released from several sources. For air quality, policy development, and environmental impact assessments, it provides a basic input. In particular, in India, where emissions come from many sectors, including transportation, industries, power plants, biomass burning, and residential cooking, accurate emission inventories are absolutely vital to understanding pollution sources and developing effective mitigation strategies (Guttikunda & Calori, 2013).

To estimate emissions, the following equation is used:

$$E_p = \mathrm{EF}_\mathrm{p} * A \tag{12.1}$$

where E_p is the total emissions in tons per year for pollutant p, EF_p is the respective emission factor, and A is the activity data.

Two primary methodologies are used in the development of emission inventories: the Approach from the bottom up, which projects emissions using individual source emission factors and activity data. Transport emissions, for example, are computed using vehicle kilometers traveled (VKT), fuel consumption rates, and emission factors obtained from actual observations (Pandey & Venkataraman, 2014). This method generates high-resolution estimations but requires large-scale ground-level data collection. The top-down method estimates emissions using aggregated national or regional fuel usage and emission variables. Emissions are commonly derived using satellite observations and national energy statistics. However, the method may lack spatial precision. Hybrid methodologies are often utilized to strike a mix between accuracy and data availability, combining both approaches. City-scale emissions, for instance, might rely on bottom-up data, whereas national inventories employ top-down estimates modified with ground observations. To increase model accuracy, emission inventories must consider changes in emissions over place and time. Urban air quality modeling requires high-resolution inventories (1 × 1 km), although lower resolutions (10 × 10 km) would be acceptable for regional research (Guttikunda & Calori, 2013). Geographic information systems and satellite remote sensing are among the geospatial methods used to enhance spatial distribution. Emissions change hourly, daily, and seasonally depending on things like agricultural burning, transportation congestion, and industry activity. For instance, the burning of agricultural debris in Punjab and Haryana peaks in October and November, greatly influencing Delhi's winter air quality (Jain et al., 2014). Temporal scaling elements help to explain these fluctuations in air quality simulations.

12.4 Case Study: Air Quality in Delhi

12.4.1 Introduction

Public health and environmental sustainability are seriously threatened in metropolitan settings by air pollution. High amounts of PM ($PM_{2.5}$) and gaseous pollutants, including nitrogen oxides (NO_x), sulfur dioxide (SO_2), and carbon monoxide (CO), routinely affect Delhi, the capital of India. With an emphasis on three main points, (1) prediction of $PM_{2.5}$ and gaseous pollutants using computational techniques, (2) source apportionment of pollutants using advanced modeling approaches, and (3) an evaluation of the effectiveness of policy interventions using scenario analysis, this part presents a case study in Delhi.

12.4.2 Urban Environment $PM_{2.5}$ and Gaseous Pollution Prediction

The design of appropriate mitigation plans depends on the accurate prediction of air pollutants. DL and ML methods have recently been extensively embraced for $PM_{2.5}$ and gaseous pollution prediction in urban environments. These models use real-time sensor data, emission inventories, and a variety of input parameters, including meteorological variables (such as temperature, humidity, and wind speed).

Reasonable accuracy pollution concentrations have been predicted using RF, support vector regression, and gradient boosting techniques, statistically and ML-based models. LSTM networks and convolutional neural networks (CNNs) have shown a better ability to capture temporal fluctuations in air pollution levels. In complicated metropolitan settings, hybrid models, integration of physical dispersion models with ML-based forecasting methods, have improved predictability.

12.4.3 Results

Capturing pollution patterns, ML-based models beat conventional regression-based methods (Gupta et al., 2013; Zhang et al., 2023). Predictive capacity is greatly improved by including satellite remote sensing data. Variations in seasonality and diurnal effect model performance call for flexible frameworks. Developing sensible air quality control plans depends on knowing where the sources of pollution are. Source apportionment methods assess contributions from many emission sources using receptor and dispersion models. Analyzing ambient air samples helps PMF and CMB models to determine source contributions (Jain et al., 2014; Guttikunda & Calori, 2013; Yadav et al., 2023). Urban pollution transport and dispersion are simulated by Gaussian plume models and Lagrangian particle dispersion models. Chemical speciation data is combined with atmospheric transport models using hybrid approaches, including ML-assisted source apportionment.

The principal causes of Delhi's $PM_{2.5}$ are that about 40% of Delhi's $PM_{2.5}$ concentration comes from vehicle emissions, so they are the primary source. About 18% of the overall levels of $PM_{2.5}$ are related to industrial activity, including emissions from power plants and manufacturing sites. Particularly in peri-urban regions, biomass burning and home fuel consumption add more than 15% and aggravate winter season air quality problems. Furthermore, roughly 27% of the contributing aerosols and long-range transport are secondary aerosols, therefore underscoring the impact of atmospheric chemical processes and regional pollution transport on Delhi's air quality. The effect of policy intervention on air quality in Delhi has several policies in place to help reduce air pollution. By modeling several regulatory and technological initiatives, computational scenario analysis offers an understanding of their efficiency. Important policies and their effects, including Odd-Even Vehicle Rationing: Although temporary declines in $PM_{2.5}$ levels (~5–10%) are observed, the overall impact is still limited due to emissions from other sources. A significant decrease in SO_2 emissions caused by the closure of coal-based power plants results in reduced secondary sulfate generation. Targeted actions have helped to reduce episodic extreme pollution episodes related to biomass burning in the winter months. Bharat Stage VI Emission Standards: Although the increasing number of vehicles balances total gains, vehicle emissions will be reduced.

12.4.4 Final Thought and Future Approaches

Computational approaches applied in air quality management provide a solid basis for estimating pollution levels, source identification, and evaluation of the effectiveness of the policy. Important guidance for the following studies and policy development consists of the following: development of real-time air quality monitoring networks and interaction with AI-based forecast systems (Govardhan et al., 2024); improving the enforcement of emission criteria for both automotive and industrial sources; improvement of infrastructure supporting public transport and support of electric mobility; and strengthening regional cooperation to regulate the movement of transboundary pollutants. This case study lays the basis for evidence-based policymaking and emphasizes the need for computational modeling in tackling problems of urban air pollution problems.

12.5 Uncertainties and Limitations in Computational Models

Notwithstanding developments in computer air pollution modeling, the complexity of atmospheric processes, data quality, and model assumptions still cause uncertainty. In regulatory and policy contexts, these ambiguities affect the accuracy, dependability, and applicability of the model. Predictive performance and error minimization depend on a methodical approach to model validation, sensitivity analysis, and data integration.

12.5.1 Model Performance Assessment and Validation

Determining the dependability of computational air pollution models is critically on model validation. It compares in situ measurements (Zhang et al., 2023), satellite retrievals, and real-world data from ground-based monitoring stations with model predictions. Performance evaluation makes frequent use of statistical measures, including RMSE, MAE, correlation coefficients, and bias analysis (Boylan & Russell, 2006). However, differences between observed and simulated values usually result from uncertainty in emission inventories, climatic input, chemical reaction processes, and spatial-temporal resolution (Hogrefe et al., 2001).

Sensitivity analysis aids in the identification of important model parameters that significantly affect output variance. This mechanism helps prioritize efforts at data cleansing and increase model robustness (Koo et al., 2020). Monte-Carlo simulations, Bayesian inference, and ensemble modeling are among the often-used uncertainty quantification methods to project the spectrum of possible results and related confidence intervals (Cohan et al., 2011). Eulerian models, including CMAQ and WRF-Chem, show uncertainty arising from reaction rate constants, vertical mixing schemes, and cloud chemistry interactions (Appel et al., 2013). Likewise, Gaussian and Lagrangian models encounter questions about dispersion coefficients, turbulence parameters, and deposition rates (Draxler & Hess, 1998). Improving model predictability and guaranteeing reliable air quality assessments depend on addressing these ambiguities (Hanna et al., 2001). Computational models depend on ground-based sensor networks, remote sensing data, reanalysis products, and multisector emission inventories, among other datasets. Perfect data integration is hampered by variations in spatial and temporal resolution, measurement techniques, and data preparation methods (van Donkelaar et al., 2010). Large-scale coverage is provided by satellite-based air quality retrievals, but they also demand bias correction and confirmation against ground-based observations (Xue et al., 2020). Model inputs are similarly unclear when emission factors, industrial activity logs, and real-time traffic emissions vary. To solve these difficulties and improve model accuracy, advanced data assimilation techniques, including ML-based gap filling, data fusion methods, and real-time calibration, are under growing investigation. Improving the dependability of computer models in air pollution evaluation, policymaking, and intervention techniques requires addressing these uncertainties and constraints. Future developments in hybrid modeling methodologies, AI-driven model correction, and high-density atmospheric simulations will further refine predictive skills and lower air quality modeling (Schultz et al., 2021).

12.5.2 Future Directions and Emerging Trends

Advancements in data collection, AI, and real-time simulation methods are driving even more advances in computational air pollution modeling. Rising technologies, including digital twins, edge computing, AI-driven forecasting, and the Internet of Things (IoT), are changing predictive modeling and air quality monitoring. These advances increase real-time decision-making and spatial and temporal resolution and allow proactive pollution management techniques (Schultz et al., 2021). By offering high-frequency real-time pollution measures at fine spatial scales, the explosion of IoT-enabled air quality sensors has transformed atmospheric pollution monitoring (Morawska et al., 2018). Applied in dense metropolitan networks, these low-cost sensors enhance satellite-based observations and conventional regulatory monitoring stations. Still, problems persist, including data dependability, sensor drift, and calibration discrepancies (Kumar et al., 2015). By allowing on-site data processing, lowering latency, and improving computational efficiency, Edge computing solves these constraints. Real-time air quality mapping and exposure assessment are much improved by combining IoT-based networks with AI algorithms, therefore enabling innovative city projects and early warning systems (Mead et al., 2013).

12.5.3 AI-Driven Real-Time Forecasting Models

Leveraging large datasets from ground-based, satellite, and meteorological sources, ML and DL models have become increasingly potent tools for real-time air pollution forecasting (Zhang et al., 2023). Constrained by uncertainty in emission inventories and meteorological inputs, traditional numerical models, including WRF-Chem and CMAQ, call for large computer resources. To increase predictive accuracy, AI-based models, which employ data-driven techniques such as RNNs, CNNs, and transformer architectures, also leverage at a fraction of the computing cost of conventional models; these models clearly detect nonlinear interactions, capture intricate spatio-temporal pollution patterns, and offer real-time forecasts (Hersbach et al., 2020). Furthermore, hybrid methods that combine AI with physics-based models improve interpretability and resilience in air quality (Xing et al., 2020).

12.5.4 Role of Digital Twins in Air Pollution Management

A transforming tool in air pollution control is digital twins, virtual duplicates of real-world systems. Digital twins give dynamic and interactive representations of urban air quality situations (Tao et al., 2018) by combining real-time sensor data, high-resolution simulations, and AI-driven analytics. By using scenario-based modelling made possible by these systems, legislators can evaluate before implementing the effects of urban planning policies, pollution control measures, and climate change mitigation initiatives. By always updating simulations with real data, digital twins also enable adaptive management by enhancing decision-making processes in response to changing pollution conditions (Liu et al., 2015). Digital twin systems combined with edge computing and IoT improve real-time monitoring and intervention capacity even more, therefore indicating a major change toward predictive and prescriptive environmental management (Lu et al., 2021).

12.5.5 Conclusion

The integration of real-time data collection, AI-driven predictive analytics, and dynamic digital twin simulations will define computational air pollution modeling in the future. These new methods provide better computational efficiency, increased forecast accuracy, and more flexibility to accommodate challenging metropolitan settings. Advancements in IoT, AI, and digital twins will be especially important in determining next-generation air quality control plans as they develop, therefore guaranteeing sustainable and data-driven decision-making for the betterment of the planet.

12.6 Conclusion

The integration of computer techniques has revolutionized the discipline of air pollution modeling and prediction. From conventional emission inventories to sophisticated AI-based predictive models, this chapter has offered a thorough investigation of the several computational methods applied in air pollution research. We have discussed how hybrid approaches, CFD, numerical modeling, and ML methods could capture the complexity of air pollution dynamics. We also underlined how remote sensing and high-resolution data sources might help to increase model accuracy and real-time monitoring. An important lesson from this chapter is that every computational technique has advantages and drawbacks. While ML techniques provide the benefit of adaptability and high predicted accuracy with enough training data, emission inventory methods remain fundamental for regulatory uses and policy decisions. Urban air quality studies benefit greatly from CFD modeling since it offers an important understanding of microscale pollution distribution. Advances in processing power and data availability enable the integration of several techniques, therefore reflecting the direction of air pollution modeling. Several research gaps and possibilities still exist, despite these developments. The demand for more large-scale, high-quality datasets to adequately train AI-based models presents one of the main difficulties. Although ML presents interesting predictive power, its reliance on large datasets limits its generalizability in many geographical and environmental settings. Moreover, the interpretability and openness of AI models have to be improved if we are to increase their adoption in legislative environments. Integration of emission inventories with real-time observational data and computational models is another important research void. This integration can improve the accuracy of forecasting and support the prompt implementation of the legislative intervention. Furthermore, given that their influence on air quality and radiative forcing is still under active investigation, the depiction of aerosol dynamics and their optical characteristics in models calls for further attention. Future studies should concentrate on creating hybrid modeling techniques using both AI-driven and physics-based methodologies to increase predictive capacity. In addition, converting computational developments into practical air quality control techniques depends on multidisciplinary cooperation among atmospheric scientists, data scientists, and legislators. In essence, although computational approaches have greatly increased air pollution research, ongoing efforts are needed to improve models, improve data integration, and improve predicted accuracy. Dealing with these difficulties will open the path for more strong and practical insights, so improving air quality control and policy development.

Acknowledgments The authors would like to express their sincere gratitude to Dr. Akhilesh Kumar Yadav for his guidance and support during the preparation of this chapter. They are also thankful to the Technical Review Committee for their valuable comments and suggestions, which helped improve the manuscript. Our appreciation extends to the Department of Civil Engineering and the Department of Electrical Engineering, Indian Institute of Technology Delhi, for providing the necessary academic environment and facilities to carry out this work. Finally, they thank all colleagues who contributed indirectly through insightful discussions and encouragement.

References

Arya, S. P. (1999). *Air pollution meteorology and dispersion* (Vol. 310). Oxford University Press.

Appel, K. W., Pouliot, G. A., Simon, H., Sarwar, G., Pye, H. O. T., Napelenok, S. L., Akhtar, F., & Roselle, S. J. (2013). Evaluation of the Community Multiscale Air Quality (CMAQ) model version 5.0 against observations from 2006 to 2009. *Geoscientific Model Development, 6*(2), 179–215. https://doi.org/10.5194/gmd-6-179-2013

Bisht, D. S., Tiwari, S., Dumka, U. C., Srivastava, A. K., Safai, P. D., Ghude, S. D., Chate, D. M., Rao, P. S. P., Ali, K., Prabhakaran, T., Panickar, A. S., Soni, V. K., Attri, S. D., Tunved, P., Chakrabarty, R. K., & Hopke, P. K. (2016). Tethered balloon-born and ground-based measurements of black carbon and particulate profiles within the lower troposphere during the foggy period in Delhi India. *Science of the Total Environment, 573*, 894–905. https://doi.org/10.1016/j.scitotenv.2016.08.185

Blocken, B. (2015). Computational fluid dynamics for urban physics: Importance scales possibilities limitations and ten tips and tricks towards accurate and reliable simulations. *Building and Environment, 91*, 219–245. https://doi.org/10.1016/j.buildenv.2015.02.015

Britigan, N., Alshawa, A., & Nizkorodov, S. A. (2006). Quantification of ozone levels in indoor environments generated by ionization and ozonolysis air purifiers. *Journal of the Air & Waste Management Association, 56*(5), 601–610. https://doi.org/10.1080/10473289.2006.10464467

Central Pollution Control Board. (2022). *Annual report 2022–23*. Central Pollution Control Board (CPCB), Ministry of Environment, Forest and Climate Change, Government of India, CPCB_AnnualReport_2022-23.

Cimorelli, A. J., Perry, S. G., Venkatram, A., Weil, J. C., Paine, R. J., Wilson, R. B., Lee, R. F., Peters, W. D., & Brode, R. W. (2005). AERMOD: A dispersion model for industrial source applications. Part I: General model formulation and boundary layer characterization. *Journal of Applied Meteorology, 44*(5), 682–693.

Cohan, D. S., & Napelenok, S. L. (2011). Air quality response modeling for decision support. *Atmosphere, 2*(3), 407–425. https://doi.org/10.3390/atmos2030407

Crippa, M., Guizzardi, D., Muntean, M., Schaaf, E., Dentener, F., Van Aardenne, J. A., Monni, S., Doering, U., Olivier, J. G. J., & Pagliari, V. (2018). Gridded emissions of air pollutants for the period 1970–2012 within EDGAR v4. 3.2. *Earth System Science Data, 10*(4), 1987–2013.

Draxler, R. R., & Hess, G. D. (1998). An overview of the HYSPLIT_4 modelling system for trajectories, dispersion, and deposition. *Australian Meteorological Magazine, 47*(4), 295–308.

Gousseau, P., Blocken, B., & van Heijst, G. J. F. (2011). CFD simulation of pollutant dispersion around isolated buildings: On the role of convective and turbulent mass fluxes in the prediction accuracy. *Journal of Hazardous Materials, 194*, 422–434. https://doi.org/10.1016/j.jhazmat.2011.08.008

Govardhan, G., Ghude, S. D., Kumar, R., Sharma, S., Gunwani, P., Jena, C., Yadav, P., Ingle, S., Debnath, S., Pawar, P., Acharja, P., Jat, R., Kalita, G., Ambulkar, R., Kulkarni, S., Kaginalkar, A., Soni, V. K., Nanjundiah, R. S., & Rajeevan, M. (2024). Decision Support System version 1.0 (DSS v1.0) for air quality management in. *Delhi India Geoscientific Model Development, 17*(7), 2617–2640. https://doi.org/10.5194/gmd-17-2617-2024

Gupta, P., Khan, M. N., da Silva, A., & Patadia, F. (2013). MODIS aerosol optical depth observations over urban areas in Pakistan: Quantity and quality of the data for air quality monitoring. *Atmospheric Pollution Research, 4*(1), 43–52.

Guttikunda, S. K., & Calori, G. (2013). A GIS based emissions inventory at 1 km× 1 km spatial resolution for air pollution analysis in Delhi, India. *Atmospheric Environment, 67*, 101–111.

Hanna, S. R., Egan, B. A., Purdum, J., & Wagler, J. (2001). Evaluation of the ADMS, AERMOD, and ISC3 dispersion models with the OPTEX, Duke Forest, Kincaid, Indianapolis and Lovett field datasets. *International Journal of Environment and Pollution, 16*(1–6), 301–314. https://doi.org/10.1504/IJEP.2001.000626

Hersbach, H., Bell, B., Berrisford, P., Hirahara, S., Horányi, A., Muñoz-Sabater, J., et al. (2020). The ERA5 global reanalysis. *Quarterly Journal of the Royal Meteorological Society, 146*(730), 1999–2049. https://doi.org/10.1002/qj.3803

Hogrefe, C., Rao, S. T., Kasibhatla, P., Kallos, G., Tremback, C. J., Hao, W., Olerud, D., Xiu, A., McHenry, J., & Alapaty, K. (2001). Evaluating the performance of regional-scale photochemical modeling systems: Part I—Meteorological predictions. *Atmospheric Environment, 35*(24), 4159–4174. https://doi.org/10.1016/S1352-2310(01)00182-0

Holmes, N. S., & Morawska, L. (2006). A review of dispersion modelling and its application to the dispersion of particles: An overview of different dispersion models available. *Atmospheric Environment, 40*(30), 5902–5928.

Jain, N., Bhatia, A., & Pathak, H. (2014). Emission of air pollutants from crop residue burning in India. *Aerosol and Air Quality Research, 14*(1), 422–430. https://doi.org/10.4209/aaqr.2013.01.0031

Kalnay, E., Kanamitsu, M., Kistler, R., Collins, W., Deaven, D., Gandin, L., Iredell, M., Saha, S., White, G., & Woollen, J. (2018). *The NCEP/NCAR 40-year reanalysis project. P. Vol1_146-Vol1_194 in renewable energy*. Routledge.

Klimont, Z., Kupiainen, K., Heyes, C., Purohit, P., Cofala, J., Rafaj, P., Borken-Kleefeld, J., & Schöpp, W. (2017). Global anthropogenic emissions of particulate matter including black carbon. *Atmospheric Chemistry and Physics, 17*(14), 8681–8723.

Koo, H., Iwanaga, T., Croke, B. F. W., Jakeman, A. J., Yang, J., Wang, H.-H., Sun, X., Lü, G., Li, X., Yue, T., Yuan, W., Liu, X., & Chen, M. (2020). Position paper: Sensitivity analysis of spatially distributed environmental models—A pragmatic framework for the exploration of uncertainty sources. *Environmental Modelling & Software, 134*, 104857. https://doi.org/10.1016/j.envsoft.2020.104857

Krotkov, N. A., McLinden, C. A., Li, C., Lamsal, L. N., Celarier, E. A., Marchenko, S. V., Swartz, W. H., Bucsela, E. J., Joiner, J., & Duncan, B. N. (2016). Aura OMI observations of regional SO2 and NO2 pollution changes from 2005 to 2015. *Atmospheric Chemistry and Physics, 16*(7), 4605–4629.

Kumar, P., Morawska, L., Martani, C., Biskos, G., Neophytou, M., Di Sabatino, S., Bell, M., Norford, L., & Britter, R. (2015). The rise of low-cost sensing for managing air pollution in cities. *Environment International, 75*, 199–205. https://doi.org/10.1016/j.envint.2014.11.019

Kumar, A., Patil, R. S., Dikshit, A. K., & Kumar, R. (2017). Application of AERMOD for short-term air quality prediction with forecasted meteorology using WRF model. *Clean Technologies and Environmental Policy, 19*(7), 1955–1965. https://doi.org/10.1007/s10098-017-1379-0

Kurokawa, J., & Ohara, T. (2020). Long-term historical trends in air pollutant emissions in Asia: Regional emission inventory in Asia (REAS) version 3. *Atmospheric Chemistry and Physics, 20*(21), 12761–12793.

Lateb, M., Meroney, R. N., Yataghene, M., Fellouah, H., Saleh, F., & Boufadel, M. C. (2016). On the use of numerical modelling for near-field pollutant dispersion in urban environments—A review. *Environmental Pollution, 208*, 271–283. https://doi.org/10.1016/j.envpol.2015.07.039

Liu, L., Oza, S., Hogan, D., Perin, J., Rudan, I., Lawn, J. E., Cousens, S., Mathers, C., & Black, R. E. (2015). Global, regional, and national causes of child mortality in 2000–13, with projections to inform post-2015 priorities: An updated systematic analysis. *The Lancet, 385*(9966), 430–440.

Lu, Y., Maharjan, S., & Zhang, Y. (2021). Adaptive edge association for wireless digital twin networks in 6G. *IEEE Internet of Things Journal, 8*(22), 16219–16230. https://doi.org/10.1109/JIOT.2021.3098508

Nieuwstadt, F. T. M. (1991). Lectures on air pollution modeling. edited by A. Venkatram and JC Wyngaard. American meteorological society, 1988. 390 Pp. *Journal of Fluid Mechanics, 229*, 690–691.

Mead, M. I., Popoola, O. A. M., Stewart, G. B., Landshoff, P., Calleja, M., Hayes, M., Baldovi, J. J., McLeod, M. W., Hodgson, T. F., Dicks, J., Lewis, A., Cohen, J., Baron, R., Saffell, J. R., & Jones, R. L. (2013). The use of electrochemical sensors for monitoring urban air quality in low-cost high-density networks. *Atmospheric Environment, 70*, 186–203. https://doi.org/10.1016/j.atmosenv.2012.11.060

Morawska, L., Thai, P. K., Liu, X., Asumadu-Sakyi, A., Ayoko, G., Bartonova, A., Bedini, A., Chai, F., Christensen, B., Dunbabin, M., Gao, J., Hagler, G. S. W., Jayaratne, R., Kumar, P., Lau, A. K. H., Louie, P. K. K., Mazaheri, M., Ning, Z., Motta, N., Mullins, B., Rahman, M. M., Ristovski, Z., Shafiei, M., Tjondronegoro, D., Westerdahl, D., & Williams, R. (2018). Applications of low-cost sensing technologies for air quality monitoring and exposure assessment: How far have they gone? *Environment International, 116*, 286–299. https://doi.org/10.1016/j.envint.2018.04.018

Paatero, P., & Hopke, P. K. (2003). Discarding or downweighting high-noise variables in factor analytic models. *Analytica Chimica Acta, 490*(1–2), 277–289.

Pandey, A., & Venkataraman, C. (2014). Estimating emissions from the Indian transport sector with on-road fleet composition and traffic volume. *Atmospheric Environment, 98*, 123–133. https://doi.org/10.1016/j.atmosenv.2014.08.039

Pu, B., & Qiu, Y. (2016). Emerging trends and new developments on urban resilience: A bibliometric perspective. *Current Urban Studies, 4*(01), 36.

Schultz, M. G., Betancourt, C., Gong, B., Kleinert, F., Langguth, M., Leufen, L. H., Mozaffari, A., & Stadtler, S. (2021). Can deep learning beat numerical weather prediction? *Philosophical Transactions of the Royal Society A: Mathematical Physical and Engineering Sciences, 379*(2194), 20200097. https://doi.org/10.1098/rsta.2020.0097

Scire, J. S., Robe, F. R., Fernau, M. E., & Yamartino, R. J. (2000). *A user's guide for the calmet meteorological model* (p. 37). Earth Tech.

Seinfeld, J. H., & Pandis, S. N. (2016). *Atmospheric chemistry and physics: From air pollution to climate change*. John Wiley & Sons.

Srishti Jain, S. K., Sharma, T. K. M., & Saxena, M. (2018). Source apportionment of PM in Delhi India using PCA/APCS UNMIX and PMF. *Particuology, 37*, 107–118. https://doi.org/10.1016/j.partic.2017.05.009

Tao, F., Cheng, J., Qi, Q., Zhang, M., Zhang, H., & Sui, F. (2018). Digital twin-driven product design manufacturing and service with big data. *The International Journal of Advanced Manufacturing Technology, 94*(9–12), 3563–3576. https://doi.org/10.1007/s00170-017-0233-1

Tominaga, Y., & Stathopoulos, T. (2013). CFD simulation of near-field pollutant dispersion in the urban environment: A review of current modeling techniques. *Atmospheric Environment, 79*, 716–730. https://doi.org/10.1016/j.atmosenv.2013.07.028

Turner, J. C., Oakes, P. J., Alexander Haslam, S., & McGarty, C. (1994). Self and collective: Cognition and social context. *Personality and Social Psychology Bulletin, 20*(5), 454–463.

U.S. Environmental Protection Agency. (2024). *FY 2023 EPA annual performance report (APR) evidence and evaluation summaries*. https://www.epa.gov/system/files/documents/2024-03/fy23-evidence-evaluation-summaries.pdf

van Donkelaar, A., Martin, R. V., Brauer, M., Kahn, R., Levy, R., Verduzco, C., & Villeneuve, P. J. (2010). Global estimates of ambient fine particulate matter concentrations from satellite-based aerosol optical depth: Development and application. *Environmental Health Perspectives, 118*(6), 847–855. https://doi.org/10.1289/ehp.0901623

Xing, J., Li, S., Jiang, Z., Wang, S., Ding, D., Dong, Z., Zhu, Y., & Hao, J. (2020). Deep learning for prediction of the air quality response to emission changes. *Environmental Science & Technology, 54*(14), 8589–8600. https://doi.org/10.1021/acs.est.0c02923

Veefkind, J., Pepijn, I., Aben, K., McMullan, H., De Förster, J., Vries, G. O., Jacques Claas, H. J., Eskes, J. F. D. H., & Kleipool, Q. (2012). TROPOMI on the ESA sentinel-5 precursor: A GMES mission for global observations of the atmospheric composition for climate, Air quality and ozone layer applications. *Remote Sensing of Environment, 120*, 70–83.

Xue, L., Hou, P., Zhang, Z., Shen, M., Liu, F., & Yang, L. (2020). Application of systematic Strategy for agricultural non-point source pollution control in Yangtze river Basin, China. *Agriculture, Ecosystems & Environment, 304*, 107148.

Yadav, A. K. (2015). Elemental composition and source apportionment of suspended particulate matters and health risk assessment in mining and nonmining areas of Odisha, India. *Journal of Hazardous, Toxic, and Radioactive Waste, 19*(3), 04014037.

Yadav, A. K., Sahoo, S. K., Dubey, J. S., Kumar, A. V., Pandey, G., & Tripathi, R. M. (2019). Assessment of particulate matter, metals of toxicological concentration, and health risk around a mining area, Odisha, India. *Air Quality, Atmosphere & Health, 12*(7), 775–783.

Yadav, A. K., Sahoo, S. K., Kumar, A. V., Shirin, S., Jamal, A., Patra, A. C., Dubey, J. S., Thakur, V. K., Lenka, P., Singh, S., Jha, V., & Tripathi, R. M. (2023). Identifying size-segregated particulate matter ($PM_{2.5}$, PM_{10} and SPM) sources in an industrial town of India. *Aerosol Science and Engineering, 7*(4), 455–473.

Yadav, V., Yadav, A. K., Singh, V., & Singh, T. (2024). Artificial neural network an innovative approach in air pollutant-prediction for environmental applications, A review. *Results in Engineering, 22*, 102305.

Yadav, A. K., Gupta, K., Bodhale, R., & Phuleria, H. C. (2025). Metals and trace element associated health risk in size-segregated household dust of an industrial town in India. *Air Quality, Atmosphere & Health*. https://doi.org/10.1007/s11869-025-01794-5

Zhang, Z., Cheng, Y., Liang, L., & Liu, J. (2023). The measurement of atmospheric black carbon: A review. *Toxics, 11*(12), 975.

Part III
Emerging Strategies and Innovations

Chapter 13
Harnessing Biodiversity for Air Pollution Mitigation Through Nature-Based Strategies and Bioresource Utilization

Kailash Chandra Joshi and Prem Ranjan

13.1 Introduction

Air pollution is an important environmental issue that affects human health, natural systems, and climate. It comes from different sources, including industries, vehicles, farming activities, and deforestation. These activities release pollutants such as particulate matter $(PM)_{2.5}$, $PM_{10,}$ NO_x, SO_2, and CO_2, which decrease air quality and cause health problems (Yadav, 2021; Yadav et al., 2019, 2023). According to the World Health Organization (WHO), nearly 99% of the population lives in areas where the air pollutant concentration is above according to the WHO air pollutant concentration guidelines (Table 13.1). Almost seven million people die early every year due to air pollution; low- and middle-income countries are mostly affected by this (World Health Organization (WHO), 2024). In India, the situation is even more serious. Studies show that, in 2019, about 1.67 million deaths were caused by long-term exposure to polluted air. The impact of that is also on the economy—India lost almost $36.8 billion in the same year due to health issues caused by air pollution (Pandey et al., 2021). But it not only affects health but also harms the environment, reduces crop production, and plays a role in climate change by increasing extreme weather events.

Nature-based solutions (NbSs) are best for tackling these issues. These methods use natural processes to absorb, filter, and reduce air pollution. Phytoremediation is when plants remove these pollutants from the air and soil and help clean the environment. Some plants also act as natural filters by trapping dust and harmful particles in their leaves. Similar to plant-based biofuels, biodiesel offers cleaner alternatives to traditional fuels and reduces emissions (Table 13.2). Promoting green spaces, adopting cleaner energy, and improving industrial policies can reduce air

K. C. Joshi · P. Ranjan (✉)
Centre For Biodiversity Conservation and Management, G.B. Pant National Institute of Himalayan Environment, Kosi-Katarmal, Almora, India

A. K. Yadav (ed.), *Computational Approach for Air Pollution*,
https://doi.org/10.1007/978-3-032-09215-1_13

Table 13.1 Guidelines for air quality for pollutants (World Health Organization (WHO), 2021 guideline)

Pollutant	Averaging time	2005 guideline (μg/m^3)	2021 guideline (μg/m^3)
$PM_{2.5}$	Annual	10	5
	24-h	25	15
PM_{10}	Annual	20	15
	24-h	50	45
Ozone (O_3)	Peak season	–	60
	8-h	100	100
NO_2	Annual	40	10
	24-h	–	25
SO_2	24-h	20	40
CO	24-h (mg/m^3)	–	4

Table 13.2 Species that are good for phytoremediation

Category	Species	Function	References
Trees	*Ficus religiosa* (Peepal)	High CO_2 absorption, PM removal	Tiwari and Malviya (2021)
	Azadirachta indica (neem)	Antimicrobial, air-purifying	Gawroński et al. (2018)
	Mangifera indica (mango)	SO_2 absorption, heavy metal tolerance	Sharma, et al. (2019)
Grasses and shrubs	*Chrysopogon zizanioides* (vetiver)	Heavy metal stabilization	Nakbanpote et al. (2024)
	Eichhornia crassipes (water hyacinth)	Absorbs heavy metals in water bodies	Malik (2007)
Algae and mosses	Spirulina spp.	Heavy metals absorption	Rangsayatorn et al. (2002)
	Lichens and bryophytes	Bio-monitors for air quality	Conti and Cecchetti (2001)

pollution and create healthier living conditions. This chapter examines how nature-based strategies can help manage air pollution and improve environmental quality.

13.1.1 Types of Pollutants

Generally, air pollutants are classified into two categories: the first one is the primary pollutant directly emitted from the source, and the second one is the secondary pollutant emitted by reaction with the primary pollutant. However, according to the WHO, five types of pollutants affect human health more:

PM: The diameter of the PM is 2.5 and 10 μm. The composition of the PM is dust particles, minerals, water, nitrate, ammonia, sodium, sulfate, and chloride. The sources of PM are pollen, sea spray, windblown dust from erosion, agricultural spaces, roadways, and mining operations, and another source, the burning of polluting fuel.

Carbon monoxide (CO) is a colorless and odorless gas that is very harmful and dangerous to human health. It is produced by the combustion of carbonaceous fuels such as coal, fuel wood, and motor vehicles, which are other contributors to CO in the ambient air.

Ozone ($O_{3)}$: There are generally two types; the first is stratospheric ozone, which is beneficial for all and protects the Earth from harmful UV radiation. However, tropospheric ozone is a secondary pollutant formed by reacting with NOx, volatile organic compounds (VOCs), and sunlight, which is a photochemical smog, mainly in urban and industrial regions.

Nitrogen Oxides (NO_x): NO_x refers to highly reactive gases. Nitrogen dioxide (NO_2) is a reddish-brown gas that is a strong oxidant and is soluble in water. In NO_x, nitric oxide (NO) and NO_2 are the main contributors to air pollution.

Sulfur Dioxide ($SO_{2)}$: It is naturally odorless, colorless, acidic, and easily soluble in water. SO_2 is one of the main air pollutants and contributes to the environmental and health problems produced by burning fossil fuels. It is also responsible for secondary pollutant formation, such as sulfate, aerosols, and fine PM (2.5).

Formaldehyde, lead, polycyclic aromatic hydrocarbons (PAHs), and Radon are harmful air pollutants that pollute the environment and harm our health. Formaldehyde is used in building materials, furniture, household products, and vehicle emissions. Long-term exposure can cause respiratory problems and even cancer. Lead was once found in gasoline and industrial processes and is still in the air from industrial emissions and old infrastructure. It is especially harmful to children, impacting brain development and general health. PAHs are from fossil fuels, wood, and tobacco. They are carcinogenic and can cause severe lung and heart problems over time. Radon is a naturally occurring gas that is radioactive in nature. It seeps into homes from the ground and can accumulate in unventilated spaces, increasing the risk of lung cancer. These pollutants degrade the air and our health, so we must reduce exposure and increase air pollution controls (Figs. 13.1, 13.2 and 13.3).

13.1.2 Overview of Global Air Pollution Trends and Impact Statistics

In urban areas of developing countries, outdoor air pollution led to have caused 4.14 million premature deaths in 2019, with the main hotspots in South Asia, sub-Saharan Africa, and the Middle East. A study was conducted to analyze air pollution trends in 13,160 urban areas worldwide from 2000 to 2019. Their findings indicate that $PM_{2.5}$ levels declined slightly by 0.2% per year, but 65% of cities still showed rising pollution levels, particularly in regions such as South Asia, the

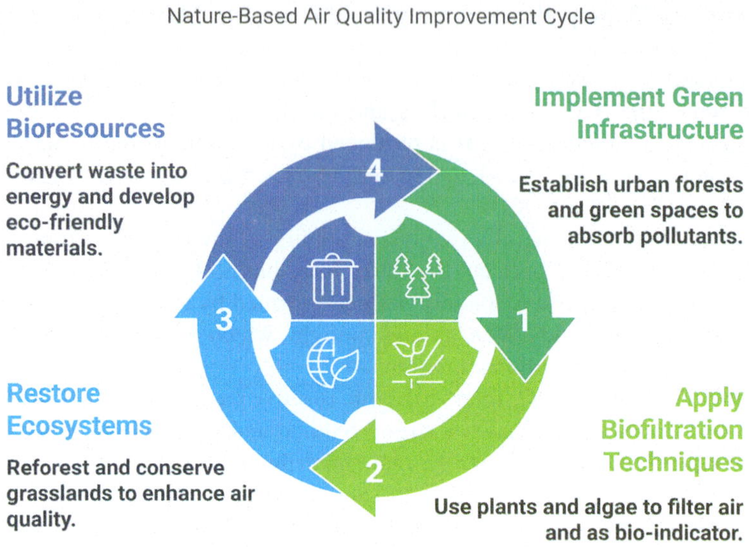

Fig. 13.1 Nature-based air pollution control (Prepared by Author from the software)

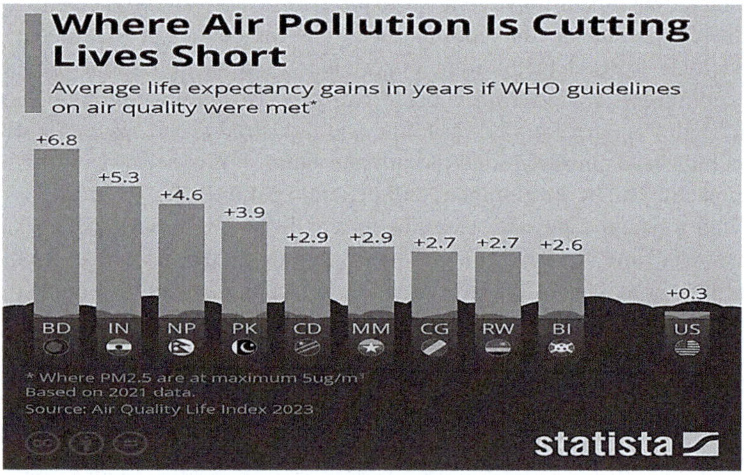

Fig. 13.2 $PM_{2.5}$ concentration trend (Source: Statista n.d.)

Middle East, and sub-Saharan Africa. Furthermore, NO_2 levels increased by 0.4% annually in 71% of cities, with worsening conditions observed in high-population regions. Exposure to ground-level ozone (O_3), another major concern, increased at 89% of the monitoring stations, largely due to lower NO_x titration and climate-related factors (Sicard et al., 2023). Urban populations are now increasingly exposed to higher O_3 levels, contributing to increased respiratory diseases and premature mortality. Exposure to NO_2 contributes significantly to pediatric asthma cases

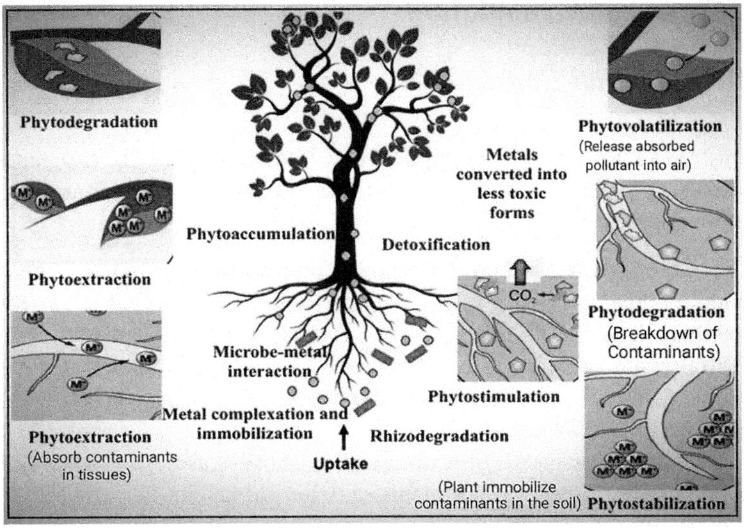

Fig. 13.3 Mechanisms of phytoremediation (Source: Khanna et al., 2023)

worldwide, particularly in urban areas with heavy traffic congestion. Air pollution remains a severe crisis in India, with $PM_{2.5}$ and PM_{10} levels exceeding WHO safety limits by more than 100% in cities such as Delhi, Mumbai, Lahore, and sub-Saharan Africa. The rapid increase in the population of urban places, along with industrial activities, has led to a rise in environmental pollution, and the levels of SO_2 and NO_2 emissions are still hazardous.

The NCAP, also known as the National Clean Air Program, is trying to preserve the Air Quality Index under control by aiming for a decline in $PM_{2.5}$ and PM_{10} levels by 20–30% by 2024, with a concentration on emissions reduction along with quality air surveillance (Sharma, Kumar, Shahzad, et al., 2019). Meanwhile, Europe and North American countries have been able to reduce NO_2 and PM emissions from vehicles through obligations or regulations on yearly automobile fuel economy standards and using more renewable resources for energy, as well as limiting the industry and raising the level of air quality in numerous cities. Improvement in air quality remains a challenge, particularly in urban areas. In South Asia and Africa, air pollution monitoring systems are still lacking, significantly hampered by any attempt to control it. Human health and the environment are the main victims of the dramatic decline in air quality resulting from industrial development, the expansion of cities, transportation, and agricultural activities. The WHO reports that around 99% of the world's population resides in areas with high levels of pollution. This level of exposure affects millions of lives, leading to early mortality and severe economic and environmental effects (Manisalidis et al., 2020). Therefore, there is a need for nature-based strategies and policies that target the synergistic impact caused by urban development, transportation, and industrial emissions.

13.2 Causes of Air Pollution

13.2.1 Industrial Emissions

The industrial sector releases the highest air pollutants, PM, SO_2, NO_2, and VOCs. Power plants, oil refineries, and the chemical and manufacturing industries contribute to these emissions.

- Studies have shown that coal-fired power plants and steel industries emit huge amounts of SO_2 and NO_x, major precursors to acid rain and smog formation (Choudhary & Garg, 2013).
- Industrial zones such as those in China, India, and the United States have been identified as the main pollution hotspots where air quality exceeds WHO limits (Appannagari, 2017).
- The release of toxic gases, *for example*, the Bhopal gas tragedy (1984) in India, highlights how industrial pollutants are hazardous to the environment and human health.

13.2.2 Urbanization and Deforestation

Urban growth at a rapid rate and deforestation have primarily been responsible for increasing levels of air pollution. For infrastructure construction, tree felling and forest cutting decrease the natural ability to absorb CO_2 and generate oxygen.

- Land use changes and increasing deforestation have promoted and accelerated the loss of natural air purifiers, causing more dust, smoke, and heat-trapping pollutants in urban areas (Appannagari, 2017).
- The increase in construction activities generates significant dust pollution, contributing to the emission of PM, PM_{10} and $PM_{2.5}$. Studies have shown that urbanization-driven pollution is responsible for nearly 25% of the total PM burden in developing nations (Manisalidis et al., 2020).

13.2.3 Transportation and Vehicle Exhaust

The transportation industry is the most significant contributor to CO, NO_x, PM, and benzene, which are all dangerous to human health and the environment. According to studies, vehicle emissions are responsible for almost 80% of urban air pollution in megacities such as Delhi, Beijing, and Los Angeles (Manisalidis et al., 2020).

- Older vehicles and the continued use of diesel and petrol-based fuels emit a lot of black carbon, which is a major contributor to air pollution and climate change.

- High-traffic cities such as New Delhi and Mumbai have recorded alarmingly high NO_2 and $PM_{2.5}$ levels, resulting in poor air quality and increased respiratory issues.

13.2.4 Agricultural Burning and Biomass Combustion

Burning of crop residues and biomass combustion in agricultural activities emit fine particulate $PM_{2.5}$, black carbon, methane (CH_4), and greenhouse gas (GHG).

- In South Asian countries such as India, Nepal, and Pakistan, stubble burning is a seasonal activity that causes severe air pollution, especially during winter (Choudhary & Garg, 2013).
- Biomass combustion for domestic cooking and heating in rural areas aggravates indoor and outdoor air pollution.

13.3 Impacts of Air Pollution

13.3.1 Human Health Effects

Air pollution has a severe and widespread impact on human health, affecting the respiratory, cardiovascular, and nervous systems.

13.3.1.1 Respiratory Diseases

Prolonged exposure to $PM_{2.5}$, PM_{10}, and NO_2 causes asthma, chronic obstructive pulmonary disease, bronchitis, and lung infections.

- Studies have shown that residents of highly polluted cities such as Delhi, Beijing, and Mexico City experience an increased risk of lung disease due to exposure to air pollution.
- WHO statistics reveal that 30% of lung cancer deaths worldwide are caused by air pollution, making it a significant environmental carcinogen (Choudhary & Garg, 2013).

13.3.1.2 Cardiovascular Effects

Air pollutants such as CO and NO_x are responsible for high blood pressure, heart attacks, strokes, and premature deaths.

- Studies have discovered that long-term exposure to $PM_{2.5}$ raises the threat of cardiovascular death by 24% (Manisalidis et al., 2020).

13.3.1.3 Nervous System Effects

Air pollution negatively affects the nervous system, leading to cognitive impairment, neuroinflammation, and neurodegenerative diseases such as Alzheimer's and Parkinson's.

- Cognitive decline and neurodegenerative diseases: Pollutants such as $PM_{2.5}$ and heavy metals cross the blood–brain barrier, triggering oxidative stress and neurodamage, promoting cognitive decline.
- Neurodevelopmental disorders in Children: Prenatal and early childhood exposure to pollutants such as $PM_{2.5}$ and NO_2 leads to the risk of Attention-Deficit/Hyperactivity Disorder (ADHD), learning disabilities, and lower IQ levels.
- Mental health effects: Long exposure to air pollution has been associated with depression, anxiety, and increased brain inflammation, leading to psychiatric illnesses.

13.3.2 Environmental Impact

13.3.2.1 Climate Change

- CO_2, CH_4, and black carbon contribute to global warming and temperature rise (Appannagari, 2017).
- Air pollution accelerates glacial melting in the Himalayas, leading to severe climate-related disasters.

13.3.2.2 GHG Buildup

- NO_2, SO_2, and CO emissions lead to acid rain, which harms forests, soil fertility, and aquatic ecosystems (Choudhary & Garg, 2013).
- Depletion of O_3 due to air pollution increases the exposure to harmful UV radiation, causing skin cancers and ecosystem disruption.

13.3.3 Economic Consequences

13.3.3.1 Healthcare Costs

- The rise in air pollution-related illnesses places a substantial financial burden on healthcare systems around the world.
- In developing countries, air pollution is responsible for billions of dollars in healthcare costs (Manisalidis et al., 2020).

13.3.3.2 Loss of Agricultural Productivity

- Crop yields are negatively affected by O_3 and PM, thus reducing food production overall.
- Studies estimate that air pollution lowers wheat and rice yields by almost 15% in the most polluted areas (Choudhary & Garg, 2013).

13.4 NbSs for Air Pollution Mitigation

13.4.1 Concept and Importance of NbS in Environmental Management

To address the challenges posed in the previous section, NbS proposes the management of natural ecosystems that can provide solutions to problems such as air pollution. Such solutions stem from an ecosystem-based management principle that relies on natural and semi-natural areas for essential activities such as pollutant filtration, carbon sequestration, and climatic moderation. International Union for Conservation of Nature (IUCN) states that NbS strategies can be applied across urban and rural landscapes to enhance the environment and meet biodiversity targets (IUCN, 2020).

Globally, air pollution is one major issue that increases the incidence of many respiratory and cardiovascular diseases and damages the surrounding ecosystem. The expansion of urban areas and the establishment of industries have only exacerbated the challenge by raising the levels of many pollutants, such as PM, NOx, SO_2, and VOCs. These traditional technological solutions usually involve the operation of air filtration power plants and the installation of catalytic converters. These are also expensive and energy-intensive.

NbS provides an alternative air quality management that is more comprehensive and cost-effective. NbS usage of NbS enables the inclusion of green infrastructure, such as urban forests, green roofs, and wetlands, which additionally aid in reducing temperature and water pollution and improve public health (Kabisch et al., 2017).

13.4.2 Role of Ecosystems in Air Purification

Green vegetation provides valuable services, such as a filter that removes unwanted air pollutants. Ecosystems defend the air from contamination by NOx, SO_2, and O_3 through different actions. Air phytoremediation is the natural mechanism by which an ecosystem passes the gaseous, heterogeneous substance through the plant parts and the soil microbial community and is transformed to settle.

Most of the ecosystem services performed by nature that relate to the air originate from the uptake of different airborne contaminants. Urban forests, especially trees of various species and Dryopteris filix mas, help isolate air SOx compounds and plants through their pores; stems undergo very slow neutralization, which changes their damaging impact. Furthermore, vegetation cover prevents the emission of fine PM, which is a source of respiratory problems for people. Urban trees are vital in reducing the concentrations of several air pollutants, particularly in one area of the densely populated city. Street tree research in London has shown that street trees reduce NO_2 by 15% (Pugh et al., 2012).

Another essential function of ecosystem services in terms of air quality sustainability is to transform pollutants into less harmful products by microbial breakdown. Some soil microbes, especially in root-associated zones, help transform dangerous contaminants into less harmful substances. This is best seen in green spaces and wetlands where microbial activity peaks.

Moreover, the regulation of temperature and the moderation of pollution distribution are additional functions of ecosystems. The urban heat island effect, where cities experience higher heat caused by human activities and infrastructure, exacerbates air pollution by slowing the dispersion of pollutants. Vegetation helps mitigate this effect, and plants counter it by providing shade and evaporative cooling areas around plants. Reducing the accumulation of pollutants improves atmospheric circulation and overall air quality.

13.4.3 Advantages of NbS Over Conventional Technological Interventions

NbS options are comparatively more advantageous than scrubbers and agent catalytic converters due to their lower cost. Unlike NbS, which is based on existing natural systems that require little maintenance, technological solutions need considerable financial resources for installation, maintenance, and energy consumption. For example, urban forests and wetlands, forms of green infrastructure, are self-sustainable for an extended period and do not need interventions that cost a lot of money (Kabisch et al., 2017). Other reasons why NbS is advantageous are its lower biodiversity, self-sustained urban beauty, noise pollution, and improved mental health. NbS offered in green structures lowers stress and enhances brain function, which is more conducive to living (IUCN, 2020).

In addition to this, NbS facilitates a balanced method of carbon dioxide capture. Unlike anthropogenic methods that only scrub pollutants, NbS actively removes carbon dioxide (CO_2) from the air. This slows the rate of climate change and improves air quality. Urban forests and wetlands serve as carbon pools, reducing the level of GHG and helping to reduce the effects of climate change. NbS is a helpful remedy for public health protection, which is another notable benefit. Insufficient air quality is one of the leading causes of chronic respiratory and heart disease, especially in urban areas. NbS helps reduce the concentrations of pollution, thus lowering the chances of asthma, lung infections, and other respiratory disorders. Furthermore, green spaces are beneficial for health because they encourage walking, jogging, and other forms of exercise (World Health Organization (WHO), 2016).

13.5 Bioremediation and Phytoremediation by Plants

Bioremediation is the procedure used to clean the environment's pollution using microorganisms or plants to improve it. Using biological processes is less costly than physical or chemical remediation methods. Bioremediation, for example, refers to exhaustive spore cleaning of environmental contaminants, including heavy metals, hydrocarbons, and industrial waste chemicals (Gentry et al., 2004).

Meanwhile, phytoremediation is a branch of bioremediation that relies solely on plants to remove, modify, or contain environmental contaminants. Its filtration efficiency of $PM_{2.5}$ and PM_{10}, super-heavy metals, and harmful gases such as SO_2 and NOx is why it has become admired (Pilon-Smits, 2005). Phytoremediation, as a result, improves air quality and overall environmental quality in urban and industrial settings (Nowak et al., 2014).

13.5.1 Mechanisms of Bioremediation by Plants

13.5.1.1 Microbial Interactions in the Rhizosphere

The area of soil around the plant root is called the rhizosphere. Some organisms in the soil ecosystems around the root are Pseudomonas, Bacillus, and Mycorrhizal fungi. The Bacillus and Pseudomonas species help in the degradation of organic pollutants and heavy metals by enzyme secretion, increasing bioavailability and aiding in the amelioration of pollution processes (Kuiper et al., 2004).

13.5.1.2 Enzymatic Breakdown

Various plants synthesize certain enzymes, for example, peroxidases, laccases, and dehalogenases, which catalyze the breakdown of complex, harmful chemicals into more straightforward, less harmful chemicals. These processes are crucial in the detoxification of pesticides, hydrocarbons, and industrial solvents (Macek et al., 2000).

13.5.1.3 Bioaccumulation and Detoxification

For bioaccumulation of harmful metals such as lead (Pb), cadmium (Cd), and arsenic (As), some plant species such as *Brassica juncea*, Indian mustard, and the sunflower Helianthus annuus have shown great promise in controlling the spread of these harmful metals in the environment.

13.5.2 Mechanisms of Phytoremediation for Air Pollution

Phytofiltration: Plants trap PM ($PM_{2.5}$ and PM_{10}) on their leaf surfaces, reducing atmospheric pollution. Leaf microstructures, including trichomes and waxes, play a crucial role.

Phytovolatilization: Some plant species absorb pollutants and release them into the air in a less toxic form. For example, *Populus species* (poplar) volatilize mercury (Hg) and selenium (Se).

Phytostabilization: Plants prevent airborne dispersal of heavy metals and toxins by stabilizing them in the soil. Species such as *Chrysopogon zizanioides* (Vetiver) and *Salix* spp. (Willow) are commonly used in contaminated sites.

13.6 Urban Green Spaces Role in Air Pollution Mitigation

Urban green spaces work as natural air filters, reducing pollutants and improving urban air quality. Green roofs and vertical gardens are highly effective in absorbing CO_2 and PM, cooling urban environments, and reducing pollution (Speak et al., 2012). Tree belts and buffer zones in industrial areas significantly reduce pollutant dispersion, improving air quality for nearby communities. Wetlands and riparian vegetation contribute to air purification by improving moisture levels and trapping contaminants.

13.6.1 *Case Studies and Government Initiatives in India*

Multiple urban green initiatives in India have illustrated the use of phytoremediation and biofiltration to control air pollution. The Aravalli Biodiversity Park in Delhi and the Sanjay Gandhi National Park in Mumbai have shown that tree canopy and vegetation cover help lower the concentration of PM in the air (Sierra-Porta et al., 2023).

India has implemented some measures to strengthen air pollution mitigation policies by combining them with urban forestry and phytoremediation policies. The NCAP, which started in 2019, seeks to achieve a 20–30% reduction of $PM_{2.5}$ and PM_{10} concentration by 2024 by greening urban spaces, building green belts, and afforestation (Jha et al., 2025). Likewise, the Smart Cities Mission facilitates the incorporation of green infrastructure in urban development and promotes tree growth, rooftop gardens, and pollution-absorbing green walls. In addition, India is applying the Miyawaki forest model, where fast-growing native forests are planted in several cities such as Mumbai, Chennai, and Bengaluru to mitigate air pollution.

13.7 Bioindicators: Plants as Pollution Monitors

Bioindicators are living organisms whose presence indicates changes in the environment. Therefore, they are ideal assistants to gauge air, water, and soil pollution. Common bioindicators, plants, lichens, and mosses are used most frequently to monitor air pollution due to their sensitivity to airborne contaminants such as SO_2, NOx, O_3, heavy metals, and $PM_{2.5}$ and PM_{10}. Natural methods of monitoring pollution are more reliable than engineered systems because they are more expensive and economical. Such organisms efficiently evaluate pollution patterns, accumulating and reflecting pollution changes over time (Conti & Cecchetti, 2001).

13.7.1 *Role of Bioindicator Species in Detecting Pollution Levels*

Lichens and mosses have apparent characteristics that distinguish them as bioindicators due to their response to pollution in urban and industrial environments. Lichens are structures formed by the symbiosis of a fungus and an alga that extract nutrients and toxins from the atmosphere. Their presence in a particular area helps determine the quality of air. Lichens sensitive to pollution, such as Usnea and Ramalina, suggest clean air, while Xanthoria and Phaeophyscia, which are tolerant to pollution, indicate a filthy environment. Their absence negatively impacts the area. Lichens are vulnerable to pollutants such as roots and cuticles, so they lack protective structures, and as such, their assimilation of biogenic SO_2, NOx, and heavy metals makes them suitable for monitoring industrial emissions.

Among bioindicators, mosses are another enjoyable group because, like lichens, they can absorb pollutants from the atmosphere. Mosses do not have roots and are dependent solely on precipitation for water and nutrients. Specialist mosses accomplish this by accumulating heavy metals, such as Pb, Cd, and mercury (Hg), to aid in the scientific examination of pollution deposition. This includes species such as *Hypnum cupressiforme* and Sphagnum, whose great ability to accumulate toxic metals gives reason to believe in the occurrence of industrial pollution and vehicle emissions (Bargagli, 1998).

13.8 Role of Bioresources: Biofuels and Biodiesel in Pollution Reduction

Biodiesel and biofuels are renewable energy sources from plant materials, including vegetable oils, algae, and agricultural waste. Such bioresources provide an ecologically cleaner alternative to fossil fuels by reducing GHG emissions and air pollution. Unlike fuels derived from petroleum, biofuels contain much less CO_2, NO_2, SO_2, and PM. This makes biofuels necessary and a priority for sustainable energy transitions and the management of climate change (Demirbas, 2009).

13.8.1 Role of Plant-Based Biofuels and Biodiesel in Reducing Fossil Fuel Dependency

Biofuels, namely, bioethanol and biodiesel, are crafted from biological sources such as sugarcane, maize, soybeans, palm oil, and jatropha. These fuels are essential for diversifying energy sources, decreasing dependency on fossil fuels, reducing carbon emissions, and improving energy security. Bioethanol, one of the most widely used substitutes for fossil fuels, is produced by the fermentation of sugar-containing crops, including sugarcane, maize, and wheat. It is commonly mixed with gasoline to reduce CO_2 and CO emissions. Countries such as Brazil and the United States have successfully implemented fuel blends of E10 (10% ethanol) and E85 (85% ethanol), leading to significant air pollution (Goldemberg, 2008). Biodiesel, another crucial biofuel, is derived from vegetable oils (soybean, palm, canola) and cooking oils through a chemical process called transesterification, resulting in a sulfur-free, biodegradable alternative to petroleum diesel. Compared to conventional diesel, biodiesel reduces SO_2 emissions by nearly 100%, CO emissions by 50%, and PM by 47%, making it a cleaner and more sustainable energy source (Knothe, 2010). In addition to plant-based biofuels, algae-based biofuels have emerged as a promising solution because of their high lipid content, rapid growth rate, and ability to utilize wastewater and carbon capture technologies. Unlike traditional crops, algae do not compete with food production and can grow in non-arable land, making them a highly sustainable energy source. The U.S. Energy Department has reported that algae-based biodiesel has the potential to reduce CO_2 emissions by 68% compared to fossil fuels, positioning it as a viable alternative in the global energy transition (Chisti, 2007).

13.8.2 India's National Bio-Energy Mission

India has actively promoted biofuel production as part of its National Bio-Energy Mission, focusing on biodiesel derived from Jatropha and Pongamia plants. These non-edible oilseed crops have been identified as sustainable feedstocks that do not compete with food production of food and can be cultivated in degraded or wasteland. The Indian government has set an ambitious target of 20% ethanol blending (E_2O) by 2025, which is expected to significantly reduce crude oil imports, improve energy security, and lower transport-related emissions by approximately 15%. This initiative aligns with India's commitment to reduce GHG emissions under the Paris Agreement and to create economic opportunities in rural areas through the cultivation and processing (Parvez et al., 2025).

13.9 Challenges and Future Prospects

Despite its effectiveness, phytoremediation faces challenges such as long remediation timeframes, variability in pollutant uptake between species, and environmental constraints. Climate, soil composition, and contamination levels significantly affect phytoremediation efficiency. Modern technologies such as genetic engineering and synthetic biology aim to increase plants' pollutant absorption and tolerance of pollutants from plants to resolve these shortcomings.

Policymakers must consider incorporating bioremediation strategies into urban development and industrial emission control frameworks to obtain better public health and environmental benefits. Implementing biofiltration systems, urban forest initiatives, and extensive afforestation activities will be vital to resolving the air pollution challenge in India. By integrating plant-based and microbial bioremediation with biofiltration, city authorities can significantly reduce air pollution, improve public health, and foster positive changes in sustainable urban development.

13.10 Conclusion

Air pollution has become a pressing issue around the world, affecting populations' health, biodiversity, and climate. Industrialization, the movement of people to urban areas, transportation, and agricultural practices have caused an increase in emissions, leading to a marked decline in air quality. This chapter has examined a number of nature-centered approaches and bioresource processes to aid in the reduction of air pollution and highlighted phytoremediation, bioremediation, bioindicators, and biofuels. NbS, such as urban green spaces, tree planting, and ecosystem restoration, actively reduces environmentally harmful emissions by capturing and metabolizing toxins. Similarly, bioethanol, biodiesel, and algae-based fuels act as bioresources, replacing fossil fuels and significantly reducing GHG. Understanding how vegetation aids in the control of pollution is necessary because it helps in the mitigation of $PM_{2.5}$, PM_{10} emissions, NOx, SO_2, and even CO_2. Bioresources known as lichens and mosses serve as bioremediation indicators, which allow them to continuously control pollution at a minimal expense, making them ideal for environmental monitoring. These approaches serve the purpose of improving air quality and preserving biodiversity and climate, which is the changing nature of the world.

India's National Bio-Energy Mission and the NCAP illustrate how the nation is committed to using biological methods for air pollution management. The emphasis on afforestation, ethanol blend, and urban green infrastructure shows how policy initiatives can enhance natural mitigation processes. In addition to their effectiveness, extended remediation time frames, contaminant species selection, and climate fluctuations remain complex problems. Increasing the efficacy of pollutant removal from plants and microbes through genetic engineering, synthetic biology, and bioengineering of biohazard plants and microbes is necessary to address these challenges. In conclusion, nature-based strategies are pragmatic and sustainable strategies to reduce pollution. Incorporating phytoremediation, biofuels, and urban greening into industrial and municipal planning can vastly improve air quality and restore environmental balance, allowing governments and policymakers to act with a more significant impact. Together with technological advances and further research, these strategies guarantee a better, healthier, and cleaner environment for future generations.

Acknowledgments The authors are very grateful to the Govind Ballabh Pant National Institute of Himalayan Environment, Kosi-Katarmal, Almora, for providing support and valuable suggestions for completing this chapter.

References

Appannagari, R. R. (2017). Environmental pollution causes and consequences: A study. *North Asian International Research Journal of Social Science & Humanities, 3*(8), 151–161.

Bargagli, R. (1998). *Trace elements in terrestrial plants. An ecophysiological approach to biomonitoring and biorecovery* (Vol. 1, pp. 1–324). Springer-Verlag.

Chisti, Y. (2007). Biodiesel from microalgae. *Biotechnology Advances, 25*(3), 294–306.

Choudhary, M. P., & Garg, V. (2013). Causes, consequences and control of air pollution. In *All India seminar on methodologies for air pollution control*. MNIT.

Conti, M. E., & Cecchetti, G. (2001). Biological monitoring: Lichens as bioindicators of air pollution assessment—A review. *Environmental Pollution, 114*(3), 471–492.

Demirbas, A. (2009). Progress and recent trends in biodiesel fuels. *Energy Conversion and Management, 50*(1), 14–34.

Gawroński, P., Jensen, P. E., Karpiński, S., Leister, D., & Scharff, L. B. (2018). Pausing of chloroplast ribosomes is induced by multiple features and is linked to the assembly of photosynthetic complexes. *Plant Physiology, 176*(3), 2557–2569.
Gentry, T., Rensing, C., & Pepper, I. A. N. (2004). New approaches for bioaugmentation as a remediation technology. *Critical Reviews in Environmental Science and Technology, 34*(5), 447–494.
Goldemberg, J. (2008). The Brazilian biofuels industry. *Biotechnology for Biofuels, 1*(1), 6.
IUCN. (2020). *Nature-based solutions for climate change mitigation and adaptation*. International Union for Conservation of Nature and Natural Resources.
Jha, N., Pandey, A., Khan, U., Gupta, A., & Gour, C. P. (2025). The role of public in mitigating air pollution in Delhi (2004-2024): A critical review. *Environmental Claims Journal*. https://doi.org/10.1080/10406026.2025.2537092
Kabisch, N., Korn, H., Stadler, J., & Bonn, A. (2017). *Nature-based solutions to climate change adaptation in urban areas: Linkages between science, policy and practice*. Springer.
Khanna, K., Kohli, S. K., Kaur, R., Handa, N., Bakshi, P., Sharma, P., Ohri, P., & Bhardwaj, R. (2023). Reconnoitering the efficacy of plant growth promoting rhizobacteria in expediting phytoremediation potential of heavy metals. *Journal of Plant Growth Regulation, 42*(10), 6474–6502.
Knothe, G. (2010). *The biodiesel handbook* (2nd ed.). Taylor & Francis.
Kuiper, I., Lagendijk, E. L., Bloemberg, G. V., & Lugtenberg, B. J. (2004). Rhizoremediation: A beneficial plant-microbe interaction. *Molecular Plant-Microbe Interactions, 17*(1), 6–15.
Macek, T., Macková, M., & Káš, J. (2000). Exploitation of plants for the removal of organics in environmental remediation. *Biotechnology Advances, 18*(1), 23–34.
Malik, A. (2007). Environmental challenge Vis a Vis opportunity: The case of water hyacinth. *Environment International, 33*(1), 122–138.
Manisalidis, I., Stavropoulou, E., Stavropoulos, A., & Bezirtzoglou, E. (2020). Environmental and health impacts of air pollution: A review. *Frontiers in Public Health, 8*, 14.
Nakbanpote, W., Jitto, P., Taya, U., & Prasad, M. N. V. (2024). Chrysopogon zizanioides (vetiver grass) for abandoned mine restoration and phytoremediation: Cogeneration of economical products. In *Bioremediation and bioeconomy* (pp. 385–417). Elsevier.
Nowak, D. J., Hirabayashi, S., Bodine, A., & Greenfield, E. (2014). Tree and forest effects on air quality and human health in the United States. *Environmental Pollution, 193*, 119–129.
Pandey, A., Brauer, M., Cropper, M. L., Balakrishnan, K., Mathur, P., Dey, S., Turkgulu, B., Kumar, G. A., Khare, M., Beig, G., & Gupta, T. (2021). Health and economic impact of air pollution in the states of India: The global burden of disease study 2019. *The Lancet Planetary Health, 5*(1), e25–e38.
Parvez, M., Lal, S., Khan, O., Ahmad, M., Lal Meena, S., & Salvi, B. L. (2025). Sustainable regulatory framework for ethanol and biodiesel blending with petro-fuels in India to meet fuel demand: A review of biofuel policies. *Biofuels, 16*(6), 638–651.
Pilon-Smits, E. (2005). Phytoremediation. *Annual Review of Plant Biology, 56*(1), 15–39.
Pugh, T. A., MacKenzie, A. R., Whyatt, J. D., & Hewitt, C. N. (2012). Effectiveness of green infrastructure for improvement of air quality in urban street canyons. *Environmental Science & Technology, 46*(14), 7692–7699.
Rangsayatorn, N., Upatham, E. S., Kruatrachue, M., Pokethitiyook, P., & Lanza, G. R. (2002). Phytoremediation potential of spirulina (Arthrospira) platensis: Biosorption and toxicity studies of cadmium. *Environmental Pollution, 119*(1), 45–53.
Sharma, A., Kumar, V., Shahzad, B., Tanveer, M., Sidhu, G. P. S., Handa, N., Kohli, S. K., Yadav, P., Bali, A. S., Parihar, R. D., & Dar, O. I. (2019). Worldwide pesticide usage and its impacts on ecosystem. *SN Applied Sciences, 1*(11), 1446.
Sharma, R., Kumar, R., Sharma, D. K., Son, L. H., Priyadarshini, I., Pham, B. T., Tien Bui, D., & Rai, S. (2019). Inferring air pollution from air quality index by different geographical areas: Case study in India. *Air Quality, Atmosphere & Health, 12*(11), 1347–1357.

Sicard, P., Agathokleous, E., Anenberg, S. C., De Marco, A., Paoletti, E., & Calatayud, V. (2023). Trends in urban air pollution over the last two decades: A global perspective. *Science of the Total Environment, 858*, 160064.

Sierra-Porta, D., Solano-Correa, Y. T., Tarazona-Alvarado, M., & de Villavicencio, L. A. N. (2023). Linking PM_{10} and $PM_{2.5}$ pollution concentration through tree coverage in urban areas. *CLEAN, 51*(5), 2200222.

Speak, A. F., Rothwell, J. J., Lindley, S. J., & Smith, C. L. (2012). Urban particulate pollution reduction by four species of green roof vegetation in a UK city. *Atmospheric Environment, 61*, 283–293.

Statista. (n.d.). *Average life expectancy gains if air pollution rules met*. https://www.statista.com/chart/30841/average-life-expectancy-gains-if-air-pollution-rules-met/

Tiwari, A., & Malviya, N. (2021). Phytopharmacological evaluation of hydroalcoholic extract of bark of pongamia glabra vent. For anti-obesity activity and effect on lipid profile and hepatic enzymes. *Journal of Advanced Scientific Research, 12*(4), 155–162.

World Health Organization (WHO). (2016). *Urban green spaces and health: A review of evidence*. World Health Organization.

World Health Organization (WHO). (2021). *WHO global air quality guidelines: Particulate matter ($PM_{2.5}$ and PM_{10}), ozone, nitrogen dioxide, Sulfur dioxide and carbon monoxide*. World Health Organization. https://www.who.int/publications/i/item/9789240034228

World Health Organization (WHO). (2024). *Ambient (Outdoor) air pollution and health*. https://www.who.int/news-room/fact-sheets/detail/ambient-(outdoor)-air-quality-and-health

Yadav, A. K. (2021). Human health risk assessment in opencast coal mines and coal-fired thermal power plants surrounding area due to inhalation. *Environmental Challenges, 3*, 100074.

Yadav, A. K., Sahoo, S. K., Dubey, J. S., Kumar, A. V., Pandey, G., & Tripathi, R. M. (2019). Assessment of particulate matter, metals of toxicological concentration, and health risk around a mining area, Odisha, India. *Air Quality, Atmosphere & Health, 12*(7), 775–783.

Yadav, A. K., Sahoo, S. K., Kumar, A. V., Shirin, S., Jamal, A., Patra, A. C., Dubey, J. S., Thakur, V. K., Lenka, P., Singh, S., Jha, V., & Tripathi, R. M. (2023). Identifying size-segregated particulate matter ($PM_{2.5}$, PM_{10} and SPM) sources in an industrial town of India. *Aerosol Science and Engineering, 7*(4), 455–473.

Chapter 14 Digital Pathways to Cleaner Air: Reducing Construction-Related CO_2 Through Building Information Modeling, Digital Twins, and Artificial Intelligence

Fanny E. Berigüete Alcántara, José S. Santos Castillo, Aneuris De Los Santos Melo, and César E. Arguedas Garro

14.1 Introduction

The construction industry is one of the largest global emitters of carbon dioxide (CO_2) as well as other greenhouse gases (GHGs), contributing 37% of global emissions and 34% of global energy consumption, according to the United Nations Environment Programme (UNEP, 2022). GHG emissions are quantified in their CO_2 equivalent as gas released into the atmosphere; the release of each 1 kg of GHG in its CO_2 equivalent, per square meter (m^2) of useful interior surface has the capacity to absorb infrared radiation from the Sun and heat the atmosphere for a period of 50–100 years (De Wolf et al., 2023; Sturman & Quénol, 2023). The Intergovernmental Panel on Climate Change (IPCC) notes that each kilogram of GHGs released per square meter of usable surface area absorbs thermal radiation, increasing global warming according to the Global Warming Potentials defined by Myhre et al. (2013). In the European Union, for example, 19% of energy in 2022 came from solid fossil fuels, generating toxic GHGs such as methane (CH_4) and nitrous oxide

F. E. Berigüete Alcántara (✉)
Department of Architectural Technology, Universitat Politècnica de Catalunya UPC, Barcelona, Spain
e-mail: fanny.esther.beriguete@upc.edu

J. S. Santos Castillo
University Research Institute for Sustainability Science and Technology, Universitat Politècnica de Catalunya UPC, Barcelona, Spain

A. D. L. S. Melo
School of Architecture and Urbanism, Universidad Central del Este UCE, San Pedro de Macoris, Dominican Republic

C. E. A. Garro
School of Architecture, Universidad de Costa Rica UCR, San José, Costa Rica

A. K. Yadav (ed.), *Computational Approach for Air Pollution*,
https://doi.org/10.1007/978-3-032-09215-1_14

(N_2O), in contrast to countries such as Costa Rica or Norway, with almost 100% renewable energy mixes (United Nations, 2025).

In this context, digital technologies such as building information modeling (BIM), digital twins (DTs), and artificial intelligence (AI) emerge as key tools to mitigate emissions and improve environmental quality. BIM allows the integration of emissions data (kg CO_2 eq) of materials, based on standards such as UNE-EN 15804:2012+A2 (AENOR, 2020), optimizing the selection of sustainable alternatives such as recycled timber or low-impact concrete (Eastman et al., 2011). DTs, through Internet of Things (IoT) sensors, monitor energy consumption and pollutants in real time, while AI with predictive life cycle analysis (LCA) identifies strategies to reduce the environmental footprint (Zhang et al., 2023). These capabilities are supported by databases. One case is the indexation of CO_2 records in the Instituto Valenciano de la Edificación (IVE, 2024), which indexes material emissions for BIM (Arguedas Garro et al., 2020) and AI processes, linking digitalization of the information of building materials and sustainability (Arguedas Garro, 2021).

In this chapter, it is identified that BIM, DT, and AI tools are useful to obtain criteria on the ecological impact of buildings by processing CO_2 emissions data. Projects such as Gemeentehuis Brummen, for example, identify a 50% decrease in emissions of GHGs based on Design for Disassembly (DfD), where the 6D BIM simulation tool was used to reduce environmental impact data (Figl et al., 2019). With the application of technologies, there is also the challenge of interoperability and the costs of data production and collection. In a current context, where climate change can be felt in the urbanized world, this study identifies the importance of sustainable construction related to the use of environmental impact data in relation to the use of digital technologies, public health, and the Sustainable Development Goals (SDGs) to reduce air pollution.

14.2 Digital Technologies for Sustainable Construction

14.2.1 Building Information Modeling

BIM is a collaborative work methodology that enables the integrated development of building and infrastructure projects through a model that unifies information, encompassing the entire lifecycle from planning to operation and decommissioning (cradle, grave, and reuse). BIM is different from traditional approaches, characterized by more fragmented processes. BIM enables collaboration between stakeholders through the development of established objectives and the unification of data that make up a project's process. The main objective of BIM is to ensure the production of data that facilitates the exchange of information among construction professionals, such as designers, engineers, builders, and others. Adopting this technological tool allows for the establishment of sustainable objectives in project management and planning, specifically in the organization of execution. The use of

BIM optimizes the planning and production process, making efficient information management more cost-effective (Abanda et al., 2025; Baracho et al., 2025).

BIM has transformed the Architecture, Engineering, Construction & Operations (AECO) sector by transcending the barriers of traditional 2D drawing. This methodology facilitates the generation of 3D models enhanced with detailed data on all building elements, fostering interdisciplinary collaboration and improving efficiency at all phases of the project. In addition to representing the geometry, BIM models also incorporate metadata with information on the physical, technical, and functional properties of building components, which favors informed decision-making (Berigüete et al., 2024). For its effective implementation, BIM is regulated by international regulations; ISO 19650 is one of the most recognized and fundamental standards (Abanda et al., 2025).

ISO 19650 was developed based on the British standard BS 1192, as well as the PAS 1192-2 specification, which establishes a universal standard for data management in building projects (ISO, 2019). According to the BSI Group (2023), this regulation is described as an "international standard for managing information throughout the life cycle of a built asset using BIM." It is structured in several ways, encompassing information security concepts and principles that are adaptable to various national contexts. Using the regulation allows workgroups to align processes with improved performance in project execution, expanding interoperability in data mapping. The following paragraphs identify how the standard contributes to the selection of sustainable materials in relation to optimized design, with the goal of reducing the footprint, the minor environmental impact of construction.

ISO 19650 regulates the quality of information generation and management processes, helping designers and data developers organize sustainability strategies into stages and layers. When developing a model using the BIM standard, decisions are made in the project layers, with groups working on processes. The tool allows one to establish, for example, criteria integration processes related to sustainable criteria such as energy efficiency, durability, and environmental impact. The project lifecycle stages can be presented for analysis, improvement, and prediction of future project performance. In its 6D version, BIM provides technology that allows simulating energy scenarios based on the predicted performance of analyzed materials according to their use, providing metadata that identifies their actual properties (Li et al., 2024). An example of this is by specifying low-impact building materials, such as recycled wood or lower-impact concrete, like earth-based concrete. Performance can be assessed, and the ecological footprint of a project can be optimized according to the use of its materials.

Various studies have corroborated the importance of using BIM to optimize the design strategy in its planning phase. Afzal et al. (2023) highlight the technological use of BIM tools to automate structural analysis tasks and design strategies, with the goal of achieving performance sustainability in project management. The benefit of this technological approach lies in the data interoperability algorithms that integrate 3D models of architecture, engineering, facilities, sustainable systems, and material resources. This synchronization facilitates interdisciplinary work between the various technicians involved in the project, thus promoting a more organized approach.

BIM platforms allow the incorporation of environmental indicators into construction materials and systems, which allows for the evaluation of a project's material components based on their environmental impact. An example of this is the relationship between the use of BIM technology and certification criteria such as Building Research Establishment Environmental Assessment Methodology (BREEAM) or Leadership in Energy and Environmental Design (LEED) (Mercader Moyano et al., 2019). The application of BIM technology is useful in research related to construction, energy efficiency, and environmental impact.

14.2.2 Digital Twins

DTs are virtual models that represent real physical objects or systems, combining real-time data collected through IoT sensors, with the aim of simulating, controlling, and optimizing their behavior, improving efficiency, and resource management (Berigüete et al., 2024; Robles et al., 2023). DTs have become an essential element in connecting the physical and digital worlds, ranging from individual processes to complex urban environments (De Los Santos Melo & Beriguete Alcántara, 2024).

One of the most compelling uses of DTs lies in their integration with the IoT, particularly in the development of smart cities. These urban environments take a holistic approach, with the aim of improving everyday life through advanced digital tools. Sensors and connected devices gather real-time data on traffic patterns, energy usage, air quality, and other vital indicators (Elias Bibri et al., 2025; Matei & Cocoşatu, 2024). DTs use this information to simulate the functioning of different urban systems, allowing for informed, data-driven decisions. In doing so, they help improve efficiency and sustainability while supporting emission reduction through more accurate monitoring. Furthermore, DTs can promote greater civic engagement by making information more accessible and encouraging shared decision-making about resource use. However, the implementation of DTs in smart cities must go hand in hand with a commitment to sustainability and social inclusion. Achieving this requires both public and private sectors to invest in the development of sophisticated modeling and simulation tools, incorporating emerging technologies such as the metaverse, extended reality, and generative AI (Matei & Cocoşatu, 2024). These tools are capable of dynamically interpreting multisensor environmental data within urban contexts, offering large-scale solutions to cut emissions and make better use of resources (Rastogi et al., 2025; Robles et al., 2023). The convergence of DTs, IoT, and other cutting-edge technologies is reshaping entire sectors, paving the way for autonomous, intelligent environments—from individual buildings to entire urban infrastructures (Berigüete et al., 2024).

14.2.3 Artificial Intelligence

AI stands out as an innovative technology in the AECO sector by replicating human reasoning and managing large volumes of data for purposes such as learning, anticipation, and optimization. Unlike conventional analytical methods, AI tackles complex problems in an automated and reliable manner, reducing the reliance on human expertise and maximizing project success (Pan & Zhang, 2023). In the context of construction, AI powers the analysis of massive data collected throughout a project's lifecycle, identifying patterns and trends that improve design and management. Integrated with BIM, it enables predictive analytics and energy efficiency simulations through visual programming algorithms and generative design, improving the productivity and accuracy of models. In addition, metadata-enriched BIM models feed into DTs, which, interfaced with the IoT, provide real-time operational responses to reduce energy costs and emissions. This integration also automates the detection of interdisciplinary conflicts, predicts risks, and optimizes schedules, contributing to safer and more efficient buildings (Pan & Zhang, 2023).

14.2.4 Synergies Between BIM, DT, and IA

The convergence of BIM, DT, and AI is revolutionizing the AECO industry by combining process automation, data operability, and predictive asset maintenance, with a clear focus on sustainability. This synergy enables generative design aimed at reducing ecological impact and optimizing resources and processes throughout the project lifecycle (Table 14.1).

By integrating BIM as an information base, DT for real-time monitoring, and AI for analysis and prediction, a holistic approach is created that improves energy efficiency and reduces pollution (Akbari et al., 2024). Allied technologies such as blockchain reinforce this complementarity by providing security and transparency in data-driven decision-making, facilitating more sustainable and efficient projects. This integration not only minimizes emissions and waste but also transforms construction into a more resilient sector in the face of environmental challenges.

14.3 Methodology

A mixed methodology combining theoretical research, secondary data analysis, and case study evaluation is employed, integrating qualitative and quantitative approaches.

Table 14.1 Comparison of BIM, DT, and IA

Aspect	BIM	DT	AI
Objectives	Centralize all design information for construction and operation	Simulate assets in real time	Automate, learn, and optimize decisions
Phases	Planning, design, construction, and operation	Design, construction, operation, and maintenance	Design, construction, and operation
Benefits	Facilitates collaboration and data management throughout the building's life cycle	Enables real-time data-driven decisions and analysis of energy performance and environmental impact	Saves time, provides accurate predictions, and reduces human errors
Data type	Quantitative, qualitative, geometric, and technical	BIM metadata with IoT and virtual replica	Quantitative, qualitative, generative, and automated metrics
Sustainability	Enables modeling of sustainable materials, simulation of energy scenarios in design, and waste management for certifications like LEED	Identifies inefficiencies, reduces emissions through real-time monitoring, and optimizes resources such as water and energy	Designs eco-friendly solutions, predicts consumption and emissions, and optimizes materials using advanced algorithms

14.3.1 Scope Definition

The research focuses on CO_2 emissions in the construction sector and the potential role of BIM, DT, and IA. Key questions were asked: How do these technologies optimize processes to reduce CO2 releases? What evidence supports its overall effectiveness? How are they integrated to maximize benefits? What regional factors affect their adoption? These guided the theoretical and practical analysis.

14.3.2 Data Collection and Organization

Three main sources were used:

- Review of the literature: scientific articles, regulations (ISO 19650), and IPCC and WHO reports.
- Secondary data: International Energy Agency (IEA) and World Bank statistics on emissions and technology.
- Case studies: Projects with BIM, DT, and IA, selected by impact and scale, classified by region.

The information collected was organized into thematic categories aligned with the chapter structure: enabling technologies (Sect. 14.2), current state of the sector

(Sect. 14.4), case study analysis (Sect. 14.5), and assessment of benefits and challenges (Sect. 14.6).

14.3.3 Analysis and Synthesis

The analysis included the following:

- Qualitative: Impact of BIM (materials), DT (IoT), and IA (LCA).
- Comparative: CO_2 reduction in digitized projects.
- Practical: Implications for policy and research.

Patterns and synergies between technologies were identified.

14.3.4 Validation and Limitations

Triangulation with academic, governmental, and technical sources was applied, prioritizing reliable data. Limitations include regional variability and a lack of standardization in implementation.

14.4 Current State of Digitalization and Air Quality in Construction

14.4.1 Geographical Overview of Emissions and Vulnerabilities

The construction sector faces the challenge of reducing its ecological impact through the integration of technology and sustainable strategies. Although major economies account for most of the emissions, developing countries lack effective infrastructure and regulation, which exacerbates their vulnerability. Digitization and real-time monitoring emerge as key tools in the transition to more sustainable cities.

The projected demographic growth, which could reach 10.9 billion by 2100, will intensify resource consumption and CO_2 emissions, which have increased steadily since 1960 (Telemetro, 2013). Regarding the steady expansion in CO_2 emissions since 1960, the IPCC has documented this increase in several of its reports. For example, in the 2007 Fourth Assessment Report, CO_2 emissions increased by approximately 80% between 1970 and 2004 from 21 to 38 gigatons (IPCC, 2008). This population growth increases the demand for transport, energy, and building materials, sectors that contribute significantly to the global carbon footprint.

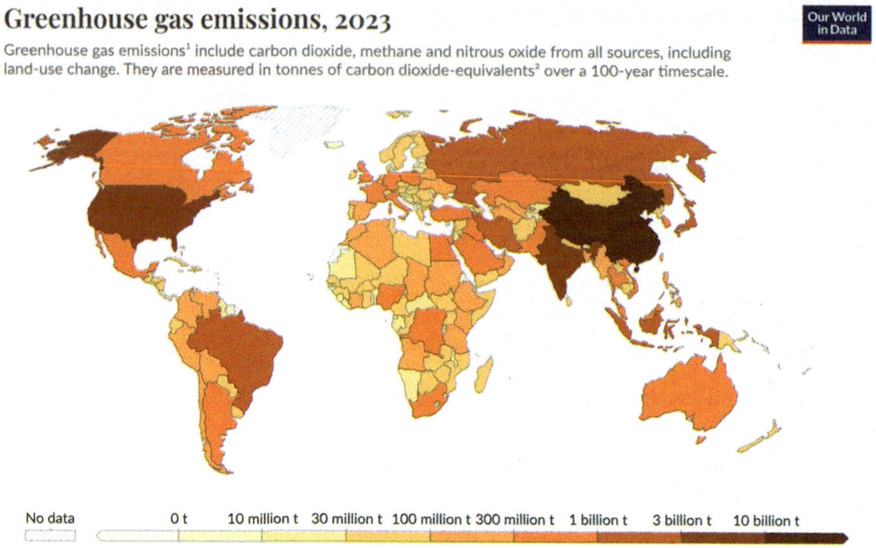

Fig. 14.1 Global distribution of greenhouse gas emissions in 2023. World map showing greenhouse gas emissions by country in 2023. Countries with the highest emissions, such as China and the United States, are shown in darker shades, indicating their greater contribution to the global carbon footprint. Source: Jones et al. (2024)

In this context, the building sector is responsible for a large share of CO_2 emissions due to its high energy consumption and the production of carbon-intensive materials. The implementation of strategies such as LCA in sustainable materials has proven to be effective in reducing their environmental impact (Lu et al., 2017). Furthermore, embedded energy in buildings and transport infrastructure is a key source of pollution in industrialized countries such as the United Kingdom (Kolosz & Grant-Muller, 2015). The main emitting economies reflect this reality. In absolute global terms up to 2023, the three countries with the greatest contribution of anthropogenic CO_2 are China, the United States, and India (Fig. 14.1). In terms of sectoral analysis of the cumulative global total, in 2021, the building industry, due to its energy consumption, emitted around 3–4 gigatons of CO_2-eq, while the manufacturing and construction sector generated approximately 7–8 gigatons. If emissions associated with transport, logistics, and waste management are considered, as well as the share of power and heat generation are considered, the total impact of this industry could be even higher, reaching between 10 and 12 gigatons of CO_2-eq globally (Fig. 14.2).

Within construction, China leads in material consumption, accounting for 50% of global cement and steel use, equivalent to 1.5 billion tons of CO_2 annually (IEA, 2022). In the United States, the construction sector is responsible for 37% of national emissions, while in India, rapid urbanization contributes 22% of the total (Global Carbon Project, 2023). In contrast, developing countries face greater challenges due

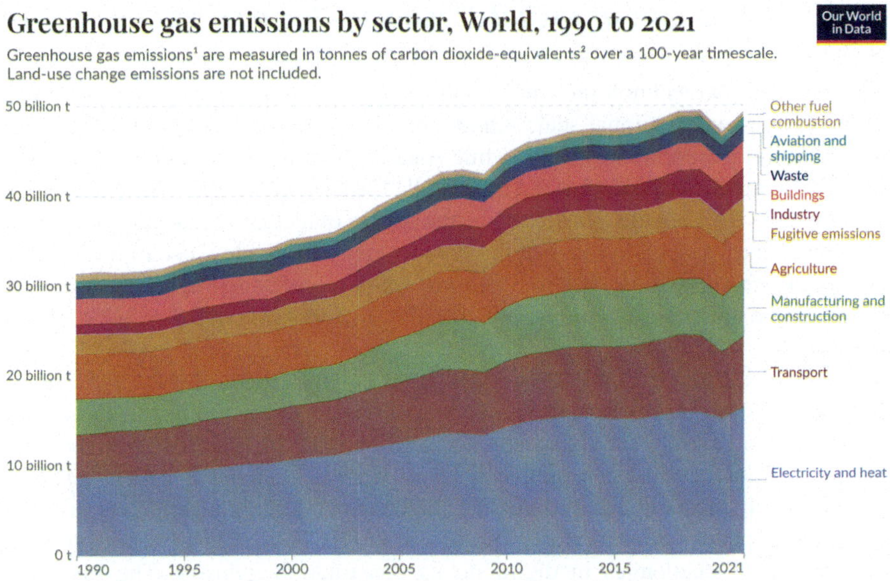

Fig. 14.2 Growth in greenhouse gas emissions by sector (1990–2021). Graph illustrating the growth of global greenhouse gas emissions from 1990 to 2021, broken down by sector. Electricity and heat, transport, and industry stand out as the main contributors to the increase in emissions. Source: Jones et al. (2024)

to outdated infrastructure and increased exposure to natural disasters. While in Europe and Japan, regulations such as the EPBD have reduced the sector's emissions by 20% in the last decade (EEA, 2022), in Latin America and Southeast Asia, 60% of buildings still do not comply with energy efficiency standards (UNEP, 2022). However, initiatives such as bamboo construction in Colombia or the integration of solar panels in the Philippines show significant progress in reducing the environmental impact. In addition, Nairobi has been able to reduce costs and emissions through the implementation of BIM, while the TDs facilitate urban planning that is more adapted to natural disasters. In parallel, air pollution from construction sector emissions poses a direct threat to public health. The World Health Organization (WHO) estimates that every year, seven million people die from exposure to pollutants such as $PM_{2.5}$ and NO_x, which aggravate respiratory and cardiovascular diseases (WHO, 2023). More polluted urban areas, such as Delhi, have reported an increase in hospitalizations (de Bont et al., 2024), mainly affecting the most vulnerable groups, such as children and the elderly (UNICEF, 2022). In this situation, digitization is an enabling tool to reduce emission volumes and address health crises. Real-time measurement of air quality and energy performance indicators in the city enables the early identification of risks, facilitating the design of specific strategies for a healthier and more resilient built environment.

14.4.2 Geography of Digital Technology Implementation

The use of digital technologies in the construction sector presents a panorama with particularities on a regional scale. On the one hand, Europe and the United States stand out in the adoption of BIM, while Asia leads in the implementation of DTs, especially Japan, South Korea, and the Middle East. However, the level of economic development does not always determine the degree of technological adoption. High-income countries, such as those in the Persian Gulf, show a low use of these, while some emerging countries, such as Singapore, show significant progress. This reveals that strategic priorities and public policies play a crucial role in the sector's innovation efforts.

Undoubtedly, the evidence shows that digital technologies contribute to the reduction of emissions and pollutants. Regions that have adopted the use of these technologies report significant improvements in air quality compared to those that have not yet adopted them. For example, 97% of Europe's air has become cleaner in recent years in most regions, progress attributed in part to digitalization in the design and management of construction projects (eSMARTCITY, 2025). In contrast, Latin America faces challenges in digitizing its construction industry. The Economic Commission for Latin America and the Caribbean has identified technological backwardness as a factor limiting regional development, stressing that digital transformation is crucial to overcoming structural barriers. The low investment in digital tools restricts the region's optimization potential in the AECO sector when developing strategies to mitigate environmental problems such as air pollution (ONU-CEPAL, 2024). In this regard, the Development Bank of Latin America (CAF) points out the urgent need to modernize the regional infrastructure to provide a way out of economic stagnation. The digitization of electrical and transportation systems, for example, is a major opportunity to take advantage of the technological revolution as a lever to improve air quality in urban environments (Cota, 2021). Digital technologies, by optimizing project management and their capacity for environmental monitoring, result in work with a lower impact on air quality. They are very useful tools to reduce environmental risks; however, while some regions of the planet have advanced in their incorporation, others remain behind, which indirectly implies unequal impacts on the atmosphere, triggering public health problems. This situation makes it very necessary for governments to initiate plans and investments for the digitization of construction and promote policies that promote sustainable construction practices to mitigate pollution and positively impact the welfare of the population.

14.5 Case Studies

14.5.1 Case Study 1: Brummen Town Hall

The headquarters building of the Brummen Town Hall (the Netherlands), projected in 2013 by RAU Architects, is a 2.200 m^2 building structure conceived for temporary use with a projected lifespan of 20 years. This means that it is a type of

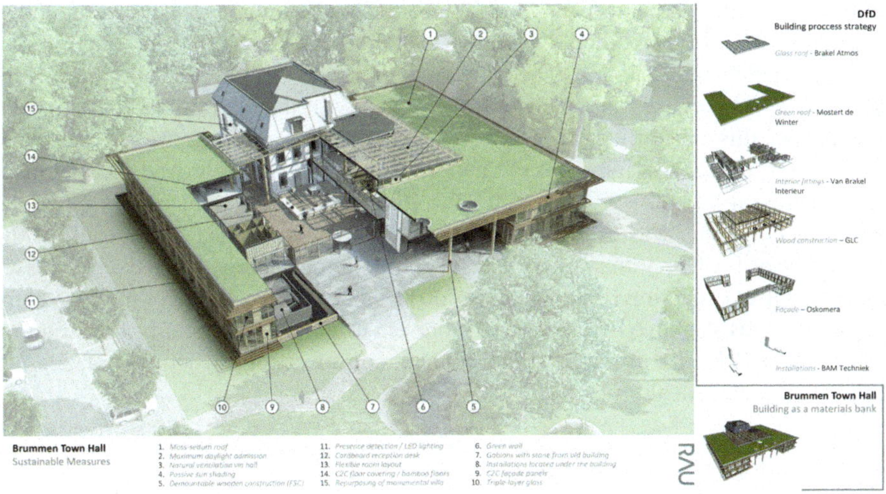

Fig. 14.3 Composition Brummen Town Hall. The diagram presents an explanatory axonometric view of the city government building, highlighting its sustainability strategies and the DfD approach. This highlights the concept of the building as a "material bank," underscoring the potential to dismantle and reuse the structure in the future, in accordance with the principles of the circular economy in architecture. Source: RAU Architects (2016)

architecture conceived according to DfD principles or, in other words, an assembly architecture. DfD promotes that components can be reused or recycled at the end of their cycle (Bragança et al., 2024) (Fig. 14.3) by adopting a materials passport that indexes in detail each building element by materiality, quantity, origin, manufacture, assembly, and so on. To this end, the adoption of BIM, using Autodesk Revit, and according to ISO 19650, was essential to develop and coordinate a modular DfD design—a project based on the use of sustainable resources, such as recycled wood, reusable steel, and pre-fabricated components with low environmental impact (Asif et al., 2024; RAU Architects, 2016). This BIM model involved a design based on efficiency, a detailed inventory of materials, and clear and orderly logistics that were generated, obtaining a 20% decrease in waste and significantly reducing emissions associated with construction (Li et al., 2024).

In addition, the use of AI facilitated early design decisions by reinforcing project sustainability through predictive LCA. AI algorithms evaluated the ecological impact of various solutions and material ratios, recommending options with lower carbon and water footprints, such as partial replacement of concrete with recycled wood panels (Zhang et al., 2023). Such simulations, integrated with the BIM model, facilitated material selection and design to generate long-term energy performance scenarios to ensure compliance with efficiency standards (Płoszaj-Mazurek & Ryńska, 2024). Furthermore, the adoption of dry construction techniques (a particular feature of DfD) was managed through BIM to optimize resource use and emissions derived from wet processes, aligning with European Union sustainability goals (Figl et al., 2019). Compared with a building of conventional construction with very similar characteristics, the Gemeentehuis Brummen presents a reduction of around 50% of embodied CO_2 emissions, equivalent to 600 tons, thanks to a design based on the selection of recycled materials and constructive efficiency (Figl et al., 2019). Likewise, the emissions associated with the operation were expected to be low; this is verified by an energy consumption, around 50 $kWh/m^2/year$, achieved by passive design thanks to natural ventilation and improved thermal insulation (Płoszaj-Mazurek & Ryńska, 2024). Furthermore, the reduction of NOx and other air pollutants was estimated to be between 15% and 20%, as a result of the lack of need for heavy machinery and the reduction of energy-intensive processes during project execution (Eastman et al., 2011). This is an example of how the combination of BIM and AI can drive a more environmentally friendly construction model, a benchmark for advocating impact-conscious construction practices to improve people's health.

14.5.2 Case Study 2: Breathe London

London, one of Europe's most densely populated capitals, is facing serious environmental challenges due to urban sprawl, traffic congestion, and high industrial emissions. Despite efforts carried out through plans and policies to mitigate pollution, levels of nitrogen dioxide (NO_2) and fine particulate matter ($PM_{2.5}$) remain high, repeatedly exceeding air quality thresholds established by the WHO (C40 Cities, 2019). These conditions compromise the health of Londoners, increasing year-on-year the cases of respiratory and cardiovascular conditions, in turn stressing the city's ability to adapt to the challenges of climate change and undermining its environmental resilience. A prominent example is the Breathe London project, launched in 2019, which deployed more than 100 low-cost sensors on street furniture and Google Street View vehicles to collect real-time data on air pollutants (Fig. 14.4). Integrating this information into a digital model of the city has made it possible to visualize the spatial and temporal distribution of air quality, identify hotspots, and assess the impact of policies such as the Ultra Low Emission Zone (ULEZ) (C40 Cities, 2019; Ittner et al., 2021).

The results have been significant: hyperlocal monitoring has allowed for optimizing urban strategies, refining transport planning, and boosting the adoption of nature-based solutions. For example, the expansion of the ULEZ in London has resulted in a 27% reduction in nitrogen dioxide (NO_2) and a 31% in particulate matter $PM_{2.5}$ in areas previously affected by high pollution. Additionally, the number of polluting vehicles entering the area daily has decreased by 13,500, representing a reduction of 38% (De Miguel, 2023; Mayor of London, 2019). The integration of real-time data into virtual models not only improves decision-making and the effectiveness of public policies but also contributes to the protection of the health and well-being of the population, demonstrating the potential of digital technologies to address environmental challenges in urban settings.

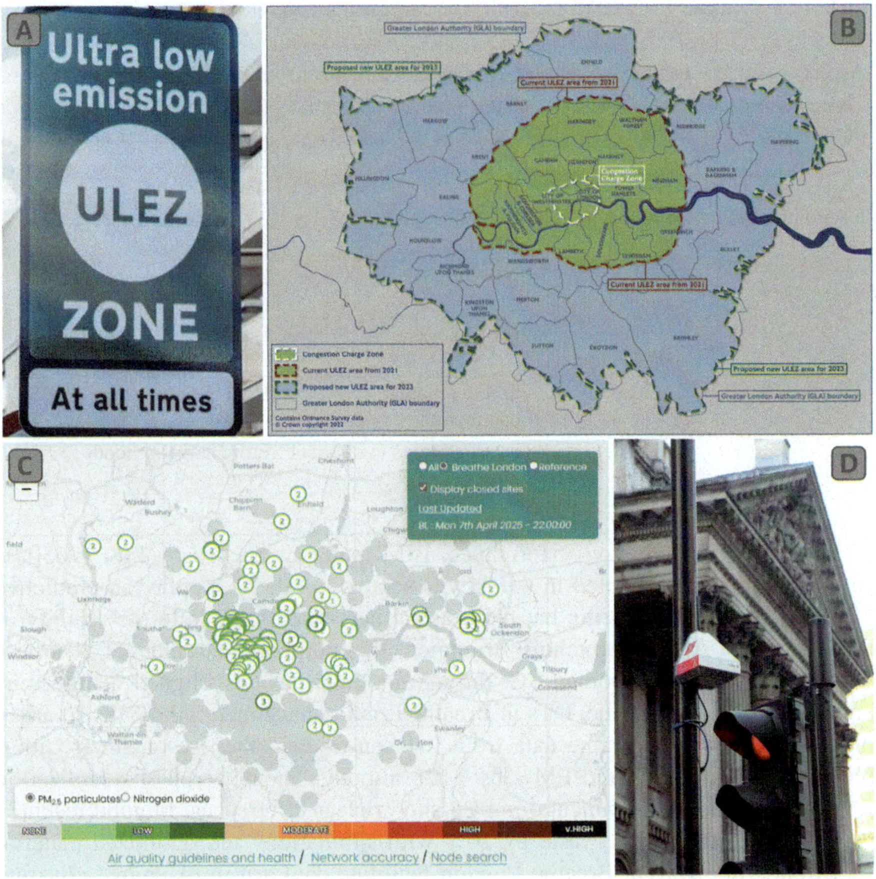

Fig. 14.4 Breathe London composition: Emissions control and air quality monitoring. (**a**) Signage informing drivers of ULEZ restrictions, (**b**) map illustrating its progressive expansion from implementation to extension in 2023, (**c**) screenshot shows Breathe London's monitoring network, whose sensor system feeds a DT for environmental management and data-driven decision making, and (**d**) IoT sensor embedded in street furniture collects real-time air quality data. Source: (**a**, **b**) BBC (2023); (**c**) Environmental Defense Fund (2020); and (**d**) Imperial College London (2024)

14.5.3 Comparative Analysis

The integration of BIM and AI in Gemeentehuis Brummen and the use of DT in Breathe London represent complementary approaches that optimize construction and urban planning. BIM and AI refine design and sustainable building, while DT improves operational management and urban resilience. Their technological performance, environmental impact, adaptability, and quantitative benefits, together with lessons learned, reveal their potential and challenges (Table 14.2).

Table 14.2 Comparison of BIM, IA, and DT in construction and urban planning

Aspect	Factor	BIM and AI (Gemeentehuis Brummen)	DTs (Breathe London)
Technological performance	Technological maturity	Global implementation in the AECO sectors, ISO 19650 standard	Rapid integration into urbanism via IoT and smart cities, low global standardization
Environmental impact	Emission reduction	50% reduction in embodied CO_2	27% reduction in NO_2, 31% reduction in $PM_{2.5}$
	Energy efficiency	50 kWh/m^2/year	Optimized urban transport
	Resource optimization	20% less waste	Data-driven planning
Adaptability	Replicability	High in modular construction	High in cities with IoT infrastructure
	Scalability	From buildings to districts	From neighborhoods to regions

In terms of technology performance, BIM maturity, backed by ISO 19650, is strong and globally adopted in the AECO sectors, while AI, applied in predictive LCA analytics, is advancing but requires further validation due to the scarcity of large datasets. At Gemeentehuis Brummen, this combination achieves 20% less waste and 50% less embodied CO_2 through modular designs, aligned with the circular economy. Meanwhile, DTs in Breathe London have matured rapidly in urban planning thanks to real-time data from IoT sensors and the rise of Smart Cities, reducing NO_2 by 27% and $PM_{2.5}$ by 31%, although their overall standardization remains low. The environmental impact shows clear differences. In good resource use, BIM and AI reduce waste by 20% with simulations and optimized material selection, while DTs at Breathe London facilitate data-driven planning, adjusting traffic and emissions in real time to indirectly optimize resources. Emission reductions reflect a static approach at Gemeentehuis Brummen (50% less embodied CO_2) versus a dynamic one at Breathe London (27% less NO_2, 31% less $PM_{2.5}$). In energy efficiency, BIM and AI achieve 50 kWh/m^2/year, a competitive standard in buildings, while Breathe London improves urban transport, although without a specific kWh metric.

Adaptability also varies depending on the context. The replicability of BIM and AI is high in modular construction, as in DfD, and is amenable to application to bridges or roads according to local demands, provided that software and data exist. DTs, replicable in cities with IoT infrastructure (e.g., London and Singapore), apply to mobility or waste management, but their use is limited in cities with little digitization. In scalability, BIM and AI can encompass sustainable planning districts, while DTs scale from neighborhoods to metropolitan regions with robust digital infrastructure, integrating buildings and urban environments. The lessons learned reinforce these observations. Data integration combines the design optimization of BIM and AI (3D modeling and predictive analytics) with the dynamic decisions of

DTs, suggesting that their union could bridge the gap between design and operational management for sustainable outcomes. Digital infrastructure is a key factor: DTs rely on sensors and IoT, a challenge in underdigitized cities, while BIM and AI require less real-time infrastructure, highlighting the need to advance urban digitization. Sustainability, at its core, reduces the environmental footprint with clear goals—less waste and CO_2 in Gemeentehuis Brummen, better air quality in Breathe London—requiring specific indicators from the outset. Future synergy between BIM and DT would integrate building and urban planning, assessing materials and long-term impacts to improve cities.

14.6 Benefits, Challenges, and Prospects

The introduction and increasing use of BIM, DT, and AI are redefining the paradigms of sustainability and efficiency in the AECO sector, creating a whole new construction culture based on a more rational and efficient holistic approach. This becomes more evident as we see how these technologies are increasingly being used in a synergistic framework. A whole process of profound transformation is taking place in the way we think of construction. Although certain facts must be stated, these tools not only reduce environmental impact but also face significant challenges in their implementation and open up promising prospects for a more sustainable sector.

According to McKinsey and Company (2022), the use of cutting-edge technologies would help reduce construction emissions by up to 20% by 2030, favoring net-zero emissions targets. However, the World Bank Group (2024) warns that affordable access to the Internet and digital skills are the key to harnessing these innovations in developing countries. In the same vein, the United Nations Development Program (UNDP, 2023) notes that digitization can advance the SDGs by promoting the strengthening of cities' resilience. This analysis explores the tangible benefits, current constraints, and future opportunities provided by these technologies, as well as their relationship with the SDGs.

BIM, DT, and AI purposefully address carbon dioxide emissions in the AECO sector by enhancing the execution of project workflows. These technologies improve the precision of structural and operational systems, delivering distinct environmental improvements through the following:

- Reduced carbon footprint: Optimization of materials and construction processes through BIM and IA enables waste to be minimized and environmental impact to be reduced.
- Improved energy efficiency: DT enables real-time monitoring of energy consumption, facilitating emission reduction strategies.
- Resource optimization: AI and BIM improve the selection of sustainable materials and promote a circular economy in construction.

- Reduction of air pollutants: Digitization allows modeling scenarios and implementing strategies to mitigate the emission of fine particles and polluting gases.
- Better urban planning and design: Through advanced simulations, these technologies can predict environmental impacts and improve urban resilience to climate change.

Despite the benefits, the adoption of these technologies faces multiple challenges:

- Interoperability: The lack of unified standards limits the integration between platforms and software from different vendors.
- High initial costs: Implementation of BIM, DT, and IA requires significant investment in training, hardware, and software.
- Resistance to change: The transition from traditional methods to digital processes faces cultural and organizational barriers.
- Lack of clear regulations: In some countries, the absence of specific regulations for the use of these technologies hinders their mass adoption.
- Technology gap in developing countries: The lack of access to technological resources limits the implementation of digital solutions in regions with limited infrastructure.

While the potential benefits and challenges of BIM, DTs, and AI have already been discussed, looking to the future of digitalization in the construction sector, there is every indication that significant developments are on the horizon that will further contribute to reducing emissions and boosting sustainability, among them:

- Quantum computing will enable faster and more accurate simulations to optimize sustainable building planning.
- Further integration of AI with BIM and DT: The evolution of predictive algorithms will facilitate automated decision-making to reduce emissions.
- Development of global standards: The creation of international standards will ensure the interoperability and uniform adoption of these technologies.
- Expansion of IoT and smart sensors: Real-time data collection will improve energy efficiency and resource management in buildings.
- Sustainable financing models: New economic strategies and public policies will incentivize the adoption of these technologies in emerging markets.

Finally, digital technologies applied to construction can contribute to the achievement of the SDGs (John, 2025). The SDGs most impacted by BIM, DT, and AI include the following:

- SDG 9 (Industry, Innovation, and Infrastructure): The integration of digital technology fosters more efficient and resilient infrastructures.
- SDG 11 (Sustainable Cities and Communities): Data-driven urban planning improves air quality and the sustainability of cities.
- SDG 12 (Responsible Consumption and Production): Optimizing materials and reducing waste through BIM and AI supports the circular economy.
- SDG 13 (Climate Action): Reducing emissions in construction and monitoring real-time data contribute to effective climate change mitigation strategies.

In addition, these technologies also indirectly promote other key SDGs.

- SDG 3 (Health and Well-Being): Reducing pollutants in construction improves air quality and reduces negative impacts on public health.
- SDG 7 (Affordable and Clean Energy): Integrating renewable energy into BIM and DT models allows for more efficient energy use.
- SDG 14 (Underwater Life) and SDG 15 (Life of Terrestrial Ecosystems): Optimizing the use of materials and reducing waste contributes to minimizing environmental impacts on marine and terrestrial ecosystems.
- SDG 17 (Partnerships to Achieve the Goals): Digitization fosters global collaboration through data sharing and the development of international standards in sustainable construction.

The adoption of BIM, DT, and AI in construction not only promises to reduce emissions and optimize resources, as highlighted in benefits and perspectives, but also invites reflection on the remaining challenges, such as interoperability and the technology gap, that hold back their global reach. Beyond sustainability, these technologies improve air quality and human health by mitigating pollutants, an impact that aligns with the SDGs and, according to McKinsey and Company (2022), "could reduce emissions from the building sector by up to 20% by 2030 through strategic digitization." Thus, its effective implementation becomes essential in an increasingly urbanized and vulnerable world.

14.7 Conclusions

The integration of BIM, DT, and AI represents a significant advance toward the reduction of CO_2 emissions and other pollutants in the construction industry, as demonstrated through the cases analyzed, such as Gemeentehuis Brummen and Breathe London. These examples show that BIM and AI optimize the design and selection of sustainable materials, achieving reductions of up to 50% in embodied emissions, while DTs, by monitoring real-time data, decrease pollutants such as NO_2 and $PM_{2.5}$ in urban environments by significant percentages (27% and 31%, respectively). This synergy not only improves energy efficiency and resource management but also promotes innovative practices such as design for assembly and resilient urban planning, aligning the sector with SDGs 11 and 13 in particular.

However, the widespread adoption of these technologies faces critical challenges, including limited interoperability between platforms, high upfront costs, and the technology gap between developed and developing economies. Despite these barriers, future prospects are promising, with advances such as quantum computing and greater IoT integration that could boost the predictive capabilities and scalability of these tools. This chapter underlines that digitization is not just a technical solution but also a comprehensive strategy to link environmental sustainability with public health, reducing air pollution and its associated impacts, such as respiratory diseases. In a global context of urbanization and climate change, BIM, DT, and

AI emerge as essential pillars to transform construction into a more sustainable and resilient sector, inviting international collaboration to ensure its equitable and effective implementation.

Acknowledgments The authors sincerely thank our affiliated institutions and others who have indirectly participated in this study.

References

Abanda, F. H., Balu, B., Adukpo, S. E., & Akintola, A. (2025). Decoding ISO 19650 through process modelling for information management and stakeholder communication in BIM. *Buildings, 15*, 431. https://doi.org/10.3390/buildings15030431

AENOR. (2020). UNE-EN 15804:2012+A2: Sostenibilidad en la construcción Declaraciones ambientales de producto-Reglas de categoría de producto básicas para productos de construcción.

Afzal, M., Li, R. Y. M., Ayyub, M. F., Shoaib, M., & Bilal, M. (2023). Towards BIM-based sustainable structural design optimization: A systematic review and industry perspective. *Sustainability, 15*(20), 15117. https://doi.org/10.3390/su152015117

Akbari, S., Sheikhkhoshkar, M., Rahimian, F. P., El Haouzi, H. B., Najafi, M., & Talebi, S. (2024). Sustainability and building information modelling: Integration, research gaps, and future directions. *Automation in Construction, 163*, 105420. https://doi.org/10.1016/j.autcon.2024.105420

Arguedas Garro, C., Serrano Lanzarote, B., & Navarro Escudero, M. (2020). Co2 emissions emissions in own materials from vernacular architecture. *The International Archives of the Photogrammetry, Remote Sensing and Spatial Information Sciences, XLIV*, 195–202. https://doi.org/10.5194/isprs-archives-XLIV-M-1-2020-195-2020

Arguedas Garro, C. E. (2021). *Estudio de datos relativos a emisiones de CO2 en materiales de construcción como estrategia para la rehabilitación de fachadas con bajo impacto ambiental.* Universitat Politècnica de València. https://doi.org/10.4995/Thesis/10251/171021

Asif, M., Naeem, G., & Khalid, M. (2024). Digitalization for sustainable buildings: Technologies, applications, potential, and challenges. *Journal of Cleaner Production, 450*, 141814. https://doi.org/10.1016/j.jclepro.2024.141814

Baracho, R. M. A., da Silva Santiago, L. G., Mendoza Assumpção Silva, A. T., Magalhães Vidigal, M., Joaquim, & Franco Porto, M. (2025). Building information modeling (BIM) for planning and construction. *Architecture, Structures and Construction, 5*(6). https://doi.org/10.1007/s44150-024-00119-x

BBC. (2023). *ULEZ expansion: London emission zone "extremely unfair" for Kent drivers*. https://www.bbc.com/news/uk-england-kent-64541632

Berigüete, F. E., Santos, J. S., & Rodriguez Cantalapiedra, I. (2024). Digital revolution: Emerging technologies for enhancing citizen engagement in urban and environmental management. *Land, 13*(11), 1921. https://doi.org/10.3390/land13111921

Bragança, L., Griffiths, P., Askar, R., Salles, A., Ungureanu, V., Tsikaloudaki, K., Bajare, D., & Zsembinszki, G. (2024). *Circular economy design and management in the built environment: A critical review of the state of the art.* Springer. https://doi.org/10.1007/978-3-031-73490-8

BSI Group. (2023). *Little book of BIM*. https://www.bsigroup.com/es-ES/iso-19650/

C40 Cities. (2019). *Mayor launches world's largest air quality monitoring network in London.* https://www.c40.org/news/mayor-launches-world-s-largest-air-quality-monitoring-network-in-london/

Cota, I. (2021). *América Latina necesita digitalizar sus infraestructuras para detener el rezago económico, según la CAF. El País.* https://elpais.com/economia/2021-07-22/america-latina-necesita-digitalizar-sus-infraestructuras-para-detener-el-rezago-economico-caf.html de Bont, J., Krishna, B., Stafoggia, M., Banerjee, T., Dholakia, H., Garg, A., Ingole, V., Jaganathan, S., Kloog, I., Lane, K., Mall, R. K., Mandal, S., Nori-Sarma, A., Prabhakaran, D., Rajiva, A., Tiwari, A. S., Wei, Y., Wellenius, G. A., Schwartz, J., & Ljungman, P. (2024). Ambient air pollution and daily mortality in ten cities of India: A causal modelling study. *Lancet Planetary Health, 8*, e433–e440. https://doi.org/10.1016/S2542-5196(24)00114-1

De Los Santos Melo, A., & Beriguete Alcántara, F. E. (2024). Maximizing learning potential: Embracing the power of digital twins in architectural and construction education of the twenty-first century. In A. J. Bienvenido-Huertas, D. Hoz-Torres, & M. L. A. Aguilera (Eds.), *Teaching innovation in architecture and building engineering* (pp. 477–493). Springer. https://doi.org/10.1007/978-3-031-59644-5_27

De Miguel, R. (2023). *La contaminación en Londres se desploma en año y medio gracias a la ampliación de su zona de bajas emisiones. El País*. https://elpais.com/clima-y-medio-ambiente/2025-03-07/la-contaminacion-en-londres-se-desploma-en-ano-y-medio-gracias-a-la-ampliacion-de-su-zona-de-bajas-emisiones.html

De Wolf, C., Cordella, M., Dodd, N., Byers, B., & Donatello, S. (2023). Whole life cycle environmental impact assessment of buildings: Developing software tool and database support for the EU framework level(s). *Resources, Conservation and Recycling, 188*, 106642. https://doi.org/10.1016/j.resconrec.2022.106642

Eastman, C., Teicholz, P., Sacks, R., & Liston, K. (2011). *BIM handbook: A guide to building information modeling for owners, managers, designers, engineers and contractors*. John Wiley & Sons Ltd.

EEA. (2022). *Building renovation: Where circular economy and climate meet*. https://www.eea.europa.eu/publications/building-renovation-where-circular-economy/building-renovation-where-circular-economy

Elias Bibri, S., Huang, J., Omar, O., & Kenawy, I. (2025). Synergistic integration of digital twins and zero energy buildings for climate change mitigation in sustainable smart cities: A systematic review and novel framework. *Energy and Buildings, 333*, 115484. https://doi.org/10.1016/j.enbuild.2025.115484

Environmental Defense Fund. (2020). *Hyperlocal monitoring*. https://globalcleanair.org/innovative-air-quality-monitoring/hyperlocal-monitoring/

eSMARTCITY. (2025). *El 97% de las regiones europeas han mejorado la calidad del aire en los últimos años*. https://www.esmartcity.es/2025/03/04/97-regiones-europeas-han-mejorado-calidad-aire-ultimos-anos?utm

Figl, H., Ilg, M., & Battisti, K. (2019). 6D BIM—terminal: Missing link for the design of CO2-neutral buildings. *IOP Conference Series Earth and Environmental Science, 323*, 012104. https://doi.org/10.1088/1755-1315/323/1/012104

Global Carbon Project. (2023). *CO_2 emissions by country*. Global Carbon Atlas. https://www.globalcarbonatlas.org/en/CO2-emissions

IEA. (2022). *Global energy review: CO2 emissions in 2021*. https://www.iea.org/reports/global-energy-review-co2-emissions-in-2021-2

Imperial College London. (2024). *Breathe London*. https://www.breathelondon.org/

IPCC. (2008). Emissions of long-lived GHGs. In *Climate change 2007: Synthesis report*. Intergovernmental Panel on Climate Change. https://www.ipcc.ch/site/assets/uploads/2018/02/ar4_syr_full_report.pdf

ISO. (2019). *Better building with new International Standards for BIM*. https://www.iso.org/news/ref2364.html

Ittner, C., Fonseca, E., & Douglas, F. (2021). *The Breathe London Blueprint: How cities can use hyperlocal air pollution monitoring to support their clean air goals*. https://globalcleanair.org/blueprint/

IVE. (2024). *Base de datos de construcción 2024*. https://bdc.f-ive.es/BDC24/1

John, M. (2025). *The Routledge handbook of global sustainability education and thinking for the 21st century* (1st ed.). Routledge. https://doi.org/10.4324/9781003171577

Jones, M. W., Peters, G. P., Gasser, T., Andrew, R. M., Schwingshackl, C., Gütschow, J., Houghton, R. A., Friedlingstein, P., Pongratz, J., & Le Quéré, C. (2024). Annual greenhouse gas emissions including land use [dataset]. In *National contributions to climate change 2024.2. Our world in data*. https://doi.org/10.5281/zenodo.14054503

Kolosz, B., & Grant-Muller, S. (2015). Extending cost—benefit analysis for the sustainability impact of inter-urban intelligent transport systems. *Environmental Impact Assessment Review, 50*, 167–177. https://doi.org/10.1016/j.eiar.2014.10.006

Li, J., Liu, Z., Han, G., & Demian, P. (2024). The relationship between artificial intelligence (AI) and building information Modeling (BIM) technologies for sustainable building in the context of smart cities. *Sustainability, 16*(24), 10848. https://doi.org/10.3390/su162410848

Lu, H. R., El Hanandeh, A., & Gilbert, B. P. (2017). A comparative life cycle study of alternative materials for Australian multi-storey apartment building frame constructions: Environmental and economic perspective. *Journal of Cleaner Production, 166*, 458–473. https://doi.org/10.1016/j.jclepro.2017.08.065

Matei, A., & Cocoşatu, M. (2024). Artificial internet of things, sensor-based digital twin urban computing vision algorithms, and blockchain cloud networks in sustainable smart city administration. *Sustainability, 16*(16), 6749. https://doi.org/10.3390/su16166749

Mayor of London. (2019). *ULEZ reduces 13,500 cars daily & cuts toxic air pollution by a third*. https://www.london.gov.uk/press-releases/mayoral/ulez-reduces-polluting-cars-by-13500-every-day.

McKinsey & Company. (2022). *The net-zero transition: What it would cost, what it could bring*. https://www.mckinsey.com/capabilities/sustainability/our-insights/the-net-zero-transition-what-it-would-cost-what-it-could-bring

Mercader Moyano, M. P., Camporeale, P. E., & Cózar-Cózar, E. (2019). Evaluación de impacto ambiental mediante la introducción de indicadores a un modelo bim de vivienda social. *Revista Hábitat Sustentable, 9*(2), 78–93. https://doi.org/10.22320/07190700.2019.09.02.07

Myhre, G., Shindell, D., Bréon, F.-M., Collins, W., Fuglestvedt, J., Huang, J., Koch, D., Lamarque, J.-F., Lee, D., Mendoza, B., Nakajima, T., Robock, A., Stephens, G., Takemura, T., & Zhang, H. (2013). Anthropogenic and natural radiative forcing. In T. F. Stocker, D. Qin, G.-K. Plattner, M. Tignor, S. K. Allen, J. Boschung, A. Nauels, Y. Xia, V. Bex, & P. M. Midgley (Eds.), *Climate change 2013: The physical science basis. Contribution of Working Group I to the Fifth Assessment Report of the Intergovernmental Panel on Climate Change* (pp. 659–740). Cambridge University Press.

ONU-CEPAL. (2024). *CEPAL advierte sobre el rezago tecnológico en América Latina y el Caribe y resalta la transformación digital como clave para superar las trampas del desarrollo*. https://www.cepal.org/es/notas/cepal-advierte-rezago-tecnologico-america-latina-caribe-resalta-la-transformacion-digital-como?utm

Pan, Y., & Zhang, L. (2023). Integrating BIM and AI for smart construction management: Current status and future directions. In *Archives of computational methods in engineering* (Vol. 30). Springer. https://doi.org/10.1007/s11831-022-09830-8

Płoszaj-Mazurek, M., & Ryńska, E. (2024). Artificial intelligence and digital tools for assisting low-carbon architectural design: Merging the use of machine learning, large language models, and building information. *Energies, 17*(12), 2997. https://doi.org/10.3390/en17122997

Rastogi, U., Kumar, S., Maurya, S. P., & Yadav, A. K. (2025). Air pollution forecasting using genetic algorithm and LSTM deep learning technique. *Aerosol Science and Engineering*. https://doi.org/10.1007/s41810-025-00343-y

RAU Architects. (2016). *Gemeentehuis Brummen*. https://www.rau.eu/portfolio/gemeentehuis-brummen/

Robles, J., Martín, C., & Díaz, M. (2023). Computers in industry OpenTwins: An open-source framework for the development of next-gen compositional digital twins. *Computers in Industry, 152*, 104007. https://doi.org/10.1016/j.compind.2023.104007

Sturman, A., & Quénol, H. (2023). *Climate change: Impacts and adaptation at regional and local* (1st ed.). OUP Oxford.

Telemetro. (2013). *ONU revisa proyección de población mundial: 10.900 millones en 2100.* https://www.telemetro.com/internacionales/2013/06/13/onu-proyeccion-poblacion-mundial-millones/1862203.html?utm

UNDP. (2023). *Three ways digital transformation accelerates sustainable and inclusive development.* https://www.undp.org/blog/three-ways-digital-transformation-accelerates-sustainable-and-inclusive-development

UNEP. (2022). *2022 global status report for buildings and construction. Towards a zero-emissions, efficient and resilient buildings and construction sector.* https://www.unep.org/resources/publication/2022-global-status-report-buildings-and-construction

UNICEF. (2022). *Air pollution: A threat to children's rights. How air pollution impacts children's lives and the action we need.* https://www.unicef.org/stories/air-pollution-threat-childrens-rights

United Nations. (2025). *2022 energy balances.* United Nations.

WHO. (2023). *WHO ambiente air quality database, 2022 update: status report.* https://iris.who.int/bitstream/handle/10665/368432/9789240047693-eng.pdf?sequence=1

World Bank Group. (2024). *Digital progress and trends report 2023.* https://doi.org/10.1596/978-1-4648-2049-6

Zhang, Y., Koon Teoh, B., Wu, M., Chen, J., & Zhang, L. (2023). Data-driven estimation of building energy consumption and GHG emissions using explainable artificial intelligence. *Energy, 262*, 125468. https://doi.org/10.1016/j.energy.2022.125468

Chapter 15
Investigating the Relationship Between Urbanization and Air Pollution: New Insights From Developed Countries

Ali Gökhan Gölçek, Fındık Özlem Alper, and Ali Eren Alper

15.1 Introduction

Urbanization constitutes a significant force behind economic and environmental transformations in modern societies. Currently, more than 57% of the global population lives in urban settlements, with ratios greater than 80% in developed nations (World Bank, 2025), which testifies to its far-reaching effects. Although urbanization is traditionally linked to economic growth and better living standards, its impact on the environment and sustainability remains controversial. Air pollution, for example, is particularly a pressing issue in heavily urbanized environments, with significant contributions to public health as well as the environment. Critical pollutants such as carbon dioxide (CO_2) and $PM_{2.5}$ are the main environmental and health indicators and represent concerns related to global warming and respiratory illness. CO_2 is a gas that occurs naturally due to the combustion of fossil fuels. CO_2, one of the main gases contributing to atmospheric heat retention, plays a significant role in climate dynamics (Florides & Christodoulides, 2009). Another pollutant indicator used in the study is $PM_{2.5}$. $PM_{2.5}$ consists of airborne particles smaller than 2.5 μm, which are known to originate from various industrial and vehicular sources (Tucker, 2000; Yadav et al., 2023). Although numerous variables influence the chemical makeup of $PM_{2.5}$, it is mainly composed of inorganic elements such as sulfate, nitrate, and ammonium, in addition to a variety of organic substances (Westervelt et al., 2016). Therefore, deciphering the multifaceted nexus between air pollution and urbanization is invaluable in the successful pursuit of sustainable development targets.

A. G. Gölçek (✉) · A. E. Alper
Department of Public Finance, Niğde Ömer Halisdemir University, Niğde, Turkey
e-mail: aligokhangolcek@ohu.edu.tr

F. Ö. Alper
Department of Economics, Niğde Ömer Halisdemir University, Niğde, Turkey

A. K. Yadav (ed.), *Computational Approach for Air Pollution*,
https://doi.org/10.1007/978-3-032-09215-1_15

Developed countries, whose high incomes are matched by sophisticated technology platforms and regulatory requirements for protecting the environment, provide a unique environment in which to investigate the environmental effects of urbanization. However, in these nations, the influence of urban expansion on pollution levels is often outweighed by the forces of economic growth and rising consumption. The G-7 countries have a large share of the world's emissions and are leaders in environmental policy. Investigating the systematic impact of urbanization on pollutants such as CO_2 and $PM_{2.5}$ in the G-7 countries can therefore be useful in informing environmental policy at local and global levels. The possibility that urbanization might exacerbate or reduce air pollution is the primary motivation for this study. This research is guided by several key inquiries, including (1) In what ways does urbanization influence CO_2 emissions and $PM_{2.5}$ pollution in G-7 countries? (2) How is this relationship shaped by factors such as economic prosperity and global integration? (3) Does the environmental impact of urbanization vary according to the socioeconomic structures of individual countries? These questions aim to unravel the multidimensional nature of urbanization's link to air pollution. Based on these inquiries, the study proposes the following hypotheses:

(H1) Urbanization reduces air pollution in developed countries by enhancing access to clean energy sources and public transportation systems.
(H2) Higher income levels increase air pollution by encouraging energy-intensive consumption and individual resource use.
(H3) Globalization exerts a mitigating effect on air pollution through the diffusion of technological innovations and international cooperation.

These hypotheses are grounded in the urban and economic dynamics of advanced economies.

To the best of our knowledge, panel data methods have not yet been extensively applied to analyze how urbanization affects CO_2 and $PM_{2.5}$ levels over the long run within a consistent group of developed nations like the G-7. This gap represents a significant limitation in understanding the environmental consequences and underscores the need for empirical evidence to guide policymakers. Existing research has largely focused on heterogeneous country groups or developing economies, often overlooking the distinct conditions of advanced nations. By concentrating on the G-7, this study aims to address this gap through a dual analysis of panel-wide and country-specific effects, examining the impact on air pollution. Additionally, modeling CO_2 and $PM_{2.5}$ separately allows the complex structure of air pollution to unravel.

The present analysis offers a novel lens on the environmental consequences of urbanization, proposing a framework that addresses not only its economic ramifications but also its ecological dimensions. In practical terms, with the high contribution rate to global emissions and leadership in sustainable development policy, this study could provide tangible recommendations for urban design and environmental policy. The impact on air pollution reaches beyond environmental needs, involving other broader issues such as public health, energy policy, and international cooperation. Therefore, the outcomes are expected to have implications on decision-making both nationally and internationally. The study examines the effects arising from urban growth on air pollution with a dataset covering 1990–2022 across the G-7 economies, employing a panel data approach. The results are evaluated both at the panel level and country-by-country, shedding light on the heterogeneous nature of the environmental effects. By exploring this relationship, the study positions itself as a vital step toward shaping a sustainable future.

15.2 Literature

Urban air pollution has become an increasing environmental issue in both developing and developed nations. Various theoretical frameworks have been proposed to conceptualize the association between urbanization and environmental quality (Chatti & Majeed, 2022; Yadav & Jamal, 2018). Among the leading theoretical approaches, ecological modernization theory interprets urbanization as an indicator of social progress, encompassing not only economic development but also institutional and structural transformations aimed at reducing environmental harm. The theory implies that environmental pressures intensify during the early and middle stages of development, particularly when income growth is prioritized over environmental safeguards. However, as sustainability awareness grows, societies can decouple environmental harm from economic expansion by adopting cleaner technologies and transitioning toward knowledge- and service-based economies. Accordingly, the theory outlines a non-linear trajectory in which environmental degradation tends to rise during early development phases, followed by a decline linked to greater technological adoption and heightened environmental awareness. The second theoretical approach, urban environmental transition theory, maintains that environmental issues arise unavoidably as urbanization advances. This theory asserts that the nature of environmental problems is a function of the level of development. In low-income environments, resource limitations pose challenges, including inadequate access to safe drinking water and insufficient sanitation infrastructure. As economies expand, these issues diminish, leaving room for industrial emissions and pollution. In wealthier urban centers, these are mitigated through policy reforms and technological progress, yet new challenges emerge due to high consumption levels. The more resource-intensive lifestyles, transport systems, and infrastructure of wealthier cities create greater environmental stress, especially in terms of carbon emissions and energy demand.

A third approach, complete city theory, emphasizes spatial planning. It suggests that high-density urban settlement can produce economies of scale in public utilities and transport, which can minimize environmental pressure. Compact cities that are well-planned are seen as a practical response to unchecked urban sprawl. Based on these conceptual foundations, empirical studies that examine how urban development interacts with environmental outcomes generally fall into four categories. First, several studies argue that urbanization worsens environmental quality by intensifying demands on infrastructure, energy, and transportation, thus disrupting ecological balance and complicating waste management. Second, a group of studies finds that urban growth can lead to environmental benefits by encouraging cleaner energy use and more efficient resource management. Third, a number of studies describe a nonlinear relationship, showing that environmental impacts of urbanization vary with time and context. Fourth, some research does not establish any clear relationship, citing inconsistencies in methodologies, countries studied, and time frames (Ahmed et al., 2019; Chatti & Majeed, 2022).

A significant portion of the empirical literature associates the growth of urban populations with increased emissions of CO_2 and $PM_{2.5}$, supporting the view that urbanization contributes to environmental deterioration. For instance, Poumanyvong and Kaneko (2010) explored data from 99 countries with different income levels covering 1975–2005; Hanif (2018) focused on sub-Saharan Africa from 1995–2015; Doğan et al. (2019) investigated selected high- and low-income countries from 1971 to 2014; Khoshnevis Yazdi and Dariani (2019) analyzed 18 Asian countries between 1980 and 2014. Other examples include Majeed and Tauqir (2020) in 156 countries (1990–2014); Aslan et al. (2021) on Türkiye (1960–2015); Chatti and Majeed (2022) on 94 countries (1998–2016); and studies by Taiwo-Onifade et al. (2021), Udemba and Keleş (2022), Sahoo and Sethi (2022), Abdi (2023), Warsame et al. (2023), Nuţă et al. (2024), and Wang et al. (2024). Taken together, these studies highlight a robust and direct association that links urbanization with increased environmental strain.

In contrast, another group of studies identifies a beneficial effect of urbanization on environmental quality. Saidi and Mbarek (2017) evaluated 19 developing countries (1990–2013); Xu et al. (2019) examined China (2005–2015); Lv and Xu (2019) covered 55 middle-income nations; Chen et al. (2020) assessed six ASEAN countries (1995–2020). Several other investigations suggest that the impact of urbanization on pollution follows a nonlinear path. Bekhet and Othman (2017) found this pattern in Malaysia (1971–2015); Irfan and Shaw (2017) in South Asia (1978–2011); Ahmed et al. (2019) in Indonesia (1971–2014); and Dong et al. (2020) across 126 nations (1990–2016). More recently, Kocoglu et al. (2022) studied 15 developing countries (1995–2015) and Talib et al. (2022) assessed 66 countries with varied income brackets (1990–2016), identifying inverted U-shaped relationships. A smaller set of studies did not find a significant or consistent association. Sadorsky (2014) reviewed 16 developing economies (1971–2009); Shaheen et al. (2019) focused on Pakistan (1972–2014); and Liu et al. (2023) analyzed China (1995–2020). Although CO_2 emissions are commonly used as a benchmark for environmental degradation, some scholars propose using the ecological footprint instead. However, even with this broader measure, the findings remain inconclusive. For example, Wang and Dong (2019) studied 14 Sub-Saharan nations (1990–2014); Nathaniel (2021) reviewed Indonesia (1970–2014); and Taiwo-Onifade et al. (2021) examined the E7 (1970–2016). Other relevant contributions include Rafique et al. (2022) on highly industrialized nations (1980–2017); Mehmood (2024) on G-11 countries (1990–2020); Danish et al. (2020) on BRICS (1992–2016); Ansari et al. (2021) on 22 high-renewable countries (1991–2016); and Ullah et al. (2023) on Türkiye

(1970–2018). Some of these studies link urbanization with higher ecological footprints, while others find the opposite.

15.3 Dataset, Model, and Methodology

15.3.1 Dataset

This study investigates how urban population levels, per capita income, and globalization affect CO_2 emissions and $PM_{2.5}$ concentrations in G-7 countries using annual data from 1990 to 2022. Two separate models are utilized to explore these relationships. Table 15.1 describes the descriptions and data sources used in the analysis.

15.3.2 Analytical Framework for Panel Estimation

The panel data methodology integrates both cross-sectional and time-series dimensions, enhancing estimation reliability by reducing problems such as multicollinearity and omitted variable bias (Hsiao, 2022). The econometric specifications employed in this study are grounded in widely accepted panel data methodologies, as demonstrated in prior works by Wang et al. (2021), Musah et al. (2021), and Ansari et al. (2022):

$$\ln CO2_{i,t} = \beta_0 + \beta_1 \ln POP_{i,t} + \beta_2 \ln Y_{i,t} + \beta_3 \ln GLB_{i,t} + \varepsilon_{i,t} \tag{15.1}$$

$$\ln PM2.5_{i,t} = \beta_0 + \beta_1 \ln POP_{i,t} + \beta_2 \ln Y_{i,t} + \beta_3 \ln GLB_{i,t} + \varepsilon_{i,t} \tag{15.2}$$

Table 15.1 Descriptive statistics

Variable	Definition	Source	Mean	Std. Dev.	Min.	Max.
$\ln CO_2$	Million tons of carbon dioxide	World Bank, World Development Indicators	6.663	0.883	5.527	8.680
$\ln PM_{2.5}$	Mean annual exposure (micrograms per cubic meter)	World Bank, World Development Indicators	2.609	0.337	1.828	3.369
lnPOP	Urban population (% of total population)	World Bank, World Development Indicators	4.356	0.0726	4.200	4.521
lnY	GDP per capita (constant 2015 US Dollar)	World Bank, World Development Indicators	10.527	0.188	10.226	11.071
lnGLB	Total (economic, social, political) globalization	KOF globalization index	4.367	0.086	4.023	4.481

In these equations, *i* and *t* represent the specific cross-sectional unit and time indices, as commonly adopted in panel data literature (Peng et al., 2021). Each variable included in the analysis is logarithmically transformed to normalize its distributions and reduce heteroskedasticity.

Model 1 focuses on CO_2 emissions (in million tons) as the primary dependent variable, while in Model 2, it is $PM_{2.5}$ concentration (measured in micrograms per cubic meter). These represent air quality indicators where lower values denote improved conditions.

The central explanatory variable in both models is the share of people residing in cities relative to the total population. This variable is expected to be negatively associated with pollutant levels, as urban concentration can facilitate infrastructure improvements and the adoption of cleaner technologies (Liu et al., 2017, 2021; Ribeiro et al., 2019). The models also include income per person (adjusted to 2015 USD) along with an index measuring globalization, which functions as the role of control variables.

15.3.3 Econometric Approach

A four-stage methodology is followed: testing for cross-sectional dependence (CSD), unit root testing for stationarity, panel cointegration testing, and estimation of long-run coefficients.

15.3.3.1 Cross-Sectional Dependency

Panel datasets are often influenced by shocks that propagate between units. To assess such dependencies, this study uses the Breusch-Pagan LM, Pesaran, Friedman, and Frees tests. These tests help determine whether disturbances affecting one country also affect others in the panel.

15.3.3.2 Stationarity Testing

The stationarity of the variables is verified by Pesaran (2007) using the cross-sectional augmented Dickey-Fuller test. This method incorporates cross-sectional means and their first differences to handle interdependencies across units. The test regression is

$$\Delta Y_{it} = \alpha_i + \rho_i^* Y_{it-1} + d_0 \overline{Y}_{t-1} + d_1 \Delta \overline{Y}_t + \varepsilon_{it} \tag{15.3}$$

The panel-level CIPS statistic is calculated as

$$CIPS = \frac{1}{N}\sum_{i=1}^{N} CADF_i \tag{15.4}$$

15.3.3.3 Panel Cointegration Testing

Following the approach of Gengenbach et al. (2016) (GUW), this study uses the GUW test for panel cointegration. This error correction-based approach is appropriate under CSD and heterogeneity and is formulated as follows.

The following equation is derived from Gengenbach et al. (2016), capturing the error-correction representation under CSD:

$$\Delta y_i = d\delta_{y,xi} + \alpha_{yi} y_{i,-1} + \omega_{i,-1}\gamma_i + \upsilon_i \pi_i + \varepsilon_{y,xi} = \alpha_{yi} y_{i,-1} + g_i^d \lambda_i + \varepsilon_{y,xi} \tag{15.5}$$

15.3.3.4 Long-Run Coefficient Estimation

To estimate long-term parameters, the study adopts the Common Correlated Effects estimator, originally introduced by Pesaran (2006). The approach filters out latent global shocks by incorporating cross-sectional averages of the variables. Results are aggregated using the Common Correlated Effects Mean Group (CCEMG) estimator.

This econometric framework allows the assessment to reflect both global interlinkages and national-level dynamics when examining the extended effects of urbanization on air quality.

15.4 Findings

15.4.1 Results of CSD and Unit Root Test

Table 15.2 reports the results of the CSD tests conducted for all variables in the panel. Data show moderate variation in CO_2 emissions and $PM_{2.5}$ concentrations in the G-7 countries. Urban population shares and globalization indices exhibit relatively consistent values, while income per capita shows a wider dispersion.

The results of the stationarity test, derived using Pesaran (2007), can be found in Table 15.3. According to the findings, none of the variables exhibit stationarity in their level forms; however, they attain stationary properties once transformed into first differences. Thus, it is concluded that all model variables exhibit first-order integration and are represented by I(1).

The results of the GUW cointegration test are presented in Table 15.4. The findings indicate evidence supporting a long-term relationship between the variables included in the analysis.

Table 15.2 CSD test results

Tests	Test statistics	Probability	Result
Model 1			
Breusch-Pagan Lagrange CSD TEST	230.666	0.000	Cross-sectionally depended
Pesaran CSD test	7.424	0.000	Cross-sectionally depended
Friedman CSD test	81.262	0.000	Cross-sectionally depended
Frees CSD test	1.998	0.000	Cross-sectionally depended
Model 2			
Breusch-Pagan Lagrange CSD test	333.858	0.000	Cross-sectionally depended
Pesaran CSD test	9.328	0.000	Cross-sectionally depended
Friedman CSD test	94.521	0.000	Cross-sectionally depended
Frees CSD test	2.514	0.000	Cross-sectionally depended

Table 15.3 Stationarity test results

Variable	t-bar	Probability	Result
At level			
InCO_2	−2.229	0.103	Non-stationary
InPM2.5	−1.603	0.691	Non-stationary
InPOP	−0.645	0.999	Non-stationary
InY	−1.389	0.865	Non-stationary
InGLB	−2.291	0.075	Non-stationary
First difference			
InCO_2	−4.370	0.000	Stationary
InPM2.5	−3.485	0.000	Stationary
InPOP	−3.004	0.000	Stationary
InY	−3.827	0.000	Stationary
InGLB	−3.822	0.000	Stationary

Table 15.4 Cointegration test results

d.y	Coefficient	T-bar statistic	p-value	Result
Model 1				
y(*t*–1)	−0.721	−3.442	≤0.01	Cointegrated
Model 2				
y(*t*–1)	−0.867	−3.357	≤0.01	Cointegrated

After establishing the cointegrated relationship, the long-run coefficients for both models are estimated using the CCEMG approach, with results presented in Tables 15.5 and 15.6.

For Model 1, the long-term coefficients suggest that, throughout the panel, CO_2 emissions show a negative correlation with the proportion of the population located in urban settlements and the globalization index but are positively related to per

Table 15.5 Long-run estimations (Model 1)

Panel (a): country-specific results	lnPOP	lnY	lnGLB
United States	1.646* (0.044)	0.545* (0.040)	0.263 (0.741)
United Kingdom	−2.745* (0.003)	−0.864* (0.036)	−0.531* (0.040)
Canada	−1.074* (0.001)	1.130* (0.007)	0.993 (0.184)
France	7.774** (0.054)	2.439* (0.000)	−1.119** (0.064)
Germany	−8.344* (0.004)	0.699* (0.004)	−1.518* (0.032)
Italy	0.065** (0.076)	1.528* (0.000)	−1.059 (0.265)
Japan	−1.410* (0.023)	0.772* (0.002)	−0.070 (0.821)
Panel (b): group results	−0.584* (0.050)	0.893* (0.019)	−0.434** (0.094)

Note: * and ** denote significance at 5% and 10% levels, respectively

Table 15.6 Long-run estimations (Model 2)

Panel (a): country-specific results	lnPOP	lnY	lnGLB
United States	2.497 (0.392)	1.161* (0.042)	−0.407 (0.715)
United Kingdom	−1.096** (0.054)	−0.621 (0.272)	0.489 (0.671)
Canada	−3.389** (0.096)	0.276 (0.671)	−0.225 (0.823)
France	−6.557** (0.068)	1.401 (0.122)	−1.063 (0.418)
Germany	8.467* (0.043)	0.120* (0.021)	0.048 (0.960)
Italy	−0.648 (0.704)	0.130* (0.008)	0.327 (0.752)
Japan	−3.147* (0.000)	0.639** (0.061)	1.139* (0.008)
Panel (b): group results	−0.553** (0.063)	0.444** (0.087)	0.044 (0.869)

Note: * and ** denote significance at 5% and 10% levels, respectively

capita income. Specifically, when the urbanization and globalization indices rise by 1%, CO_2 emissions decrease by approximately 0.584% and 0.434%, respectively. On the contrary, an increase of 1% in income per person is linked to an estimated 0.893% increase in CO_2 emissions. In Model 1, the country-level results indicate that CO_2 emissions tend to decrease with increasing urban population in the United Kingdom, Canada, Germany, and Japan, while an opposite pattern—namely, a positive association—is observed in the United States, France, and Italy. For Model 2, the long-run coefficients reveal that, across the panel, $PM_{2.5}$ concentrations are negatively related to the proportion of the population residing in urban areas, but positively related to per capita income. There is no statistically significant relationship between $PM_{2.5}$ and the globalization index. To be precise, a 1% increase in urban population is associated with a 0.553% decline in $PM_{2.5}$ levels, whereas a similar increase in per capita income corresponds to a 0.444% increase in $PM_{2.5}$ concentration.

In Model 2, the country-specific analysis shows that the relationship between urban population and $PM_{2.5}$ is negative in the United Kingdom, Canada, France, and Japan but positive in Germany. There is no significant statistical relationship in the United States and Italy.

15.5 Conclusions

At the panel level, the analysis reveals that urbanization has a mitigating effect on both CO_2 emissions and $PM_{2.5}$ concentrations, while rising income levels contribute positively to pollution. Globalization appears to lower CO_2 emissions while exhibiting no statistically significant effect on $PM_{2.5}$. Country-specific results reveal notable heterogeneity: Urbanization decreases air pollution in the United Kingdom, Canada, and Japan, but increases it in the United States and France. These findings partially support the theoretical frameworks and hypotheses in the literature, while also presenting unexpected outcomes. The observation that urbanization generally reduces CO_2 and $PM_{2.5}$ levels aligns with Hypothesis *H1* and is consistent with ecological modernization theory. In developed countries, urbanization can facilitate access to clean energy sources and leverage economies of scale, such as public transportation, to explain this result. In countries such as Japan and the United Kingdom, advanced urban planning and investments in renewable energy are likely to underpin this effect. However, the increase in air pollution linked to urbanization in the United States and France is an unexpected result. In the United States, widespread suburban sprawl and extensive dependence on private cars, and in France, the continuing role of fossil fuels in energy production, could possibly account for this divergence. This would suggest that compact city theory is not a universal truth, with urban form having varying effects under varying local conditions.

The finding that per capita income increases air pollution corroborates Hypothesis H2 and aligns with urban environmental transition theory. In upper-income economies, rising income levels trigger energy-driven consumption and greater individual consumption of resources, which increase CO_2 and $PM_{2.5}$ emissions. This effect is particularly pronounced in countries such as Canada and Japan. The result that globalization reduces CO_2 emissions but does not have a significant effect on $PM_{2.5}$ partially confirms Hypothesis *H3*. The role in promoting technological diffusion and supporting international environmental agreements may contribute to CO_2 reductions; however, $PM_{2.5}$, being a more localized pollutant, appears less influenced by global processes. The heterogeneity of the findings underscores that the environmental impacts of urbanization depend on the institutional frameworks, energy policies, and urban planning approaches of the countries. For instance, the positive relationship between urbanization and $PM_{2.5}$ in Germany may reflect the persistent influence of industrial activity and coal use. On the contrary, the negative effect on both pollutants in the United Kingdom can be attributed to stringent environmental regulations and a shift toward a service-based economy. These variations provide valuable information on the reasons behind the contradictory findings in the literature. From a policy perspective, this study highlights several key points. First, in countries where urbanization reduces air pollution (e.g., Japan and the United Kingdom), investments in public transportation and renewable energy should be encouraged. Second, in nations such as the United States and France, where urbanization increases pollution, compact city designs to curb suburban sprawl and carbon-neutral transportation systems should be prioritized. Third, given the effect of per capita income that improves pollution, energy efficiency and low-carbon consumption habits should be promoted in high-income societies. Finally, while the positive impact on CO_2 can be strengthened through international cooperation and technology transfer, the treatment of local pollutants like $PM_{2.5}$ requires more targeted national policies.

Ultimately, the findings highlight that, while urban growth in developed countries offers potential for environmental improvement, it also poses notable risks to air quality. The expected result (urbanization generally reducing pollution) and unexpected findings (country-specific heterogeneity) fill gaps in the literature, offering a fresh perspective on sustainable urban development. Future research could investigate this heterogeneity through case studies that examine urban design at a micro level. Additionally, interdisciplinary studies that incorporate the health impacts of air pollution could further enrich the practical implications of these findings. This study aims to contribute to shaping policies that guide the G-7 countries toward their environmental sustainability goals.

Acknowledgments The authors received no specific funding for this research and have no conflicts of interest to declare. We would like to thank the editorial team for their guidance during the review process.

References

Abdi, A. H. (2023). Toward a sustainable development in sub-Saharan Africa: Do economic complexity and renewable energy improve environmental quality? *Environmental Science and Pollution Research, 30*(19), 55782–55798. https://doi.org/10.1007/s11356-023-26364-z

Ahmed, Z., Wang, Z., & Ali, S. (2019). Investigating the non-linear relationship between urbanization and CO_2 emissions: An empirical analysis. *Air Quality, Atmosphere & Health, 12*(8), 945–953. https://doi.org/10.1007/s11869-019-00711-x

Ansari, M. A., Akram, V., & Haider, S. (2022). A link between productivity, globalisation and carbon emissions: Evidence from emissions by coal, oil and gas. *Environmental Science and Pollution Research, 29*(22), 33826–33843. https://doi.org/10.1007/s11356-022-18557-9

Ansari, M. A., Haider, S., & Masood, T. (2021). Do renewable energy and globalization enhance ecological footprint: An analysis of top renewable energy countries? *Environmental Science and Pollution Research, 28*(6), 6719–6732. https://doi.org/10.1007/s11356-020-10786-0

Aslan, A., Altinoz, B., & Ozsolak, B. (2021). The link between urbanization and air pollution in Turkey: Evidence from dynamic autoregressive distributed lag simulations. *Environmental Science and Pollution Research, 28*(37), 52370–52380. https://doi.org/10.1007/s11356-021-14408-1

Bekhet, H. A., & Othman, N. S. (2017). Impact of urbanization growth on Malaysia CO_2 emissions: Evidence from the dynamic relationship. *Journal of Cleaner Production, 154*, 374–388. https://doi.org/10.1016/j.jclepro.2017.03.174

Chatti, W., & Majeed, M. T. (2022). Information communication technology (ICT), smart urbanization, and environmental quality: Evidence from a panel of developing and developed economies. *Journal of Cleaner Production, 366*, 132925. https://doi.org/10.1016/j.jclepro.2022.132925

Chen, T., Li, M., Luo, L., Deng, S., Zhou, R., & Chen, D. (2020). Simulating the effects of land urbanization on regional meteorology and air quality in Yangtze River Delta, China. *Applied Geography, 120*, 102228. https://doi.org/10.1016/j.apgeog.2020.102228

Doğan, B., Saboori, B., & Can, M. (2019). Does economic complexity matter for environmental degradation? An empirical analysis for different stages of development. *Environmental Science and Pollution Research, 26*(31), 31900–31912. https://doi.org/10.1007/s11356-019-06333-1

Dong, Q., Lin, Y., Huang, J., & Chen, Z. (2020). Has urbanization accelerated PM2.5 emissions? An empirical analysis with cross-country data. *China Economic Review, 59*, 101381. https://doi.org/10.1016/j.chieco.2019.101381

Florides, G. A., & Christodoulides, P. (2009). Global warming and carbon dioxide through sciences. *Environment International, 35*(2), 390–401. https://doi.org/10.1016/j.envint.2008.07.007

Gengenbach, C., Urbain, J., & Westerlund, J. (2016). Error correction testing in panels with common stochastic trends. *Journal of Applied Econometrics, 31*(6), 982–1004. https://doi.org/10.1002/jae.2475

Hanif, I. (2018). Impact of economic growth, nonrenewable and renewable energy consumption, and urbanization on carbon emissions in Sub-Saharan Africa. *Environmental Science and Pollution Research, 25*(15), 15057–15067. https://doi.org/10.1007/s11356-018-1753-4

Hsiao, C. (2022). *Analysis of panel data*. Cambridge University Press.

Irfan, M., & Shaw, K. (2017). Modeling the effects of energy consumption and urbanization on environmental pollution in South Asian countries: A nonparametric panel approach. *Quality & Quantity, 51*(1), 65–78. https://doi.org/10.1007/s11135-015-0294-x

Khoshnevis Yazdi, S., & Dariani, A. G. (2019). CO2 emissions, urbanisation and economic growth: Evidence from Asian countries. *Economic Research-Ekonomska Istraživanja, 32*(1), 510–530. https://doi.org/10.1080/1331677X.2018.1556107

Kocoglu, M., Awan, A., Tunc, A., & Aslan, A. (2022). The nonlinear links between urbanization and CO2 in 15 emerging countries: Evidence from unconditional quantile and threshold regression. *Environmental Science and Pollution Research, 29*(12), 18177–18188. https://doi.org/10.1007/s11356-021-16816-9

Liu, H., Wong, W. K., The Cong, P., Nassani, A. A., Haffar, M., & Abu-Rumman, A. (2023). Linkage among urbanization, energy consumption, economic growth and carbon emissions: Panel data analysis for China using ARDL model. *Fuel, 332*, 126122. https://doi.org/10.1016/j.fuel.2022.126122

Liu, J., Li, M., & Ding, Y. (2021). Econometric analysis of the impact of the urban population size on carbon dioxide (CO2) emissions in China. *Environment, Development and Sustainability, 23*(12), 18186–18203. https://doi.org/10.1007/s10668-021-01433-w

Liu, Y., Gao, C., & Lu, Y. (2017). The impact of urbanization on GHG emissions in China: The role of population density. *Journal of Cleaner Production, 157*, 299–309. https://doi.org/10.1016/j.jclepro.2017.04.138

Lv, Z., & Xu, T. (2019). Trade openness, urbanization and CO_2 emissions: Dynamic panel data analysis of middle-income countries. *The Journal of International Trade & Economic Development, 28*(3), 317–330. https://doi.org/10.1080/09638199.2018.1534878

Majeed, M. T., & Tauqir, A. (2020). Effects of urbanization, industrialization, economic growth, energy consumptipn and financial development on carbon emissions: An extended STIRPAT Model for heterogeneous incoe groups. *Pakistan Journal of Commerce and Social Sciences, 14*(3), 652–681. https://www.econstor.eu/bitstream/10419/224955/1/1735131601.pdf

Mehmood, U. (2024). Assessing the impacts of eco-innovations, economic growth, urbanization on ecological footprints in G-11: Exploring the sustainable development policy options. *Journal of the Knowledge Economy, 15*(4), 16849–16867. https://doi.org/10.1007/s13132-024-01759-6

Musah, M., Kong, Y., Mensah, I. A., Antwi, S. K., & Donkor, M. (2021). The connection between urbanization and carbon emissions: A panel evidence from West Africa. *Environment, Development and Sustainability, 23*(8), 11525–11552. https://doi.org/10.1007/s10668-020-01124-y

Nathaniel, S. P. (2021). Ecological footprint, energy use, trade, and urbanization linkage in Indonesia. *GeoJournal, 86*(5), 2057–2070. https://doi.org/10.1007/s10708-020-10175-7

Nuță, F. M., Sharafat, A., Abban, O. J., Khan, I., Irfan, M., Nuță, A. C., Dankyi, A. B., & Asghar, M. (2024). The relationship among urbanization, economic growth, renewable energy consumption, and environmental degradation: A comparative view of European and Asian emerging economies. *Gondwana Research, 128*, 325–339. https://doi.org/10.1016/j.gr.2023.10.023

Peng, B., Yu, J., & Zhu, Y. (2021). A heteroskedasticity robust test for cross-sectional correlation in a fixed effects panel data model. *Economics Letters, 201*, 109799. https://doi.org/10.1016/j.econlet.2021.109799

Pesaran, M. H. (2006). Estimation and inference in large heterogeneous panels with a multifactor error structure. *Econometrica, 74*(4), 967–1012. https://doi.org/10.1111/j.1468-0262.2006.00692.x

Pesaran, M. H. (2007). A simple panel unit root test in the presence of cross-section dependence. *Journal of Applied Econometrics, 22*(2), 265–312. https://doi.org/10.1002/jae.951

Poumanyvong, P., & Kaneko, S. (2010). Does urbanization lead to less energy use and lower CO_2 emissions? A cross-country analysis. *Ecological Economics, 70*(2), 434–444. https://doi.org/10.1016/j.ecolecon.2010.09.029

Rafique, M. Z., Nadeem, A. M., Xia, W., Ikram, M., Shoaib, H. M., & Shahzad, U. (2022). Does economic complexity matter for environmental sustainability? Using ecological footprint as an indicator. *Environment, Development and Sustainability, 24*(4), 4623–4640. https://doi.org/10.1007/s10668-021-01625-4

Ribeiro, H. V., Rybski, D., & Kropp, J. P. (2019). Effects of changing population or density on urban carbon dioxide emissions. *Nature Communications, 10*(1), 3204. https://doi.org/10.1038/s41467-019-11184-y

Sadorsky, P. (2014). The effect of urbanization on CO_2 emissions in emerging economies. *Energy Economics, 41*, 147–153. https://doi.org/10.1016/j.eneco.2013.11.007

Sahoo, M., & Sethi, N. (2022). The dynamic impact of urbanization, structural transformation, and technological innovation on ecological footprint and $PM_{2.5}$: Evidence from newly industrialized countries. *Environment, Development and Sustainability, 24*(3), 4244–4277. https://doi.org/10.1007/s10668-021-01614-7

Saidi, K., & Ben Mbarek, M. (2017). The impact of income, trade, urbanization, and financial development on CO_2 emissions in 19 emerging economies. *Environmental Science and Pollution Research, 24*(14), 12748–12757. https://doi.org/10.1007/s11356-016-6303-3

Shaheen, A., Sheng, J., Arshad, S., Salam, S., & Hafeez, M. (2019). The dynamic linkage between income, energy consumption, urbanization and carbon emissions in Pakistan. *Polish Journal of Environmental Studies, 29*(1), 267–276. https://doi.org/10.15244/pjoes/95033

Taiwo-Onifade, S., Gyamfi, B. A., Haouas, I., & Bekun, F. V. (2021). Re-examining the roles of economic globalization and natural resources consequences on environmental degradation in E7 economies: Are human capital and urbanization essential components? *Resources Policy, 74*, 102435. https://doi.org/10.1016/j.resourpol.2021.102435

Talib, M. N. A., Hashmi, S. H., Aamir, M., & Khan, M. A. (2022). Testing non-linear effect of urbanization on environmental degradation: Cross-country evidence. *Frontiers in Environmental Science, 10*, 971394. https://doi.org/10.3389/fenvs.2022.971394

Tucker, W. G. (2000). An overview of $PM_{2.5}$ sources and control strategies. *Fuel Processing Technology, 65-66*, 379–392. https://doi.org/10.1016/S0378-3820(99)00105-8

Udemba, E. N., & Keleş, N. İ. (2022). Interactions among urbanization, industrialization and foreign direct investment (FDI) in determining the environment and sustainable development: New insight from Turkey. *Asia-Pacific Journal of Regional Science, 6*(1), 191–212. https://doi.org/10.1007/s41685-021-00214-7

Ullah, A., Tekbaş, M., & Doğan, M. (2023). The impact of economic growth, natural resources, urbanization and biocapacity on the ecological footprint: The case of Turkey. *Sustainability, 15*(17), 12855. https://doi.org/10.3390/su151712855

Wang, A., Shan, S., Ibrahim, R. L., & Omokanmi, O. J. (2024). A new look at environmental sustainability from the lens of green policies, eco-digitalization, affluence, and urbanization: Empirical insights from BRICS economies. *Energy & Environment, 35*(8), 4195–4222. https://doi.org/10.1177/0958305X231177736

Wang, J., & Dong, K. (2019). What drives environmental degradation? Evidence from 14 Sub-Saharan African countries. *Science of the Total Environment, 656*, 165–173. https://doi.org/10.1016/j.scitotenv.2018.11.354

Wang, W.-Z., Liu, L. C., Liao, H., & Wei, Y. M. (2021). Impacts of urbanization on carbon emissions: An empirical analysis from OECD countries. *Energy Policy, 151*, 112171. https://doi.org/10.1016/j.enpol.2021.112171

Warsame, A. A., Abdi, A. H., Amir, A. Y., & Azman-Saini, W. N. W. (2023). Towards sustainable environment in Somalia: The role of conflicts, urbanization, and globalization on environmental degradation and emissions. *Journal of Cleaner Production, 406*, 136856. https://doi.org/10.1016/j.jclepro.2023.136856

Westervelt, D. M., Horowitz, L. W., Naik, V., Tai, A. P. K., Fiore, A. M., & Mauzerall, D. L. (2016). Quantifying $PM_{2.5}$ -meteorology sensitivities in a global climate model. *Atmospheric Environment, 142*, 43–56. https://doi.org/10.1016/j.atmosenv.2016.07.040

World Bank. (2025). *Urban population (% of total population)*. https://data.worldbank.org/indicator/SP.URB.TOTL.IN.ZS

Xu, S.-C., Miao, Y. M., Gao, C., Long, R. Y., Chen, H., Zhao, B., & Wang, S. X. (2019). Regional differences in impacts of economic growth and urbanization on air pollutants in China based on provincial panel estimation. *Journal of Cleaner Production, 208*, 340–352. https://doi.org/10.1016/j.jclepro.2018.10.114

Yadav, A. K., & Jamal, A. (2018). Suspended particulate matter and its management system surrounding opencast coal mines. *Environmental Quality Management, 28*(2), 123–128.

Yadav, A. K., Sahoo, S. K., Kumar, A. V., Shirin, S., Jamal, A., Patra, A. C., Dubey, J. S., Thakur, V. K., Lenka, P., Singh, S., Jha, V., & Tripathi, R. M. (2023). Identifying size-segregated particulate matter ($PM_{2.5}$, PM_{10} and SPM) sources in an industrial town of India. Aerosol Science and Engineering, 7 (4), 455–473.

Chapter 16
Air Pollution and Recent Trends in Control Strategy

Pinki Biswas, Sushil Kumar, Gavendra Pandey, Kishore Chandra Swain, Sandip Debnath, Rakesh Kumar, and Sudesh Yadav

16.1 Introduction

Air pollution is the contamination of the outdoor and indoor air through toxic solids, liquids, and gaseous substances that have complex chemical species. Recently, the World Health Organization (WHO) declared that 99% of the world's population is breathing polluted air and a major population is confined to low- and middle-income countries. Globally, more than 6.7 million premature deaths were recorded due to exposure to poor indoor and outdoor air. Premature death contributes significantly (4.2 million), followed by indoor air (3.2 million), across the globe. Chemically, air pollutants are a group of gaseous (ground-level ozone, carbon monoxide, sulfur

P. Biswas · S. Kumar (✉)
Department of Environmental Studies, Siksha-Bhavana (Institute of Science), Visva-Bharati, Santiniketan, India
e-mail: sushil.kumar@visva-bharati.ac.in

G. Pandey
Institute for the Environment, The University of North Carolina at Chapel Hill, Chapel Hill, NC, USA

K. C. Swain
Department of Agricultural Engineering, Institute of Agriculture, Visva-Bharati, Sriniketan, Santiniketan, India

S. Debnath
Department of Genetics and Plant Breeding, Institute of Agriculture, Visva-Bharati, Sriniketan, Santiniketan, India

R. Kumar
Department of Environmental Sciences, University of Jammu, Jammu, India

S. Yadav
School of Environmental Sciences, Jawaharlal Nehru University, New Delhi, India

A. K. Yadav (ed.), *Computational Approach for Air Pollution*,
https://doi.org/10.1007/978-3-032-09215-1_16

dioxide, and nitrogen dioxide) and a wide range of particle matter sizes (Ofremu et al., 2024; World Health Organization, 2021). Furthermore, other toxic gases such as benzene, toluene, xylene, and polyaromatic hydrocarbons (PAHs) pose a serious concern for human health and welfare. The suspended particles (PM) in the ambient atmosphere have a wide size range (nm to μm) emitted from multiple sources with a complex chemical composition, and their particle size is determined by sources and formation mechanisms in the ambient atmosphere. PM is chemically constituted with water-soluble inorganic ions, carbonaceous matter, major, and trace elements (Gupta et al., 2023). Daily rises in air pollutant levels in the ambient atmosphere draw the attention of the scientific community to control and reduce ambient levels across the world (Arshad et al., 2024; Ofremu et al., 2024). Recently, monitoring, source reduction through a technological intervention approach, and nature-based control strategies have been widely accepted to reduce air pollution. The stringent air quality guidelines, which take care of industrial and economic feasibility, are a positive approach to control air pollution. The direct involvement of the local and regional community through a citizen science approach is a new insight into controlling air pollution (Awewomom et al., 2024; Yadav et al., 2022).

This chapter provides a detailed overview of the status of air pollution, the multiplicity of sources, and recent trends in control strategies that are being practiced to reduce indoor and ambient air pollution. Furthermore, nature-based air pollution control strategies and a citizen science approach will provide a new long-term advantage for air pollution control strategies.

16.2 Multiplicity of Air Pollution Sources

Air pollution is a multifactor phenomenon contributed by various sources at the local, regional, and global scales. The pollutants emitted in the ambient atmosphere come from multiple sources such as combustion activities, vehicular emissions, industrial emissions, resuspension of road dust, volcanic eruption, dust storm, and so on (Roy et al., 2024; Yadav, 2015; Yadav et al., 2014, 2019, 2023, 2025) (Fig. 16.1).

The contribution of anthropogenic sources is increasing with the increasing pace of urbanization and industrialization across the globe. Industries are the leading source of energy consumption and contribute to atmospheric pollutants worldwide (Oreggioni et al., 2022). The nature and type of pollutants depend on the type of industries, such as petrochemical, metallurgical, fertilizer industries, refineries, and power stations. The key pollutants are CO_2, particulate matter (PM), NOx and SOx, elemental carbon, black carbon, PAHs, benzene, dioxins, and heavy metals (Pb, Ar, and Cd). Chemical and manufacturing industries can release large amounts of toxins, PAHs, volatile organic compounds (VOCs), and hazardous pollutants into the air. CO can also be generated during the incomplete combustion process in industries. Thermal power plants are a predominant source of gaseous and PM in the

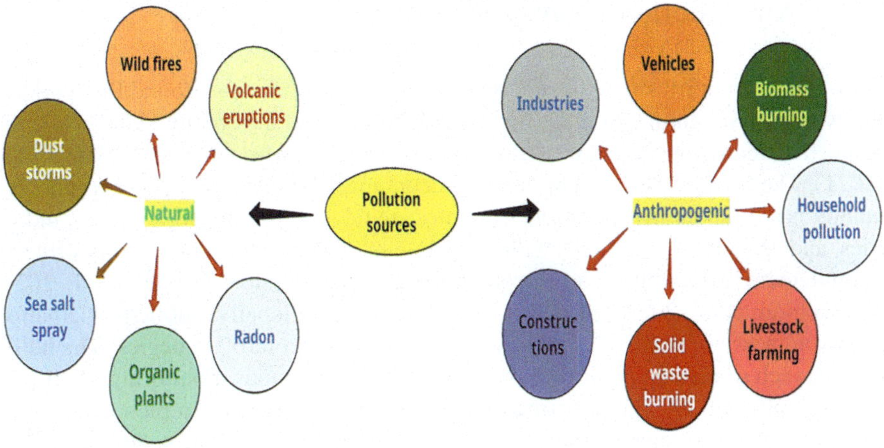

Fig. 16.1 Conceptual diagram of sources of a multiplicity of air pollution in the environment

ambient atmosphere. The transportation sector is another major source of air pollutants due to an increasing number of old vehicles, bulk engines, low fuel quality, and road conditions (Jonidi Jafari et al., 2021). NOx emission from vehicular sources also depends upon the speed limit of the vehicle, and the lowest emission lies between 70 and 90 km/h (Yadav et al., 2023). Agriculture induces activity such as sowing, plowing, harvesting, cultivating, threshing, and handling of grains emits various atmospheric pollutants. Recent studies indicate that a significant amount of ammonia (NH_3) is released into the atmosphere, which undergoes further gas-to-particle conversion and forms fine PM. The increasing application of synthetic chemical fertilizers is a potential source of NH_3 in the atmosphere in modern agricultural practices. In addition, burning crop residues is a prominent source. For instance, the Indo-Gangetic Plain over India is the potential source of various pollutants during the post-harvesting season and degrades the ambient air quality over the downwind region due to high haze pollution, tropospheric ozone generation, and severe health effects. Household air pollutants or indoor pollutants play a significant role in degrading indoor and outdoor air quality. Most of the developing countries in the South Asian region are using solid or conventional fuels for their daily cooking purposes and emit a huge amount of pollutants in an indoor environment, which results in health problems among children, the elderly, and the women population. Solid fuels are the source of various pollutants such as CO, CO_2, PM, PAHs, and so on. Other pollutants such as VOCs, benzenes, and formaldehyde are also released from building materials, paint, personal care products, air fresheners, cleaners, disinfectants, and so on. O_3 can be released from electronic devices such as printing or photocopy machines and can be present in the office environment. In addition, pollutants such as dust, asbestos, and fibers spread indoors due to any construction activity.

16.3 Air Pollution Control Strategies

Air pollution control strategies are the priority of governments around the world in the twenty-first century to meet the prescribed ambient air quality standard of the respective regulating agency of the country and the World Health Organization (2021). The holistic assessment of socio-economic loss due to exposure to poor air quality is still under investigation. The intricate relation of transport, energy, industries, and agricultural practices with air pollutant emissions provides an opportunity to implement technological advancements to control emissions (Arshad et al., 2024; Awewomom et al., 2024; Oreggioni et al., 2022). Additionally, public participation and awareness are positive approaches to air pollution control strategies. The holistic approach to managing air pollution starts with continuous monitoring of air pollutants, followed by the identification of hot spots and the implementation of stringent emission control strategies. The restriction on air pollution emission from known sources using advanced technology, followed by green belt development and community participation, is a key aspect for air pollution control strategies (Ogwu et al., 2024; Yadav et al., 2022).

16.3.1 Air Pollution Control at Sources

Industries are the key sources of PM and gaseous pollutants in the ambient atmosphere (Gupta et al., 2023; Yadav et al., 2023). The improvement of industrial infrastructure and the installation of air pollutants control devices are a preliminary approach to reducing the emission of various pollutants. The use of coal as a fuel is the source of SO_2 in the ambient atmosphere. The preliminary desulfurization of coal prior to burning, followed by the installation of a flue gas desulfurization (FGD) unit, are key steps to reduce SO_2 emission in the ambient atmosphere. Based on the requirement and level of moisture content, the FGD unit is dry or wet (WFGD) with the addition of calcium and magnesium carbonate to adsorb SO_2 prior to being released into the ambient atmosphere. A recent study reported the use of limestone, gypsum, or ammonia in FGD units in China in various power generation units. Furthermore, the use of sorbent injection in slurry or powder in upstream flue gas during the combustion activity in the fluidized bed (FB) to reduce the gaseous and PM is widely practiced. Another technique that is widely used to reduce the PM is the wet electrostatic precipitator (WESP). The use of low-NOx burners (LNBs) and selective catalytic reduction (SCR) are the two basic techniques that are widely used for the reduction of NOx emissions globally (Kim et al., 2020). The use of overfire air (OFA) technology with a combination of SCR and LNB provides better control of gaseous and PM emissions (Asif et al., 2022). Recently, a study conducted on the OFA effect of the installation of OFA on NOx emission indicates >30–40% reduction compared with LNB. The reduction of atmospheric carbon load is an ongoing challenge to achieve net-zero emissions and carbon neutrality (Table 16.1).

Table 16.1 Summary of recent studies to control and reduce pollutants through different methods

Pollutants	Methods	Efficiency	References
NOx, VOCs, CO, PM_{10}	Vehicle exhaust catalysts	NO_2 (20%), PM_{10} (10%), VOCs (30%), and CO (70%)	Jonidi Jafari et al. (2021)
NOx	SCR, low NOx burner	550–650 to 330–430 mg/Nm^3	Liu et al. (2018)
NOx	Over-fire air (OFA), oxy-fuel combustion	50%	Vodička et al. (2021)
SO_2	WFGD	95%	Grinišin et al. (2022)
CO_2	Post-combustion carbon capture adsorption	85%	Hussin and Aroua (2020)
Hg and PM	Activated carbon injection with electrostatic precipitator (ESP)	50–90% mercury and 99% PM	Wang (2020)
SO_2, NOx, Hg, PM	Activated coke	90–98% SO_2, 15–80% NOx, 90–99% Hg, 80–85% PM	Li et al. (2020)
PM	ESP	99%>	He et al. (2024)
PM	WESPs	>70% for particles 10 nm to 2.5 μm	Sadeghpour et al. (2021)
PM	Fabric filter	>99%	Kumar and Kumar (2018)

Various recent strategies to reduce the source-based carbon emissions are a priority in developed and developing countries (Awewomom et al., 2024). Three types of combustion processes, such as pre-combustion, oxyfuel combustion, and post-combustion, occur during power generation and industrial processes (Obi et al., 2025). The use of the pre-combustion process, such as integrated gasification combined cycle, is one of the advanced techniques to convert combustion products to syngas (Han et al., 2024). Furthermore, burning fuel in the oxygen-rich environment using the oxyfuel combustion technique less than 95% pure O_2 produces complete combustion as a byproduct of CO_2 with the least secondary pollutants. Fuel gases enriched with CO_2 are easily absorbed or absorbed using gas separation membranes and gas absorption membranes before being released into the ambient air. In addition, direct injection of activated carbon in flue gases significantly reduces mercury in the ambient atmosphere. The use of a multi-pollutant removal approach using activated coke and electrocatalytic oxidation is an efficient technique to reduce SO_2, NO_2, Hg, and PM in the ambient atmosphere and is widely installed by countries such as Japan and Germany (Asif et al., 2022). The vehicular emissions and transport sectors are the major contributors of atmospheric pollutants across the globe in the form of automobiles such as two-wheelers, three-wheelers, four-wheelers, light vehicles, heavy-duty vehicles, trains, and airplanes. The emission of pollutants from automobiles could be controlled through the installation of air pollution control devices, the use of good-quality fuel, or the switch to alternative fuels. The improvement of fuel quality, smooth roads, and shifting toward alternative fuels are the main strategies to reduce air pollution in the air. For instance, the complete phaseout of lower sulfur-content diesel and lead-free use of gasoline and diesel fuels are examples of reducing the atmospheric load of sulfur and lead in many cities across the world.

The use of alternative fuels such as liquefied petroleum gas or compressed natural gas (CNG) is a significant approach to reducing air pollution in ambient air. The wide use of catalytic converters or modified three-way vehicle exhaust catalytic converters is a recent technique to reduce the CO, NOx, and VOCs in ambient air. The use of diesel oxidation catalysts (DOCs) and diesel particulate filters in vehicles significantly reduces the levels of atmospheric pollutants. The use of DOCs effectively oxidizes total hydrocarbons, CO, and PM in the exhaust air. Installation of SCR in light-duty diesel vehicles significantly reduces NOx emissions in exhaust air and is widely used in Brazil, Korea, Japan, China, and the Philippines. The blending of fuel with ethanol with gasoline, CNG significantly reduces the pollutant emissions (Jonidi Jafari et al., 2021). Proper disposal of crop residue using best management practices could reduce pollutant emissions. The installation of a circulating FB gasifier and a down-draft gasifier is an efficient technique to reduce air pollutants. A rich agricultural society widely used crop residue as cooking fuel around the world due to its easy and cost-effective availability. Hence, the promotion of clean fuel and the use of crop residues for other purposes provide a sustainable solution to a clean environment. The use of crop residues in biogas production, biofertilizers, is another positive approach to air quality management.

16.3.2 Policy Framework and Legislation to Improve Air Quality

The issue of air pollution is a global threat and has severely affected human health and welfare since the preindustrial period. In the eighteenth century, atmospheric pollutant levels amplified and degraded ambient air quality under the Industrial Revolution (Ofremu et al., 2024). The sudden rise of air pollution episodes such as Los Angeles in 1943 and London smog in 1952 drew the need to make restrictions on emissions from various sources. Rising pollutant levels and their dispersion and transboundary movement during the Industrial Revolution period became a concern. To prevent exposure to air pollution, the United Kingdom (UK) launched the UK Clean Air Act (CAA) in 1956, followed by several modifications and periodic revisions. Furthermore, in 1970, the US government first established the National CAA and showed concern about air pollution. Later, the Montreal Protocol, which is an international global protocol, was established in 1987 to eradicate ozone-depleting substances like CFCs from the atmosphere. Next, in the year 1989, the Basel Convention, another international convention, was assigned to restrict the transboundary movement of hazardous air pollutants between countries. Then the Environmental Protection Agency formulated National Ambient Air Quality Standards under the CAA amendment in 1990 for the six main pollutants (CO, NO_2, SO_2, PM, Pb, and O_3), which are termed criteria pollutants are significant moves to prevent air pollution. The restriction of greenhouse gases emitted from industries and vehicles is undermined through the Kyoto Protocol, an international treaty signed in 1997 under the UN Framework Convention on Climate Change (UNFCCC). Lately, the Paris Agreement is an international global agreement that was signed in 2015 to control global warming below 1.5 °C by the suggestion of the Intergovernmental Panel on Climate Change. For instance, the Indian government formulated the Air (Prevention and Control of Pollution) Act in 1981, followed by the umbrella act or the Environment Protection Act 1986 for the betterment of the environment. The Chinese government has taken steps to improve the ambient air quality standard in 1982 and has periodically reviewed the national ambient air quality standard. The State Council of China first formulated a 5-year national Air Pollution Control and Action Plan from 2013 to 2017 to reduce the concentration of $PM_{2.5}$ and the targeted blue-sky days in urban areas (Jiang et al., 2021). In a similar manner, other countries such as Germany and Korea formulated a national electricity supply plan to reduce atmospheric pollution. Further policy formulation for sifting from traditional coal-based thermal power plants to renewable energy forward move to improve ambient air quality (Asif et al., 2022).

16.3.3 Biofiltration Is a Promising Nature-Based Solution

Natural approaches are a holistic way and a cost-effective, less maintenance, and successful eco-friendly design to reduce air pollution, as summarized in Table 16.2. It is widely admired in the current scenario to control the environmental impacts that pose the least negative impact in urban areas, cities, and megacities. Recently, the WHO has already suggested ensuring a green space per capita of 9 m^2/person. According to the European Commission, nature-based solution (NbS) can be defined as "Solutions that are inspired and supported by nature, which are cost-effective, simultaneously provide environmental, social, and economic benefits, and help build resilience" (Menon & Sharma, 2021). The International Union for Conservation of Nature to assess the effectiveness of NbS approaches globally. NbS can act as an ecosystem approach that can improve air quality, biodiversity, and human health associated with their well-being. The process of air pollution control starts by involving plants or microbes to break down pollutants and transform them into nontoxic or less-toxic components. In 1980, the National Aeronautics and Space Administration successfully established the role of plants to control the air pollution contaminants in the space shuttle (Table 16.2).

Plants are autotrophic organisms that have the ability to retain a wide range and large amounts of toxic gases, particles, and aerosols. Therefore, phytoremediation is a kind of bioremediation technology that is widely used to remove contaminants from the air. The intensive gas exchange cell and metabolic activity accumulate and absorb air pollutants and reduce the pollutants' mobility, toxicity, and volume of contaminants. The degradation of hazardous pollutants through enzymes such as catalase or dehydrogenase in urban vegetation. Lilies, microalgae, herbs, and trees

Table 16.2 Summary of recent studies to control and reduce pollutants through nature-based strategies

Pollutants	Plants	References
Formaldehyde	*Chamaedorea elegans, Nephrolepis obliterata*	Teiri et al. (2018)
PM and TVOCs, CO_2, CO	*Areca palm*	Bhargava et al. (2021)
TVOCs, PM, NOx, O_3, and BTEX	*Chamaedorea elegans, Epipremnum aureum, Ficus lyrata, Neomarica gracillis, Peperomia obtusifolia, Spathiphyllum wallisii, Syngonium podophyllum, Nephrolepis exaltata, Scheffera arboricola, Syngonium podophyllum, Spathiphyllum wallisii, Syngonium podophyllum*	Pettit et al. (2019)
VOCs, PM, HCHO, and acetone	*Spathiphyllum wallisii*	Morgan et al. (2022)
Size-segregated PM	*Nematanthus glabra, Schefflera arborcola, Nephropelis exaltata bostoniensis, Nephrolepis cordifolia, Epipremnum aureum, grass or moss, Epipremnum aureum, Chlorophytum orchidastrum, Chlorophytum comosum*	Ibrahim et al. (2021)
PM and heavy metals	*Syzigium oleina* and *Wedelia trilobata*	Rachmadiarti et al. (2019)
PM	*Nematanthus glabra, Chlorophytum comosum variegatum, philodendron Xanadu*, and *Spathiphyllum wallisii*	Paull et al. (2020)
Dust and PM	*Mangifera indica, Azadirachta indica, Ficus religiosa*	Chaudhary and Rathore (2019)

are widely used as a natural solution to mitigate the concentration of various gases, particulates, potentially toxic elements, and heavy metals (Rachmadiarti et al., 2019). Morphological properties, such as leaf number, surface area, stomata structures, large crown, and waxy surfaces, are factors that retain the pollutants.

The indoor plant is efficient for the phytoremediation of VOCs and indoor air pollutants. For instance, ornamental plants such as *Chamaedorea elegans and Nephrolepis obliterata* reduce the concentration of formaldehyde in indoor air (Teiri et al., 2018). Benzene, toluene, xylene, and ethylbenzene (BTEX) can be eliminated from the indoor environment by using *Opuntia microdasys, Schefflera arboricola, and Spathiphyllum wallisii* plant species. The combination of Ophiopogon *japonicus* with *Staphylococcus epidermis* and *Pseudomonas* spp. degrades phenol in an indoor environment. In another study, PM, TVOCs, CO_2, and CO have been significantly reduced using plants such as *Areca palm* in the indoor environment (Bhargava et al., 2021). Plant species such as *Chamaedorea elegans*, *Epipremnum aureum*, *Ficus lyrata*, *Neomarica gracillis*, *Peperomia obtusifolia*, *Spathiphyllum wallisii*, *Syngonium podophyllum*, *Nephrolepis exaltata*, *Scheffera arboricola*, *Syngonium podophyllum*, *Spathiphyllum wallisii*, and *Syngonium podophyllum* can be used to reduce TVOCs, PM, NOx, O_3, and BTEX components in residential rooms, classrooms, and indoor spaces (Pettit et al., 2019). *Spathiphyllum wallisii* can be effectively used to eliminate VOCs and all size fractions of PM, formaldehyde, and acetone (Morgan et al., 2022). The size-segregated PM has been removed using

Nematanthus glabra, Schefflera arborcola, Nephropelis exaltata bostoniensis, Nephrolepis cordifolia, Epipremnum Aureum, Grass or moss, Epipremnum aureum, Chlorophytum orchidastrum, and *Chlorophytum comosum* in indoor space (Ibrahim et al., 2021). Although phytoremediation techniques are frequently used for the indoor environment, there are several plants that can potentially reduce the outdoor air pollution. For instance, shrubs and coniferous plants such as *Parthenocissus quiquefolia and Betula pendula* are able to remove PM through adsorption on the wax surface. A study was conducted using plant species such as *Syzigium oleina* and *Wedelia trilobata* along the roadside in Indonesia to reduce PM and heavy metals (Rachmadiarti et al., 2019). Plant species such as *Nematanthus glabra, Chlorophytum comosum variegatum, Philodendron Xanadu*, and *Spathiphyllum wallisii* have accumulated PM in urban outdoor areas over Sydney and Australia (Paull et al., 2020). The hydrophilicity of the leaf, the stomatal densities, and the specific leaf area of *Muntingia calabura* enhance the deposition of PM in the streets of Hanoi, Vietnam. Plant species such as *Hebe albicans Cockayne, Hebe x youngii Metcalf, Buxus sempervirens* L., *Thymus vulgaris* L., *Cinnamomum japonicum, Loropetalum chinense*, and *Osmanthus fragrans* are used to accumulate PM from the roadside, traffic zones, the university campus, or even industrial areas. The extension of green space using *Mangifera indica, Azadirachta indica, Ficus religiosa*, and shrubs having larger canopies can capture road dust and PM effectively (Chaudhary & Rathore, 2019). Overall, green plants, including herbs, shrubs, trees, and aquatic plants, are efficient pollutant absorbers and a natural filtration unit and a promising nature-based solution in the near future (Menon & Sharma, 2021).

16.4 Future Perspective for Air Pollution Control Strategies

Air pollution control strategies are a combined effort of technological, community, and nature-based interventions to prevent exposure to poor ambient air quality (Ogwu et al., 2024). Reducing exposure to poor air quality is helpful to reduce the global burden of diseases at the local, regional, and global levels (World Health Organization, 2021). Recently, the development of an increasing number of monitoring stations to trace real-time air quality data and the instantaneous implementation of policy will be a positive approach to controlling air quality. The installation of low-cost sensors and prevention strategies for poor air quality reduces the global burden of diseases. The use of artificial intelligence and model-based monitoring will be a powerful technique to forecast poor air quality. Proper urban planning with appropriate green space will be a positive move for the management of air quality in megacities. The expansion of megacities with smart development equipped with air quality monitoring stations is a promising option for air quality management. Proper disposal of solid waste in the urban environment with zero emissions will be a positive move for air quality management. Furthermore, community awareness and participation with a citizen science approach will be a positive move to control the air pollution in the near future. Implementing stringent air pollutant emission standards for vehicles and industries will significantly move toward reducing air pollution in the near future (Oreggioni et al., 2022).

16.5 Conclusion

Air pollution control is a recent global challenge for every country in the era of economic race. To maintain the quality of the air, various countries formulated the air quality standard with periodic modifications to protect public health and welfare. Holistic control strategies of the emission of air pollutants start with the identification of multiple sources that provide positive feedback to maintain ambient air quality. The installation of recent advanced techniques in industries is a source-based reduction of air pollution. Furthermore, the adoption of new technology in vehicles is an appreciable approach to reducing air pollution in the transport sector. In addition, management of solid waste dumping and development of sustainable disposal mechanisms at the local, regional, and global levels contribute significantly to the reduction in air pollution. The community participation to avoid the burning of biomass and the burning of solid waste in open space in developing and underdeveloped areas is a significant move to reduce global air pollution. Furthermore, the use of nature-based technology to reduce waste generation is a promising solution for the near future. The development of green belts around cities and rural and industrial areas is a significant move for air pollution. The implementation of stringent air quality at the country level is a recent demand for air pollution control. Overall, only technological intervention is not a solution to air pollution. Therefore, active participation of the community in citizen science approaches is a current demand to control atmospheric pollution.

Acknowledgments The authors are thankful to the anonymous reviewers for critically reviewing the book chapter to improve the quality. The author (Pinki Biswas) is grateful to the Non-NET Ph.D. fellowship provided by Visva-Bharati, Santiniketan, for carrying out the research work.

References

Arshad, K., Hussain, N., Ashraf, M. H., & Saleem, M. Z. (2024). Air pollution and climate change as grand challenges to sustainability. *Science of the Total Environment, 928*, 172370.

Asif, Z., Chen, Z., Wang, H., & Zhu, Y. (2022). Update on air pollution control strategies for coal-fired power plants. *Clean Technologies and Environmental Policy, 24*(8), 2329–2347.

Awewomom, J., Dzeble, F., Takyi, Y. D., Ashie, W. B., Ettey, E. N. Y. O., Afua, P. E., & Akoto, O. (2024). Addressing global environmental pollution using environmental control techniques: a focus on environmental policy and preventive environmental management. *Discover Environment, 2*(1), 8.

Bhargava, B., Malhotra, S., Chandel, A., Rakwal, A., Kashwap, R. R., & Kumar, S. (2021). Mitigation of indoor air pollutants using areca palm potted plants in real-life settings. *Environmental Science and Pollution Research, 28*, 8898–8906.

Chaudhary, I. J., & Rathore, D. (2019). Dust pollution: Its removal and effect on foliage physiology of urban trees. *Sustainable Cities and Society, 51*, 101696.
Grinišin, N., Bešenić, T., Kozarac, D., Živić, M., Wang, J., & Vujanović, M. (2022). Modelling of absorption process by seawater droplets for flue gas desulfurization application. *Applied Thermal Engineering, 215*, 118915.
Gupta, S., Rajput, A., Kumar, R., Yadav, S., Verma, A., Gautam, A. S., & Gautam, S. (2023). Seasonal variation and sources of $PM_{2.5}$ bound water-soluble ions at Jammu city in the foothills of the North-Western Himalayas, India. *Geological Journal, 58*(12), 4333–4348.
Han, L., Zhao, J., Rong, N., Wang, Z., Qi, Z., Shen, Z., & Yu, H. (2024). Energy and exergy analyses of biomass IGCC power plant using calcium looping gasification with in situ CO_2 capture and negative carbon emission. *Biomass Conversion and Biorefinery, 14*(19), 23649–23666.
He, J., Chen, K., Xu, J., & Sun, Y. (2024). Urban air pollution and control. In *Reference module in earth systems and environmental sciences*. Elsevier.
Hussin, F., & Aroua, M. K. (2020). Recent trends in the development of adsorption technologies for carbon dioxide capture: A brief literature and patent reviews (2014–2018). *Journal of Cleaner Production, 253*, 119707.
Ibrahim, I. Z., Chong, W. T., Yusoff, S., Wang, C. T., Xiang, X., & Muzammil, W. K. (2021). Evaluation of common indoor air pollutant reduction by a botanical indoor air biofilter system. *Indoor and Built Environment, 30*(1), 7–21.
Jiang, X., Li, G., & Fu, W. (2021). Government environmental governance, structural adjustment and air quality: A quasi-natural experiment based on the three-year action plan to win the blue sky Defense war. *Journal of Environmental Management, 277*, 111470.
Jonidi Jafari, A., Charkhloo, E., & Pasalari, H. (2021). Urban air pollution control policies and strategies: A systematic review. *Journal of Environmental Health Science and Engineering, 19*, 1911–1940.
Kim, H. S., Kasipandi, S., Kim, J., Kang, S. H., Kim, J. H., Ryu, J. H., & Bae, J. W. (2020). Current catalyst technology of selective catalytic reduction (SCR) for NOx removal in South Korea. *Catalysts, 10*(1), 52.
Kumar, D., & Kumar, D. (2018). *Sustainable management of coal preparation*. Woodhead Publishing.
Li, Y., Zhang, X., Huangfu, L., Yu, F., Chen, Z., Li, C., & Gao, S. (2020). The simultaneous removal of SO_2 and NO from flue gas over activated coke in a multistage fluidized bed at low temperature. *Fuel, 275*, 117862.
Liu, X., Tan, H., Wang, Y., Yang, F., Mikulčić, H., Vujanović, M., & Duić, N. (2018). Low NOx combustion and SCR flow field optimization in a low volatile coal fired boiler. *Journal of Environmental Management, 220*, 30–35.
Menon, J. S., & Sharma, R. (2021). Nature-based solutions for co-mitigation of air pollution and urban heat in Indian cities. *Frontiers in Sustainable Cities, 3*, 705185.
Morgan, A. L., Torpy, F. R., Irga, P. J., Fleck, R., Gill, R. L., & Pettit, T. (2022). The botanical biofiltration of volatile organic compounds and particulate matter derived from cigarette smoke. *Chemosphere, 295*, 133942.
Obi, D., Onyekuru, S., & Orga, A. (2025). Minimizing carbon capture costs in power plants: A novel dimensional analysis framework for techno-economic evaluation of Oxyfuel combustion, pre-combustion, and post-combustion capture systems. *Energy Science & Engineering, 13*, 1749–1770.
Ofremu, G. O., Raimi, B. Y., Yusuf, S. O., Dziwornu, B. A., Nnabuife, S. G., Eze, A. M., & Nnajiofor, C. A. (2024). Exploring the relationship between climate change, air pollutants and human health: Impacts, adaptation, and mitigation strategies. *Green Energy and Resources, 3*, 100074.
Ogwu, M. C., Ro, B., & Thapa, B. (2024). Eco-friendly methods for combating air pollution. In *Sustainable strategies for air pollution mitigation: Development, economics, and technologies* (pp. 237–277). Springer Nature.
Oreggioni, G. D., Mahiques, O., Monforti-Ferrario, F., Schaaf, E., Muntean, M., Guizzardi, D., & Crippa, M. (2022). The impacts of technological changes and regulatory frameworks on

global air pollutant emissions from the energy industry and road transport. *Energy Policy, 168*, 113021.

Paull, N. J., Krix, D., Irga, P. J., & Torpy, F. R. (2020). Airborne particulate matter accumulation on common green wall plants. *International Journal of Phytoremediation, 22*(6), 594–606.

Pettit, T., Bettes, M., Chapman, A. R., Hoch, L. M., James, N. D., & Irga, P. J. (2019). The botanical biofiltration of VOCs with active airflow: Is removal efficiency related to chemical properties? *Atmospheric Environment, 214*, 116839.

Rachmadiarti, F., Purnomo, T., Azizah, D. N., & Fascavitri, A. (2019). Syzigium oleina and Wedelia trilobata for phytoremediation of lead pollution in the atmosphere. *Nature, Environment and Pollution Technology, 18*(1), 157–162.

Roy, A., Mandal, M., Das, S., Kumar, M., Popek, R., Awasthi, A., & Sarkar, A. (2024). Non-exhaust particulate pollution in Asian countries: A comprehensive review of sources, composition, and health effects. *Environmental Engineering Research, 29*(3), 230113.

Sadeghpour, A., Oroumiyeh, F., Zhu, Y., Ko, D. D., Ji, H., Bertozzi, A. L., & Ju, Y. S. (2021). Experimental study of a string-based counterflow wet electrostatic precipitator for collection of fine and ultrafine particles. *Journal of the Air & Waste Management Association, 71*(7), 851–865.

Teiri, H., Pourzamani, H., & Hajizadeh, Y. (2018). Phytoremediation of VOCs from indoor air by ornamental potted plants: A pilot study using a palm species under the controlled environment. *Chemosphere, 197*, 375–381.

Vodička, M., Hrdlička, J., & Skopec, P. (2021). Experimental study of the NOX reduction through the staged oxygen supply in the oxy-fuel combustion in a 30 kWth bubbling fluidized bed. *Fuel, 286*, 119343.

Wang, S. (2020). Near-zero air pollutant emission technologies and applications for clean coal-fired power. *Engineering, 6*(12), 1408–1422.

World Health Organization. (2021). *WHO global air quality guidelines: particulate matter ($PM_{2.5}$ and PM_{10}), ozone, nitrogen dioxide, sulfur dioxide and carbon monoxide*. World Health Organization.

Yadav, A. K. (2015). Elemental composition and source apportionment of suspended particulate matters and health risk assessment in mining and nonmining areas of Odisha, India. *Journal of Hazardous, Toxic, and Radioactive Waste, 19*(3), 04014037.

Yadav, A. K., Gupta, K., Bodhale, R., & Phuleria, H. C. (2025). Metals and trace element associated health risk in size-segregated household dust of an industrial town in India. *Air Quality, Atmosphere & Health*. https://doi.org/10.1007/s11869-025-01794-5

Yadav, A. K., Sahoo, S. K., Dubey, J. S., Kumar, A. V., Pandey, G., & Tripathi, R. M. (2019). Assessment of particulate matter, metals of toxicological concentration, and health risk around a mining area, Odisha, India. *Air Quality, Atmosphere & Health, 12*(7), 775–783.

Yadav, A. K., Sahoo, S. K., Kumar, A. V., Shirin, S., Jamal, A., Patra, A. C., Dubey, J. S., Thakur, V. K., Lenka, P., Singh, S., Jha, V., & Tripathi, R. M. (2023). Identifying size-segregated particulate matter ($PM_{2.5}$, PM_{10} and SPM) sources in an industrial town of India. *Aerosol Science and Engineering, 7*(4), 455–473.

Yadav, A. K., Sahoo, S. K., Patra, A. C., Dubey, J. S., Lenka, P., Sagar, D. V., Kumar, A. V., & Tripathi, R. M. (2014). Source identification of particulate matter and associated intake of elements through inhalation in an industrial area of Odisha, India. *Toxicological & Environmental Chemistry, 96*(3), 410–425.

Yadav, A. K., Shirin, S., Hopke, P. K., Pal, D., & Jamal, A. (2022). Process to reduce particulate matter in ambient air using bubbles of sodium palmitate. *Chemical Engineering & Technology, 45*(8), 1497–1500.

Part IV
Specialized Applications and Case Studies

Chapter 17
COVID-19 Emission Inventory and Air Quality Simulation in Madrid, Spain

Roberto San Jose, Juan L. Perez-Camanyo, and Miguel Jimenez-Gañan

17.1 Introduction

Air pollution is a global public health problem that has been a focus of attention for decades. According to the World Air Quality Report 2020, air pollution remains one of the greatest health hazards facing people around the world (Krzyzanowski et al., 2005). Among the main contributors to air pollution is road transport, especially in urban environments. Exhaust emissions from road transport are a major contributor to air pollution in urban areas, raising much concern about the effects of urban air quality on human health (Rahaman et al., 2022). In general, transport activity contributes about 30% of NO_x emissions and 20% of total particle emissions in areas with road networks and cities with a high density of population (Han & Naeher, 2006). In addition, the effects of air pollution are not limited to respiratory problems but can have wider impacts on human health, including cardiovascular disease, cancer, and neurodevelopmental problems (Cohen et al., 2017). Children, the elderly, and people with pre-existing conditions are particularly vulnerable to the effects of air pollution. Mitigation of air pollutant concentrations has therefore become a public health issue, and it is necessary to address effective measures to reduce the effects of air pollution, such as those aimed at reducing emissions from transport. This study helps to provide information on the behavior of pollutants in the context of a massive reduction in NO_x emissions, which must be taken into account if air quality is to be improved.

The coronavirus pandemic 2019 (COVID-19) had a significant impact on social and economic activity and therefore on air quality. Movement restrictions led to a reduction in anthropogenic emissions of air pollutants and greenhouse gases (Liou

R. S. Jose (✉) · J. L. Perez-Camanyo · M. Jimenez-Gañan
Environmental Software and Modelling Group, Computer Science School, Technical University of Madrid, Madrid, Spain
e-mail: roberto@fi.upm.es

A. K. Yadav (ed.), *Computational Approach for Air Pollution*,
https://doi.org/10.1007/978-3-032-09215-1_17

et al., 2023). The decrease in industrial activity and reduced road transport is the explanation for the decrease in NO_x and particulate matter (PM) in all cities during the 2020 lockdown (Le et al., 2020). In addition, significant decreases in global other emissions have been recorded due to reduced human activity during COVID-19 (Gaucher et al., 2022). The reduction in emissions implied improvements in air quality based on some pollutants in cities. This was observed from the analysis of measurements from air quality monitoring networks and satellites (Collivignarelli et al., 2020).

This episode has been described as "the largest controlled air quality experiment conducted in history" (Forster et al., 2020) that has allowed the impact of emission reductions on air quality to be assessed (Wang et al., 2020). Mobility datasets show that during the pandemic, emissions from the transport sector were significantly reduced due to stay-at-home orders (Venter et al., 2020). This has had a positive impact on some pollutants, such as NO_2, as they have been drastically reduced in many urban areas (European Space Agency (ESA), 2020), but increases in O_3 and PM concentrations were also observed in some places. This episode of emission reductions has provided a unique scientific opportunity to detect, attribute, and understand the impacts of anthropogenic emissions on air quality. Scientific studies on this large experiment will inform future policies on reducing emissions and improving air quality around the world. Decreases in emissions of nitrogen oxides (NO_x) and PM have been reported in various parts of the world. However, recent studies have found that the responses of secondary pollutants, such as ozone (O_3) and fine PM ($PM_{2.5}$), vary due to the complex nonlinear interactions present in atmospheric chemistry due to the complexity of atmospheric chemistry and how emissions of chemical compounds can have cascading effects and produce secondary pollutants (Seinfeld et al., 1998); therefore, a more complete understanding of atmospheric chemistry is needed to predict the effects of these reductions in secondary pollutants, which this chapter aims to contribute to study these effects; most researchers have used methods such as ground-based observation statistical analysis (Tobías et al., 2020) and numerical modeling analysis. Numerical models, used in this study, are more effective tools for analyzing the impact of COVID-19 restrictions on air quality because they can simulate complex atmospheric dynamics and consider multiple factors simultaneously. Currently, there are several studies in different areas with different conclusions on the change in concentrations. For example, Sulaymon et al. (2021) found that COVID-19 control measures in China led to a significant decrease in air pollutant emissions and an improvement in air quality in cities, including a reduction in NO_2 and $PM_{2.5}$ concentrations. Nakada and Urban (2020) and Dahech et al. (2022) showed a significant decrease in NO_2 levels in the air during the partial blockade in São Paulo, while O_3 concentrations increased in some areas due to the reduction of NO_x compounds. Dahech et al. (2022) indicated a significant decrease in NO_2 and $PM_{2.5}$ concentrations in Barcelona during the COVID-19 blockade, while O_3 concentrations increased in some areas due to a reduction. Javed et al. (2021) found that COVID-19 control measures in the Yangtze River Delta region led to a significant decrease in NO_2 and $PM_{2.5}$ concentrations, while O_3 concentrations increased due to the reduction of NO_x compounds.

In this study, the WRF/Chem numerical model is used to simulate air pollutants in Madrid, Spain, during COVID-19 confinement and to assess, with high spatial resolution, the impact of anthropogenic emission reductions on air quality. Numerical models have proven to be valuable tools for investigating air quality changes during the COVID-19 pandemic (Yumin et al., 2021). In this work, the use of high-resolution numerical models will provide more detailed information on the spatial distribution of air pollution, which is essential for developing and optimizing effective air pollution control strategies in the city of Madrid and other places. In this research, we have analyzed variations in a range of air quality components, also considering the impact of meteorological factors on atmospheric oxidants, which closely link NO_2 and O_3, hence the focus on these two components that are of vital importance in urban air quality. Determining to what extent the COVID-19 measures influenced air quality is challenging due to multiple factors. Beyond daily fluctuations, meteorological conditions, such as increased precipitation, can reduce pollutant concentrations independently of the effects of the closure. Additionally, seasonal variations in emissions have a direct impact on air pollution levels. Even if local emissions decrease, pollutants such as $PM_{2.5}$ and O_3 may still increase due to long-range transport, as observed in April 2020. Specifically, two simulations were carried out: a baseline with BAU emissions and one with emission reductions due to containment measures (COVID). The comparison of the two scenarios will allow an estimate of the actual impacts of the containment measures on air quality in Madrid, Spain. This modeling approach has also been adopted by other researchers, such as Wang et al. (2022). It is important to note that meteorology can play an important role in the formation, transport, deposition, and transformation of air pollution, which can affect the concentrations of air pollutants such as ozone and fine PM (Nosek et al., 2016). It is therefore necessary to use modeling approaches similar to the one mentioned above to assess the impact of abatement measures, independent of the effects of meteorology, in order to isolate the effects of emission reductions and to analyze the mechanisms responsible for the change (increase of some pollutants and decrease of other pollutants) in atmospheric concentrations from emission reductions.

17.2 Modeling Tools

In this work, two numerical modeling models widely used in atmospheric environmental research have been used. The model was the Weather Research and Forecasting Model Coupled with Chemistry (WRF/Chem), which allows implementing the bidirectional coupling of atmospheric dynamics and atmospheric chemistry (Grell et al., 2005; Skamarock et al., 2008). The WRF/Chem model was developed by the National Center for Atmospheric Research and is an advanced meteorological model that has been coupled with a chemical model to simulate atmospheric dynamics and chemistry in the atmosphere. This model has been used in numerous air quality studies and air pollutant forecasts (Binkowski & Roselle, 2003; Fujita et al., 2015; Yadav, 2015).

The WRF is driven by global meteorological data from the Global Forecast System produced by the European National Centers for Environmental Prediction (NCEP). In this study, the NCEP GDAS Final Analysis 0.25° (ds083.3) was used as the initial and boundary atmospheric conditions to simulate air quality. The initial and boundary chemical conditions were extracted from the Whole Atmosphere Community Climate Model output (Marsh et al., 2013). It was continuously integrated during the study period and updated every 6 hours to improve the precision of the results. This modeling approach has been used in several previous studies, suggesting that it is a reliable methodology for simulating atmospheric conditions and air quality in different case studies (Marsh et al., 2013; Wie & Moon, 2018).

The WRF/Chem model is set with the following parameters: Microphysics Scheme Lin (Lin et al., 1983); Longwave and Shortwave Radiation Scheme RRTMG Scheme (Iacono et al., 2008); Surface Layer Scheme Monin-obukov (Monin & Obukhov, 1954); Planetary Boundary Layer Scheme YSU (Ouermi et al., 2017); Land Surface Scheme Noah Scheme (Niu et al., 2011); Cumulus Parameterization Scheme Grell-3D Scheme (Grell & Freitas, 2014); Chemical Mechanism Scheme CBMZ Scheme (Zaveri & Peters, 1999); and Aerosol Mechanism Scheme MOSAIC Scheme (Zaveri et al., 2008). This scheme considers multiple chemical processes, such as oxidation of organic compounds, ozone formation, sulfuric acid formation, and nitrate formation. It also includes the interaction between aerosols and atmospheric gases and takes into account aerosol dynamics such as nucleation, growth, and coagulation. MOSAIC includes the simulation of multiple aerosol species, allowing the assessment of the chemical composition and optical properties of aerosols in the atmosphere.

The study used the European emission inventory CAMS-REG-AP (Granier et al., 2011) (v4.2_ry 2019) to derive anthropogenic emissions in Europe without the influence of COVID-19 blocking measures. This emission inventory provides detailed data on greenhouse gases and air pollutant emissions from anthropogenic sources in Europe. It has been widely used in air quality studies and atmospheric modeling (Granier et al., 2011). Emissions from the inventory were processed using the EMIMO tool (UPM) to produce WRF/Chem-ready emissions. EMIMO is an emission modeling tool used to calculate emissions of anthropogenic and natural sources of air pollutants on different spatial and temporal scales (San José et al., 2015).

17.3 Experiment

Three nested simulation domains have been set in the experiment. These domains allow different geographical areas to be represented with different levels of spatial detail. The outer domain (domain 1) covers most of the Iberian Peninsula and was established with a spatial resolution of 25 km. The intermediate domain (domain 2) covers the Madrid region with a spatial resolution of 5 km, and the inner domain (domain 3) covers the municipality of Madrid with a spatial resolution of 1 km. In terms of vertical resolution, 35 vertical hybrid layers were established from the surface up to a pressure of 50 hPa to represent the different altitudes and levels of the atmosphere. This nested simulation approach is often used in atmospheric modeling to represent different spatial and temporal scales while saving computational resources.

The simulation period ran from February 1, 2020, to April 12, 2020, with two sub-periods of 36 days each: pre-lockdown and lockdown. The first subperiod, which runs from 1 February to 7 March 2020, was before the lockdown measures were imposed, and normal emissions were observed. The second subperiod, from March 8 to April 12, 2020, was during the lockdown, during which a significant decrease in emissions was observed. On March 8, 2020, the first emission reductions were observed, suggesting that the lockdown measures had a significant impact on pollutant emissions. On March 15, 2020, the Spanish national blockade came into force, and on March 28, 2020, the Spanish government banned all nonessential activities. These additional restriction measures helped to further reduce the emissions of pollutants into the atmosphere. On April 13, 2020, workers in some non-essential sectors, such as construction and industry, who were not allowed to work remotely, were allowed to return to work. The relaxation of restrictions was reflected in an increase in emissions. These dates are consistent with NO2 concentration measurements in the city of Madrid; Fig. 17.1 shows the time series of the median (50th percentile) value measured by the stations of the Madrid City Council monitoring network.

Analysis showed a significant decrease in the atmospheric concentration of NO_2 during the lockdown period. To assess the effect of the measures on air quality, two emission scenarios have been prepared. The first scenario, known as BAU,

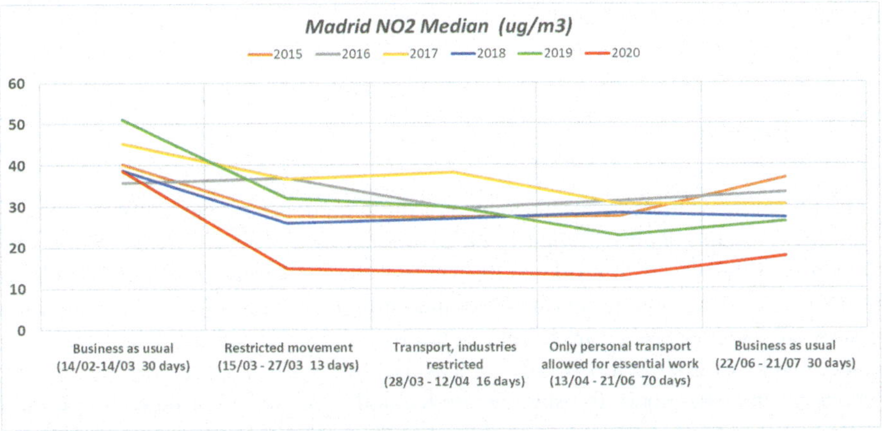

Fig. 17.1 The median of NO_2 atmospheric daily concentrations measured by Madrid monitoring network, from February 14 to June 21 for the year 2015–2020

considers the development of emissions in the absence of intervention measures. This scenario is therefore representative of what would have happened under normal conditions without locking measures. The second scenario, called the COVID simulation, takes into account the changes in emissions due to the lockdown of COVID-19. This scenario is based on a rescaling of the anthropogenic emissions inventory from March 8. Estimates of emission reductions during the lock-in period (COVID simulation) are taken from the recent article by Guevara et al. (2021), which has published the country-specific percentage reduction. In particular, specific daily reduction rates have been obtained for the following emission activities: public energy, industry, road transport, and aviation. The impact of the COVID-19 lockdown on primary emissions in Europe was quantified using a dataset of reduction factors that vary by country and day. This dataset allows modeling and assessment of air quality changes resulting from emission reductions. The reduction factors were provided at daily and country-specific resolution for the main emission sources, including power plants (energy industry), manufacturing industry, road transport, and aviation (landing and take-off cycles). The dataset covers the period from February 21, when the first regional blockade in Europe was implemented in Lombardy (Italy), to April 26, 2020. This timeframe includes 5 weeks (from March 23 to April 26) of the most stringent and relatively stable mobility and economic restrictions across Europe. The calculated reduction factors were integrated with the European emission inventory of the Copernicus Atmosphere Monitoring Service, using adjusted time profiles to estimate time-dependent emission reductions per country and pollutant sector. During the most stringent lock-in phase, average emission reductions across the EU-30 (EU-28 plus Norway and Switzerland) were estimated at −33% for NO_x, −8% for volatile organic compounds other than methane (NMVOC), −7% for SO_x, and −7% for $PM_{2.5}$. For most pollutants, more than 85% of the total reduction was attributed to road traffic, with the exception of SO_x. In the countries with the most stringent restrictions, such as Italy, France, and Spain, the reductions were even more pronounced, reaching −50% for NO_x, −14% for NMVOCs, −12% for SO_x, and − 15% for $PM_{2.5}$.

17.4 Results

In this case, to evaluate the performance of these models, simulated hourly concentrations were compared with hourly observational data provided by the monitoring networks of the city of Madrid and the Madrid region. A total of 48 monitoring stations were used to obtain the observed and simulated data. To evaluate the performance of the models, a single representative value per pollutant and hour was calculated by averaging the 48 values measured by the stations and the 48 simulated values in the cells where the stations are located. The statistical indicators used to evaluate the performance of the models were the normalized mean bias (NMB, Eq. (17.1)), the root mean square error (RMSE, Eq. (17.2)), and the squared correlation coefficient (R^2, Eq. (17.3)). The squared correlation coefficient is used to measure the relationship between observed and simulated data, with values closer to 1 indicating a better relationship. The mean normalized skewness measures the average difference between the simulated and observed values, normalized by the observed values, with values closer to 0 indicating a better simulation. The mean squared error measures the average squared difference between the simulated and observed values, with values closer to 0 indicating a better simulation. It is important to bear in mind that the average concentration of the grid cell in which the station is located may not be representative of the measurement at a control location. If a station is located near a pollution source, such as a road or a factory, it is possible that the results will be higher than at a location a few hundred meters away from the source. Remember that the results we are comparing have a spatial resolution of 1 km; hence, it is also a good option to average the values of the 48 stations to obtain a representative value:

$$\mathrm{NMB} = \frac{\sum_{i=1}^{N}(M_i - O_i)}{\sum_{i=1}^{N} O_i} \times 100 \tag{17.1}$$

$$\mathrm{RMSE} = \sqrt{\frac{1}{N}\sum_{i=1}^{N}(M_i - O_i)^2} \tag{17.2}$$

$$R^2 = \left[\frac{\sum_{i=1}^{N}\left(M_i - \overline{M}\right)\left(O_i - \overline{O}\right)}{\sqrt{\sum_{i=1}^{N}\left(M_i - \overline{M}\right)^2 \sum_{i=1}^{N}\left(O_i - \overline{O}\right)^2}}\right]^2 \tag{17.3}$$

Overall, we can say that, based on the results of the performance assessment, both models capture the magnitude and temporal evolution of the four key air pollutants reasonably well. NMBs remain below 30%, indicating good accuracy in the data simulation. Furthermore, the R^2 values are above 0.5 in most cases, indicating a good correlation between the observed and simulated data. In particular, during the lockdown period, the models showed a better reproduction of the observed NO_2 values, suggesting that the emission reduction estimates made by Guevara et al. (2021) are accurate. NMBs range from 18% to 9% for the WRF/Chem model, while the R^2 values range from 0.67 to 0.72. This would confirm that the emission reduction estimates for the lockdown period made by Guevara et al. (2021) are very accurate. In the pre-lockdown period, both models showed a slight overestimation of NO_2 levels and an underestimation of O_3 levels, indicating that there may be uncertainty in NOx emissions. For particulate simulation, the WRF/Chem model shows a higher accuracy. This result suggests that the MOSAIC aerosol scheme used in WRF/Chem is more accurate under the meteorological conditions of the studied period. In general terms, both models reproduce the trend of NO_2 and O_3

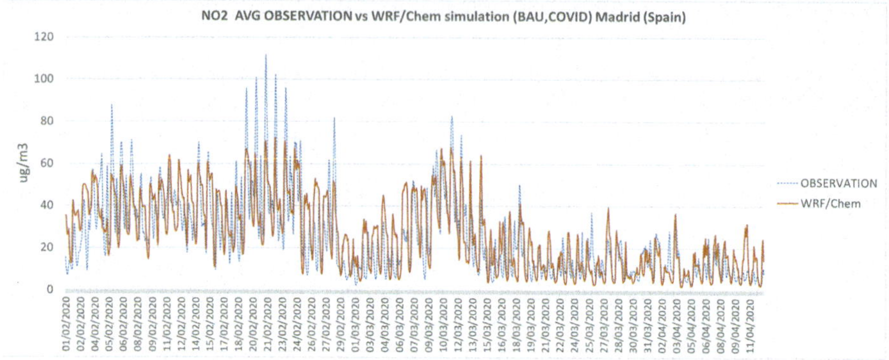

Fig. 17.2 NO_2 time series simulated for the WRF/Chem (orange) model, compared to the observations (blue) for pre-lockdown and lockdown periods in Madrid, Spain, for the average values of the Madrid monitoring stations

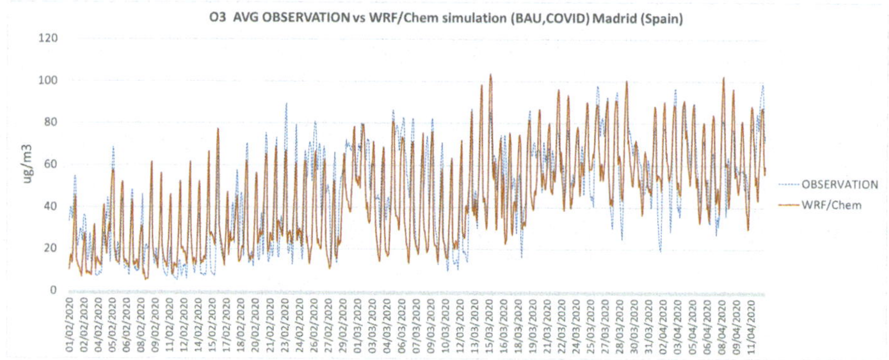

Fig. 17.3 O_3 time series simulated for the WRF/Chem (orange) model, compared to the observations (blue) for pre-lockdown and lockdown periods in Madrid, Spain, for the average values of the Madrid monitoring stations

concentrations very well with R^2 above 0.7. Figures 17.2 and 17.3 show hourly time series of NO_2 and O_3 measured and simulated with the WRF/Chem model, respectively, for Madrid, Spain, during the two periods of the case study.

The role of primary nitrogen dioxide emissions (Keuken et al., 2009) in the evolution of NO_2 and O_3 concentrations in a traffic environment explains the decrease in atmospheric NO_2 concentration after the measures taken to address the COVID-19 pandemic and the corresponding increase in O_3 concentrations. The agreement between the simulated NO_2 and O_3 values and the observations during both periods can be seen very well in the figures above and corresponds to the values in Table 17.1. Figures 17.4 and 17.5 illustrate the effects of the lockdown measures on average O_3 and NO_2 concentrations during the period from March 8 to April 12, comparing the

Table 17.1 Performance of the model simulations

	NMB (%)		RMSE (μg/m³)		R^2 [0,1]	
Model	Pre-lockdown	Lock-down	Pre-lockdown	Lock-down	Pre-lockdown	Lock-down
NO_2	18	9	14	9	0.67	0.72
O_3	−14	−2	15	13	0.75	0.78
PM_{10}	18	−18	14	20	0.67	0.46
$PM_{2.5}$	27	24	11	6	0.53	0.57

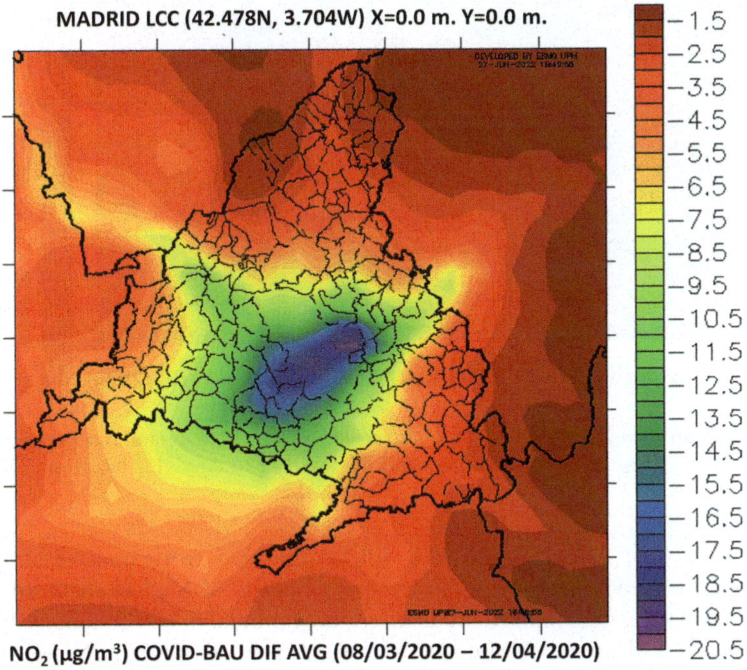

Fig. 17.4 Map showing the differences in average surface NO_2 concentrations (μg/m³) between COVID and BAU scenarios for the period from March 8 to April 12, based on WRF/Chem simulation

COVID and BAU simulations (absolute differences). Figure 17.4 highlights the impact of emission reductions on NO_2 levels, with the most significant decrease observed in Madrid, mainly due to reduced traffic emissions. The reductions were most pronounced in urban areas, especially along major highways and densely populated areas. As NO levels decrease, nocturnal O_3 depletion becomes less efficient, resulting in higher O_3 concentrations at night, especially near urban centers and main roads. Figure 17.5 reveals an increase in O_3 concentrations of up to 20 μg/m³

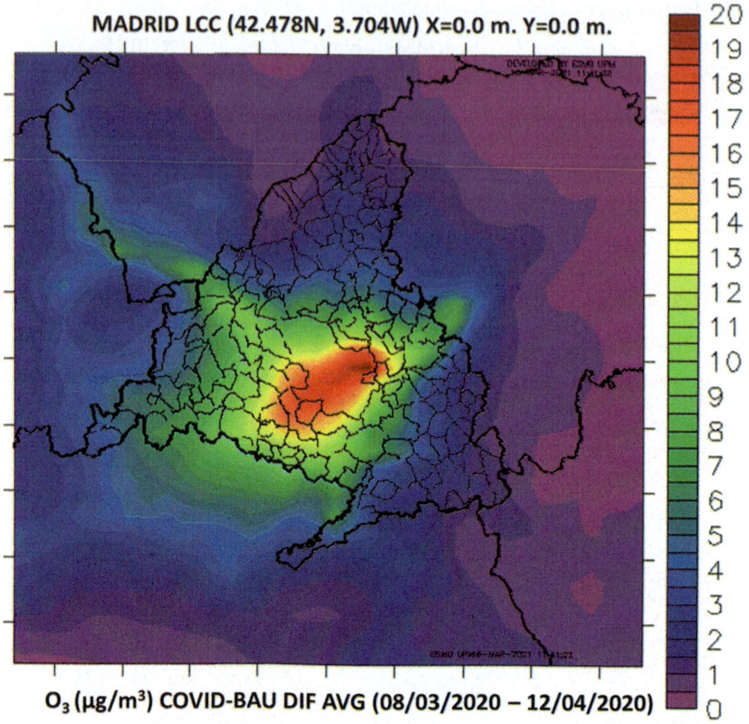

Fig. 17.5 Map showing the differences in average surface O_3 concentrations (μg/m^3) between COVID and BAU scenarios for the period from March 8 to April 12, based on WRF/Chem simulation

in the Center of Madrid, precisely where emission reductions—mainly from traffic—were greatest. On the contrary, areas outside of Madrid show little or no impact, indicating that emission reductions had minimal influence outside the city.

Figure 17.6 illustrates the impact of closure measures on daily NO_2 concentrations on March 30, 2020, the day when all nonessential activities were suspended. Compared to a normal Monday under standard conditions, NO_2 levels show a significant reduction of up to 20.5%. The spatial distribution of these changes is determined by the prevailing wind patterns, with a northeast–southwest direction influencing the dispersion of pollutants throughout the region.

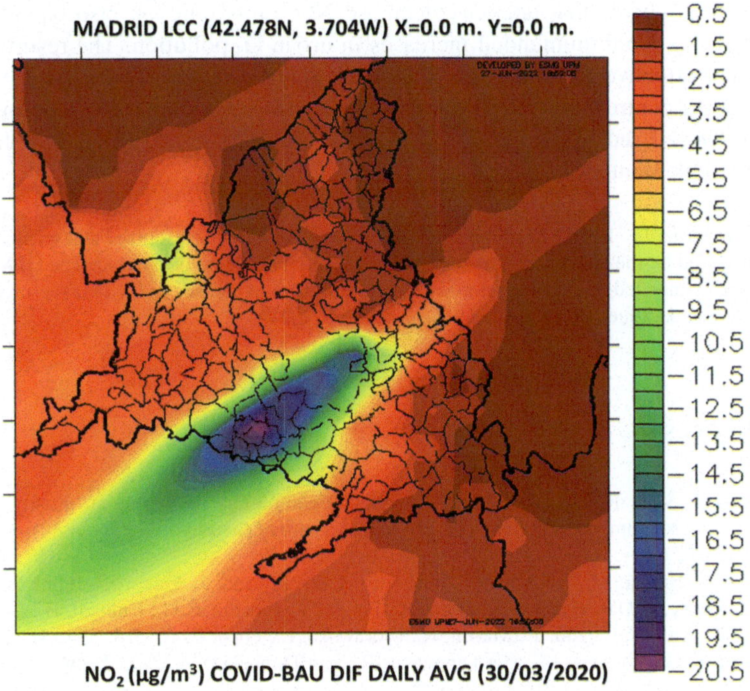

Fig. 17.6 Map showing the differences in average surface NO_2 concentrations (µg/m^3) between COVID and BAU scenarios for the period from March 30, based on WRF/Chem simulation

17.5 Conclusion

This study takes advantage of the COVID-19 lockdown as a unique large-scale experiment to analyze the effects of substantial emission reductions, especially from the transport sector, on air pollution. To assess the impact of blocking measures on air quality in the Community of Madrid (Spain), two simulations were carried out: one reflecting actual emission reductions during blocking (COVID simulation) and the other assuming BAU conditions without restrictions (BAU simulation). These simulations were carried out using two air quality models. The WRF/Chem results for the lockdown period, based on the adjusted emission inventory, showed better agreement with surface observations than those of the pre-lock-in period, where the BAU inventory was applied.

The results indicate that the models successfully captured the magnitude and temporal evolution of the main air pollutants, with performance parameters within the expected ranges. Emission reductions, mainly from the transport sector, significantly altered air quality in Madrid, leading to a marked decrease in NO_2 concentrations and a secondary increase in O_3 levels. Comparison between BAU and COVID scenarios shows a significant reduction in NO_x and a corresponding increase in ozone. The experience of COVID-19 closure underlines the need for emission reduction strategies that not only decrease NO_x emissions but also balance VOC reductions to avoid unintended increases in urban O_3 pollution. The results underscore that effective air quality improvement requires a coordinated and balanced approach to control multiple pollutants. Furthermore, air quality modeling tools, such as those used in this study, can help simulate possible mitigation strategies prior to implementation.

Acknowledgments The UPM authors acknowledge the computer resources and technical assistance provided by the Centro de Supercomputación y Visualización de Madrid. The UPM authors thankfully acknowledge the computer resources, technical expertise, and assistance provided by Red Española de Supercomputación.

References

Binkowski, F. S., & Roselle, S. J. (2003). Models-3 community multiscale air quality (CMAQ) model aerosol component 1 model description. *Journal of Geophysical Research-Atmospheres, 108*, 1409.

Cohen, A. J., Brauer, M., Burnett, R., Anderson, H. R., Frostad, J., Estep, K., Balakrishnan, K., Brunekreef, B., Dandona, L., Dandona, R., Feigin, V., Freedman, G., Hubbell, B., Jobling, A., Kan, H., Knibbs, L., Liu, Y., Martin, R., Morawska, L., Pope, C. A., 3rd, Shin, H., Straif, K., Shaddick, G., Thomas, M., van Dingenen, R., van Donkelaar, A., Vos, T., Murray, C. J. L., & Forouzanfar, M. H. (2017). Estimates and 25-year trends of the global burden of disease attributable to ambient air pollution: An analysis of data from the global burden of diseases study 2015. *Lancet, 389*(10082), 1907–1918. https://doi.org/10.1016/S0140-6736(17)30505-6

Collivignarelli, M. C., Abbà, A., Bertanza, G., Pedrazzani, R., Ricciardi, P., & Carnevale Miino, M. (2020). Lockdown for Covid-2019 in Milan: What are the effects on air quality? *Science of The Total Environment., 732*, 139280.

Dahech, S., Abdmouleh, M. A., & Lagmiri, S. (2022). Spatiotemporal variation of air quality (PM and no2) in southern Paris during COVID-19 lockdown periods. *Atmosphere, 13*, 289.

European Space Agency (ESA). (2020). *Nitrogen dioxide concentrations over Europe*. Retrieved April 17, 2023, from https://www.esa.int/Applications/Observing_the_Earth/Copernicus/Sentinel-5P/Nitrogen_dioxide_concentrations_over_Europe

Forster, P. M., Forster, H. I., Evans, M. J., Gidden, M. J., Jones, C. D., Keller, C. A., Lamboll, R. D., Quéré, C. L., Rogelj, J., Rosen, D., Schleussner, C.-F., Richardson, T. B., Smith, C. J., & Turnock, S. T. (2020). Current and future global climate impacts resulting from covid-19. *Nature Climate Change, 10*, 913–919.

Fujita, E. M., Campbell, D. E., Stockwell, W. R., Saunders, E., Fitzgerald, R., & Perea, R. (2015). Projected ozone trends and changes in the ozone-precursor relationship in the south coast Air Basin in response to varying reductions of precursor emissions. *Journal of the Air & Waste Management Association, 66*, 201–214.

Gaucher, Y., Tanaka, K., Ciais, P., & Boucher, O. (2022). Limited impact of COVID-19 recovery packages on near-term CO2 emissions pathways. *Environmental Research Communications, 4*, 101006.

Granier, C., Bessagnet, B., Bond, T., D'Angiola, A., Denier van der Gon, H., Frost, G. J., Heil, A., Kaiser, J. W., Kinne, S., Klimont, Z., Kloster, S., Lamarque, J.-F., Liousse, C., Masui, T., Meleux, F., Mieville, A., Ohara, T., Raut, J.-C., Riahi, K., Schultz, M. G., Smith, S. J., Thompson, A., van Aardenne, J., van der Werf, G. R., & van Vuuren, D. P. (2011). Evolution of anthropogenic and biomass burning emissions of air pollutants at global and regional scales during the 1980–2010 period. *Climatic Change, 109*, 163–190.

Grell, G. A., & Freitas, S. R. (2014). A scale and aerosol aware stochastic convective parameterization for weather and air quality modeling. *Atmospheric Chemistry and Physics, 14*, 5233–5250.
Grell, G. A., Peckham, S. E., Schmitz, R., McKeen, S. A., Frost, G., Skamarock, W. C., & Eder, B. (2005). Fully coupled "online" chemistry within the WRF model. *Atmospheric Environment, 39*, 6957–6975.
Guevara, M., Jorba, O., Soret, A., Petetin, H., Bowdalo, D., Serradell, K., Tena, C., Denier van der Gon, H., Kuenen, J., Peuch, V.-H., & Pérez García-Pando, C. (2021). Time-resolved emission reductions for atmospheric chemistry modelling in Europe during the COVID-19 lockdowns. *Atmospheric Chemistry and Physics, 21*, 773–797.
Han, X., & Naeher, L. P. (2006). A review of traffic-related air pollution exposure assessment studies in the developing world. *Environment International, 32*, 106–120.
Iacono, M. J., Delamere, J. S., Mlawer, E. J., Shephard, M. W., Clough, S. A., & Collins, W. D. (2008). Radiative forcing by long-lived greenhouse gases: Calculations with the AER radiative transfer models. *Journal of Geophysical Research, 113*, 9944.
Javed, Z., Tanvir, A., Wang, Y., Waqas, A., Xie, M., Abbas, A., Sandhu, O., & Liu, C. (2021). Quantifying the impacts of covid-19 lockdown and spring festival on air quality over Yangtze River Delta region. *Atmosphere, 12*, 735.
Keuken, M., Roemer, M., & van den Elshout, S. (2009). Trend analysis of urban NO2 concentrations and the importance of direct NO2 emissions versus ozone/Nox equilibrium. *Atmospheric Environment, 43*, 4780–4783.
Krzyzanowski, M., Kuna-Dibbert, B., & Schneider, J. (2005). *Health effects of transport-related air pollution*. World Health Organization. Regional Office for Europe.
Le, T., Wang, Y., Liu, L., Yang, J., Yung, Y. L., Li, G., & Seinfeld, J. H. (2020). Unexpected air pollution with marked emission reductions during the COVID-19 outbreak in China. *Science, 369*, 702–706.
Lin, Y.-L., Farley, R. D., & Orville, H. D. (1983). Bulk parameterization of the snow field in a cloud model. *Journal of Climate and Applied Meteorology, 22*, 1065–1092.
Liou, Y.-A., Vo, T.-H., Nguyen, K.-A., & Terry, J. P. (2023). Air quality improvement following covid-19 lockdown measures and projected benefits for environmental health. *Remote Sensing, 15*, 530.
Marsh, D. R., Mills, M. J., Kinnison, D. E., Lamarque, J.-F., Calvo, N., & Polvani, L. M. (2013). Climate change from 1850 to 2005 simulated in CESM1(WACCM). *Journal of Climate, 26*, 7372–7391.
Monin, A. S., & Obukhov, A. M. (1954). Basic Laws of turbulent mixing in the surface layer of the atmosphere. *Contributions to Geophysics and Geodesy, 24*, 163–187.
Nakada, L. Y. K., & Nakada Urban, R. C. (2020) COVID-19 pandemic: impacts on the air quality during the partial lockdown in São Paulo state, Brazil. Sci. Total Environ., 730 p. 139087. https://doi.org/10.1016/j.scitotenv.2020.139087
Niu, G.-Y., Yang, Z.-L., Mitchell, K. E., Chen, F., Ek, M. B., Barlage, M., Kumar, A., Manning, K., Niyogi, D., Rosero, E., Tewari, M., & Xia, Y. (2011). The community Noah land surface model with multiparameterization options (Noah-MP): 1. model description and evaluation with local-scale measurements. *Journal of Geophysical Research, 116*, D12.
Nosek, Š., Kukačka, L., Kellnerová, R., Jurčáková, K., & Jaňour, Z. (2016). Ventilation processes in a three-dimensional street canyon. *Boundary-Layer Meteorology, 159*, 259–284.
Ouermi, T. A. J., Knoll, A., Kirby, R. M., & Berzins, M. (2017). Optimization strategies for WRF single-moment 6-class microphysics scheme (WSM6) on Intel microarchitectures. In *2017 Fifth International Symposium on Computing and Networking (CANDAR)*.
Rahaman, S., Jahangir, S., Chen, R., & Kumar, P. (2022). Restricted anthropogenic activities and improved urban air quality in China: Evidence from real-time and remotely sensed datasets using air quality zonal modeling. *Atmosphere, 13*, 961.
San José, R., Pérez, J. L., Balzarini, A., Baró, R., Curci, G., Forkel, R., Galmarini, S., Grell, G., Hirtl, M., Honzak, L., Im, U., Jiménez-Guerrero, P., Langer, M., Pirovano, G., Tuccella, P., Werhahn, J., & Žabkar, R. (2015). Sensitivity of feedback effects in CBMZ/mosaic chemical mechanism. *Atmospheric Environment, 115*, 646–656.

Seinfeld, J. H., Pandis, S. N., & Noone, K. (1998). atmospheric chemistry and physics: From air pollution to climate change. *Physics Today, 51*, 88–90.

Skamarock, W. C., Klemp, J. B., Dudhia, J., Gill, D. O., Barker, D. M., Duda, M. G., & Powers, J. G. (2008). *A description of the Advanced Research WRF version 2. NCAR Technical Note NCAR/TN-468+ STR, 113.*

Sulaymon, I. D., Zhang, Y., Hopke, P. K., Zhang, Y., Hua, J., & Mei, X. (2021). Covid-19 pandemic in Wuhan: Ambient air quality and the relationships between criteria air pollutants and meteorological variables before, during, and after lockdown. *Atmospheric Research, 250*, 105362.

Tobías, A., Carnerero, C., Reche, C., Massagué, J., Via, M., Minguillón, M. C., Alastuey, A., & Querol, X. (2020). Changes in air quality during the lockdown in Barcelona (Spain) one month into the SARS-COV-2 epidemic. *Science of The Total Environment., 726*, 138540.

Venter, Z. S., Aunan, K., Chowdhury, S., & Lelieveld, J. (2020). Covid-19 lockdowns cause global air pollution declines with implications for public health risk. *Environmental Sciences, 117*, 18984–18990.

Wang, T., Du, H., Zhao, Z., Zhang, J., & Zhou, C. (2022). Impact of meteorological conditions and human activities on air quality during the COVID-19 lockdown in Northeast China. *Frontiers in Environmental Science, 10*, 877268.

Wang, Y., Wen, Y., Wang, Y., Zhang, S., Zhang, K. M., Zheng, H., Xing, J., Wu, Y., & Hao, J. (2020). Four-month changes in air quality during and after the COVID-19 lockdown in six megacities in China. *Environmental Science & Technology Letters, 7*, 802–808.

Wie, J., & Moon, B.-K. (2018). Impact of the western North Pacific subtropical high on summer surface ozone in the Korean peninsula. *Atmospheric Pollution Research, 9*, 655–661.

Yadav, A. K. (2015). Elemental composition and source apportionment of suspended particulate matters and health risk assessment in mining and nonmining areas of Odisha, India. *Journal of Hazardous, Toxic, and Radioactive Waste, 19*(3), 04014037.

Yumin, L., Shiyuan, L., Ling, H., Ziyi, L., Yonghui, Z., Li, L., Yangjun, W., & Kangjuan, L. (2021). The casual effects of covid-19 lockdown on air quality and short-term health impacts in China. *Environmental Pollution, 290*, 117988.

Zaveri, R. A., Easter, R. C., Fast, J. D., & Peters, L. K. (2008). Model for simulating aerosol interactions and chemistry (mosaic). *Journal of Geophysical Research, 113*, 8782.

Zaveri, R. A., & Peters, L. K. (1999). A new lumped structure photochemical mechanism for large-scale applications. *Journal of Geophysical Research-Atmospheres, 104*, 30387–30415.

Chapter 18
Intuitionistic Fuzzy Multi-Criteria Decision-Making Technique for the Selection of Property

Vijay Kumar

18.1 Introduction

A beautiful property in a good locality is a fantasy for everyone, also in a pollution-free environment on the lap of Mother Earth. It is a fundamental requirement for every living being in the universe, the selection of which is a complicated process and needs some sort of robust decision-making because of the involvement of certain factors. The requirements of an individual vary with time, but a property in a clean environment is of utmost priority to everyone. Real estate companies offer a variety of options to customers; it is very difficult to decide on the right choice. A robust decision support system is required to get the right property that covers all the related factors. The factors that influence the selection of properties are vague and have a great deal of ambiguity. Some of the key factors considered in this study are budget, average air quality index (AQI) of the locality, developers' reputation, basic amenities, and so on. In other words, an individual needs a property in a pollution-free environment with all basic facilities with good construction quality within their budget. In the real estate context, the features of a flat can be divided into three groups, such as structural attributes, nearby services, and environmental resources. Due to urbanization, clean air is the key selling feature in the real estate sector. Many real estate developers highlight this feature in their project details and make their projects costly, and sometimes the cost is beyond the reach of a common man. Many researchers have validated in their research that air quality is one of the important components for property buyers, regardless of size and other features.

In real-life situations, we encounter deterministic and non-deterministic phenomena. Out of this, non-deterministic phenomena face more challenges in the study of decision-making processes. The study of non-deterministic phenomena

V. Kumar (✉)
Department of Applied Sciences, School of Engineering and Technology, Manav Rachna International Institute of Research and Studies, Faridabad, India

A. K. Yadav (ed.), *Computational Approach for Air Pollution*,
https://doi.org/10.1007/978-3-032-09215-1_18

was taken up by statistics if the occurrences of patterns in the process are stable. In real-life applications, a large amount of imprecise and vague information has been generated, which is beyond the scope of existing mathematical models. To handle situations that contain vague and precise information, Zadeh (1965) proposed the concept, which is the generalization of classical set theory, known as fuzzy set theory, that has the potential to deal with incomplete and imprecise information. Therefore, fuzzy set theory is very popular among the community of researchers working in the domains of decision-making. The generalization of fuzzy sets was introduced by Atanassov (1986) as an intuitionistic fuzzy set (IFS), which contains a degree of membership and non-membership grades, respectively. In special cases, IFS describes the fuzzy nature more comprehensively than fuzzy sets. In the era of technology, many decision support systems have been developed that help to solve routine decision-making problems easily under certain favorable or unfavorable circumstances. Decision-support tools provide support to the decision-makers that allow the discipline of multi-criteria decision-making (MCDM) to grow in every sphere of the universe. Many domains such as medical diagnosis, decision theory, pattern recognition, and so on have been handled well using fuzzy sets and their generalizations because of the presence of uncertainty in the input information. IFSs have emerged as a potential theory for dealing with real-life problems in uncertain situations. In this chapter, a decision support system has been proposed, which supports an individual to select the better property in uncertain situations. A mathematical model for the selection of properties with a primary focus on a pollution-free environment using the generalized fuzzy information theory measure of Tsallis.

18.2 Air Quality

Air quality is difficult to define and varies from place to place. Poor air quality has health implications, including respiratory problems, heart disease, and other auxiliary health problems. Previously, poor air quality was only confined to industrial towns and densely populated areas, but due to urbanization, it is almost everywhere, and sometimes during the winter season, it is worse (Yadav et al., 2022). Many respiratory diseases such as asthma, bronchitis, and other lung diseases perpetuate cardiovascular diseases, cognitive function, and mental health, affecting overall well-being, including headaches, irritation of the eyes, nose, throat, and so on. Pregnant women can have adverse birth outcomes. It affects outdoor activities and travel and worsens the work and home environment.

The AQI is the key indicator for understanding the quality of air in which we live in. It is a numerical value that indicates the levels of air pollution and its hazardous implications for health. With the increasing health concerns, understanding AQI is essential. Its range varies from 0 to 500, and its levels indicate better and worse air quality in that region. The AQI index is divided into six categories, and each category is assigned a different color, as shown in Fig. 18.1. In 2014, AQI was launched in India under the Swachh Bharat Mission for the general public to provide

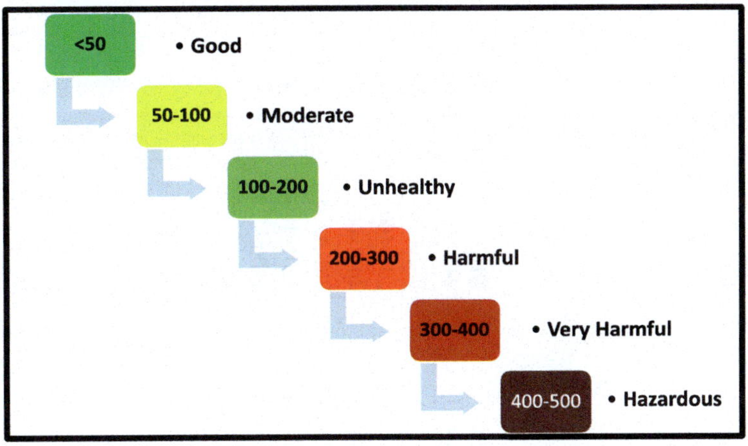

Fig. 18.1 Levels of AQI. (Source: Air Quality Standards, Central Pollution Control Board, New Delhi)

real-time reporting and forecasting of air quality in that region. In addition, it is a tool to communicate the actual state of air quality to people so that they may understand the status of air quality and take necessary action. Satisfactory AQI levels become the best choice for the property in that locality.

The United States has regulation of air quality since the 1970s, and a separate act has been established as the Clean Air Act. Under this act, air emission regulations were established to protect public health from hazardous air pollutants. Air pollution comprises gases and solid particles that change the characteristics of the air we breathe. In other words, air contains harmful pollutants, such as particulate matter (PM10), particulate matter ($PM_{2.5}$), nitrogen dioxide (NO_2), sulfur dioxide (SO_2), carbon monoxide (CO), ozone (O_3), ammonia (NH_3), and lead (Pb), which cannot be detected by the human eye. Much of such pollution spreads through road/construction dust, vehicular exhaust, burning of garbage/crop residues, industrial emissions, thermal power plant exhausts, chemical fumes, a spray of insecticides/pesticides, and so on. Air quality affects everyone, but mostly children, pregnant women, old people, traffic policemen, auto-rickshaw drivers, street vendors, and so on. In this chapter, the relationship between housing prices and air pollution in the locality has been addressed. The AQI category of a particular area at a given time depends on the level of pollutants present in the air, given in Table 18.1.

Individuals who have the ability to pay a hefty price to buy property in the vicinity of the area with better air quality are willing to pay a high price to buy property in the area. Therefore, real estate developers must keep this point in mind while offering property opportunities to the public. Making decisions in real-life situations is very crucial for everyone. From the available set of alternatives, selecting the best alternative is a million-dollar task. All the domains that exist in the universe such as optimization, pattern recognition, engineering, medicine, image processing, and so on require decision-making. In real-life situations, the information available

Table 18.1 Breakpoint for AQI according to the pollutants

AQI category (range)	PM_{10} 24 h	$PM_{2.5}$ 24 h	NO_2 24 h	O_3 8 h	CO 8 h (mg/m^3)	SO_2 24 h	NH_3 24 h	PB 24 h
Good (0–50)	0–50	0–30	0–40	0–50	0–1.0	0–40	0–200	0–0.5
Satisfactory (51–100)	51–100	31–60	41–80	51–100	1.1–2.0	41–80	201–400	0.6–1.0
Moderate (101–200)	101–250	61–90	81–180	101–168	2.1–10	81–380	401–800	1.1–2.0
Poor (201–300)	251–350	91–120	181–280	169–208	10.1–17	381–800	801–1200	2.1–3.0
Very poor (301–400)	351–430	121–250	281–400	209–748	17.1–34	801–1600	1201–1800	3.1–3.5
Severe (401–500)	430^+	250^+	400^+	748^+	34^+	1600^+	1800^+	3.5^+

Source: Ahuja, Astha (2019): Report of the National AQI by Central Pollution Control Board, New Delhi

is uncertain or imprecise; therefore, the intervention of MCDM-based computational methods is necessary to make effective decision-making. Under such situations, fuzzy theory gives remarkable results. Among these, IFS theory plays a vital role in tackling imprecise or vague situations in special cases and helps in the development of information-theoretic measures for decision-making purposes.

18.3 Literature Review

Ridker and Henning (1967) established a relationship between air quality and housing prices and used regression models to investigate whether air pollution has a significant impact on housing prices. After this, many researchers participated in this and conducted various studies to show that there is an impact of air pollution on property prices. Chay and Greenstone (2005) analyzed property prices in cities where pollution levels are above the government norms. They found that housing prices increase in places with good air quality and that follow all the compliance with environmental regulations. Choi et al. (2023) recently conducted a study to show that degraded air quality has a direct and long-term impact on housing prices in that locality. They studied the housing prices of Los Angeles City during and after the Aliso Canyon gas leak tragedy within an 8 km radius of the gas well and experienced an additional 8–13% drop in the prices, even after the air quality was restored. On the same note, Borja-Urbano et al. (2021) studied the effect of air pollution on urban housing prices in almost all polluted cities of Ecuador, which impact prices statistically significant with a decrease in value from about 1% to 3%. Zou et al. (2022) conducted a study in China using machine learning techniques to investigate the impact of air pollution on housing prices in Shanghai and found that the impact is very low. Cai et al. (2024) conducted studies on the same parameters in China and found that for housing, high-income groups pay an extra amount for cleaner air localities.

The South Korean housing market on air pollution has been studied by Lu and Lee (2022) using the two-stage spatial model, which found that about a 1% increase in air pollution levels in residential localities causes a decrease in the value by about 0.5% in the surroundings of the locality. Wang (2006) investigated in their studies that there is a positive correlation between air quality and residential property prices in large cities with people having additional paying capabilities for air quality, wherein small- and medium-sized cities have a negative correlation for the same situation. Researchers such as Miłuch and Kopczewska (2024) and Tang and Niemeier (2021) investigated in their studies the impact of air quality on housing prices. They conducted studies in some cities in Poland, California, and so on to study economic significance. In recent years, Asif (2024) discussed that the growth of real estate depends on environmental factors, including air quality and a lush green environment. Poor air quality can significantly affect real estate prices, and places with poor AQI levels have slower property sales. Air is the driving force to maintain a disease-free routine at all stages of life.

According to the hedonic theory of price, lack of environmental amenities poses a negative impact on housing prices, and the rest of the feature variables remain constant. Kim and Yoon (2019) studied housing prices in Seoul while looking for the concentration of PM10 annually and the average change in air quality between 2015 and 2016. The authors studied 1 lakh housing transactions and investigated how air quality impacts property value. From these studies, it is observed that a potential home buyer is more sensitive to air quality when the differences between two locations are more substantial than within a city. Researchers used linear logarithmic, spatial error models, and spatial autoregressive models to study the relationship between real estate prices and air pollution. According to Picard et al. (2010), every home has four qualities, including the physical appearance of the property, demography, accessibility, and environmental quality. Researchers such as Vijay et al. (2016), Smith and Huang (1995) , Wei et al. (2012), and Kumar and Jain (2018) used fuzzy sets and IFS in their research for decision-making in various domains of their studies.

18.4 Information-Theoretic Measures

We would start with a probabilistic approach and then take up fuzzy and then intuitionistic fuzzy approaches.

18.4.1 Probability Approach

Entropy is a quantitative measure of randomness in a system, which is used to calculate uncertainty. The term entropy was coined by Shannon (1948), which gave birth to a separate discipline, namely, information theory (Ash, 2012). Shannon provides mathematical insights into the concepts of information by demonstrating the proofs of the general results along with their consequences. After that, entropy has been extensively applied in different domains of science and technology, including quantum computing, communication theory, image processing, decision-making, and so on. Shannon entropy has generalizations proposed by Renyi (1961), Tsallis (1988), Havrda and Charvát (1967), Pal and Pal (1989), and so on. These generalized entropy measures have applications in various domains of study such as pattern recognition, image extraction, feature selection, decision-making, and so on.

18.4.2 Fuzzy Approach

Zadeh (1965) introduced the concept of fuzzy sets to handle uncertain problems. Since its inception, it has been applied to the problems of decision-making and deal with imprecise and uncertain information gathered by humans. In real-life applications, information is available in both qualitative and quantitative forms. Fuzzy sets have the inherent potential to effectively handle both types of information and to develop a human-friendly ecosystem of rules. This can be seen in a medical diagnosis in which imprecise and uncertain information is available, which is required to establish a relationship between symptoms and diseases. Fuzzy sets deal with such situations very well by assigning some degree of incompleteness to the diagnosis process and making decisions by addressing certain factors, which may be overlooked by the doctor during the diagnosis process. In such situations, many fuzzy-based models intervene and can help establish certain relationships such as the variation of important symptoms, the pattern of symptoms of different diseases, the relationship between the diseases and their stages, and initial and final diagnosis. Fuzzy-based models are useful when the diseases are in the initial stage, which can be iterated and configured according to the requirement. Sanchez (1993, 1996) proposed a similar type of relationship that represents the knowledge base by establishing a fuzzy max–min relationship between the symptoms and the disease. Fuzzy theory provides promising results in uncertain situations, irrespective of any domain.

18.4.3 Intuitionistic Fuzzy Approach

IFS is the generalization of the fuzzy set that has an additional function, known as the non-membership function, to fuzzy sets that have only a membership function. It has the advantage over fuzzy sets to deal with real-life problems in special cases. Many scholars have researched entropy measures based on IFS. Burillo and Bustince (1996) introduced the first axiom of IF entropy. It was a distance measure to calculate the degree of intuitionism associated with IFS. Szmidt and Kacprzyk (2001a) extended the definition of IF entropy, which was the extension of (De Luca & Termini, 1993), and proposed a new entropy measure for IFS. For the past many years, researchers have been busy using generalized fuzzy sets for decision-making purposes. To cite a few, Verma and Sharma (2013), Mishra (2016), Wei et al. (2012), Jianqiang and Zhong (2009), Liu and Ren (2014), Bai (2013), Boran et al. (2009), Burillo and Bustince (1996), Das et al. (2016), Gani et al. (2011), Guorong (2011), Hung and Chen (2009), Kulak et al. (2015), Kumar and Jain (2018), Kumar et al. (2012), Szmidt and Kacprzyk (2001b), Verma and Sharma (2012), Wei et al. (2012), Wang (2006), Wang and Zhang (2008), Wang and Wang (2012), Wang et al. (2016), Wei and Ye (2010), Yager (2001, 2004, 2008, 2009), and Zhang (2014) introduced various forms of generalized fuzzy measures in various uncertain situations for the purpose of better decision-making.

18.5 Decision-Making

In decision-making, the selection of the best alternative from a finite number of available alternatives is made for a subsequent course of action. Decision-making is not an activity, but an art that is essential to solving real-life problems. Researchers in various disciplines have shown great interest in solving real-world problems related to decision-making by using various techniques. In real-life situations, usually, ambiguous or imprecise information is available, and in such cases, fuzzy and IFS theories have emerged as powerful tools. Solving such real-life situations flavored with information measures in fuzzy and IFS environments marks the beginning of a new discipline as fuzzy information theory. Information-theoretical measures have been deployed to solve many real-life situations in the fuzzy and IF domains to deal with ambiguous situations. For this, we must suggest a correct mathematical model that acts as a tool to solve real-life problems. Researchers have developed many decision-making techniques to handle such real-life situations. Decision-making is a vital activity to make decisions based on judgment and is a cognitive process. One of the favorable alternatives has been selected or the alternatives among the available to get better decision-making. In the deterministic problem-solving process, decision-makers take the situation as an optimization problem, whereas in nondeterministic problems, the problem-solving concept contains uncertainty, and making decisions under uncertainty is quite challenging. In these situations, a lot of incomplete information makes the process of evaluating the criteria of alternatives. In such a situation, IFS theory can model and handle vague information and situations very well. There are varieties of situations in which decision-making can be performed such as individual decision-making, multiperson decision-making, multistage decision-making, and MCDM. Of these, MCDM is used frequently in real-life situations and can handle incompleteness problems.

18.6 Multi-Criteria Decision-Making

MCDM is a sub-domain of decision-making that develops decision support systems to evaluate performance criteria. It encompasses the domains of science and technology to solve real-world problems. For more than six decades, MCDM has grown significantly, providing real-time support to decision-makers to rank alternatives according to their order of preference by classifying them into smaller groups. MCDM techniques have been used across the domains of science and technology by various researchers, but MCDM techniques under a fuzzy environment provide promising results. MCDM problems are those in which there is a relationship between the set of alternatives against the given set of factors with the objective to rank the alternatives according to the order of preferences. A set of domain experts is involved in the process to provide initial input in terms of linguistic variables.

These variables are the basis for setting up the decision framework to rank alternatives to get the performance rating of alternatives. The goal of the whole process is to generate meaningful information from the available incomplete/uncertain information to provide a better solution.

18.7 Basic Preliminaries

18.7.1 Crisp Set

The crisp set is a classical set. A crisp set θ of C is defined as follows:

$$\exists \quad f_\theta(c) : C \to \{0, 1\},$$

where C is the universal set and

$$f_\phi(t) = \begin{Bmatrix} 1, & \text{if } t \in \phi \\ 0, & \text{if } t \notin \phi \end{Bmatrix}.$$

In a classical set, an element is either present in the set or not.

18.7.2 Fuzzy Set

Zadeh (1965) proposed the generalization of the classical set, in which elements have a degree of membership. Lotfi Asker Zadeh is a mathematician and computer scientist, and a professor of computer science at the University of California. [[In mathematics, a set can be thought of as any collection of distinct things considered as a whole.

A fuzzy set ξ is defined in a discrete universe of discourse $C = \{c_1, \ldots, c_n\}$ as follows: $\xi = \{< c, \mu_\xi(c) > : c \in C\}; C \to [0, 1]$.

where

1. μ_ξ is the membership function;
2. $\mu_\xi(c)$ is called the membership grade of $c \in C$ in ξ.

Fuzzy sets have many generalizations. IFSs are one such generalization that has been used much for application purposes to deal with real-world situations.

18.7.3 Intuitionistic Fuzzy Sets

Atanassov (1986) an IFS ψ in $C = \{c_1, \ldots, c_n\}$is given as follows

$$\psi = \{< c, | \mu_\psi(c), | \nu_\psi(c) >, | c \in C\}; C \to [0, 1]$$

where

1. μ_ψ and ν_ψ are the membership and non-membership functions of set ψ;
2. $\pi_\psi(c) = 1 - \mu_\psi(c) - \nu_\psi(c)$ is the intuitionistic index or hesitation index of c in ψ.

In the limiting case, if $\pi_\psi(c) = 0$, IFS → FS.
In this work, IFSs have been used for the purpose of decision-making.

18.7.4 Intuitionistic Fuzzy Relation

Based on Biswas (1997), if ψ is an IFS of C, the max-min composition of the IFR R ($U \to V$) with ψ is an IFS ζ of V denoted by $\zeta = R \circ \psi$ and has membership function $\mu_\zeta(y) = \max_{c \in C} \{\min [\ \mu_\psi(c),\ \mu_R(c,v)\ ]\}$ and non-membership function $\nu_\zeta(y) = \min_{c \in C} \{\max [\ \nu_\psi(c),\ \nu_R(c,v)\ ]\}, \forall\ v \in V$.

18.7.5 Tsallis Measure of Entropy

Tsallis (1988) introduced a non-extensive measure of entropy as a generalization of the Boltzmann–Shannon–Gibbs theory. For a random variable X with a probability mass function $p = (p_1, \ldots, p_n)$, the Tsallis measure of entropy is defined as follows:

$$H_\alpha^{(T)}(p) = \frac{1}{1-\alpha}\left[\sum_{i=1}^{N} (p_i)^\alpha - 1\right]$$

where $\alpha > 0$ and is called as non-extensive index.

The IF version of the Tsallis measure of entropy is defined as follows:

$$H_\alpha^{(T)}(p) = \frac{1}{1-\alpha}\left[\sum_{i=1}^{N}\left[(\mu_i)^\alpha + (\nu_i)^\alpha + (\pi_i)^\alpha\right] - 1\right], \quad \alpha > 0.$$

Tsallis entropy is the limiting case of Shannon entropy for $\alpha \to 1$.

In this chapter, the Tsallis measure in the IF environment has been used to select the best property among the available properties under the defined parameters.

18.8 Relation Between Factors and Brands

For a particular locality, air quality with an AQI of less than 100 is the top priority for the customer to buy property. Purchasing a home is a complex and challenging task and thus requires effective decision-making skills. The proposed decision support system provides a tool to buyers by which they can select the best property based on certain factors, with the main focus on AQI according to their needs and desires. A case study has been developed on the basis of the information about customers/buyers, available projects offered by various real estate developers, and their factors that affect the purchase of the property with assigned degrees of membership, non-membership, and hesitation index. A composition among sets has been taken using the IF max–min relation. The mathematical model for the selection of property using the Tsallis fuzzy information theory measure has been proposed to select property in a clean environment. Using this model, a buyer can assess a need and desire to buy their house and study various real estate developers without even having sound knowledge about it. In addition, buyers can cross-check the results with trusted real estate developers.

18.9 Methodology

Before buying any property, customers must analyze their needs and desires to select the best location according to the AQI. He should fix certain factors that are responsible for the selection of the property and analyze those factors available in various existing real estate developers. In this way, they should fix the real estate developers and the related factors available in their projects, which could help them in the decision-making process. To achieve the objectives of the study for the selection of property according to the given facts, the following are the steps:

- Establish the IF relationship between customers and factors that affect the selection for the purchase of property by assigning some membership and non-membership grades, given in Table 18.2.
- Establish the IF relationship between factors and the available information about the projects offered by various real estate developers by assigning some membership and non-membership grades, given in Table 18.3.

Table 18.2 Membership and non-membership values of customers and factors

	f_1		f_2		f_3		f_4		f_5	
Q	μ	ν	μ	ν	μ	ν	μ	ν	μ	ν
c_1	0.7	0.2	0.6	0	0.3	0.7	0.5	0.2	0	0.6
c_2	0	0.6	0.5	0.3	0.6	0.1	0	0.6	0	0.9
c_3	0.5	0.2	0.7	0.2	0	0.5	0.2	0.3	0.3	0.4
c_4	0.4	0	0.3	0.5	0.2	0.3	0.8	0.1	0.5	0.3

Table 18.3 Membership and non-membership values of factors and the selected real estate developers

R	RE_1		RE_2		RE_3		RE_4		RE_5	
	μ	ν	μ	ν	μ	ν	μ	ν	μ	ν
f_1	0.7	0.1	0.3	0.1	0.4	0.4	0.2	0.6	0.2	0.7
f_2	0.4	0.5	0.8	0	0.5	0.3	0.3	0.3	0	0.6
f_3	0.1	0.5	0.2	0.7	0.6	0.2	0.5	0.3	0.2	0.4
f_4	0.3	0.2	0.5	0.3	0.6	0.1	0.7	0.2	0.4	0.3
f_5	0.2	0.5	0.1	0.6	0.2	0.7	0.3	0.6	0.9	0

Table 18.4 Membership and non-membership values of customers and selected real estate developers

T	RE_1		RE_2		RE_3		RE_4		RE_5	
	μ	ν	μ	ν	μ	ν	μ	ν	μ	ν
c_1	0.7	0.2	0.6	0	0.5	0.2	0.5	0.2	0.4	0.3
c_2	0.4	0.5	0.5	0.3	0.6	0.2	0.5	0.3	0.2	0.4
c_3	0.5	0.2	0.7	0.2	0.5	0.3	0.3	0.3	0.3	0.3
c_4	0.4	0.1	0.5	0.1	0.6	0.1	0.7	0.2	0.5	0.3

Table 18.5 Decision values of customers and the selected real estate developer for $\alpha = 0.1$

T	RE_1	RE_2	RE_3	RE_4	RE_5
c_1	1.7896	**0.9585**	1.8566	1.8566	1.8729
c_2	**1.8220**	1.8566	1.8365	1.8566	1.8625
c_3	1.8566	**1.7896**	1.8566	1.8729	1.8729
c_4	1.8220	1.8220	1.8123	**1.7896**	1.8566

Table 18.6 Customer decision values and the selected real estate developer for $\alpha = 0.3$

T	RE_1	RE_2	RE_3	RE_4	RE_5
c_1	1.4525	**0.8822**	1.6088	1.6088	1.6476
c_2	**1.5330**	1.6088	1.5600	1.6088	1.6234
c_3	1.6088	**1.4525**	1.6088	1.6476	1.6476
c_4	1.5330	1.5330	1.5085	**1.4525**	1.6088

- The max–min and min–max composition of the IF relations (Table 18.4) has been used to establish the relationship between the customers and the given real estate developers, given in Table 18.4.
- To handle uncertainty and perform better decision-making, the Tsallis entropy measure in an IF environment has been used for different values of the parameter α, as explained in Tables 18.5, 18.6, 18.7, 18.8, 18.9, 18.10, 18.11, and 18.12.

The customer with the minimum weight has been taken up for decision-making. However, if there is a tie in weight elements, both types of developers can be considered for decision-making and the flow chart of the methodology (Fig. 18.2).

Table 18.7 Decision values of customers and the selected real estate developer for $\alpha = 0.5$

T	RE_1	RE_2	RE_3	RE_4	RE_5
c_1	1.2002	**0.8141**	1.4041	1.4041	1.4558
c_2	**1.3116**	1.4041	1.3380	1.4041	1.4242
c_3	1.4041	**1.2002**	1.4041	1.4558	1.4558
c_4	1.3116	1.3116	1.2771	**1.2002**	1.4041

Table 18.8 Decision values of customers and the selected real estate developer for $\alpha = 0.7$

T	RE_1	RE_2	RE_3	RE_4	RE_5
c_1	1.0090	**0.7531**	1.2341	1.2341	1.2919
c_2	**1.1388**	1.2341	1.1588	1.2341	1.2575
c_3	1.2341	**1.0090**	1.2341	1.2919	1.2919
c_4	1.1388	1.1388	1.0980	**1.0090**	1.2341

Table 18.9 Decision values of customers and the selected real estate developer for $\alpha = 0.9$

T	RE_1	RE_2	RE_3	RE_4	RE_5
c_1	0.8623	**0.6983**	1.0919	1.0919	1.1515
c_2	**1.0016**	1.0919	1.0129	1.0919	1.1169
c_3	1.0919	**0.8623**	1.0919	1.1515	1.1515
c_4	1.0016	1.0016	0.9572	**0.8623**	1.0919

Table 18.10 Decision values of customers and the selected real estate developer for $\alpha = 1.1$

T	RE_1	RE_2	RE_3	RE_4	RE_5
c_1	0.7483	**0.6490**	0.9725	0.9725	1.0308
c_2	**0.8907**	0.9725	0.8934	0.9725	0.9978
c_3	0.9725	**0.7483**	0.9725	1.0308	1.0308
c_4	0.8907	0.8907	0.8448	**0.7483**	0.9725

Table 18.11 Decision values of customers and the selected real estate developer for $\alpha = 1.3$

T	RE_1	RE_2	RE_3	RE_4	RE_5
c_1	0.6584	**0.6046**	0.8714	0.8714	0.9268
c_2	**0.7996**	0.8714	0.7948	0.8714	0.8962
c_3	0.8714	**0.6584**	0.8714	0.9268	0.9268
c_4	0.7996	0.7996	0.7536	**0.6584**	0.8714

Table 18.12 Decision values of customers and the selected real estate developer for $\alpha = 1.5$

T	RE_1	RE_2	RE_3	RE_4	RE_5
c_1	0.5865	**0.5645**	0.7854	0.7854	0.8368
c_2	**0.7037**	0.7854	0.7127	0.7854	0.8092
c_3	0.7854	**0.5865**	0.7854	0.8368	0.8368
c_4	0.7237	0.7237	0.6786	**0.5865**	0.7854

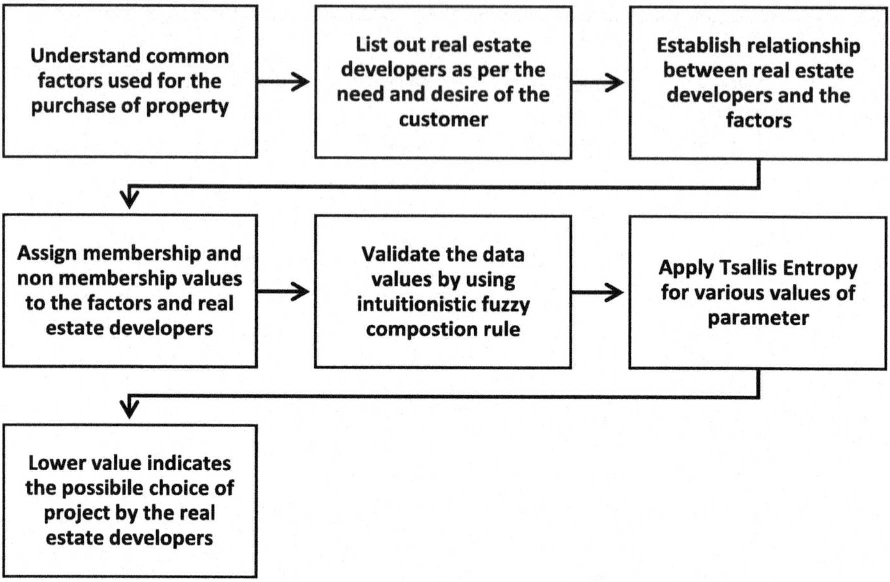

Fig. 18.2 Flow chart of the methodology

18.10 Case Study: Selection of Property from the Available Real Estate Developers

A case study of the factors that characterize each customer and the factors that characterize the available real estate developers has been framed as follows:

Consider the following notations:

1. Factors: F = {Budget, Security, Average AQI of the locality, Developer Reputation, Basic Amenities}
2. Customers: $C = \{ c_1, c_2, c_3, c_4\}$
3. Real Estate Developers: $RE = \{ RE_1, RE_2, RE_3, RE_4, RE_5\}$
4. IF relation between customers and factors: $Q(C \to F)$
5. IF relation between factors and the available real estate developers: $R(F \to RE)$
6. IF relation between the customers and available real estate developers: $T(C \to RE)$

18.11 Conclusion

From Tables 18.5, 18.6, 18.7, 18.8, 18.9, 18.10, 18.11, and 18.12, it is concluded that the customers c_1 and c_3 can purchase a property offered by the real estate developer RE_2. Customer c_2 can purchase a property offered by the real estate developer RE_3, and customer c_4 can purchase a property offered by the real estate developer

RE_4. All the decision values have shown the same trend, and the minimum value is the decision value. This method can help the common man in selecting the property as per the AQI. It not only helps the customer in the selection of property but also helps them in the selection of trusted real estate developer projects. The proposed model can also be applied to other decision-making problems.

Acknowledgments I would like to thank the referees for their helpful remarks.

References

Ahuja, A. (2019). *Air pollution: What is air quality index, how is it measured and its health impact*. Swachh Guide.

Ash, R. B. (2012). *Information theory*. Courier Corporation.

Asif, M. (2024). The impact of air quality on real estate demand. https://www.linkedin.com/pulse/impact-air-quality-real-estate-demand-muhammad-asif-epl7f/

Atanassov, K. T. (1986). Intuitionistic fuzzy sets. *Fuzzy Sets and Systems, 20*(1), 87–96.

Bai, Z. Y. (2013). An interval-valued intuitionistic fuzzy TOPSIS method based on an improved score function. *The Scientific World Journal, 2013*(1), 879089.

Biswas R. (1997) Intuitionistic fuzzy relations. *Bull. Sous. Ens. Flous.Appl. (BUSEFAL)*, Vol. 70, pp. 22–29.

Boran, F. E., Genç, S., Kurt, M., & Akay, D. (2009). A multi-criteria intuitionistic fuzzy group decision making for supplier selection with TOPSIS method. *Expert Systems with Applications, 36*(8), 11363–11368.

Borja-Urbano, S., Rodríguez-Espinosa, F., Luna-Ludeña, M., & Toulkeridis, T. (2021). Valuing the impact of air pollution in urban residence using hedonic pricing and geospatial analysis, evidence from Quito, Ecuador. *Air, Soil and Water Research, 14*, 11786221211053277.

Burillo, P., & Bustince, H. (1996). Entropy on intuitionistic fuzzy sets and on interval-valued fuzzy sets. *Fuzzy Sets and Systems, 78*(3), 305–316.

Cai, Y., Smit, M. J., & Helbich, M. (2024). Economic effects of air quality on housing prices: evidence from Beijing, China. *Journal of Housing and the Built Environment, 39*(2), 885–908. https://doi.org/10.1007/s10901-023-10108-z

Chay, K. Y., & Greenstone, M. (2005). Does air quality matter? Evidence from the housing market. *Journal of Political Economy, 113*(2), 376–424.

Choi, A., Kim, P., & Park, A. (2023). The effects of air quality on housing prices: Evidence from the Aliso canyon gas leak. *Journal of Housing Economics, 62*, 101966.

Das, S., Guha, D., & Dutta, B. (2016). Medical diagnosis with the aid of using fuzzy logic and intuitionistic fuzzy logic. *Applied Intelligence, 45*, 850–867.

De Luca, A., & Termini, S. (1993). A definition of a nonprobabilistic entropy in the setting of fuzzy sets theory. In *Readings in fuzzy sets for intelligent systems* (pp. 197–202). Morgan Kaufmann.

Gani, A. N., Sritharan, N., & Kumar, C. A. (2011). Weighted average rating (WAR) method for solving group decision making problem using an intuitionistic trapezoidal fuzzy hybrid aggregation (ITFHA) operator. *International Journal of Pure and Applied Sciences and Technology, 6*(1), 54–61.

Guorong, X. (2011). Models for multiple attribute decision making with intuitionistic trapezoidal information. *International Journal of Advancement in Computing Technology, 3*(6), 21–25.

Havrda, J., & Charvát, F. (1967). Quantification method of classification processes. Concept of structural α-entropy. *Kybernetika, 3*(1), 30–35.

Hung, C. C., & Chen, L. H. (2009). A multiple criteria group decision making model with entropy weight in an intuitionistic fuzzy environment. In *Intelligent automation and computer engineering* (pp. 17–26). Springer.

Jianqiang, W., & Zhong, Z. (2009). Aggregation operators on intuitionistic trapezoidal fuzzy number and its application to multi-criteria decision making problems. *Journal of Systems Engineering and Electronics, 20*(2), 321–326.

Kim, S. G., & Yoon, S. (2019). Measuring the value of airborne particulate matter reduction in Seoul. *Air Quality, Atmosphere & Health, 12*, 549–560.

Kulak, O., Goren, H. G., & Supciller, A. A. (2015). A new multi criteria decision making approach for medical imaging systems considering risk factors. *Applied Soft Computing, 35*, 931–941.

Kumar, V., Bharti, I., & Sharma, Y. K. (2012). Fuzzy diagnosis procedure of the types of glaucoma. *International Journal of Applied Information Systems, 1*(6), 42–45.

Kumar, V., & Jain, S. (2018). Alternate procedure for the diagnosis of malaria via intuitionistic fuzzy sets. In *Nature inspired computing: Proceedings of CSI 2015* (pp. 49–53). Springer.

Liu, M., & Ren, H. (2014). A new intuitionistic fuzzy entropy and application in multi-attribute decision making. *Information, 5*(4), 587–601.

Lu, J., & Lee, Y. (2022). Spatial effects of air pollution on the housing market: Evidence from South Korea. *Journal of Real Estate Research, 44*(1), 131–150. https://doi.org/10.1080/08965803.2021.2008109

Miłuch, O., & Kopczewska, K. (2024). Fresh air in the city: The impact of air pollution on the pricing of real estate. *Environmental Science and Pollution Research, 31*(5), 7604–7627.

Mishra, A. R. (2016). Intuitionistic fuzzy information measures with application in rating of township development. *Iranian Journal of Fuzzy Systems, 13*(3), 49–70.

Pal, N. R., & Pal, S. K. (1989). Object-background segmentation using new definitions of entropy. *IEE Proceedings E (Computers and Digital Techniques), 136*(4), 284–295.

Picard, N., Antoniou, C., & De Palma, A. (2010). *Econometric models. Sustain City* (Vol. 2). Working paper.

Renyi, A. (1961). On measures of entropy and information. In *Proceedings of the fourth Berkeley symposium on mathematical statistics and probability, volume 1: contributions to the theory of statistics* (Vol. 4, pp. 547–562). University of California Press.

Ridker, R. G., & Henning, J. A. (1967). The determinants of residential property values with special reference to air pollution. *The Review of Economics and Statistics, 49*, 246–257.

Sanchez, E. (1993). Solutions in composite fuzzy relation equations: Application to medical diagnosis in Brouwerian logic. In *Readings in fuzzy sets for intelligent systems* (pp. 159–165). Morgan Kaufmann.

Sanchez, E. (1996). Truth-qualification and fuzzy relations in natural languages, application to medical diagnosis. *Fuzzy Sets and Systems, 84*(2), 155–167.

Shannon, C. E. (1948). A mathematical theory of communication. *The Bell System Technical Journal, 27*(3), 379–423.

Smith, V. K., & Huang, J. C. (1995). Can markets value air quality? A meta-analysis of hedonic property value models. *Journal of Political Economy, 103*(1), 209–227.

Szmidt, E., & Kacprzyk, J. (2001a). Entropy for intuitionistic fuzzy sets. *Fuzzy Sets and Systems, 118*(3), 467–477.

Szmidt, E., & Kacprzyk, J. (2001b). Intuitionistic fuzzy sets in some medical applications. In *International conference on computational intelligence* (pp. 148–151). Springer.

Tang, M., & Niemeier, D. (2021). How does air pollution influence housing prices in the bay area? *International Journal of Environmental Research and Public Health, 18*(22), 12195.

Tsallis, C. (1988). Possible generalization of Bolzmann-Gibbs statistics. *Journal of Statistical Physics, 52*, 479–487.

Verma, R., & Sharma, B. D. (2012). On generalized intuitionistic fuzzy divergence (relative information) and their properties. *Journal of Uncertain Systems, 6*(4), 308–320.

Verma, R., & Sharma, B. D. (2013). Exponential entropy on intuitionistic fuzzy sets. *Kybernetika, 49*, 114–127.

Vijay, K., Hari, A., & Kiran, P. (2016). Intuitionistic trapezoidal fuzzy prioritized weighted geometric operator: an algorithm for the selection of suitable treatment for lung cancer. In *Proceedings of fifth international conference on soft computing for problem solving: SocProS 2015* (Vol. 1, pp. 605–615). Springer.

Wang, J. (2006). Multi-criteria decision-making approach with incomplete certain information based on ternary AHP. *Journal of Systems Engineering and Electronics, 17*(1), 109–114.
Wang, J., & Wang, P. (2012). Intuitionistic linguistic fuzzy multi-criteria decision-making method based on intuitionistic fuzzy entropy. *Control Decision, 27*, 1694–1698.
Wang, J. Q., & Zhang, Z. (2008). Programming method of multi-criteria decision-making based on intuitionistic fuzzy number with incomplete certain information. *Control and Decision, 23*(10), 1145–1148.
Wang, L. E., Liu, H. C., & Quan, M. Y. (2016). Evaluating the risk of failure modes with a hybrid MCDM model under interval-valued intuitionistic fuzzy environments. *Computers & Industrial Engineering, 102*, 175–185.
Wei, C. P., Gao, Z. H., & Guo, T. T. (2012). An intuitionistic fuzzy entropy measure based on trigonometric function. *Control and Decision, 27*(4), 571–574.
Wei, G. (2012). Hesitant fuzzy prioritized operators and their application to multiple attribute decision making. *Knowledge-Based Systems, 31*, 176–182.
Wei, P., & Ye, J. (2010). Improved intuitionistic fuzzy cross-entropy and its application to pattern recognitions. In *2010 IEEE international conference on intelligent systems and knowledge engineering* (pp. 114–116). IEEE.
Yadav, A. K., Shirin, S., Emmanouil, C., & Jamal, A. (2022). Effect of seasonal and meteorological variability of air pollution in Singrauli coalfield. *Aerosol Science and Engineering, 6*, 61–70.
Yager, R. R. (2001). Fusion of multi-agent preference orderings. *Fuzzy Sets and Systems, 117*(1), 1–12.
Yager, R. R. (2004). OWA aggregation over a continuous interval argument with applications to decision making. *IEEE Transactions on Systems, Man, and Cybernetics, Part B (Cybernetics), 34*(5), 1952–1963.
Yager, R. R. (2008). Prioritized aggregation operators. *International Journal of Approximate Reasoning, 48*(1), 263–274.
Yager, R. R. (2009). Prioritized OWA aggregation. *Fuzzy Optimization and Decision Making, 8*, 245–262.
Zadeh, L. A. (1965). Fuzzy sets. *Information and Control, 8*, 338–353.
Zhang, Z. (2014). Intuitionistic trapezoidal fuzzy prioritized operators and their application to multiple attribute group decision making. *British Journal of Mathematics & Computer Science, 4*(14), 1951.
Zou, G., Lai, Z., Li, Y., Liu, X., & Li, W. (2022). Exploring the nonlinear impact of air pollution on housing prices: A machine learning approach. *Economics of Transportation, 31*, 100272. https://doi.org/10.1016/j.ecotra.2022.100272

Chapter 19
Probabilistic and Computational Health Risk Modeling for Assessment of Air Pollution Exposure

Deep Chakraborty, Apurba Koley, Surajit Karmakar, Vidya Shanmugam, Nitu Gupta, Chandrani Sinharoy, Srinivasan Balachandran, Krishnendu Mukhopadhyay, and Sankar Sambandam

19.1 Introduction

Fossil fuel combustion, industrial discharges, wildfires, and the use of solid household fuel led to air pollution. Air pollution remains one of the most significant environmental health adversities, with millions of premature deaths attributed annually. Computational and probabilistic modeling has now formally become a method for measuring exposure risks, either directly or used to plan abatement tools. The current world situation regarding air pollution is a major public health issue, and the vast majority of the world's population is exposed to unsafe levels of air pollution (Marcantonio et al., 2021; Rentschler & Leonova, 2023; Yadav, 2021; Yu et al., 2023). Almost 94% (~7.3 billion) of the world's population is subject to unsafe levels of particulate matter (PM)2.5, and most of the time, exceed the most recent World Health Organization (WHO) 2021 guideline, 5 μg/m^3. Air pollution is a major concern, and many within low- and middle-income countries experience the highest exposure rates, for instance, regions including East Asia and the Pacific and South Asia have unsafe exposure rates of almost 95% and 99%, respectively, likely

D. Chakraborty (✉) · V. Shanmugam · K. Mukhopadhyay · S. Sambandam
Department of Environmental Health Engineering, Sri Ramachandra Faculty of Public Health, Sri Ramachandra Institute of Higher Education and Research (DU), Chennai, India
e-mail: deepchakraborty@sriramachandra.edu.in

A. Koley · C. Sinharoy · S. Balachandran
Department of Environmental Studies, Siksha-Bhavana (Institute of Science), Visva-Bharati, Santiniketan, India

S. Karmakar
CSIR-National Environmental Engineering Research Institute, Kolkata Zonal Centre, Kolkata, India

N. Gupta
Department of Environmental Science, Tezpur University, Tezpur, Assam, India

A. K. Yadav (ed.), *Computational Approach for Air Pollution*,
https://doi.org/10.1007/978-3-032-09215-1_19

due to technological and economic disadvantages when compared with alternative energy sources. Exposure to unsafe levels of $PM_{2.5}$ can increase mortality rates by 4% and hazardous exposure (>35 μg/m^3) by 24%. World Health Organization (2021) stated that since most air pollution is modifiable as it comes from a variety of vulnerable sources—particularly combustion emissions—they hope "that the economic recovery following this pandemic serves as an opportunity for world leaders to prioritize and speed up air quality improvements. In acknowledging the urgency driving this global call to action and the ability of government, business, and individuals to make even faster transitions to sustainably reduce air pollution emissions and improve air quality globally. Air pollution is more at risk for global public health than high blood pressure, other than high blood pressure being the leading modifiable risk factor" (Institute for Health Metrics and Evaluation, 2024). The general health impacts of air pollution are complex and enormous. Many millions of research studies on the health impacts of air pollution suggest that exposure to PM air pollution can lead to significant impacts from a health perspective. Specifically, studies have linked fine particles in air pollution with respiratory problems, cardiovascular disease, pulmonary diseases (asthma, chronic obstructive pulmonary disease (COPD), immune response function and imbalance, oxidative imbalance, and modified genes and epigenetics, leading to instances of lung cancer (Chen et al., 2017; Falcon-Rodriguez et al., 2016; Hime et al., 2018; Sokhi et al., 2021; Thurston et al., 2017). Pollutants such as NO_2, SO_2, O_3, and CO pose serious health risks and are tracked by air quality services in countries around the world. Children aged 0–5 years are the most vulnerable population to the health effects of air pollution: It is estimated that air pollution is responsible for more than 700,000 deaths every year in children under the age of 5 years worldwide, mostly due to household pollution in low-income areas (Health Effects Institute (HEI), 2024).

19.1.1 Status of the Global Burden of Diseases Due to Air Pollution

To assess the health of the population, you need to consider the mortality and morbidity of the population. This summation of mortality and morbidity is called the burden of disease, and it is measured using a comprehensive summary metric called "Disability Adjusted Life Years" (DALYs) (Murray & Lopez, 2013). A DALY is equivalent to loss of year of full health (WHO, 2020). In 1993, the first results of the Global Burden of Disease, Injuries, and Risk Factors Study (GBD) were published by Health Metrics and Evaluation (IHME) (Roser et al., 2021). Subsequently, the "Disease Burden Unit" was created in 1998 by the WHO and worked alongside IHME in publishing comprehensive reports (annually) on 371 diseases and injuries for 204 countries and territories (Murray, 2022; GBD (Global Burden of Disease), 2019). From 1990 to the present time, there has been a clear shift in paradigm from communicable to non-communicable diseases, caused by a number of risk factors, such as high blood pressure, tobacco smoking (including second-hand smoking), and household air pollution from solid fuels (Lajoie, 2015; Lim et al., 2010). Air pollution is the second leading risk factor for death. In 2021, 8% of the total disease burden was caused by PM air pollution (the leading contributor among risk factors). Notably, 48% of COPD is attributed to air pollution, and 34% of preterm births were attributed to exposure to air pollution (Global Burden of Disease (GBD), 2021). $PM_{2.5}$ air pollution is the leading cause of the burden of disease. Chronic exposure to $PM_{2.5}$ is associated with heart disease, lung cancer, pneumonia, and type 2 diabetes. The burden of air pollution varies globally. The regions with the highest ambient $PM_{2.5}$ are Southeast Asia, East Asia, and Oceania; South Asia; and North Africa and the Middle East. The most susceptible population to air pollution is children under 5 years. Children are affected more than adults due to their developing organs and lower immunity (Health Effects Institute (HEI), 2024).

19.1.2 Status of Burden of Disease Due to Air Pollution in India

Countries in South Asia and East, West, Central, and Southern Africa have the highest burden of disease linked to air pollution, with India recording a massive 1.24 million deaths in 2017 (Health Effects Institute (HEI), 2024; GBD MAPS Working Group, 2018). India is stated to have the highest level of outdoor air pollution in the world. Research has shown that 99.9% of the Indian population lives in areas that have exceeded the WHO Air Quality Guideline of 10 μg/m^3 for $PM_{2.5}$ (GBD MAPS Working Group, 2018). Some of the main contributors (to various degrees) of ambient $PM_{2.5}$ are residual biomass burning, coal burning, anthropogenic dust (related to human activities), agricultural burning, and brick production. The level of air pollution varies from state to state. States with low sociodemographic index (SDI), such as Bihar, Uttar Pradesh, Rajasthan, and Jharkhand, have the highest levels of air pollution (ambient and household). In 2017, 28% of global DALYs were attributed to air pollution, with 11% of premature deaths in people younger than 70 years due to the same (Balakrishnan et al., 2017). In India, 38% of cardiovascular disease and diabetes are attributed to air pollution. Along with this, ischemic heart disease, stroke, COPD, and lung cancer are also common health impacts of air pollution (Balakrishnan et al., 2017). In 2019, the economic loss from morbidity and death attributable to air pollution was high in India, consuming 1.36% form India's GDP (Pandey et al., 2019). The total healthcare cost in 2019 was estimated to be $103.7 billion. With air pollution responsible for 11.5% of the disease burden, the crude estimate of healthcare cost expenditure due to air pollution would be $11.9 billion. As stated previously, due to SDI, there was a threefold difference in economic loss (due to premature death and morbidity) between states, with Northern states recording a higher economic loss (National Health Accounts Estimates for India, 2019).

19.1.3 Difference Between Epidemiological and Modeling Techniques

Using epidemiological techniques provides empirical evidence to establish causal associations between air pollution and various health outcomes of air pollution, inherently quantifying risks related to air pollution exposure (Kirwa et al., 2021). Short-term air pollution studies are those that examine the short-term effects of air pollution exposure, such as air pollution levels on a day-to-day basis. They are also known as time-series studies. Long-term air pollution studies use cross-sectional, case-control, and cohort designs. Data from air pollution epidemiological studies are obtained from existing registers and routinely collected air pollution and health data (Anderson et al., 2007). The epidemiological study laid the foundation for quantifying the attributable burden of air pollution. Newer studies have started to incorporate modeling of air pollution exposure (Forastiere et al., 2024). Air pollution and atmospheric models are used to better understand and investigate the atmospheric environment and also the impact of air pollutants on health (San José et al., 2008; Yadav, 2015, 2022; Yadav et al., 2019; Yadav & Hopke, 2020). The objective of this chapter is to explore the current state of health burden due to air pollution and different modeling approaches of air pollution and health effects through a bibliometric analysis. Additionally, the chapter provides a brief overview of various computational and probabilistic modeling techniques, exposure assessment methods, and health impact models commonly used in air quality studies.

19.2 Current Scenario in Air Pollution Modeling: A Bibliometric Analysis

19.2.1 Literature

To understand the status of the air pollution modeling scenario worldwide, the published literature was mined using the open-access Web of Science database (Gupta, et al., 2024a; Koley et al., 2024a). This chapter exported a total of 774 papers utilizing the specific keywords TS = [("air pollution" OR "airborne pollutants" OR "atmospheric pollution" OR "particulate matter" OR "$PM_{2.5}$" OR "PM10" OR "NOx" OR "SOx" OR "volatile organic compounds" OR "heavy metals" OR "black carbon") AND ("human health risk" OR "risk assessment" OR "health impact" OR "toxicological risk") AND ("computational modeling" OR "probabilistic modeling" OR "Bayesian model" OR "Monte Carlo simulation" OR "machine learning" OR "artificial intelligence" OR "exposure modeling" OR "stochastic modeling" OR "deterministic modeling")].

19.2.2 Visualization of Similarity

These exported data were incorporated into the VOS viewer software for analysis based on the visualization of keyword similarity (Banerjee et al., 2024; Gupta et al., 2025; Gupta et al., 2024b; Koley et al., 2023; Laha et al., 2024), in which the threshold of keywords was selected to be 7, showing 148 co-occurring keywords with 6 clusters and a link strength of 2575. Figure 19.1a depicts the visualization of the keyword similarity network based on co-occurrence weight (Gupta et al., 2024c; Gupta 2024d; Koley et al., 2024b; Koley et al., 2024c), and Fig. 19.1b illustrates the visualization of the keyword similarity network based on the average publication year.

Cluster 1 (red): Air pollution modeling for health risk assessment is basically used to explore the exposure risk to hazardous chemicals and their associated human health impact. Long-term exposure to pollutants such as PM (PM_{10}, $PM_{2.5}$, and ultrafine particles), nitrogen dioxide, PAHs, and ozone has been associated with respiratory problems and an increase in death rates. Recently, black carbon has been known to be a significant source of air pollution linked to cardiovascular and lung disorders. Machine learning (ML), deep learning, and neural networks are increasingly being used to forecast air pollution levels and evaluate health hazards. Epidemiological studies help to demonstrate links between pollution and illnesses such as asthma and respiratory infections. Pollutant concentrations are used to estimate the possible health risk utilizing exposure modeling and risk assessment methodologies. Modeling such as neural networks, uncertainty, deep learning, regression modeling, and ML is used in air pollution modeling to assess health risk impacts and predict exposure-related outcomes.

Cluster 2: The modeling of health risks in air pollution includes several elements such as hazardous metals (Pb, Cd, arsenic, mercury, and copper) and their bioaccumulation in human populations. MC simulation and probabilistic risk assessment can evaluate noncarcinogenic and carcinogenic hazards from exposure to pollutants such as street dust and heavy metal pollution. MC simulations were extensively

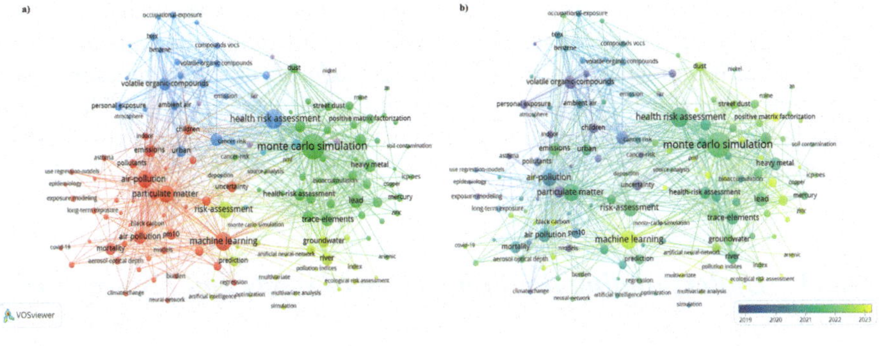

Fig. 19.1 Co-occurrence network visualization of keywords. (**a**) Network colored by clusters representing research keyword themes. (**b**) Overlay visualization showing temporal evolution

used during health risk analysis. The source study and a good matrix factorization find pollutant traits and possible environmental hazards. Studies on soil pollution and geochemistry help clarify heavy metal distribution in urban settings. Accurate risk assessment is enhanced by speciation methods like ICP-OES, which also help identify pollutants. Meta-analysis guides policy and mitigation plans through data integration to assess long-term health effects.

Cluster 3: Modeling of health risks in air pollution focuses on volatile organic compounds (VOCs) such as benzene, toluene, and benzene, toluene, ethylbenzene, and xylenes (BTEX), which are common in urban and occupational settings. Exposure to microenvironmental inhalation, such as in indoor and outdoor air, contributes to an increased probability of developing cancer and other health effects. MC simulation and health risk assessment methods can be used to help estimate the potential effects of emissions and individual exposure. Understanding the dispersal of pollution in the atmosphere can help mitigate exposure risks for workers and the general public as well.

Cluster 4: Air pollution health risk modeling adopts artificial intelligence, neural networks, and MC simulation to assess and quantify the level of exposure and estimate future health impacts. The use of multivariate analysis techniques and pollution indices also helps to identify sources of contamination and determine metal pollution concentrations. Ecological risk assessment examines the consequences of contamination on groundwater sources and rivers, which is essential to the safety of public health. Geographic information system (GIS) technology to better geographically identify pollution hotspots, for more locationally targeted treatment. The use of positive matrix factorization and optimization methods will help strengthen source attribution to optimally improve health risk assessments. Together, these approaches help us understand potential human health hazards in order to implement appropriate pollution management interventions.

Clusters 5 and 6: The health risk modeling in air pollution evaluates carcinogenicity using exposure to heavy metals such as chromium and nickel, which have been associated with lung cancer and oxidative stress. The Incremental Lifetime Cancer Risk statistic represents the long-term cancer risk based on exposure to inhalation. Inhaled contaminants can be deposited, and their patterns of deposition affect human health; thus, it is important to also analyze non-carcinogenic hazards using hazard quotients. Sensitivity analysis aids in the identification of significant factors that influence health outcomes, thus enhancing model accuracy. Uncertainty analysis provides robust risk assessment by accounting for data variability. These techniques help us understand pollution-related cancer risks and promote public health initiatives (Fig. 19.1).

19.2.3 Understanding the Publications Trends and Forecasting

The topic of computational and probabilistic human health risk modeling for air pollution exposure has received a tremendous increase in research papers in recent years. The trend was forecasted using interpolation to fill missing gaps, with a 95%

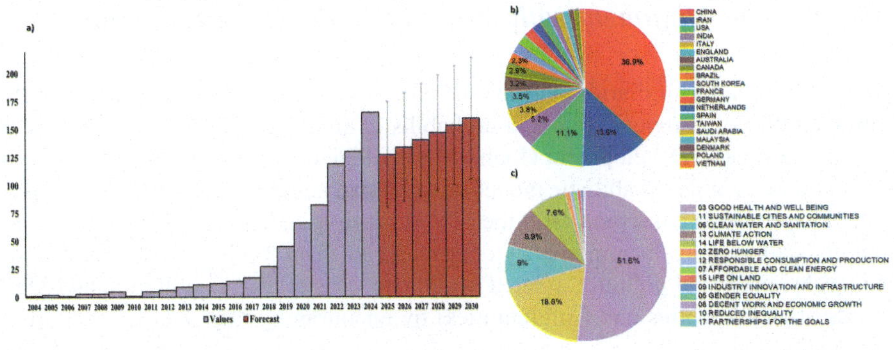

Fig. 19.2 Trends and distribution of research. (**a**) Annual publication output from 2004 to 2023 with a forecast up to 2030, including 95% confidence intervals. (**b**) Country-wise contribution to research publications. (**c**) Distribution of publications mapped to relevant SDGs

confidence interval, extending projections until 2030 (Fig. 19.2a). Before 2015, research efforts were modest, with fewer than 20 papers each year. However, a significant increase was observed from 2018 to 2024, reaching 165 publications by 2024. Forecasting indicates that the number of publications will continue to rise, with projections of around 160 by 2030. There is an increase in the growing relevance of air pollution risk assessment, which includes advanced computational approaches and probabilistic modeling. The increased use of MC simulations and ML technologies has increased the accuracy of risk assessment. Future research should focus more on multidisciplinary cooperation to improve exposure assessment and mitigation techniques.

19.2.4 Scientific Production Worldwide

To understand the global positioning of air pollution and risk assessment modeling, the distribution of scientific publications was analyzed. The main contributing countries include China (371), Iran (137), the United States (112), India (52), Italy (38), England (35), Australia (32), and Canada (29). Figure 19.2b depicts the country-wise contribution in scientific production relevant to this topic, each with a minimum of 20 publications included in this analysis. The findings indicate that China dominates this research domain, followed by Iran and the United States. As air pollution and health risk assessment remain critical global concerns, future research efforts should emphasize greater international collaboration. Strengthening cooperative scientific publications will enhance knowledge exchange and innovative solutions to mitigate air pollution risks.

19.2.5 The Alignment with Sustainable Development Goals

The study found a similarity between the current objective of this chapter, namely, compatibility of computational and probabilistic human health risk modeling with many Sustainable Development Goals (SDGs) depicted in Fig. 19.2c. The topic of computational and probabilistic modeling of human health risk as a result of air pollution exposure is consistent with the SDGs, since it addresses environmental and public health issues. It immediately contributes to SDG 3 (Good Health and Well-being) by analyzing pollution-related health concerns related to pollution and SDG 11 (Sustainable Cities and Communities) by encouraging improvements in urban air quality.

"Good Health and Well Being" (SDG 3) emerged as the most investigated issue, with 638 papers focusing on how air pollution affects human health. "Sustainable Cities and Communities" (SDG 11) follows with 233 research studies that emphasize the relevance of urban air quality. Research on "Clean Water and Sanitation" (SDG 6) and "Climate Action" (SDG 13) reveals an increasing concern about environmental contamination. Furthermore, research on "Life Below Water" (SDG 14) and "Zero Hunger" (SDG 2) demonstrates the link between pollution, food security, and marine ecosystems. Little research on SDGs such as "Gender Equality" (SDG 5) and "Partnerships for the Goals" (SDG 17) highlights the need for more multidisciplinary collaboration.

19.3 Computational Modeling Approaches

Computational models integrate air quality data, physiological responses, and health impact assessments to estimate potential risks. The key types of models include:

Deterministic models: Deterministic models rely on fixed parameters and assumptions to establish a direct link between pollutant exposure and health effects.

19.3.1 Air Quality Index Models

The classification of air quality index (AQI) models developed between 1960 and 2021 can be broadly categorized into two main groups: Single-pollutant-based AQI models and multi-pollutant-based AQI models. Single-pollutant-based AQI models utilize either the linear interpolation method or the segmented linear interpolation method. In the segmented linear interpolation method, models such as PSI (using the maximum operator), ASI, DAQx, and LAQx are commonly used. Multipollutant-based AQI models are further divided into three subcategories that are Pollutant

Aggregated Models, Spatial (multisite) Aggregated Models, and Health-based AQI Models (Kumar, 2022).

19.3.2 Air Pollution Deterministic Index Model

The Air Pollution Deterministic Index Model (APDIM) is a type of deterministic approach used to develop pollution indices for assessing air quality. This model is typically structured around a function α, a dimensionless number representing air quality conditions. APDIM calculates α by taking the weighted values of the individual pollutant parameters measured at various spatial points and comparing them with the air quality standard. One of the main advantages of this model is that α is a nondimensional value (i.e., not expressed in ppm or mg/m^3), making it universally applicable. The resulting pollution indices have the corresponding fixed values that, in a way similar to certain levels of air quality, each by a color code that is readily understandable to the general public (Akhtar & Shahkhan, 2018; Garanrd, 1996).

19.3.3 Dose–Response Models

It applies known toxicological thresholds to make an estimate of the probability of health effects.

19.3.3.1 Poisson Log-Linear Model

In many communities, time-series studies of the effect of PM on mortality, researchers infer daily mortality counts using the commonly used additive (Poisson) log-linear model that treats daily counts as independent Poisson random variables with a mean that varies over time, *mt* for day *t*, expressed as follows:

$$\log(mt) = \text{confounders}_t + \beta.PMt$$

In the model, confounders_t includes other variables associated with daily mortality, PM_t is the exposure to PM for day t, and β is the change in mortality associated with PM (Roberts & Martin, 2007). The model assumes a linear dose–response relationship and usually employs either a single-day or multiday moving average of PM as the exposure indicator; however, the model can be enriched in an effort to allow for nonlinearity in the dose–response relationship by using more flexible functional forms from the literature including piecewise linear functions, as has been done in a number of studies (Kim et al., 2004; Smith et al., 2000).

19.3.3.2 Poisson Regression Models

Generalized additive Poisson regression models were used to explore the relationship between city air pollution and mortality to control for potential overdispersion. This model can take advantage of including smooth nonparametric functions to explore potential nonlinear relationships between daily mortality and the variables of weather, seasonal trends, and potentially others. We assume a linear function relating daily deaths to NO_2 concentrations; the model for each specific city can be written as follows:

$$\log\left(E\left(Y^c\right)\right) = \beta^c NO_2 + \sum_{i=1}^{\varphi} S_i^c\left(Z_i^c\right)$$

The dependent variable Y^c is the daily number of deaths in city c, and $E(Y^c)$ is the expected value of Y^c. β^c is the coefficient for the logarithmic relative risk associated with NO_2 in city c. Independent variables Z^c_i are non-pollutant covariates, and $S^c_i(\cdot)$ is a smooth function of non-pollutant covariates (Cleveland & Devlin, 1988; Hastie & Tibshirani, 1990; Schwartz, 1994). To create these smooths, we used locally estimated scatterplot smoothing, which has proven to be common in time series studies of health effects of mortality or morbidity (Samoli & Touloumi, 2003).

19.3.4 Eulerian and Lagrangian Dispersion Models

19.3.4.1 Eulerian Dispersion Models

This model assists in the simulation of the long-range dispersion of PM and nitrogen oxides (NO_x) in cities (Kumar & Sharan, 2010) and the prediction of thermal plume behavior from power plants in coastal seas (Baptista et al., 1984). Eulerian models solve the advection-diffusion equation on a stationary gridded surface and determine pollutant concentrations over time. The governing equation for crosswind-integrated steady-state concentrations is as follows:

$$U(z)\partial x / \partial C = \partial / \partial z\left(Kz(x,z)\partial C / \partial z\right)$$

where $U(z)$ is the vertical wind profile, Kz is the vertical eddy diffusivity, and $C(x,z)$ is the pollutant concentration (Kumar & Sharan, 2010). Ultimately, numerical solutions will break the advection-diffusion equation into parts for diffusion and parts for diffusion and then advection for finite difference or finite element methods (Baptista et al., 1984; Byun et al., 2003). For example, Eulerian models have various turbulence parameterizations, such as vertical eddy diffusivity tensor Dij, as it is a function of atmospheric stability and surface flux exchanges (Byun et al., 2003).

19.3.4.2 Lagrangian Dispersion Models

Lagrangian models serve to investigate the spread of pollutants among buildings in urban canyons, where intricate flow structures are prominent (Kaplan & Dinar, 1996). Lagrangian particle dispersion models (LPDMs) are pivotal in footprint computation in atmospheric chemistry and greenhouse gas flux determination (Fillola et al., 2023). LPDMs simulate the transport of pollutants by employing stochastic differential equations to model particle movement. The position of each particle is updated in response to the mean wind field and a stochastic term that simulates turbulent fluctuations:

$$x(t+\Delta t) = x(t) + u_{\text{mean}}(x,t)\Delta t + u'(t)\Delta t$$

where $x(t)$ is the particle's position vector at time t and $u_{\text{mean}}(x,t)$ is the mean wind speed at the particle's position. The $u'(t)$ indicates turbulent fluctuation, usually treated as a random variable with statistical characteristics obtained from the theory of atmospheric turbulence (Hegarty et al., 2013; Rossi, 2014).

As per Rossi (2014), the stochastic turbulent term is usually expressed in terms of a Markov process (Langevin equation):

$$du_i' = -u_i' / TL\, dt + \sqrt{}2\sigma^2 / TL_d W_i$$

In this, TL is the Lagrangian timescale, σ^2 is the turbulent velocity variance, and dWi is a Wiener process (Gaussian white noise).

19.3.5 *Probabilistic Models*

The probabilistic approach is employed in various methods, including structural equation modeling, Bayesian networks (BNs), fuzzy logic, or fuzzy cognitive maps, and in combination with fault tree analysis methods. The probabilistic approach plays an important role in expert systems and knowledge representation (Khodabakhshian et al., 2023; Wee et al., 2015). Probabilistic modeling is used to quantify the uncertainties associated with health risk assessment. Quantitative modeling is more realistic in estimating the impact from exposure to avoid uncertainties than deterministic models. Probabilistic modeling is especially adept at capturing variability across real-world circumstances and is often applied in environmental health studies as health risk assessment for pollutants such as $PM_{2.5}$ and NO_2 (Whittemore, 2010).

Quantifying uncertainty in risk estimation uses numerous computational and probabilistic approaches that equally consider variability due to stochastic processes, limited data, and uncertainty due to model errors. New methods augment previous approaches that may utilize MC simulations with improved efficiencies and closure of standard Bayesian frameworks to provide a probabilistic rationale, and ML for scalable uncertainty analysis. A summary of the main methods from recent literature is provided below:

19.3.6 MC Simulations

Multilevel MC (MLMC) reduces computational costs by hierarchically combining low-accuracy/low-cost simulations with fewer high-accuracy/high-cost evaluations. This control variates method redistributes computational budgets in order to reduce variance in estimates (Zhang, 2021). Multifidelity MC extends MLMC by combining outputs from models of different fidelity (e.g., low-fidelity physics models combined with high-fidelity simulations) to reduce convergence (Zhang, 2021). Multimodel MC remedies imprecise probabilities due to limited datasets by merging multiple MC loops into one loop through multimodel inference and importance sampling (Zhang, 2021). Extended MC involves interdependencies among risks, valuing uncertainties according to their cooccurrence relationships and not as independent variables (Ostadi & Abbasi Harofteh, 2022). Use probability distributions to create multiple exposure scenarios and estimate a range of potential health effects. Markov chain MC samples posterior distributions to measure parameter uncertainty in high-dimensional spaces, allowing for robust probabilistic updates (Khadeer & Supriya, 2024).

19.3.7 Bayesian Networks

BNs are a basic technique for representing and reasoning about uncertainty in probabilistic models and can be viewed as a graph representation of an uncertain complex system. They consist of nodes (random variables) and directed edges (conditional dependencies), which combine to form a directed acyclic graph, encoding the joint probability distribution over a collection of variables (Kote et al., 2018).

A BN expresses the joint probability distribution $P(X_1, X_2,...,X_n)$ as a product of conditional probabilities:

$$P(X_1,X_2,\ldots,X_n)=\prod_{n}^{i=1}P(X_i \mid Pa(X_i))$$

where $Pa(X_i)$ is the notation for the parents of X_i in the DAG (Cheng et al., 2001; Kote et al., 2018). This factorization exploits conditional independence assumptions about the model to allow BNs to be very successful at representing high-dimensional distributions.

19.3.8 Agent-Based Models

Agent-based models (ABMs) simulate the interaction of independent "agents" in a designated environment, thereby generating emergent group behaviors (Monti et al., 2023; Wilensky & Rand, 2015). Setting theoretical exploration into a bounded domain is well established and generates higher learning about understanding how assumptions made at the individual level led to patterns that are observable at other levels (Railsback & Grimm, 2019). As Axelrod (1997) in opposition to induction that explores patterns based on existing data, ABM gives you a framework by which to build your intuitive understanding of complex systems. In the context of probabilistic modeling, where assumptions are made on the basis of an agent's action, an agent's interactions, and relationship to environments, introducing stochastic variation into the model makes it possible to study a complex emergent phenomenon that arises as the result of microlevel ambiguity. Every ABM can be expressed in the form:

$Z_t \sim P_t(Z_t | \Theta, Z\tau < t)$where Z_t is the set of variables of interest within the system at time t. Θ is the parameter that governs the behavior of the model, and P_t represents a probability function; however, it is implicitly defined through the specific rules and interactions of the model. t is the discrete time step, and Z_τ is the history of the variables within a system prior to time t.

19.3.9 ML Approaches

ML employs various methods to construct mathematical models that allow predictions based on prior information or data. Presently, it finds application across numerous domains, such as powering recommendation engines, filtering emails, automatically tagging content on platforms like Facebook, identifying images, and processing spoken language (Carbonell et al., 1983). ML addresses the problem of creating computer programs that can improve their performance on their own by adding experience (Jordan & Mitchell, 2015; Zhang et al., 2021). Within building energy system modeling, ML approaches have shown value in a number of domains, not limited to: (1) energy demand forecasting (Fan et al., 2017), (2) system operation and control optimization (Peng et al., 2018), (3) identification and diagnosis (Kim & Katipamula, 2018), and (4) urban-level modeling and analysis (Nosratabadi et al., 2019).

19.3.9.1 Supervised Learning

Supervised learning is a mainstay in ML, where the objective is to learn some function fθ parameterized by θ, that maps inputs x to outputs y. Generally, this is done by minimizing a loss function L on the data$\{(x_i,y_i)\}_{i=1}{}^N$:

$$\min_{\Theta} 1/N\sum_{i=1}^{n} L\left(f_\theta\left(x_i\right), y_i\right)$$

As an example, for linear regression, $f\theta(x) = \theta Tx$, and L is the mean squared error:

$$L\left(f_\theta\left(x_i\right), y_i\right) = \left(y_i - \theta^T x_i\right)^2$$

In practice, this optimization is solved via gradient descent or a variant of it, particularly in deep learning (Jain, 2024; Weinan, 2019).

19.3.9.2 Deep Learning and Neural Networks

Deep neural networks are multilayered function approximators. A nonlinear activation function is applied after a linear transformation in each layer:

$$h^{(l)} = \sigma\left(W^{(l)} h^{(l-1)} + b^{(l)}\right)$$

where σ is a non-linear activation (such as sigmoid or ReLU), $W^{(l)}$ and $b^{(l)}$ are weights and biases, and $h^{(l)}$ is the result of layer l. Backpropagation calculates gradients for optimization, and a loss function is used to compare the final output with the target (Jain, 2024; Weinan, 2019).

19.3.10 Exposure Assessment Techniques in Modeling

19.3.10.1 Remote Sensing

Remote sensing is a key technique for detecting and measuring different air pollutants by using multiple sensors and image devices. Such technologies can be employed to provide extensive coverage and data for air quality assessment on satellites, aircraft, drones, and ground stations (Zhou et al., 2023). Two advantages of remote sensing through satellites are global coverage and the ability to monitor long-term trends of large-scale air pollution (Stratoulias et al., 2024). The concentration and distribution of atmospheric contaminants can be quantified through instruments such as spectrometers and imaging sensors onboard satellites (Tikader et al., 2024). There are two types of satellites, one geostationary and the other polar orbiting. Both satellites are important for monitoring air quality and providing valuable information for investigation and regulation. Geostationary satellites offer high-resolution temporal data for observing current pollutions in real time, and polar orbiting satellites are beneficial to know the global and long-term trend of air pollution (Holloway et al., 2021; Lahoz et al., 2012). Various sensors and spectral bands are employed in remote sensing, all serving a distinct purpose and providing a unique view into the dynamics of the Earth. They include visible and near-infrared bands, thermal infrared bands, shortwave infrared bands, light detection and ranging (LiDAR) sensors, radio decision and ranging (Radar) sensors, microwave sensors, and hyperspectral sensors (Panda et al., 2016; Sikakwe, 2023). Aerial remote sensing has advantages such as flexibility, high spatial resolution, accessibility to hard terrains, and near-real-time monitoring capacity that make it effective in dynamic situations or emergency situations (Dube et al., 2022). It also possesses a variety of equipment such as various cameras, LiDAR, Radar, and thermal sensors to detect, map, and monitor various air pollution phenomena (Bansod et al., 2017; Zhou et al., 2023). Ground-based remote sensing is the collection of information using sensors mounted on the surface of the Earth. The technique offers accurate localized data for a variety of purposes, including air quality assessment, observation of air pollutants (NOx, SOx, ozone, and PM) (Verma et al., 2023). This type of technique comprises weather stations, GPS receiver, LiDAR, and Radar for data gathering the data related to air pollution (Sharma & Yadav, 2023).

19.3.10.2 GIS and Spatial Analysis Technique

The GIS is critical in mapping, analyzing, and abating air pollution due to its visualization and spatial analysis tools (Franch-Pardo et al., 2020). GIS helps to enable targeted mitigation by locating the sources of pollution in air pollution control by mapping their location. It generates real-time air quality by blending satellite imagery data, weather data, and air quality sensor stations (Ahamad et al., 2022). This enables it to identify hotspots of pollution, as well as track changes over time and successful assessment of air pollution abatement and control policy that assists policymakers in decision-making (Matejicek, 2005). GIS technologies are commonly applied to map air pollution because of their ability to perform spatial analysis and fuse spatial data from numerous sources (Tabunschik et al., 2023). Some of the GIS tools and software utilized include ArcGiS, QGIS, Google Earth Engine, GRASS GIS, ENVI, and inverse distance weighting (IDW). These GIS tools and software offer robust capabilities in mapping of air pollution, analyzing spatial trends, and visualizing data, all vital components for monitoring and researching air quality (Tikader et al., 2024). To examine the spatial distribution of pollutants, identify hotspots, and assess their impacts on the environment and human health, spatial analytic methods are crucial in air quality studies (Adza et al., 2022). The shared methods and analyses employed in the estimation of pollution variation, concentration, and detection of spatial patterns are IDW, Spatial join Combine dataset, Spatial autocorrection analysis, Buffer analysis, Hotspot analysis, Kriging, Spline, and geographically weighted regression (Tikader et al., 2024).

19.3.10.3 Land Use Regression Model

Recently, due to its ability to anticipate local variations in air pollution concentrations, the land use regression (LUR) method has become popular for studying the exposure of an individual to ambient air pollution (Morley & Gulliver, 2018). In an example of multivariate regression, LUR uses characteristics such as traffic, land use, population density, terrain, and other geographical criteria as independent variables and the observed levels of some pollutant as the dependent variable (Hoek et al., 2008). LUR models have been applied to satisfactorily model annual mean concentrations of SO_2, NO_x, NO_2, PM_{10}, $PM_{2.5}$, elemental carbon, shoot content, and VOCs (Liu et al., 2019; Zhang et al., 2018). Briggs et al. (1997) developed the LUR model within the Small Area Variations in Air Quality and Health study. Initially, it was referred to simply as regression mapping. By creating a linear regression model from a limited number of air pollution monitoring sites and potential predictor variables collected by GIS technology from a wide variety of land use types, this technique reproduces the concentrations of air pollutants (AZMi et al., 2023). These models were then applied to a number of unsampled locations in the study area, known as "virtual monitoring sites." The method initially saw its greatest application in Europe, but it has seen an increase in modern literature in studies applied worldwide. Although LUR techniques have often been effectively used to simulate the geographical variance of ambient air pollution, the approach has a number of drawbacks. The ability of LUR models to distinguish between the effects of certain priority pollutants is limited. The very localized changes in concentration that could occur close to sources might not be represented by LUR models (Yang et al., 2017). Typically, the model is made to forecast the total concentration alone rather than a particular source-related component of the overall concentration (He et al., 2018). Another drawback of current LUR models is that monitoring campaigns are frequently only covered for a limited period of time, making it impossible to precisely calculate exact concentrations that can be compared to air quality standards (Hoek et al., 2008).

19.3.11 Personal Exposure Models

Direct or indirect methods can be used to measure personal exposure. While exposure levels are measured stationarily or by models in the indirect technique, exposure levels are ascertained on an individual basis in the direct approach (using a biological marker or personal sampler) (Cordier et al., 2023). Passive samplers are widely used devices in personal sampling. They work on the basis of a gas's passive diffusion, and Fick's equation of diffusion can be used to determine the concentration in air (Bai et al., 2020). This sampler is mostly used for NO_2, but it is also used for CO, SO_2, formaldehyde, O_3, ammonia, and VOC (Kim et al., 2023). The main problem with this device is its lack of accuracy in data. There is another device called a real-time monitor for gaseous pollutants, but it takes a comparatively long time for detection (Monn, 2001). It is also used to detect different sizes of particles based on the principle of light scattering (Chen et al., 2018). Another use of this device is to detect combustion aerosols using the principle of a photoemission aerosol sensor (Wallace, 2005). Biological markers are of two types: exposure markers and effect markers. An effect marker is a preclinical indication of irregularities that may potentially be part of a medical diagnostic (Halim, 2011). On the other hand, the level of the analyte that has crossed the human borderline is reflected in an exposure marker (Rietjens et al., 2018). Biological markers are considered to be valid when they have definite background variability, known drug kinetics, and biological relevance (Whittaker et al., 2022). The advantage of using biomarkers is that all exposure pathways are incorporated and all exposures are integrated over time, but the basic limitation of these markers is that it is impossible to differentiate between the exposure pathways. In several epidemiological studies, ambient air monitoring networks have been used to gather data of exposure (Concas et al., 2021). It has online monitors that provide constant data with enough time resolution. The accuracy and precision level of these monitors are normally good. But the operation cost of such a type of network is quite expensive. The microenvironmental model is another approach to measuring personal exposure. It is a reliable and user-friendly method. A "chunk of air space with homogeneous pollutant concentration" is what Duan (1982) referred to as a "microenvironment" (ME). MEs are illustrative of either an indoor or an outdoor place. In an epidemiological study, it is viable to use group-MEs, which combine comparable MEs into ME categories (Monn, 2001).

19.3.12 Hybrid Models

The flexibility of hybrid models to combine the best characteristics of numerous approaches has rendered them an effective system for monitoring air pollution, therefore enhancing adaptability, precision, and robustness (Zou et al., 2009). At the very least, in a hybrid way, this type of modeling is useful in cases where human exposure to air pollution needs to be replicated and is a rational and cost-effective mode of combining the merits of various models or monitoring data sources (Ozkaynak et al., 2013). In the context of epidemiology, these hybrid models can link regional to personal monitoring, air dispersion to regional or personal monitoring (Pantusheva et al., 2022), or regional grid with local plume models and urban/local models with regional models (Nikmo et al., 1999). Another recent attempt is to better represent total ambient air pollution concentrations from all sources in all their complexity through model combination. In this regard, Community Multiscale Air Quality and the American Meteorological Society have been used (Appel et al., 2017; Askariyeh et al., 2017). The combinations of neural network plus genetic algorithm and kriging plus genetic are another two models used to monitor several air pollutants (Zhalehdoost & Taleai, 2022). The regression model is combined with the geostatistical model of IDW and also used to monitor the pollution concentrations (Abdullah & Barua, 2022). The advantages of this model are that it provides higher data accuracy, effectively works across different spatial scales, using a predictive algorithm, it can fill any existing gaps, and it can be modified and tuned for specific regions and for particular pollution types (Zou et al., 2009). There are a number of challenges in implementing hybrid models and interpreting their findings. Like, sometimes, regional and personal monitoring of data is hard to acquire. Integration of different exposure models is difficult occasionally (Ryan et al., 2007). Differences in measurement scales are other issues related to hybrid models (Wilson & Zawar-Reza, 2006).

19.3.13 Health Impact Modeling

Computational models estimate the potential health impacts of exposure to air pollution through the following:

19.3.13.1 Epidemiological Modeling

Epidemiological models use statistical correlations between exposure to air pollution and health outcomes to estimate population-level risks. For instance, long-term exposure to PM and nitrogen dioxide (NO_2) has been associated with increased mortality and morbidity, with studies attributing significant annual mortality burdens to current pollution levels (Manisalidis et al., 2020). First, these models used regression analysis to identify the mixing of variables such as socioeconomic characteristics that may influence the effects of pollution. In Washington, United States, these investigators demonstrated that environmental covariates such as temperature and humidity hardly influenced influenza-associated hospitalization estimates, with implications for the robustness of classical ecologic models (Somayaji et al., 2020). However, problems arise when attempting to tease apart multipollutant exposure effects and compensate for latency periods during chronic disease development (Ghio et al., 2018).

19.3.13.2 Physiological Modeling

Physiological models simulate how biological systems respond to pollutants. They include toxicological pathways. For example, air pollution disrupts iron balance using humic-like substances. This disruption leads to systemic inflammation and fibrosis. These mechanisms are also seen with cigarette smoke (Ghio et al., 2018). Pollutants can activate the neuroendocrine system, which worsens metabolic and immune system problems. This contributes to issues such as hypertension and diabetes. These models help explain why some individuals are more susceptible, especially in populations with pre-existing allostatic load, or cumulative stress load (Kodavanti, 2019). Although these models provide valuable information about the mechanisms, they usually depend on animal studies or laboratory data. This reliance limits the degree to which findings can be applied to humans (Ghio et al., 2018).

19.3.13.3 ML Approaches

ML algorithms improve risk prediction by identifying complex patterns in different datasets. Federated learning frameworks suggest combining real-time environmental, health, and socioeconomic data to predict AQIs and forecast hospitalizations (Neo et al., 2022). Studies comparing ML techniques, such as support vector regression and artificial neural networks (ANNs), found that ANNs are better at predicting PM concentrations with minimal error. Wearable devices are also useful for monitoring health in real time. They help link pollution exposure to individual respiratory outcomes (Mun & Cho, 2022). Challenges include protecting data privacy in federated systems and simplifying models for clinical use (Neo et al., 2022).

19.4 Challenges and Future Direction

Air quality modeling encounters various challenges that effectively limit its assessment and management of pollution-related health risks. One major problem is the lack of multiple contaminant approaches. Most studies and models examine contaminants individually, although real-world exposures consist of complex mixtures. This is tied to a limited focus on pollutant mixtures, which weakens understanding of cumulative health effects. There is also a need for stronger and more integrated tools in environmental epidemiology to better assess the health impacts of various pollutants. Another key challenge is the weak connection between air quality and population health. More research is required to strengthen this link. Individual exposure is often underestimated because it relies on collective data from multiple monitoring stations, which may not represent actual levels of personal exposure. Assessing exposure accurately is difficult because it requires high-resolution data and complex modeling techniques. Additionally, there is limited knowledge about how air pollutants move, which is crucial to predict their spread and health impacts. Air pollution models struggle with integration, especially when it comes to effectively combining risk assessment, geographical information systems, and other applications. There is insufficient understanding of how to apply current models. Usually lacking the precision and accuracy required to assist in risk management and decisions by local authorities on time. Building integrated multipollutant models that use high-resolution exposure spatial data, real-time performance monitoring, and more defined epidemiological methods could improve the development of next-generation air quality models. The use of interactions between pollutants in the assessment of individual exposures should be improved by improving predictive capacity to offer opportunities for public health action, anticipation, and policy action.

19.5 Conclusion

This overview emphasizes the important role of air pollution modeling as a key tool in assessing and managing air quality. Modeling is essential when measurements are missing or inadequate. It helps to evaluate different sources of pollution and predict pollution levels or review possible scenarios. We have looked at both deterministic and stochastic-statistical modeling methods. Deterministic models are especially useful in practical situations. When calibrated and used correctly, these models provide clear and objective source–receptor relationships. This makes them vital for effective air quality management and policy development.

Acknowledgments Deep Chakraborty wants to acknowledge the Department of Environmental Health Engineering, Sri Ramachandra Faculty of Public Health, Sri Ramachandra Institute of Higher Education and Research, for giving the opportunity to write this book chapter, and Apurba Koley wants to acknowledge the Bio-energy laboratory, Department of Environmental Studies, Siksha-Bhavana (Institute of Science), Visva-Bharati, for providing necessary research support.

References

Abdullah, S., & Barua, D. (2022). Combining geographical information system (GIS) and machine learning to monitor and predict vegetation vulnerability: An empirical study on Nijhum Dwip, Bangladesh. *Ecological Engineering, 178*, 106577.

Adza, W. K., Hursthouse, A. S., Miller, J., & Boakye, D. (2022). Exploring the combined association between road traffic noise and air quality using QGIS. *International Journal of Environmental Research and Public Health, 19*(24), 17057.

Ahamad, F., Bhutiani, R., & Ruhela, M. (2022). Environmental quality monitoring using environmental quality indices (EQI) geographic information system (GIS) and remote sensing: A review. In *GIScience for the sustainable management of water resources* (pp. 331–348). Apple Academic Press.

Akhtar, F., & Shahkhan, S. (2018). Air pollution deterministic index Modeling: Application in Quetta, Pakistan. *IOP Conference Series: Materials Science and Engineering, 414*(1), 012014.

Anderson, L. H., Martinson, B. C., Hall, K. M., & Duncan, I. G. (2007). *Critical review of stochastic simulation literature and applications for health actuaries*. Society of Actuaries.

Appel, K. W., Napelenok, S. L., Foley, K. M., Pye, H. O., Hogrefe, C., Luecken, D. J., & Young, J. O. (2017). Description and evaluation of the community multiscale air quality (CMAQ) modeling system version 5.1. *Geoscientific Model Development, 10*(4), 1703–1732.

Askariyeh, M. H., Kota, S. H., Vallamsundar, S., Zietsman, J., & Ying, Q. (2017). AERMOD for near-road pollutant dispersion: Evaluation of model performance with different emission source representations and low wind options. *Transportation Research Part D: Transport and Environment, 57*, 392–402.

Axelrod, R. (1997). The dissemination of culture: A model with local convergence and global polarization. *Journal of Conflict Resolution, 41*(2), 203–226.

Azmi, W. N. F. W., Pillai, T. R., Latif, M. T., Koshy, S., & Shaharudin, R. (2023). Application of land use regression model to assess outdoor air pollution exposure: A review. *Environmental Advances, 11*, 100353.

Bai, Y., Huo, L., Zhang, Y., Liu, J., Shao, H., Wu, C., & Guo, Z. (2020). A spatial fractional diffusion model for predicting the characteristics of VOCs emission in porous dry building material. *Science of the Total Environment, 704*, 135342.

Balakrishnan, et al. (2017). The impact of air pollution on deaths, disease burden, and life expectancy across the states of India: The global burden of disease study. *The Lancet Planetary Health, 3*(1), e26–e39.

Banerjee, S., Gupta, N., Pramanik, K., Gope, M., GhoshThakur, R., Karmakar, A., & Balachandran, S. (2024). Microbes and microbial strategies in carcinogenic polycyclic aromatic hydrocarbons remediation: A systematic review. *Environmental Science and Pollution Research, 31*(2), 1811–1840.

Bansod, B., Singh, R., Thakur, R., & Singhal, G. (2017). A comparision between satellite based and drone based remote sensing technology to achieve sustainable development: A review. *Journal of Agriculture and Environment for International Development (JAEID), 111*(2), 383–407.

Baptista, A. E. D. M., Adams, E. E., & Stolzenbach, K. D. (1984). *Eulerian-Lagrangian analysis of pollutant transport in shallow water*. MIT Energy Laboratory.

Briggs, D. J., Collins, S., Elliott, P., Fischer, P., Kingham, S., Lebret, E., Pryl, K., Van Reeuwijk, H., Smallbone, K., & Van Der Veen, A. (1997). Mapping urban air pollution using gis: A regression-based approach. *International Journal of Geographical Information Science, 11*(7), 699–718.

Byun, D. W., Lacser, A., Yamartino, R., & Zannetti, P. (2003). Eulerian dispersion models. In P. Zannetti (Ed.), *Air quality modeling: Theories, methodologies, computational techniques, and available databases and software*. Air & Waste Management Association.

Carbonell, J. G., Michalski, R. S., & Mitchell, T. M. (1983). An overview of machine learning. *Machine Learning, 1*, 3–23.

Chen, D., Liu, X., Han, J., Jiang, M., Xu, Y., & Xu, M. (2018). Measurements of particulate matter concentration by the light scattering method: Optimization of the detection angle. *Fuel Processing Technology, 179*, 124–134.

Chen, G., Zhang, W., Li, S., Zhang, Y., Williams, G., Huxley, R., & Guo, Y. (2017). The impact of ambient fine particles on influenza transmission and the modification effects of temperature in China: A multi-city study. *Environment International, 98*, 82–88.

Cheng, J., Grainer, G., Kelly, J., Bell, D. A., & Lius, W. (2001). *Learning bayesian networks from data: An information-theory based approach, 2001*.

Cleveland, W. S., & Devlin, S. J. (1988). Robust locally-weighted regression and smoothing scatterplots. *Journal of the American Statistical Association, 74*, 829–836.

Concas, F., Mineraud, J., Lagerspetz, E., Varjonen, S., Liu, X., Puolamäki, K., & Tarkoma, S. (2021). Low-cost outdoor air quality monitoring and sensor calibration: A survey and critical analysis. *ACM Transactions on Sensor Networks (TOSN), 17*(2), 1–44.

Cordier, S., Warembourg, C., & Stewart, P. A. (2023). Exposure assessment. In *Handbook of epidemiology* (pp. 1–33). Springer.

Duan, N. (1982). Models for human exposure to air pollution. *Environment International, 8*(1–6), 305–309.

Dube, T., Shekede, M. D., & Massari, C. (2022). Remote sensing for water resources and environmental management. *Remote Sensing, 15*(1), 18.

Falcon-Rodriguez, C. I., Osornio-Vargas, A. R., Sada-Ovalle, I., & Segura-Medina, P. (2016). Aeroparticles, composition, and lung diseases. *Frontiers in Immunology, 7*, 3.

Fan, C., Xiao, F., & Zhao, Y. (2017). A short-term building cooling load prediction method using deep learning algorithms. *Applied Energy, 195*, 222–233.

Fillola, E., Santos-Rodriguez, R., Manning, A., O'Doherty, S., & Rigby, M. (2023). A machine learning emulator for Lagrangian particle dispersion model footprints: A case study using NAME. *Geoscientific Model Development, 16*(7), 1997–2009.

Forastiere, F., Orru, H., Krzyzanowski, M., et al. (2024). The last decade of air pollution epidemiology and the challenges of quantitative risk assessment. *Environmental Health, 23*, 98. https://doi.org/10.1186/s12940-024-01136-5

Franch-Pardo, I., Napoletano, B. M., Rosete-Verges, F., & Billa, L. (2020). Spatial analysis and GIS in the study of COVID-19. A review. *Science of the Total Environment, 739*, 140033.

Garanrd, K. (1996). *Environmental engineering* (pp. 936–939). McGraw-Hill International.

GBD (Global Burden of Disease). (2019). Global burden of 369 diseases and injuries in 204 countries and territories, 1990–2019: A systematic analysis for the Global Burden of Disease Study 2019. *Lancet, 396*(10258), 1204–1222. https://doi.org/10.1016/S0140-6736(20)30925-9

GBD MAPS Working Group. (2018). *Burden of disease attributable to major air pollution sources in India. Special report 21*. Health Effects Institute.

Ghio, A. J., Soukup, J. M., & Madden, M. C. (2018). The toxicology of air pollution predicts its epidemiology. *Inhalation Toxicology, 30*(9–10), 327–334.

Global Burden of Disease (GBD). (2021). *Risk factor factsheets*. Institute for Health Metrics and Evaluation. https://www.healthdata.org/research-analysis/diseases-injuries/factsheets-overview/about-risk-factor

Gupta, N., Banerjee, S., Koley, A., Basu, A., Gogoi, N., Hoque, R. R., & Balachandran, S. (2024a). Fungal strategies for the remediation of polycyclic aromatic hydrocarbons. In *Bioremediation for sustainable environmental cleanup* (pp. 86–108). CRC Press.

Gupta, N., Koley, A., Banerjee, S., Ghosh, A., Hoque, R. R., & Balachandran, S. (2024b). Nanomaterial-mediated strategies for enhancing bioremediation of polycyclic aromatic hydrocarbons: A systematic review. *Hybrid Advances, 7*, 100315.

Gupta, N., Banerjee, S., Koley, A., Bharali, P., GhoshThakur, R., Hoque, R. R., & Balachandran, S. (2024c). Strategies for remediation of polycyclic aromatic hydrocarbons in contaminated soil: A systematic review and bibliometric analysis. *Applied Soil Ecology, 204*, 105688.

Gupta, N., Koley, A., Saha, A., Hoque, R. R., & Balachandran, S. (2024d). Microbial fuel cells: Bifunctionalized approach for wastewater treatment and energy recovery innovation. In *Application of microbial technology in wastewater treatment and bioenergy recovery* (pp. 465–506). Springer Nature.

Gupta, N., Koley, A., Banerjee, S., Show, B. K., Ghosh, A., Banerjee, S., & Balachandran, S. (2025). Microbial strategies of chrysene remediation. In *Challenges and sustainable solutions in bioremediation* (pp. 34–53). CRC Press.

Halim, A. B. (2011). Biomarkers in drug development: A useful tool but discrepant results may have a major impact. In *Drug discovery and development-present and future* (Vol. 4, pp. 401–424). IntechOpen.

Hastie, T., & Tibshirani, R. (1990). *Generalized additive models*. Chapman and Hall.

He, B., Heal, M. R., & Reis, S. (2018). Land-use regression modelling of intra-urban air pollution variation in China: Current status and future needs. *Atmosphere, 9*(4), 134.

Health Effects Institute (HEI). (2024). *State of global air 2024. Special report*. Health Effects Institute.

Hegarty, J., Draxler, R. R., Stein, A. F., Brioude, J., Mountain, M., Eluszkiewicz, J., & Andrews, A. (2013). Evaluation of Lagrangian particle dispersion models with measurements from controlled tracer releases. *Journal of Applied Meteorology and Climatology, 52*(12), 2623–2637.

Hime, N. J., Marks, G. B., & Cowie, C. T. (2018). A comparison of the health effects of ambient particulate matter air pollution from five emission sources. *International Journal of Environmental Research and Public Health, 15*(6), 1206.

Hoek, G., Beelen, R., De Hoogh, K., Vienneau, D., Gulliver, J., Fischer, P., & Briggs, D. (2008). A review of land-use regression models to assess spatial variation of outdoor air pollution. *Atmospheric Environment, 42*(33), 7561–7578.

Holloway, T., Miller, D., Anenberg, S., Diao, M., Duncan, B., Fiore, A. M., & Zondlo, M. A. (2021). Satellite monitoring for air quality and health. *Annual Review of Biomedical Data Science, 4*(1), 417–447.

Institute for Health Metrics and Evaluation. (2024). Air pollution. In *The global burden of disease*. University of Washington.

Jain, V. (2024). Mathematics & techniques related to machine learning (ML). *International Journal for Multidisciplinary Research (IJFMR), 6*(4), 1.

Jordan, M. I., & Mitchell, T. M. (2015). Machine learning: Trends, perspectives, and prospects. *Science, 349*(6245), 255–260.

Kaplan, H., & Dinar, N. (1996). A Lagrangian dispersion model for calculating concentration distribution within a built-up domain. *Atmospheric Environment, 30*(24), 4197–4207.

Khadeer, D. A., & Supriya, M. A. (2024). Bayesian inference methods for uncertainty quantification in data science: Taming the complexity. *International Journal of Novel Research and Development, 9*(3), f674–f679.

Khodabakhshian, A., Puolitaival, T., & Kestle, L. (2023). Deterministic and probabilistic risk management approaches in construction projects: A systematic literature review and comparative analysis. *Buildings, 13*(5), 1312.

Kim, S. J., Lee, H. Y., Lee, S. J., & Choi, S. D. (2023). Passive air sampling of VOCs, O3, NO2, and SO2 in the large industrial city of Ulsan, South Korea: spatial–temporal variations, source identification, and ozone formation potential. *Environmental Science and Pollution Research, 30*(60), 125478–125491.

Kim, S. Y., Lee, J. T., Hong, Y. C., Ahn, K. J., & Kim, H. (2004). Determining the threshold effect of ozone on daily mortality: An analysis of ozone and mortality in Seoul, Korea, 1995–1999. *Environmental Research, 94*, 113–119.

Kim, W., & Katipamula, S. (2018). A review of fault detection and diagnostics methods for building systems. *Science and Technology for the Built Environment, 24*(1), 3–21.

Kirwa, K., Szpiro, A. A., Sheppard, L., Sampson, P. D., Wang, M., Keller, J. P., Young, M. T., Kim, S. Y., Larson, T. V., & Kaufman, J. D. (2021). Fine-scale air pollution models for epidemiologic research: Insights from approaches developed in the multi-ethnic study of atherosclerosis and air pollution (MESA air). *Current Environmental Health Reports, 8*(2), 113–126. https://doi.org/10.1007/s40572-021-00310-y

Kodavanti, U. P. (2019). Susceptibility variations in air pollution health effects: Incorporating neuroendocrine activation. *Toxicologic Pathology, 47*(8), 962–975.

Koley, A., Ghosh, A., Banerjee, S., Gupta, N., Thakur, R. G., Show, B. K., & Balachandran, S. (2024a). Phytoremediation of wastewater discharged from paper and pulp, textile and dairy industries using water hyacinth (*Eichhornia crassipes*). In *Bioremediation for sustainable environmental cleanup* (pp. 238–261). CRC Press.

Koley, A., GhoshThakur, R., Das, K., Gupta, N., Banerjee, A., Show, B. K., & Balachandran, S. (2024b). Growth dynamics and nutrient removal from biogas slurry using water hyacinth. *Sustainability, 16*(11), 4450.

Koley, A., Gupta, N., Singh, A., GhoshThakur, R., & Balachandran, S. (2024c). Microalgae-based biofuel for sustainable bioenergy production. In *Application of microbial technology in wastewater treatment and bioenergy recovery* (pp. 359–384). Springer.

Koley, A., Mukhopadhyay, P., Gupta, N., Singh, A., Ghosh, A., Show, B. K., & Balachandran, S. (2023). Biogas production potential of aquatic weeds as the next-generation feedstock for bioenergy production: A review. *Environmental Science and Pollution Research, 30*(52), 111802–111832.

Kote, N., Biba, M., & Canaj, E. (2018). Bayesian networks: A state-of-the-art survey. *Learning, 9*, 30.

Kumar, P. (2022). A critical evaluation of air quality index models (1960–2021). *Environmental Monitoring and Assessment, 194*(5), 1–45.

Kumar, P., & Sharan, M. (2010). An analytical model for dispersion of pollutants from a continuous source in the atmospheric boundary layer. *Proceedings of the Royal Society A: Mathematical, Physical and Engineering Sciences, 466*(2114), 383–406.

Laha, T., Gupta, N., Pal, M., Koley, A., Masto, R. E., Hoque, R. R., & Balachandran, S. (2024). Chemical speciation and health risk assessment of potentially toxic elements in playground soil of bell metal commercial town of eastern India. *Environmental Geochemistry and Health, 46*(11), 453.

Lahoz, W. A., Peuch, V. H., Orphal, J., Attié, J. L., Chance, K., Liu, X., & Amraoui, L. E. (2012). Monitoring air quality from space: The case for the geostationary platform. *Bulletin of the American Meteorological Society, 93*(2), 221–233.

Lajoie, J. (2015). *Understanding the measurement of global burden of disease*. National Collaborating Centre for Infectious Disease. https://nccid.ca/publications/understanding-the-measurement-of-global-burden-of-disease/

Lim, S., et al. (2010). A comparative risk assessment of burden of disease and injury attributable to 67 risk factors and risk factor clusters in 21 regions, 1990–2010: A systematic analysis for the global burden of disease study. *The Lancet, 380*(9859), 2224–2260.

Liu, T., Xiao, J., Zeng, W., Hu, J., Liu, X., Dong, M., Wang, J., Wan, D., & Ma, W. (2019). A spatiotemporal land-use-regression model to assess individual level long-term exposure to ambient fine particulate matters. *MethodsX, 6*, 2101–2105.

Manisalidis, I., Stavropoulou, E., Stavropoulos, A., & Bezirtzoglou, E. (2020). Environmental and health impacts of air pollution: A review. *Frontiers in Public Health, 8*, 14.

Marcantonio, R., Javeline, D., Field, S., & Fuentes, A. (2021). Global distribution and coincidence of pollution, climate impacts, and health risk in the Anthropocene. *PLoS One, 16*(7), e0254060.

Matejicek, L. (2005). Spatial modelling of air pollution in urban areas with GIS: A case study on integrated database development. *Advances in Geosciences, 4*, 63–68.

Monn, C. (2001). Exposure assessment of air pollutants: A review on spatial heterogeneity and indoor/outdoor/personal exposure to suspended particulate matter, nitrogen dioxide and ozone. *Atmospheric Environment, 35*(1), 1–32.

Monti, C., Pangallo, M., De Francisci Morales, G., & Bonchi, F. (2023). On learning agent-based models from data. *Scientific Reports, 13*(1), 9268.

Morley, D. W., & Gulliver, J. (2018). A land use regression variable generation, modelling and prediction tool for air pollution exposure assessment. *Environmental Modelling & Software, 105*, 17–23.

Mun, E., & Cho, J. (2022). Review of internet of things-based artificial intelligence analysis method through real-time indoor air quality and health effect monitoring: focusing on indoor air pollution that are harmful to the respiratory organ. *Tuberculosis Respiratory Disease, 86*(1), 23.

Murray, C. J., & Lopez, A. D. (2013). Measuring the global burden of disease. *The New England Journal of Medicine, 369*(5), 448–457. https://doi.org/10.1056/NEJMra1201534

Murray, C. J. L. (2022). The global burden of disease study at 30 years. *Nature Medicine, 28*, 2019–2026. https://doi.org/10.1038/s41591-022-01990-1

National Health Accounts Estimates for India. (2019). *Ministry of Health and Family Welfare*. https://nhsrcindia.org/sites/default/files/2021-06/FINAL%20National%20Health%20Accounts%202016-17%20Nov%202019-for%20Web%20%281%29.pdf

Neo, E. X., Hasikin, K., Mokhtar, M. I., Lai, K. W., Azizan, M. M., Razak, S. A., & Hizaddin, H. F. (2022). Towards integrated air pollution monitoring and health impact assessment using federated learning: A systematic review. *Frontiers in Public Health, 10*, 851553.

Nikmo, J., Tuovinen, J. P., Kukkonen, J., & Valkama, I. (1999). A hybrid plume model for local-scale atmospheric dispersion. *Atmospheric Environment, 33*(27), 4389–4399.

Nosratabadi, S., Mosavi, A., Keivani, R., Ardabili, S., & Aram, F. (2019). State of the art survey of deep learning and machine learning models for smart cities and urban sustainability. In *International conference on global research and education* (pp. 228–238). Springer International Publishing.

Ostadi, B., & Abbasi Harofteh, S. (2022). A novel risk assessment approach using Monte Carlo simulation based on co-occurrence of risk factors: A case study of a petrochemical plant construction. *Scientia Iranica, 29*(3), 1755–1765.

Ozkaynak, H., Baxter, L. K., Dionisio, K. L., & Burke, J. (2013). Air pollution exposure prediction approaches used in air pollution epidemiology studies. *Journal of Exposure Science & Environmental Epidemiology, 23*(6), 566–572.

Panda, S. S., Rao, M. N., Thenkabail, P. S., Misra, D., & Fitzgerald, J. P. (2016). Remote sensing systems—Platforms and sensors: Aerial, satellite, UAV, optical, radar, and LiDAR. In *Remote sensing handbook* (Vol. I, pp. 3–86). CRC Press.

Pandey, et al. (2019). Health and economic impact of air pollution in the states of India: The global burden of disease study. *The Lancet Planetary Health, 5*(1), e25–e38.

Pantusheva, M., Mitkov, R., Hristov, P. O., & Petrova-Antonova, D. (2022). Air pollution dispersion modelling in urban environment using CFD: A systematic review. *Atmosphere, 13*(10), 1640.

Peng, Y., Rysanek, A., Nagy, Z., & Schlüter, A. (2018). Using machine learning techniques for occupancy-prediction-based cooling control in office buildings. *Applied Energy, 211*, 1343–1358.

Railsback, S. F., & Grimm, V. (2019). *Agent-based and individual-based modeling: A practical introduction*. Princeton University Press.

Rentschler, J., & Leonova, N. (2023). Global air pollution exposure and poverty. *Nature Communications, 14*(1), 4432.

Rietjens, I. M., Dussort, P., Günther, H., Hanlon, P., Honda, H., Mally, A., & Eisenbrand, G. (2018). Exposure assessment of process-related contaminants in food by biomarker monitoring. *Archives of Toxicology, 92*, 15–40.

Roberts, S., & Martin, M. A. (2007). A distributed lag approach to fitting non-linear dose–response models in particulate matter air pollution time series investigations. *Environmental Research, 104*(2), 193–200.

Roser, M., Ritchie, H., & Spooner, F. (2021). *Burden of disease* published online at OurWorldinData.org. https://ourworldindata.org/burden-of-disease

Rossi, D. (2014). Techniques for Lagrangian modelling of dispersion in geophysical flows.

Ryan, P. B., Burke, T. A., Cohen Hubal, E. A., Cura, J. J., & McKone, T. E. (2007). Using biomarkers to inform cumulative risk assessment. *Environmental Health Perspectives, 115*(5), 833–840.

Samoli, E., & Touloumi, G. (2003). Zanobetti A, et alInvestigating the dose-response relation between air pollution and total mortality in the APHEA-2 multicity project. *Occupational and Environmental Medicine, 60*, 977–982.

San José, R., Baklanov, A., Sokhi, R. S., Karatzas, K., & Pérez, J. L. (2008). Air quality modeling. *Encyclopedia of ecology*, 111–123.

Schwartz, J. (1994). Eneralized additive models in epidemiology. In *Invited papers. 17th international biometric conference* (pp. 55–80). International Biometric Society.

Sharma, K., & Yadav, S. (2023). Air quality monitoring using geospatial technology and field sensors. In *Geospatial analytics for environmental pollution modeling: Analysis, control and management* (pp. 91–118). Springer.

Sikakwe, G. U. (2023). Mineral exploration employing drones, contemporary geological satellite remote sensing and geographical information system (GIS) procedures: A review. *Remote Sensing Applications: Society and Environment, 31*, 100988.

Smith, R. L., Spitzner, D., Kim, Y., & Fuentes, M. (2000). Threshold dependence of mortality effects for fine and coarse particles in Phoenix, Arizona. *Journal of the Air & Waste Management Association, 50*, 1367–1379.

Sokhi, R. S., Moussiopoulos, N., Baklanov, A., Bartzis, J., Coll, I., Finardi, S., & Kukkonen, J. (2021). Advances in air quality research–current and emerging challenges. *Atmospheric Chemistry and Physics Discussions, 2021*, 1–133.

Somayaji, R., Neradilek, M. B., Szpiro, A. A., Lofy, K. H., Jackson, M. L., Goss, C. H., & Ortiz, J. R. (2020). Effects of air pollution and other environmental exposures on estimates of severe influenza illness, Washington, USA. *Emerging Infectious Diseases, 26*(5), 920.

Stratoulias, D., Nuthammachot, N., Dejchanchaiwong, R., Tekasakul, P., & Carmichael, G. R. (2024). Recent developments in satellite remote sensing for air pollution surveillance in support of sustainable development goals. *Remote Sensing, 16*(16), 2932.

Tabunschik, V., Gorbunov, R., & Gorbunova, T. (2023). Unveiling air pollution in crimean mountain rivers: Analysis of sentinel-5 satellite images using google earth engine (GEE). *Remote Sensing, 15*(13), 3364.

Thurston, G. D., Kipen, H., Annesi-Maesano, I., Balmes, J., Brook, R. D., Cromar, K., & Brunekreef, B. (2017). A joint ERS/ATS policy statement: What constitutes an adverse health effect of air pollution? An analytical framework. *European Respiratory Journal, 49*(1), 1600419.

Tikader, M., Mukhopadhyay, D., & Dabhadker, K. (2024). Remote sensing and GIS in air pollution mitigation: A bibliometric review of Chhattisgarh, India. *International Journal of Environment and Climate Change, 14*(4), 174–193.

Verma, S., Gangwar, T., Singh, J., Prakash, D., & Payra, S. (2023). Urban air quality monitoring and modelling using ground monitoring, remote sensing, and GIS. In *Geospatial analytics for environmental pollution modeling: Analysis, control and management* (pp. 213–247). Cham.

Wallace, L. (2005). Real-time measurements of black carbon indoors and outdoors: a comparison of the photoelectric aerosol sensor and the aethalometer. *Aerosol Science and Technology, 39*(10), 1015–1025.

Wee, Y. Y., Cheah, W. P., Tan, S. C., & Wee, K. (2015). A method for root cause analysis with a Bayesian belief network and fuzzy cognitive map. *Expert Systems with Applications, 42*(1), 468–487.

Weinan, E. (2019). Machine learning: Mathematical theory and scientific applications. *Notices of the American Mathematical Society, 66*(11), 1813–1820.

Whittaker, C., Chesnais, C. B., Pion, S. D., Kamgno, J., Walker, M., Basáñez, M. G., & Boussinesq, M. (2022). Factors associated with variation in single-dose albendazole pharmacokinetics: A systematic review and modelling analysis. *PLoS Neglected Tropical Diseases, 16*(10), e0010497.

Whittemore, A. S. (2010). Evaluating health risk models. *Statistics in Medicine, 29*(23), 2438–2452.

WHO. (2020). Methods and data sources for global burden of disease estimates 2000–2019. https://cdn.who.int/media/docs/default-source/gho-documents/global-health-estimates/ghe2019_daly-methods.pdf?sfvrsn=31b25009_7

Wilensky, U., & Rand, W. (2015). *An introduction to agent-based modeling: modeling natural, social, and engineered complex systems with NetLogo*. MIT Press.

Wilson, J. G., & Zawar-Reza, P. (2006). Intraurban-scale dispersion modelling of particulate matter concentrations: Applications for exposure estimates in cohort studies. *Atmospheric Environment, 40*(6), 1053–1063.

World Health Organization. (2021). *WHO global air quality guidelines: particulate matter ($PM_{2.5}$ and PM_{10}), ozone, nitrogen dioxide, sulfur dioxide and carbon monoxide*. World Health Organization.

Yadav, A. K. (2015). Elemental composition and source apportionment of suspended particulate matters and health risk assessment in mining and nonmining areas of Odisha, India. *Journal of Hazardous, Toxic, and Radioactive Waste, 19*(3), 04014037.

Yadav, A. K. (2021). Human health risk assessment in opencast coal mines and coal-fired thermal power plants surrounding area due to inhalation. *Environmental Challenges, 3*, 100074.

Yadav, A. K. (2022). Characterization and effective dose of gross alpha and gross beta activities in particulate matters ($PM_{2.5}$, PM_{10}, SPM) in India. *Aerosol Science and Engineering, 6*(4), 360–369.

Yadav, A. K., & Hopke, P. K. (2020). Characterization of radionuclide activity concentrations and lifetime cancer risk due to particulate matter in the Singrauli coalfield, India. *Environmental Monitoring and Assessment, 192*, 680. https://doi.org/10.1007/s10661-020-08619-1

Yadav, A. K., Sahoo, S. K., Dubey, J. S., Kumar, A. V., Pandey, G., & Tripathi, R. M. (2019). Assessment of particulate matter, metals of toxicological concentration, and health risk around a mining area, Odisha, India. *Air Quality, Atmosphere & Health, 12*(7), 775–783.

Yang, X., Zheng, Y., Geng, G., Liu, H., Man, H., Lv, Z., & de Hoogh, K. (2017). Development of $PM_{2.5}$ and NO_2 models in a LUR framework incorporating satellite remote sensing and air quality model data in Pearl River Delta region. *China. Environmental pollution, 226*, 143–153.

Yu, W., Ye, T., Zhang, Y., Xu, R., Lei, Y., Chen, Z., & Guo, Y. (2023). Global estimates of daily ambient fine particulate matter concentrations and unequal spatiotemporal distribution of population exposure: A machine learning modelling study. *The Lancet Planetary Health, 7*(3), e209–e218.

Zhalehdoost, A., & Taleai, M. (2022). A review of the application of machine learning and geospatial analysis methods in air pollution prediction. *Pollution, 8*(3), 904–933.

Zhang, J. (2021). Modern Monte Carlo methods for efficient uncertainty quantification and propagation: A survey. *Wiley Interdisciplinary Reviews: Computational Statistics, 13*(5), e1539.

Zhang, L., Wen, J., Li, Y., Chen, J., Ye, Y., Fu, Y., & Livingood, W. (2021). A review of machine learning in building load prediction. *Applied Energy, 285*, 116452.

Zhang, Z., Wang, J., Hart, J. E., Laden, F., Zhao, C., Li, T., Zheng, P., Li, D., Ye, Z., & Chen, K. (2018). National scale spatiotemporal land-use regression model for PM2.5, PM10 and NO2 concentration in China. *Atmospheric Environment, 192*(April), 48–54.

Zhou, B., Zhang, S., Xue, R., Li, J., & Wang, S. (2023). A review of space-air-ground integrated remote sensing techniques for atmospheric monitoring. *Journal of Environmental Sciences, 123*, 3–14.

Zou, B., Wilson, J. G., Zhan, F. B., & Zeng, Y. (2009). Air pollution exposure assessment methods utilized in epidemiological studies. *Journal of Environmental Monitoring, 11*(3), 475–490.

Chapter 20
Advancing Sustainable and Inclusive Evaluation Frameworks for Health and Environmental Impact Assessment

Asma Hasan, Baby Tabassum, Mohammad Hashim, and Nagma Khan

20.1 Introduction

Integrated Environmental Health Impact Assessment (IEHIA) is an emerging multidisciplinary idea that is of growing importance in the twenty-first century. Unambiguously, global climate change is occurring; this phenomenon is changing the contexts of public health as well as the environment (Morgan, 2012). Recent studies state that, with climate change as the main cause, the interdependence of human health and environmental sustainability has been central in debates on modern global development at the end of the century. The Intergovernmental Panel on Climate Change projected a range of 1.2–6.5 °C for the average global temperature increase in its fourth assessment report. In addition to having direct and indirect effects on the environment, climate change also poses risks to human health and safety. Direct health effects may include mortality and morbidity from natural calamities such as hot and cold spells, droughts, and storms, as well as from changes in the quality of water and air resulting from changing climate factors. Conversely, indirect consequences might result from changes in environmental circumstances, such as air quality deterioration or landscape transformation, which, in turn, affect human health outcomes or change disease ecology (Holman et al., 2005; Yadav, 2021, 2022).

A. Hasan · B. Tabassum (✉) · N. Khan
Toxicology Lab, Department of Zoology, Govt. Raza P.G. College, Rampur, India

Mahatma Jyotiba Phule Rohilkhand University, Bareilly, India

M. Hashim
Mahatma Jyotiba Phule Rohilkhand University, Bareilly, India

A. K. Yadav (ed.), *Computational Approach for Air Pollution*,
https://doi.org/10.1007/978-3-032-09215-1_20

Health and environmental impact assessments (HIAs and EIAs) have been carried out as separate procedures, sometimes neglecting the intricate links among environmental stresses and public health. This separation has hindered our ability to properly handle the interconnected nature of environmental damage and health effects. While poor health may make it difficult to solve environmental issues, environmental degradation can aggravate medical difficulties. Growing awareness of these links demands a framework change in evaluation systems (Waheed et al., 2009). As climate change, biodiversity loss, pollution, and noncommunicative diseases increase, this study proposes an integrated HIA and EIA paradigm with sustainability and inclusivity. This type of strategy guarantees that policy and decision-making procedures adequately meet the demands of sensitive populations and ecosystems, enabling more complete and fair solutions to the environmental and health (E&H) problems of the present time (Hambling et al., 2011).

20.2 Genesis

Integrated Environmental Health Monitoring (IEHM) plays a crucial role in determining significant stressors in the environment, assessing their status, and determining the health effects of climate change (Lebret, 2015). At present, no one has established IEHM. The Integrated Assessment of Health Risks of Environmental Stressors in Europe project, which was funded by the European Union, looked at ways to connect and combine different types of information and technology to make IEHM work better. To perform an IEHIA, which is "an inclusive and, as far as feasible, comprehensive assessment of the risks to and impacts on human health as a result either of exposures to a defined set of environmental hazards or of the effects of policies or other interventions that operate via the ambient or living environment," IEHM gathers data that are objective and reliable. Therefore, knowledge on various forms leads to relationships among environmental health hazards is necessary. Research studies and policy choices traditionally largely dealt with single-pollutant and single-effect links; data on exposure and the influence of changes to the environment on human health were not combined (Firket, 1936). However, recently, efforts to investigate the connections between several pressures and various effects on health have been growing fast. Data shortages or the reality that different data collection techniques have diverse purposes and cannot be easily combined have hindered integrated research. IEHM's main concentration is monitoring to evaluate, assess, and comprehend environmental impacts on human health, such as chemical exposure, climate change, air pollution, and so on (Fig. 20.1); knowledge of this helps to guide decisions (Corvalan et al., 1996).

We require a structured, ongoing cause–effect chain approach to characterize interrelated E&H challenges. The intricacy of E&H issues and the interplay of many parameters at different organizational levels and sizes make the design of IEHM programs difficult. To address this, we need innovative data utilization, particularly technology data integration (Bell et al., 2004).

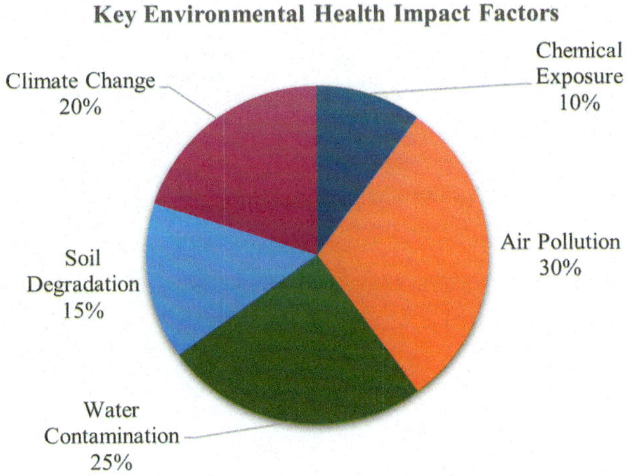

Fig. 20.1 Pie chart visualizing key environmental health impact factors

European initiatives to link E&H information include the European Community Health Indicators, the European Integrated Environment and Health Monitoring and Response System, the WHO's "Health for All" initiatives, and the ECEH's indicator work. Such systems collect data from different origins but have different objectives. An IEHM application to track and quantify environmental changes in well-being is based on this study. We also discuss the needs and frameworks based on extensive literature and data.

20.3 Criteria for Sustainable Development Goals

There is a transparent procedure followed to choose pertinent environmental health indicators (EHIs) for the Sustainable Development Goals (SDGs). Although the SDGs are mostly intended for nations and regions, their indicators can help with health in EIAs (Fig. 20.2). Examining all the SDG metrics from UNSTATS (Gulis et al., 2022) comes first. The next stage is to focus on the health-related metrics for a particular nation, such as Denmark, using accessible national data, since the SDGs cover various areas beyond health. The relevant health indicators were found using metadata from Statistics Denmark and the Institute of Health Metrics and Evaluation in Seattle, USA (Winkler et al., 2020).

The final phase is to select the health outcome indicators that fit the EIA. For this purpose, we consulted five public health specialists, experts in HIA, environmental health, and EIA. Direct and complicated are the two categories into which health outcome indicators fall. Although complicated indicators may need to be separated into more precise outcomes before being used in EIA, direct health indicators may be used immediately in evaluations.

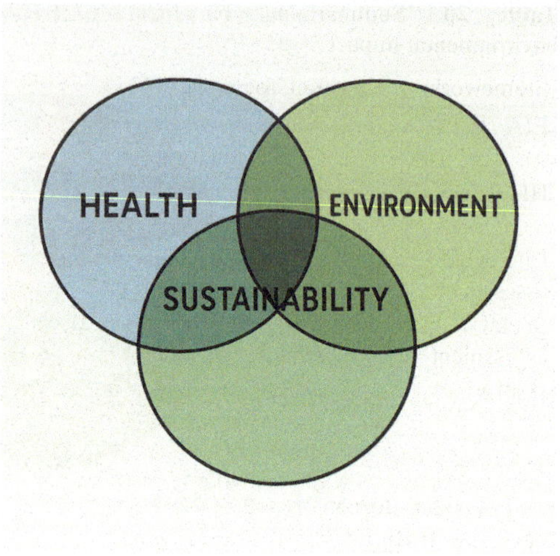

Fig. 20.2 Diagram illustrating the relationship between health, environment, and sustainability

20.4 Framework

An IEHM software should be built on a disciplined foundation to allow concurrent monitoring and interpretation (Harris et al., 2015). Frameworks offer a methodical technique to find connections or interactions between the environment and human health, therefore, helping to clarify difficult E&H scenarios. Actually, the arrangement of thoughts, notions, and ideas of an action typically follows a paradigm to make sense (Cave et al., 2020), thus facilitating the understanding of the relationships among the several components of the activity.

Solid groundwork has the following (Harris et al., 2015):

- Conceptual clarity and scope: guarantees that the framework has logical and reasonable connections and covers all main ideas.
- Flexibility to allow one to consider the problem at any level of the framework or at any component.
- Balance: The paradigm equally well handles problems with environmental or health emphasis.
- Usability: the paradigm allows a workable approach to create appropriate indications.

By establishing a connection between environmental components and health outcomes, EHI programs help develop adaptation strategies in the face of changing environmental conditions. It is crucial to have a versatile paradigm in order to include indications that describe interactions in natural, ecological, and human-made systems. A more comprehensive strategy than just identifying symptoms is to combine indicators into determinants, which might include physical, social, economic, environmental, and behavioral health considerations (Kasperson, 1988).

Table 20.1 Summarizing the main assessment frameworks related to health and environmental impact

Framework	Key focus	Application
EIA	Assessing the environmental consequences of projects	Infrastructure, industrial projects, urban planning
HIA	Assessing health outcomes of policies and projects	Public health policies, urban development
Life cycle assessment	Examining environmental effects through a product's life	Product design, industrial processes
Social impact assessment	Evaluating projects and policy social effects	Community projects, social policies
IEHIA	Combining health and environmental risk assessment	Holistic policy evaluation, multi-sectoral projects

The first step in the development of EHIs is to determine which environmental variables influence human health. A more helpful approach would be to categorize indicators into causes and outcomes rather than just providing a list of them.

It is essential to use an ecosystem health perspective while creating and using EHI. By taking an ecosystem-level view, we may learn more about the factors that affect human health and where diseases originate. Highlights the interdependence between ecosystems and people, with altered ecosystems identified as potential risk factors for the public. By including social, economic, and political elements, ecological health fortifies the connections between human health and environmental sustainability. Environmental health, health, and the environment have all contributed to the development of various frameworks. To analyze and track the effects of climate change on human health, this study compares several frameworks based on the characteristics they offer for creating EHI (illustrated in Table 20.1).

20.4.1 Pressure State Response Framework

Organisation for Economic Co-operation and Development (1993) created the pressure-state response (PSR) method to standardize environmental performance metrics in the early 1990s. Environmental indicators initially monitored the "state" of the environment or the pace of physical change in nature. This method showed that something was wrong, but it did not reveal the problem or possible solutions. Human activities, or the "pressures" that are driving change, and societal "responses" to regulate or mitigate the effects of those actions, are therefore incorporated into the paradigm. At first, environmental statistics frequently relied on straightforward one-to-one relationships between individual environmental pressures, changes, and societal responses. Despite its widespread use, the framework is continuously improving. Its roots in environmental economics prevent it from delving deeply into the relationships between causes and consequences (Kjellström, 1999). Distinguishing between "pressure" and "state" signs has been a major challenge.

We should expand the framework to tackle sustainable development more effectively. People often argue against the PSR, arguing that it oversimplifies the relationship between environmental impacts and human actions. The Australian federal government made adjustments to the paradigm by including feedback loops to improve it. More than just human health and ecosystems are now included in the pressure-state-impact-response paradigm. This model has changed over time to include larger forces and their effects. However, it has become apparent that the PSR paradigm does not adequately address the connections between changes in the environment and human health.

20.4.2 Driving Force-State-Response Framework

The Driving Force-State-Response (DSR) framework, which the United Nations Commission on Sustainable Development accepted, is an evolution of the PSR framework. Sustainable development, according to the DSR paradigm, consists of the following parts: The driving force behind sustainable development is the human actions, processes, and patterns that influence it. Another aspect is the condition of sustainable development. Third on the list are policy decisions and other reactions to variations in sustainability. Instead of "pressure," the term "driving force" better captures the integration of institutional, social, and economic components. Similarly to how social, economic, and institutional indicators can have both positive and negative impacts on sustainable development, the phrase "driving force" allows this. Indicators of the environmental status can let the public and decision-makers of the scientific findings derived in fields and laboratories. Signals have to be focused on a certain group if they are to be effective. An ideal set of indicators would be a way to minimize an enormous amount of data into a more manageable format while keeping the key insights relevant to the research issue at hand (von Schirnding, 2002).

20.4.3 Driving Force-Pressure-State-Exposure-Effect-Action Framework

World Health Organisation. International Classification of Diseases (2021) created the Driving Force-Pressure-State-Exposure-Effect-Action (DPSEEA) model in the early 1990s to assist in the establishment of environment and health indicators. Driving response (anthropogenic) refers to human activities that activate and push the processes of the environment. Usually, human activity or exploitation of the environment reflects pressure on the environment. In humans, exposure is the state in which one is in contact with the environment. In humans, impact is the result of exposure to environmental dangers, which can indicate health consequences. Action refers to policies or treatments intended to minimize or avoid health impacts; they can be incorporated at any stage of the framework. DPSEEA focuses on anthropogenic indicators more than natural and physical hazards, so it does not do a good job of showing how complex the relationships are between natural and human systems (Dannenberg, 2016). Furthermore, using the framework linearly leads to limited results (González Del Campo et al., 2020). Although the DPSEEA is designed to be an interlinking network for E&H concerns, it shows that several driving variables can cause multiple health outcomes and various exposures. DPSEEA effectively involves stakeholders in assessing exposure-health effects, enabling the identification and tracking of environmental-health indicators, and developing cost-efficient therapies throughout the causal chain (Kørnøv et al., 2020). The DPSEEA has been used for a variety of purposes, including but not limited to assessing the efficacy of adaptation and mitigation policies in the face of climatic drift; developing E&H indicators to monitor human health susceptibility; and monitoring the health consequences of climate change in Europe. With its wide-ranging uses, DPSEEA might offer insightful direction on creating an IEHM program, therefore allowing tracking and evaluation of how climatic changes affect human health.

20.4.4 Multiple Exposures-Multiple Effects Model

When it comes to the environment and children's health, the multiple exposures-multiple effects (MEME) paradigm is a significant advance on the DPSEEA model. This model highlights the interconnected nature of environmental exposures and their effects on children's health. It takes the DPSEEA architecture and streamlines it while adding to it. Differentiating between the state and pressure components can be challenging in practice. Since it is possible to evaluate exposure indicators more or less directly, the MEME paradigm groups the three components under the term "exposure." This is due to the frequent substitution of state and pressure components for actual exposure (Renn & Graham, 2005). Furthermore, the MEME model acknowledges that exposures and health consequences are impacted by contextual variables, including demography, socioeconomic development, and social circumstances. The shared characteristics allow for easy switching between the MEME and DPSEEA frameworks as needed. Recognizing contextual elements is an important part of the MEME framework's emphasis on the interconnected nature of E&H issues (Moore & Kempton, 2009). Still, it does not distinguish between factors that are close by (exposures) and those that are further away (pressure and state). When developing and implementing treatments further up the causal chain, the capacity of a paradigm to differentiate between proximal and distal causes is very valuable. Reducing unfavorable health outcomes may frequently be more cost-effective with these higher-level treatments. Consequently, the DPSEEA framework can be useful for organizing topics more formally when dealing with complex environmental health challenges, such as the health implications of climate change.

20.4.5 Framework of Environmental Public Health Indicators

The Environmental Public Health Indicator (EPHI) model was designed by the U.S. Centers for Disease Control and Prevention; it organizes indicators into four groups: danger, exposure, health impact, and action. Using concepts from Edwards and Meagher (2020), the framework integrates structural components from the DPSEEA paradigm for conserving the environment and public health with the PSR framework for sustainable development. Hazard indicators point to events or practices that could cause one to come into contact with dangerous pollutants or hazardous environments. Found in tissues or fluids, exposure indicators are biological markers that signify the presence of potentially dangerous compounds for health (Briggs, 2008). Although intervention indicators focus on programs or policies meant to reduce or avoid exposure, hazards, or health consequences, health effect indicators highlight diseases or disorders originating from environmental hazards. EPHI indicators are used to measure health status or environmental hazards. According to Moore and Kempton (2009), the best indicators are those that consistently gather data, satisfy certain requirements, and forecast the relationship between human health and environmental conditions. The EPHI framework, similar to the DPSEEA and MEME models, is based on the concept of the environmental health chain, which states that anything occurring before exposure is a potential "hazard indicator." It is less useful for planning and implementing treatments than the DPSEEA framework, but, similar to the MEME framework, it does not differentiate proximal (direct) from distal (indirect) causes.

20.4.6 Health Impact Assessment

An HIA is a method for determining how a proposed policy, program, or initiative could affect people's health, particularly those who are already at a disadvantage. The goal is to provide decision-makers and stakeholders with suggestions that optimize health outcomes while minimizing potential negative repercussions. Until now, most HIAs have been implemented on a regional or even a municipal level, with very restricted accessibility. However, past evaluations of the consequences of climate change have used HIA approaches. According to Patz et al. (2008), the HIA paradigm gives a thorough and applicable method to improve judgments about health and climate change policies. The HIA is a more all-encompassing technique for making decisions that considers health equity and may help to assess the co-benefits of activities as well as the lower of individual disease risks. To evaluate the effects of climate change on the health of Bangladeshi citizens, Nelson (2003) utilized the disability-adjusted life year index within HIA. They were able to rate and compare different dangers using this technique, demonstrating its usefulness for risk evaluation and classification. Although the HIA paradigm has been shown to be useful in assessing the health effects of climate change, as demonstrated by Nelson

(2003) and Patz et al. (2008), it is not intended for the purpose of creating indicators or determining what steps to take. Consequently, it does not provide specific steps for developing EHIs to track how climate change affects people's well-being.

20.4.7 Integrated Environmental Health Impact Assessment

IEHIA is a technique used to evaluate health-related issues resulting from environmental variables and the health consequences of policies and actions that influence the environment. Taking a comprehensive perspective on both health and the environment, this method acknowledges the complexity, interdependence, and uncertainty of the real world. It recognizes that the outside factors driving the change inside the system influence the whole system, including the health of the environment. Therefore, IEHIA investigates the effects of environmental assets and hazards in view of these outside changes (Mekouar, 2018). Combining many assessment techniques, the IEHIA framework includes HIA, comparative risk assessment, and conventional risk assessment. It employs both qualitative and quantitative techniques: the qualitative approach concentrates on framing problems and choosing suitable evaluation techniques, while the quantitative approach uses models, often in the face of insufficient data, to analyze difficult questions (Fig. 20.3).

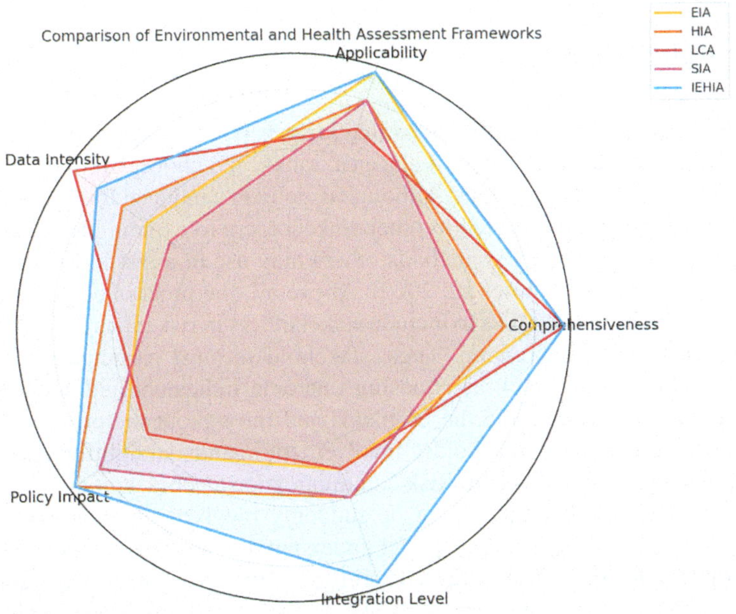

Fig. 20.3 Radar chart (Spider chart) comparing different assessment frameworks based on multiple criteria

The qualitative method is rather participatory, so stakeholder engagement becomes vital. On the contrary, the quantitative method addresses complicated multidimensional systems, usually with insufficient data. This integrated approach deals with environmental health issues' nonlinearity and multi-causality (Bhatia & Wernham, 2008). Similarly to the HIA framework, we can use the IEHIA model to assess the impact of climate change on human health. However, it does not expressly target the creation of EHIs and does not offer a clear direction for steps to take. Consequently, the IEHIA paradigm provides little help in creating EHIs to assess and track how climate change affects human health.

20.5 Limitations

An integrated framework for assessing environmental health impacts combines quantitative approaches for evaluating complicated systemic problems with a qualitative approach to identify concerns and selecting appropriate assessment methods (Bhatia & Wernham, 2008). Problems could occur with any approach. Participatory approaches, such as the qualitative one, can be challenging to manage, since they need input from stakeholders. Modeling and analyzing complicated systems are at the core of the quantitative method, but reliable findings are generally difficult due to a lack of resources (Nooteboom, 2000).

20.6 Stakeholders

Stakeholders should be included in evaluations; it is also vital to understand the several uses this helps to achieve. It clarifies, ranks, and chooses for risk managers the appropriate policies or actions. Although usually in a reduced form, it also offers data that must be communicated to other stakeholders to explain the risks, defend the actions taken, and provide methods others may use to assist in lowering those risks (González Del Campo et al., 2020). Above all, the process of generating the assessment offers opportunities to include stakeholders in risk management, thereby fostering trust and collaboration. Thus, stakeholders must participate not just in developing the process, such as choosing outcome indicators and deciding their relevance, but also in defining the problems and throughout the evaluation stage. Although the concept of stakeholder involvement sounds straightforward, implementing it is much more difficult. Risk communication teaches, among other important things, that stakeholders cannot be suddenly involved on short notice. Real meaningful engagement requires mutual understanding and trust, which takes time to develop (Ravn Boess et al., 2021). Therefore, early involvement with stakeholders is essential as constant communication helps build trust. Especially when decision-makers seek fast responses, time and money might be major obstacles (Prüss-Üstün et al., 2003). Engaging stakeholders might also be difficult given the mistrust of bureaucratic cultures within authorities or specialists. Finding people who can really represent big, varied groups can be very challenging; identifying real stakeholders is challenging enough (Prüss-Üstün et al., 2003). Additionally, variations in knowledge, communication ability, and authority among individuals might impede stakeholder participation. In local settings, these difficulties are sufficient, but they become much more difficult when handling intricate regional or global problems. Although certain approaches, such as citizen panels or interactive websites that enable more public engagement, are being developed, real, direct involvement in regional or global choices remains rare and usually lacks appropriate information (Istamto et al., 2014).

20.7 Multivariate

Systemic problems can be somewhat difficult to evaluate, particularly if they have several sources. One of the key challenges is knowing how several elements cooperate to produce hazards. When many dangerous substances are mixed, for instance, their effects could just stack up if they all behave the same way (Grilli et al., 2021). However, if they operate differently, the outcomes might be more complex, with both expected positive and negative impacts difficult to predict. Sometimes health hazards result from cooperating wholly separate agents, such as noise and air pollution, or even more complex ones, including particles in the air and electromagnetic radiation. Furthermore, elements unrelated to the environment, such as genes or behavior, can also affect health as well; for example, how exposure to radon combined with smoking raises cancer risk (Hansson & Polk, 2018). Furthermore, health hazards may not necessarily result from concurrent exposure to detrimental elements. They can also arise from exposures at several phases of life, either increasing or lowering the chance of disease later on or perhaps shielding against it. For instance, early exposure to allergens may increase or lower the chance of respiratory difficulties; exposure to air pollution before birth may lead to low birth weight and subsequent health complications. Due to this complex network of factors, it is difficult to identify the precise causes of health problems. Although each occurrence builds on the one before it, causes can be found throughout time and are sometimes molded by other elements. It becomes abundantly evident that some health problems have no "basic" cause. Rapport et al. (1998) notes that causes cannot always be clearly found and that, in many circumstances, several elements contribute rather than one. In the disciplines of toxicology and epidemiology, researchers often seek to control various risk variables in their studies to account for them in the analysis. On the other hand, when the objective is to understand the whole system of causes and how they connect to policies and health outcomes, this strategy does not perform well. This makes it rather crucial to understand the interplay among several factors. However, it is difficult to determine how to share blame for a health problem when several elements are involved. Although in some circumstances these connections may have relatively minor impacts, in extreme cases, one element may entirely

rely on another, as in some gene–environment interactions (Bois et al., 2010). Dividing the influence evenly among elements might cause an overset environment; the overall influence of the elements adds up to more than 100%. Solving this puzzle is not simple. It cannot be corrected with strict rules, such as the Bradford Hill criteria, which may ignore significant study biases. Ignoring difficult-to-measure elements does not lower uncertainty; rather, it limits the study and masks significant unresolved issues. In contrast, striving to incorporate all elements might make the process too difficult and not always produce better outcomes. The secret is to combine the complexity of the assessment with its goal. Three key factors should be taken into account: the demands of the users of the findings, the degree of uncertainty they are ready to accept, and the expenses of compiling additional data. Although there are techniques, such as value of information analysis, that can assist in defining this balance, they are not always useful for difficult tests (Rosenthal et al., 2007). Therefore, the most crucial issue is to ensure that the rational and open approach utilized to assign causes is followed. An issue's overall health burden should never be more than 100%; moreover, elements along a causative route should not be double-counted. Creating a clear causal model that specifies when it is suitable to sum effects helps to prevent this. Another problem with multicausal systems is that positive elements that could safeguard health also influence health consequences in addition to negative risk elements. By eliminating the negative consequences of hazards, these positive elements can help reduce disease or enhance health. Our conception of causality should incorporate elements that improve health, in addition to the elements causing illness. Several studies overlook the good elements, as illness generally takes center stage. However, not ignoring these elements might distort the picture of the circumstances and support bad policy judgments.

20.8 Adaptation, Change, and Time

Comparison of health outcomes in various contexts is an integral part of evaluation, as indicated before. In most cases, major life changes are required to bring about a transition from one state to another. In talks on topics such as climate change, the emphasis has typically been on government measures, such as policies, for adaptation. However, adaptability entails a tremendous deal more. During this phase, people and organizations react to both potential threats and policies and measures put in place to mitigate them. Some changes may not be immediately apparent due to systemic delays, such as the lag time between a risk's exposure and the onset of its consequences, or because the first alteration may set off a cascade of reactions (Edwards & Meagher, 2020). Thus, adaptations can persist for a long time, possibly beyond a century. Furthermore, the starting and ending points of this process do not always signify a steady or balanced condition. This is because systems undergo inherent changes as a result of age or evolution, and outside forces might intervene before the system has a chance to completely adapt. Therefore, many systems are constantly evolving. Environmental health systems are dynamic, which means that evaluations must also be. It implies that the results are contingent upon the evaluation time frame. The effects of continuous adjustments are not always captured when comparing before- and after-images of a system, which could lead to an underestimation of the real implications. Short-term evaluations do not consider the original intervention and its reactions, potentially leading to incorrect conclusions (Martens et al., 2000). Efforts to evaluate policies and products throughout their entire life, known as life-cycle evaluations, could fail to take into account the consequences that will last for generations to come (Lindner, 2011). Assessments may have to span substantially longer times, potentially hundreds or thousands of years, to capture these long-term impacts. As an example, a landfill's life-cycle analysis considered the decomposition of garbage for as long as 500 years. A Swiss researcher even extended its time horizon to 60,000 years in the hopes that another ice age would have passed by then! It is difficult to explain these lengthy timescales in the context of short-term policymaking, and they introduce significant uncertainty into the analysis. Fixed before-and-after comparisons, or static situations, have their limitations when dealing with systems that adapt and evolve. Although they can provide general answers about the objectives of various policies, they fail to provide much information on the real-world effects of attempting to accomplish those objectives (Kay et al., 2000). Therefore, adaptive scenarios, which represent the system's response to changes in input circumstances, are typically more helpful. These scenarios do not predict the eventual state but rather the potential evolution of the system over time. However, because these models are derived from forecasts or predictions, they incorporate all the inherent uncertainties in modeling complex systems. When doing environmental assessments, these uncertainties might amplify because data is often scarce, and models are simply best guesses. They take on an even greater significance when considering social systems that are reliant on individual actions. It cannot always depend on people's reactions; at certain points, they may behave in ways that contradict our expectations or the common good. It is possible for people's reactions to events to diverge far from what observers anticipate, as the "tragedy of the commons" demonstrates (Pohjola & Tuomisto, 2011).

20.9 Conclusion

Developing and applying EHIs to monitor how climate change affects human health presents a difficult task. A systematic framework is needed to explain the intricate relationships between the environment and human health. Understanding that disrupted ecosystems affect the whole population, an ecosystem health approach is efficient. This method helps to find direct as well as indirect connections between human well-being and environmental changes. However, properly reflecting the complexities of climate change's consequences on health is challenging due to variable dates, places, and non-linear interactions. Frameworks should be adaptable and customized to unique demands and goals. This study focuses on models for developing EHIs to evaluate, track, and direct actions for the health effects of climate change. Developing EHIs that track health vulnerability, develop interventions, and evaluate climate change adaptation and mitigation action might find the DPSEEA framework appropriate. EHIs that combine environmental, animal, and human data can help spot new ecological problems. This comprehensive strategy guides actions to safeguard human, animal, and environmental health and acts as an early warning system for disturbance of ecosystems. When these relationships are well characterized, interventions may be deployed earlier in the causal chain, increasing ecological health and decreasing disease loads.

Acknowledgments AH, BT, MH, and NK express their sincere appreciation to the Department of Zoology, Government Raza P.G. College, Rampur, for offering vital resources and related equipment.

References

Bell, M. L., Davis, D. L., & Fletcher, T. (2004). A retrospective assessment of mortality from the London smog episodeof 1952: The role of influenza and pollution. *Environmental Health Perspectives, 112*, 6–8.

Bhatia, R., & Wernham, A. (2008). Integrating human health into environmental impact assessment: An unrealized opportunity for environmental health and justice. *Environmental Health Perspectives, 116*, 991–1000. https://doi.org/10.1289/ehp.11132

Bois, F. Y., Jamei, M., & Clewell, H. J. (2010). PBPK modelling of inter-individual variability in the pharmacokinetics of environmental chemicals. *Toxicology, 278*, 256–267. https://doi.org/10.1016/j.tox.2010.06.007

Briggs, D. J. (2008). A framework for integrated environmental health impact assessment of systemic risks. *Environmental Health, 7*, 61.

Cave, B., Claßen, T., Fischer-Bonde, B., Humboldt-Dachroeden, S., Martín-Olmedo, P., Mekel, O., et al. (2020). Human health. In *Ensuring a high level of protectionA reference paper on addressing human health in environmental impact assessment: As per EU directive 2011/92/EU amended by 2014/52/EU*. International Association for Impact Assessment and European Public Health Association.

Corvalan, C., Briggs, D. J., & Kjellstrom, T. (1996). *Development of environmental health indicators, linkage methods for environment and health analysis, general guidelines*. UNEP, USEPA and WHO.

Dannenberg, A. L. (2016). Effectiveness of health impact assessments: A synthesis of data from five impact evaluation reports. *Preventing Chronic Disease, 13*, E84. https://doi.org/10.5888/pcd13.150559

Edwards, D. M., & Meagher, L. R. (2020). A framework to evaluate the impacts of research on policy and practice: a forestry pilot study. *Forest Policy and Economics, 114*, 101975. https://doi.org/10.1016/j.forpol.2019.101975

Firket, J. (1936). *Fog along the Meuse Valley; transactions of the faraday society*. RSC Publishing.

González Del Campo, A., Gazzola, P., & Onyango, V. (2020). The mutualism of strategic environmental assessment and sustainable development goals. *Environmental Impact Assessment Review, 82*, 106383. https://doi.org/10.1016/j.eiar.2020.106383

Grilli, N. D. M., Andrade, M. M. D., Xavier, L. Y., Santos, C. R., et al. (2021). Step by step: A participatory action-research framework to improve social participation in coastal systems. *AmbienteSociedade, 24*, e02551. https://doi.org/10.1590/1809-4422asoc20190255r1vu2021L1AO

Gulis, G., Krishnankutty, N., Boess, E. R., Lyhne, I., & Kørnøv, L. (2022). Environmental impact assessment, human health and the sustainable development goals. *International Journal of Public Health, 31*(67), 1604420. https://doi.org/10.3389/ijph.2022.1604420

Hambling, T., Weinstein, P., & Slaney, D. (2011). A review of frameworks for developing environmental health indicators for climate change and health. *International Journal of Environmental Research and Public Health, 8*(7), 2854–2875. https://doi.org/10.3390/ijerph8072854

Hansson, S., & Polk, M. (2018). Assessing the impact of transdisciplinary research: The usefulness of relevance, credibility, and legitimacy for understanding the link between process and impact. *Research Evaluation, 27*(2), 132–144.

Harris, P., Viliani, F., & Spickett, J. (2015). Assessing health impacts within environmental impact assessments: An opportunity for public health globally which must not remain missed. *International Journal of Environmental Research and Public Health, 12*, 1044–1049. https://doi.org/10.3390/ijerph120101044

Holman, I. P., Rounsevell, M. D. A., Shackley, S., Harrison, P. A., Nicholls, R. J., Berry, P. M., & Audsley, E. (2005). A regional, multi-sectoral and integrated assessment of the impacts of climate and socio-economic change in the UK-part I-methodology. *Climatic Change, 71*, 9–41. https://doi.org/10.1007/s10584-005-5927-y

Istamto, T., Houthuijs, D., & Lebret, E. (2014). Multi-country willingness to pay study on road-traffic environmental health effects: Are people willing and able to provide a number? *Environmental Health, 13*, 35.

Kasperson, R. E. (1988). The social amplification of risk a conceptual framework. *Risk Analysis, 8*, 177–187.

Kay, D., Pruss, A., & Corvalan, C. (2000). *Methodology for assessment of environmental burden of disease: Report on the ISEE session on environmental burden of disease*. World Health Organization.

Kjellström, T. (1999). Preliminary assessment of potential human health indicators for air quality. In *Technical paper no. 55 air quality*. Ministry for the Environment.

Kørnøv, L., Lyhne, I., & Davila, J. G. (2020). Linking the UN SDGs and environmental assessment: Towards a conceptual framework. *Environmental Impact Assessment Review, 85*, 106463. https://doi.org/10.1016/j.eiar.2020.106463

Lebret, E. (2015). Integrated environmental health impact assessment for risk governance purposes; Across what do we integrate? *International Journal of Environmental Research and Public Health, 13*, 71. https://doi.org/10.3390/ijerph13010071

Lindner, R. (2011). *Frameworks for assessing policy research and ACIAR'S investment in policy oriented projects in Indonesia. ACIAR impact assessment series report no. 72*. Australian Centre for International Agricultural Research. Retrieved July 30, 2023, from https://www.aciar.gov.au/sites/default/files/legacy/node/13937/ias72_pdf_72856.pdf

Martens, P., McMichael, A. J., & Patz, J. (2000). Globalization, environmental change and health. *Global Change and Human Health, 1*, 4–8. https://doi.org/10.1023/A:1011572212445

Mekouar, A. (2018). *UN environment, assessing environmental impacts-a global review of legislation* (pp. 659–660). Revue juridique de l'environnement.

Moore, M., & Kempton, P. D. (2009). A synopsis of the joint environment and human health programme in the UK. *Environmental Health, 8*, S1.

Morgan, R. K. (2012). Environmental impact assessment: The state of the art. *Impact Assessment and Project Appraisal, 30*, 5–14. https://doi.org/10.1080/14615517.2012.661557

Nelson, D. I. (2003). Health impact assessment of climate change in Bangladesh. *Environmental Impact Assessment Review, 23*, 323–341.

Nooteboom, B. (2000). Learning by interaction: Absorptive capacity, cognitive distance and governance. *Journal of Management and Governance, 4*, 69–92.

Organisation for Economic Co-operation and Development. (1993). OECD core set of indicators for environmental performance reviews: A synthesis report by the group on the state of the environment. *Environment Monographs, 83*, 1–39.

Patz, J., Campbell-Lendrum, D., Gibbs, H., & Woodruff, R. (2008). Health impact assessment of global climate change: Expanding on comparative risk assessment approaches for policy making. *Annual Review of Public Health, 29*, 27–39. https://doi.org/10.1146/annurev.publhealth.29.020907.090750

Pohjola, M. V., & Tuomisto, J. T. (2011). Openness in participation, assessment, and policy making upon issues of environment and environmental health: A review of literature and recent project results. *Environmental Health, 10*, 58. https://doi.org/10.1186/1476-069X-10-58

Prüss-Üstün, A., Mathers, C., Corvalán, C., & Woodward, A. (2003). Introduction and methods: Assessing the environmental burden of disease at national and local levels. In *Environmental burden of disease series No. 1*. World Health Organization.

Rapport, D. J., Costanza, R., & McMichael, A. J. (1998). Assessing ecosystem health. *Trends in Ecology & Evolution, 13*, 397–402. https://doi.org/10.1016/s0169-5347(98)01449-9

Ravn Boess, E., Lyhne, I., Davila, J. G., Jantzen, E., Kjellerup, U., & Kørnøv, L. (2021). Using sustainable development goals to develop EIA scoping practices: The case of Denmark. *Impact Assessment and Project Appraisal, 39*, 463–477. https://doi.org/10.1080/14615517.2021.1930832

Renn, O., & Graham, P. (2005). *White paper on risk governance towards an integrative approach*. International Risk Governance Council.

Rosenthal, J. K., Sclar, E. D., Kinney, P. K., Knowlton, K., Crauderueff, R., & Brandt-Rauf, P. W. (2007). Links between the built environment, climate and population health: Interdisciplinary environmental change research in New York city. *Annals Academy of Medicine, 36*, 834–846.

von Schirnding, Y. (2002). *Health in sustainable development planning: The role of indicators*. World Health Organisation.

Waheed, B., Khan, F., & Veitch, B. (2009). Linkage-based frameworks for sustainability assessment: Making a case for driving force-pressure-state-exposure-effect-action (DPSEEA) frameworks. *Sustainability, 1*, 441–463. https://doi.org/10.3390/su1030441

Winkler, M. S., Adongo, P. B., Binka, F., Brugger, F., Diagbouga, S., Macete, E., et al. (2020). Health impact assessment for promoting sustainable development: The HIA4SD project. *Impact Assessment and Project Appraisal, 38*(3), 225–232. https://doi.org/10.1080/14615517.2019.1694783

World Health Organisation. International Classification of Diseases. (2021). Retrieved June 7, 2021, from https://www.who.int/standards/classifications/classification-of-diseases

Yadav, A. K. (2021). Human health risk assessment in opencast coal mines and coal-fired thermal power plants surrounding area due to inhalation. *Environmental Challenges, 3*, 100074.

Yadav, A. K. (2022). Characterization and effective dose of gross alpha and gross beta activities in particulate matters ($PM_{2.5}$, PM_{10}, SPM) in India. *Aerosol Science and Engineering, 6*(4), 360–369.

Chapter 21
Environmental and Public Health Concerns in a Sustainable Future

Smrita Singh, Tanzeel Ahmed, and Ashutosh Singh Chauhan

21.1 Introduction

Air pollution is a global challenge for the environment and human health. Industrial emissions and vehicular exhaust to agricultural practices and natural phenomena cause pollution. The emissions of harmful gases, fumes, and particles into the atmosphere deteriorate air quality, causing harm to human health and ecosystems. Air pollution includes contaminants such as particulate matter (PM), ground-level ozone, nitrogen oxides, sulfur dioxide, and volatile organic compounds (VOCs) that have different sources and health problems. Despite the fact that great attention has been paid to the long-term consequences of pollution, the most dangerous pollutants are invisible in existence and isolation. Understanding this tangled web of pollutant sources is extremely important in designing mitigation strategies and protecting public health. The sources of air pollution are diverse and play a complex interplay between human activities and natural processes. A variety of pollutants, such as PM, nitrogen oxides, and sulfur dioxide, are released in combustion-related activities, such as power plants, vehicles, and industrial facilities. The emissions of dust, fumes, and chemical contaminants from industrial operations such as manufacturing, mining, and construction contribute to air pollution. The use of fertilizers, pesticides, and livestock farming in the agriculture sector generates emissions of ammonia, methane, and other pollutants that can degrade air quality. Fossil fuels in transportation are major contributors to air pollution, releasing pollutants such as

S. Singh (✉) · T. Ahmed
School of Biotechnology, IFTM University, Moradabad, India

A. S. Chauhan
Air Pollution and Stone Conservation Laboratory, Archaeological Survey of India, Agra, India

A. K. Yadav (ed.), *Computational Approach for Air Pollution*,
https://doi.org/10.1007/978-3-032-09215-1_21

nitrogen oxides, carbon monoxide, and PM. Volcanic eruptions, wildfires, and dust storms, which are natural phenomena, emit PM and gases into the atmosphere, also adding to air pollution. Tobacco smoke, gas stoves, and building materials in indoor sources can contribute to air pollution.

Exposure to air pollutants causes respiratory and cardiovascular problems, such as triggering asthma attacks, exacerbating chronic bronchitis, and increasing the risk of respiratory infections. Long-term exposure causes lung cancer, heart disease, and stroke. In children, it causes impaired lung development and increases premature mortality. Exposure to air pollution during pregnancy affects fetal development and causes low birth weight and preterm birth. Beyond human health, air pollution poses a significant threat to ecosystems and the environment. Acid rain leads to the damage of forests, lakes, and aquatic ecosystems. Ground-level ozone is reducing crop yields and impacting forest health. Furthermore, air pollution can contribute to climate change, ecological damage, and socioeconomic disruptions. Nitrogen oxides and sulfur dioxide deposition can acidify soils and water bodies, leading to detrimental effects on plants and aquatic life.

Regulatory agencies have established air quality standards to limit emissions and safeguard public health worldwide. Air quality monitoring networks track pollution levels to assess control measures. Governments are implementing policies and regulations to mitigate air pollution globally, such as promoting cleaner transportation, investing in renewable energy, and enforcing stricter emission controls for industries. The Air Act (1981) and the Environment Act (1986) promote regular air quality monitoring. Air purifiers, smog towers, and pollution-absorbing materials are being developed and deployed to reduce urban air pollution. A step toward green urban roofs and vertical gardening can help reduce air pollutants, particularly in urban areas. These green spaces absorb CO_2 and can mitigate climate change. In addition, urban green spaces contribute to the aesthetic appeal of cities and provide recreational opportunities for residents. Repurposing spaces occupied by cars into green spaces can encourage outdoor physical activity, improve air quality, reduce noise, and improve mental well-being (Roca-Barceló et al., 2024). Urban areas account for approximately 70% of global carbon emissions and more than 60% of resource use.

The transportation sector is a major driver of air pollution, climate change, and urban noise, consuming around one-third of global energy and responsible for over one-fifth of greenhouse gas emissions. Many cities lack comprehensive transportation master plans, which are essential to establish sustainable and eco-friendly urban mobility systems. Inadequate resources and planning standards result in poorly managed urban areas, haphazard transportation networks, and unplanned street layouts. Achieving the long-term goal of reducing carbon emissions is challenging without strict emissions regulations and widely accepted guidelines for a low-carbon transportation strategy. To improve urban sustainability, it is crucial to promote a shift toward less polluting modes of transportation. Urban areas account for a substantial share of global energy consumption and greenhouse gas emissions. Well-planned green spaces and coordinated efforts from governments, industries, and individuals are required for addressing air pollution (Das et al., 2024).

21.2 Methodology of Assessing Air Quality

Assessing air quality involves monitoring, modeling, and data analysis to provide a complete understanding of pollution levels and their impacts. The fixed air quality measurement stations located in designated places in rural and urban areas measure the concentration of various air pollutants in that area. Currently, various spectroscopic and remote sensing techniques that are more accurate are being used to measure the air quality of a particular area. However, these methods face limitations due to limited coverage and spatial constraints.

Air quality data along with meteorological data, for example, temperature, RH, pressure, wind speed, wind direction, and so on, are used to predict the trend of pollution levels in accordance with air quality standards. This data also helps to build air quality models to simulate the transport, dispersion, and transformation of pollutants in the atmosphere.

These air quality models help predict the pollution level of an area and prepare emission control strategies to minimize air pollution. The larger amount of air quality data along with meteorological data is useful for forecasting and tracing the pollution source and its particles in a year. It also helps in decision-making and formulation of strategies in tackling air pollution (Huang et al., 2021). The air quality models so made serve two purposes for the areas lacking air quality monitoring stations:

1. Estimation of the level of pollutants in the air.
2. Forecasting future trends of air pollution levels of that area.

These models by incorporating mathematical and numeric analysis help to predict the physical and chemical processes influencing the pollution the type of pollution of an area. Better air quality can be managed by incorporating the data in these air quality models from:

1. Air quality monitoring data (air pollutants).
2. Data from emission inventories.
3. Metrological data.

Using these models, a comprehensive approach can be taken to assess and minimize air pollution. The fixed air quality monitoring station provides accurate and high-quality data but requires higher maintenance and installation costs. Alternative low-cost and continuous monitoring systems can be used to measure the air quality of an area, but the accuracy in such cases is compromised, and they generally do not match the reference-grade instruments. Despite advancements in air quality measuring instruments, major gaps persist in data analysis and preparation of accurate models in developing regions (Süel et al., 2022; Yadav et al., 2014, 2023).

21.3 Integrating Air Quality

Nowadays, machine learning techniques are being used in air quality modeling and prediction. Machine learning models use historical data as a training set to predict future levels of pollution, pinpoint sources of pollution, and improve emission control strategies. Models developed have significant value in urban environments where various air quality factors, such as traffic patterns, industrial emissions, and meteorological conditions, are involved. In the study conducted by Méndez et al., (2023), it was observed that the combination of machine learning and conventional air quality modeling methods can enhance the accuracy and reliability of air quality assessments. This will help facilitate proactive strategies to protect public health and the environment. Advancements in deep learning, machine learning algorithms, big data technologies, and the Internet of Things are utilized to predict remote sensing technologies' capabilities and create stable and efficient models for monitoring and forecasting air pollution (Gangwar et al., 2023).

21.4 Air Pollution Sources and Types

The sources of air pollution are human-made and natural, and each contributes unique pollutants to the atmospheric mix. There are three types of air pollutants (Fig. 21.1):

- Primary
- Secondary
- Tertiary

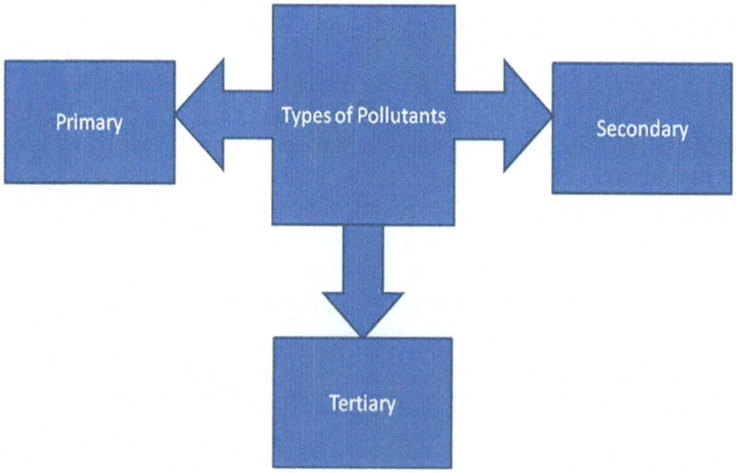

Fig. 21.1 Different types of pollutants

- Primary pollutants: These are the basic pollutants that are released directly from various sources, such as SO_2 and NO_2 from power plants and hydrocarbons, CO, and NO_2 from vehicle exhaust.
- Secondary pollutants: These are formed within the atmosphere after the release of primary pollutants through reactions with other pollutants. Examples include SO_3, H_2SO_4, and HNO_3, which can be formed by photochemical or chemical reactions or catalytic action.
- Tertiary pollutants: These are those that are formed during the reaction of secondary pollutants with other substances, resulting in compounds such as KNO_3, $NaNO_3$, and proxy acetyl nitrate.

Primary pollutants are involved in chemical processes in the atmosphere and form secondary pollutants, like ozone and secondary organic aerosols, which further deteriorate air quality. The study by Gavrilaş et al., (2022) showed that industrial activities, including manufacturing, mining, and construction, significantly contribute to air pollution by releasing various pollutants contingent on the employment of specific procedures (Fig. 21.2).

Inappropriate management and disposal of hazardous materials can result in soil and water contamination, increasing risks to ecosystems and human health. Volcanic eruptions emit significant amounts of sulfur dioxide, PM, and various gases into the atmosphere, which can affect air quality and climate at both regional and global levels. Wildfires emit substantial amounts of PM, carbon monoxide, and nitrogen oxides, thus worsening air pollution events, especially in forested regions. Dust storms, prevalent in arid and semi-arid regions, transport significant amounts of mineral dust across extensive distances, affecting air quality and visibility. Plant emissions, including VOCs, can interact with anthropogenic pollutants to produce ozone and secondary organic aerosols, thus intensifying urban air pollution. Traffic

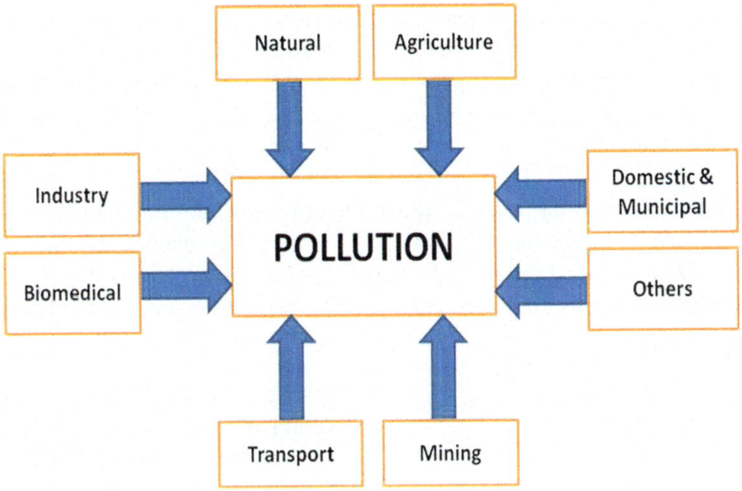

Fig. 21.2 Different sources of air pollution

is a major source of fine PM, which poses considerable health risks and contributes to climate change. Air pollutants are classified into primary and secondary types according to their formation processes. Primary pollutants include PM, a complex amalgamation of solid and liquid particles suspended in the atmosphere, recognized as a significant air pollutant with considerable health consequences. Pollutants are classified by size, where PM10 denotes particles with a diameter of 10 μm or less, and PM2.5 denotes particles with a diameter of 2.5 μm or less. Fine and ultrafine particulates arise primarily from the combustion of fossil fuels and the transformation of biogenic emissions. PM2.5 is particularly concerning because it can penetrate deeply into the respiratory system and enter the bloodstream, leading to various adverse health effects. Exposure to PM2.5 has been associated with cardiovascular, pulmonary, neurological, and psychiatric disorders. Gaseous pollutants, such as ozone, nitrogen dioxide, and VOCs, also contribute to air pollution and pose health risks. Nitrogen dioxide, often used as an indicator of air pollution, can lead to the formation of secondary pollutants such as ozone and acidic nitrate particles.

21.5 Impacts of Indoor Air Quality and Health

Indoor air quality significantly affects human health and overall well-being, as individuals spend a substantial portion of their time, particularly in urban settings, within indoor environments (Sajeev et al., 2022). These indoor spaces harbor a combination of outdoor pollutants that infiltrate through ventilation systems, as well as indoor contaminants generated from various sources, including building materials, furnishings, and occupant activities. Suboptimal indoor air quality can cause a wide range of health problems, from mild irritation to severe respiratory and cardiovascular problems. In fact, indoor air pollution is responsible for a considerable number of deaths annually. Bioaerosols, comprising PM of biological origin such as fungal spores, viruses, and bacterial cells, constitute a significant portion of indoor air pollution, contributing to approximately 5–34% of the total. Combustion byproducts, including carbon monoxide, nitrogen dioxide, and PM, are common indoor air pollutants that arise from the use of stoves, heaters, and fireplaces (Table 21.1).

Cooking activities, particularly those that involve the combustion of biomass fuels, can generate high concentrations of PM and various chemical compounds that adversely affect the respiratory and cardiovascular systems. In addition, construction materials and furniture can emit VOCs, formaldehyde, and other harmful

Table 21.1 Types of pollutants with examples

Types of pollutants	Examples
Gaseous pollutants	NO_x, CO_x, SO_x, etc.
Solid pollutants	Dust, hydrocarbons, etc.
Liquid pollutants	Acids, bases, etc.

Table 21.2 Types of pollutants and their effects on health

Category	Examples	Health effects
Biological contaminants	Mold, bacteria, viruses, pet, and pest allergens	Allergic reactions, asthma attacks, infectious diseases
VOCs	Formaldehyde, pesticides	Respiratory issues, headaches, potential carcinogenic effects
Oxides of carbon and nitrogen	CO (carbon monoxide), NO_x (nitrogen oxides)	Respiratory problems, reduced oxygen transport in blood, smog formation
PM	Dust, smoke, mold spores	Lung diseases, heart problems, reduced indoor air quality
Tobacco smoke	Secondhand smoke	Lung cancer, respiratory diseases, cardiovascular issues
Radon	Naturally occurring radioactive gas	Lung cancer risk

substances into the indoor environment. Exposure to contaminants, including VOCs, formaldehyde, and suspended dust, has been shown to adversely affect human health in newly decorated environments (Xu & Ou, 2019). Biological pollutants, such as mold, bacteria, viruses, and allergens from animals and insects, can thrive in indoor settings, particularly in humid or inadequately ventilated spaces. Contact with these biological contaminants can cause allergic reactions, asthma exacerbations, and infectious diseases. Furthermore, indoor air may be contaminated by various pollutants, including VOCs, biological contaminants, carbon and nitrogen oxides, PM, tobacco smoke, radon, mildew, formaldehyde, and pesticides (Table 21.2).

Depending on the type and concentration of pollutants, exposure time, and personal vulnerability, indoor air pollution can have a wide range of health effects. Since perceptual and cognitive processes influence how people behave and respond to environmental stresses, short-term exposure to air pollution can also have major psychological and cognitive effects. According to research, air pollution may have a negative impact on cognitive and mental health. Numerous adverse health impacts, ranging from subclinical physiological changes to severe clinical symptoms, hospital admissions, and even early mortality, have been linked to short-term and long-term exposure to ambient air pollution. Crucially, air pollutants, even at levels deemed "normal" by national standards, can still pose health risks, underscoring the need for constant vigilance and preventive measures (Wang et al., 2023). Short-term exposure to air pollutants is closely related to conditions such as chronic obstructive pulmonary disease (COPD), cough, shortness of breath, wheezing, asthma, respiratory disease, and high rates of hospitalization. Furthermore, exposure to urban air pollution in North America and Europe has been associated with increased respiratory symptoms, reduced lung function, hospitalization, and death from cardiac and respiratory diseases. The health implications of exposure to air pollution in indoor environments, such as homes, schools, and workplaces, can be significant, particularly for vulnerable populations.

21.6 Air Pollution: A Global Health Crisis

Air pollution is a major environmental health hazard that affects populations in both developed and developing nations. In 2012, the World Health Organization reported approximately seven million deaths attributable to exposure to air pollution. Ambient air pollution is increasing in urban centers, and the risk of lung diseases is particularly high in developing and low- to middle-income countries, where many people face poverty and air pollution challenges. Air pollution includes exposure to PM, ozone, nitrogen dioxide, and sulfur dioxide (Shin et al., 2023). The burden of air pollution is more severe in low- and middle-income countries, with the Southeast Asia and Western Pacific regions experiencing the largest pollution-related health impacts. Globally, around 1600 cities in 91 nations report elevated air pollution levels. According to the World Health Organization, nine out of ten people breathe air containing polluted compounds that penetrate deep into the lung tissue and the cardiovascular system. Air pollution is alarming because of its widespread impact on human health and the environment.

21.7 Results of Air Pollution Exposure

Prolonged exposure to air pollution causes chronic respiratory diseases and decreases lung function among children and the elderly. Epidemiological research has shown that there is an association between exposure to air pollutants and adverse cardiovascular events, such as heart attacks and strokes. Furthermore, studies have indicated that air pollution can contribute to the development of certain cancers, including lung cancer and bladder cancer. In 2015, ambient air pollution with fine PM accounted for significant proportions of various diseases, including 17.1% of ischemic heart disease, 14.2% of cerebrovascular disease, 16.5% of lung cancer, 21.4% of chronic obstructive pulmonary disease, and 23.6% of lower respiratory infections. Air pollution can also affect cognitive function and mental health, especially in children and the elderly. The detrimental impact of air pollution on health is well documented. Air pollution is now considered the greatest environmental threat to human health. The number of deaths attributed to air pollution substantially exceeds that of homicides and car crashes combined. Government actions to improve air quality could prevent seven million premature deaths annually and improve the overall health and economic well-being of 92% of the world's population (Table 21.3) (Luo et al., 2023).

The adverse health effects of air pollution extend beyond the respiratory and cardiovascular systems, with emerging evidence suggesting links to neurological disorders, reproductive problems, and even metabolic diseases such as diabetes. Exposure to air pollutants has been associated with various neurological disorders, including psychological disorders and neurodegenerative diseases. Research

Table 21.3 Air pollution, global threat, and government action

Aspect	Details	Source
Health impact of air pollution (2015)	– 17.1% of ischemic heart disease – 14.2% of cerebrovascular disease – 16.5% of lung cancer – 21.4% of COPD – 23.6% of lower respiratory infections	Cohen et al., (2017)
Cognitive and mental health effects	Impairs cognitive function and mental health, especially in children and the elderly	Cipriani et al., (2018)
Air pollution as a global threat	Considered the foremost environmental threat to human health	Campbell-Lendrum and Prüss-Ustün (2019)
Comparison with other causes of death	Deaths from air pollution exceed those from homicides and car crashes combined	Luo et al., (2023)
Potential benefits of government action	Could prevent seven million premature deaths annually and improve health and economic Well-being for 92% of the world's population	Luo et al., (2023)

indicates that PM and other air pollutants may contribute to neuroinflammation, oxidative stress, and ultimately cognitive decline and dementia. These pollutants can cluster together to form larger particles (Singh & Thajudeen, 2023). Air pollution poses a particular concern for children's health, as their developing lungs and immune systems are more vulnerable to the harmful effects of these pollutants. Furthermore, exposure to air pollution during pregnancy has been linked to adverse birth outcomes, such as preterm birth and low birth weight, which can have long-term consequences for the child's health and development.

Governments have the responsibility to implement measures to improve air quality. Public education on the health impacts of air pollution can help prevent exposure. Although air pollution causes diseases and disabilities, it is a preventable issue. Transitioning from fossil fuels to renewable energy sources can improve air quality and prevent millions of premature deaths annually. Aggressive air quality management programs targeted at the main sources of pollution can lower exposure to ambient air pollution and its associated disease burden for entire populations through policy action at national and subnational levels. The severity of the impact of air pollution is linked to the levels and duration of exposure. In addition, an individual's underlying health status can influence the severity of the effects from exposure to air pollution. It is imperative to target the fundamental causes of air pollution and implement evidence-based interventions to reduce emissions and protect public health. Additionally, increased exposure to air pollution is correlated with increased total healthcare expenditures among the population. Furthermore, early exposure to air pollutants can cause developmental defects in the lungs, potentially resulting in respiratory illnesses and decreased lung function later in life.

21.8 Discussion of the Detrimental Effects

Air pollution disproportionately affects children's respiratory health because of their increased vulnerability to its harmful effects. Exposure to indoor and outdoor pollution is associated with acute lower respiratory infections and chronic health problems in the long term. Recent recognition has been given to the implications of environmental exposures, particularly air pollution, during critical developmental stages, especially for children under 5 years of age (Spencer-Hwang et al., 2023). Air pollution follows the Kuznets curve (Fig. 21.3), providing an intricate relationship between development and air quality. In this curve, initially, there is a low level of air pollution, but as the development of a nation starts and demand for energy increases, leading to increased consumption of fossil fuel and rapid industrialization, a peak height is reached in the air pollution level. But in the later stages, when a nation progresses and wealth increases and shifts to cleaner and greener fuels, the level of air pollutants starts to decline, leading to a relatively cleaner environment.

The main factors contributing to indoor air pollution are the burning of solid fuel for cooking and heating purposes. This causes respiratory illness mainly in developing and low-income countries. Approximately 2.8 billion people worldwide use solid fuel, for example, wood, cow dung, and so on, as the main source of fuel. In India alone, burning of biomass contributes to approximately 15–30% of PM-2.5 emissions. Pollutants released by the burning of these solid fuels mainly affect women and young children, as they are more prone to exposure to these harmful pollutants. The main causes of exacerbating air pollution in developing countries are as follows:

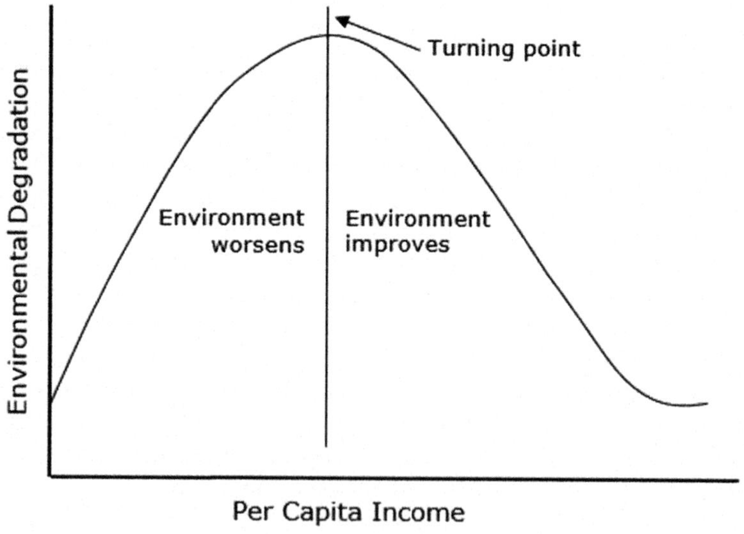

Fig. 21.3 Environmental Kuznets curve adopted from To et al., (2019)

- Rapid industrialization
- Rapid urbanization
- Improper living standards
- Lack of education
- Insufficient environmental regulation
- Lack of clean and green fuel
- Lack of modern technologies
- Older vehicles

To overcome the problem of air pollution, the government should encourage

- Use of cleaner and greener fuel
- Use of modern and safer techniques
- Use of public transport
- Educating people about air pollution
- Discarding of older vehicles
- Enforcing strict environmental regulations

21.9 Conclusion

Air pollution is a worldwide substantial risk to various respiratory and cardiovascular ailments, cancer, and early death. Children do have detrimental risk of air pollution due to their developing lungs. Comprehensive strategies to address air pollution—the reduction of emissions from transportation, industry, and residences, with the advancement of cleaner energy sources and the enforcement of more stringent environmental rules are being proposed. WHO research shows that India is adversely affected by pollution with inadequate ambient air quality. The Indian judiciary has the right to a clean and healthy environment as a basic right, highlighting the necessity of strong environmental regulation. Regulatory measures, technological advancements, and innovative ideas are crucial for addressing air pollution. Robust collaboration among governments, enterprises, and individuals to promote a cleaner, healthier, and more sustainable future for all should be initiated. Moreover, climate change is another significant issue for which environmental justice and promoting collective global responsibility are essential to reduce air pollution exposure and confront the effects of climate change. It is essential to perceive the complex relationships between air pollution and environmental elements, such as land use patterns and social disparities.

Acknowledgments The authors sincerely thank Prof. M. P. Pandey, Dr. Sanjeev Agrawal, and Prof. Vaibhav Trivedi for their support in writing this book chapter.

References

Campbell-Lendrum, D., & Prüss-Ustün, A. (2019). Climate change, air pollution and noncommunicable diseases. *Bulletin of the World Health Organization, 97*(2), 160–161.

Cipriani, G., Danti, S., Carlesi, C., & Borin, G. (2018). Danger in the air: Air pollution and cognitive dysfunction. American Journal of Alzheimer's Disease & Other Dementias, 33(6), 333–341

Cohen, A. J., Brauer, M., Burnett, R., Anderson, H. R., Frostad, J., Estep, K., Balakrishnan, K., Brunekreef, B., Dandona, L., Dandona, R., Feigin, V., Freedman, G., Hubbell, B., Jobling, A., Kan, H., Knibbs, L., Liu, Y., Martin, R., Morawska, L., Pope, C. A., & Forouzanfar, M. H. (2017). Estimates and 25-year trends of the global burden of disease attributable to ambient air pollution: An analysis of data from the global burden of diseases study 2015. *Lancet, 389*(10082), 1907–1918.

Das, S., Choudhury, M. R., Chatterjee, B., Das, P., Bagri, S. C., Paul, D., Bera, M., & Dutta, S. (2024). Unraveling the urban climate crisis: Exploring the nexus of urbanization, climate change, and their impacts on the environment and human Well-being - a global perspective. *AIMS Public Health, 11*(3), 963–1001.

Gangwar, A., Singh, S., Mishra, R., & Prakash, S. (2023). The state-of-the-art in air pollution monitoring and forecasting systems using IoT, big data, and machine learning. *Wireless Personal Communications, 130*(3), 1699–1729.

Gavrilaş, S., Ursachi, C. Ş., Perţa-Crişan, S., & Munteanu, F. (2022). Recent trends in biosensors for environmental quality monitoring. *Sensors, 22*(4), 1513.

Huang, W., Li, T., Liu, J., Xie, P., Du, S., & Teng, F. (2021). An overview of air quality analysis by big data techniques: Monitoring, forecasting, and traceability. *Information Fusion, 75*(2), 28–40.

Luo, Y., Xu, L., Li, Z., Zhou, X., Zhang, X., Wang, F., Peng, J., Cui, C., Chen, Z., & Yu, H. (2023). Air pollution in heavy industrial cities along the northern slope of the Tianshan Mountains, Xinjiang: Characteristics, meteorological influence, and sources. *Environmental Science and Pollution Research, 30*(19), 55092–55111.

Méndez, M., Merayo, M. G., & Núñez, M. (2023). Machine learning algorithms to forecast air quality: A survey. *Artificial Intelligence Review, 56*(9), 10031–10066.

Roca-Barceló, A., Rice, M. B., Nunez, Y., Thurston, G., Weinmayr, G., Straíf, K., Roscoe, C., Ebi, K. L., Andersen, Z. J., de Nazelle, A., & Negev, M. (2024). Climate action has valuable health benefits. *Environmental Epidemiology, 8*(1), e288.

Sajeev, V., Anand, P., & George, A. (2022). Indoor air pollution, occupant health, and building system controls—a COVID-19 perspective. *Air Pollution Control, 17*, 291–306.

Shin, H., Braun, D., Irene, K., & Antonelli, J. (2023). A spatial interference approach to account for mobility in air pollution studies with multivariate continuous treatments. *arXiv*. https://doi.org/10.48550/arxiv.2305.14194

Singh, A., & Thajudeen, T. (2023). A hybrid particle swarm optimization-tuning algorithm for the prediction of nanoparticle morphology from microscopic images. *Aerosol and Air Quality Research, 23*(4), 220453.

Spencer-Hwang, R., Hwang, J., Sinclair, R., Dennis, S., & Montgomery, S. (2023). Adverse health outcomes in early childhood (birth to 5 years) and ambient air pollutant exposures: a systematic review. *Air Quality, Atmosphere and Health, 16*(5), 913–944.

Süel, E., Sorek-Hamer, M., Moise, I., Pohle, M. V., Sahasrabhojanee, A., Asanjan, A. A., Arku, R. E., Alli, A. S., Barratt, B., Clark, S., Middel, A., Deardorff, E., Lingenfelter, V., Oza, N. C., Yadav, N., Ezzati, M., & Bräuer, M. (2022). What you see is what you breathe? Estimating air pollution spatial variation using street-level imagery. *Remote Sensing, 14*(14), 3429.

To, A. H., Ha, D. T.-T., Nguyen, H. M., & Vo, D. H. (2019). The impact of foreign direct investment on environment degradation: Evidence from emerging Markets in Asia. *International Journal of Environmental Research and Public Health, 16*(9), 1636.

Wang, Q., Chen, Z., Huang, W., Kou, B., & Li, J. (2023). Short-term effect of moderate level air pollution on outpatient visits for multiple clinic departments: A time-series analysis in Xi'an China. *Toxics, 11*(2), 166.

Xu, X., & Ou, S.-J. (2019). The discussion on the relationship among indoor air quality, environmental cognition and user's behaviour. *IOP Conference Series: Earth and Environmental Science, 291*(1), 012005.

Yadav, A. K., Sahoo, S. K., Kumar, A. V., Shirin, S., Jamal, A., Patra, A. C., Dubey, J. S., Thakur, V. K., Lenka, P., Singh, S., Jha, V., & Tripathi, R. M. (2023). Identifying size-segregated particulate matter ($PM_{2.5}$, PM_{10} and SPM) sources in an industrial town of India. *Aerosol Science and Engineering, 7*(4), 455–473.

Yadav, A. K., Sahoo, S. K., Patra, A. C., Dubey, J. S., Lenka, P., Sagar, D. V., Kumar, A. V., & Tripathi, R. M. (2014). Source identification of particulate matter and associated intake of elements through inhalation in an industrial area of Odisha. *India. Toxicological & Environmental Chemistry, 96*(3), 410–425.

The manufacturer's authorised representative in the EU is Springer Nature Customer Service Centre GmbH, Europaplatz 3, 69115 Heidelberg, Germany. If you have any concerns regarding our products, please contact ProductSafety@springernature.com

Printed and bound by CPI Group (UK) Ltd, Croydon, CR0 4YY

11/10/2026

02231395-0002